Climate Change Management

Series Editor

The aim of this book series is to provide an authoritative source of information on climate change management, with an emphasis on projects, case studies and practical initiatives – all of which may help to address a problem with a global scope, but the impacts of which are mostly local. As the world actively seeks ways to cope with the effects of climate change and global warming, such as floods, droughts, rising sea levels and landscape changes, there is a vital need for reliable information and data to support the efforts pursued by local governments, NGOs and other organizations to address the problems associated with climate change. This series welcomes monographs and contributed volumes written for an academic and professional audience, as well as peer-reviewed conference proceedings. Relevant topics include but are not limited to water conservation, disaster prevention and management, and agriculture, as well as regional studies and documentation of trends. Thanks to its interdisciplinary focus, the series aims to concretely contribute to a better understanding of the state-of-the-art of climate change adaptation, and of the tools with which it can be implemented on the ground.

Notes on the quality assurance and peer review of this publication

Prior to publication, the quality of the works published in this series is double blind reviewed by external referees appointed by the editor. The referees are not aware of the author's name when performing the review; the referees' names are not disclosed.

Aziza B. Karbekova
Editor

Sustainable Development of Rural Territories Based on Climate-Smart Technologies

Springer

Editor
Aziza B. Karbekova
Jalal-Abad State University named
after B. Osmonov
Manas, Kyrgyzstan

ISSN 1610-2002 ISSN 1610-2010 (electronic)
Climate Change Management
ISBN 978-3-032-20105-8 ISBN 978-3-032-20106-5 (eBook)
https://doi.org/10.1007/978-3-032-20106-5

This Springer imprint is published by the registered company Springer Nature Switzerland AG
The registered company address is: Gewerbestrasse 11, 6330 Cham, Switzerland

Introduction: Climate Change Management Based on Smart Agricultural Technologies for the Sustainable Development of Rural Areas

Climate change manifests itself everywhere, having a truly global reach, affecting all countries of the world, and an intersectoral character, leaving its imprint on every branch of the economy. However, the innovative perspective on climate change presented in this book, grounded in spatial economics, reveals a pronounced territorial specificity in the consequences of environmental degradation. These consequences are most clearly observed in rural areas due to their mono-sectoral (agricultural) production specialization and the strong dependence of productivity on environmental factors.

In the context of the accelerating Fifth Industrial Revolution, smart agricultural technologies have become crucial in addressing climate change in agriculture. These technologies represent advanced tools for automating agricultural resource use. They are also widely known as climate-smart agricultural technologies, as automation enhances the climate adaptability of agricultural production through more frequent and comprehensive phyto- and biomonitoring based on artificial intelligence; through improved controllability of smart farms, their environmental autonomy (particularly in vertical farming), and the increased flexibility of their settings within a cyber-ecological system enabled by the Internet of Things; as well as through the advancement of agricultural production technologies using machine learning.

The problem lies in the fact that although climate-smart agricultural technologies are most needed in rural areas due to their agricultural specialization, they are being implemented most intensively in urban areas, where the most favorable organizational and technological conditions exist: advanced telecommunications infrastructure, venture investment, and a digitally skilled workforce. As a result, climate-smart agricultural technologies are actively adopted to establish forms of agricultural production within city limits that were previously inaccessible to megacities. While this increases the availability of locally produced natural agricultural food for urban areas, it limits opportunities for the socio-economic development of rural territories and accelerates an already critical process of urbanization.

The above circumstances underscore the relevance of in-depth scientific research into the dissemination of climate-smart agricultural technologies in rural areas as a means of accelerating their economic growth and development. The existing literature studied the topic of climate management in agriculture based on smart technologies in considerable detail. This issue is widely covered in numerous publications. However, a key shortcoming of prior research is that this topic has been examined with a narrow focus on agricultural production (i.e., from the perspective of agricultural entrepreneurship) while its territorial location has largely been overlooked.

This has resulted in a gap in the literature, reflected in the limited study and uncertainty surrounding the specific features and prospects of climate-smart modernization of agricultural production in rural areas as a distinct economic and geographical environment. This book addresses that gap. It is written with the aim of developing a scientific concept of climate change management in rural areas based on smart agricultural technologies, systematizing international experience in this field, and proposing applied solutions to improve such management in support of the sustainable development of rural territories.

The book is founded on an original scientific concept of climate-resilient development of rural areas in the context of the digital economy. Its scientific novelty lies in the fact that it is the first to incorporate the territorial context of climate-smart modernization of agriculture. The book's theoretical significance stems from its expansion of existing views on climate-smart management of agricultural entrepreneurship by offering an innovative interpretation of it as an actor within the spatial economy.

Thus, the book has formed a systemic vision of climate-smart agriculture as an integrated whole comprising an environmental and geographical dimension (linked to combating climate change in rural areas), an economic and managerial dimension (revealing organizational models and management frameworks for addressing climate change in agricultural entrepreneurship), and a technological dimension (explaining the role of climate-smart agricultural technologies).

The practical significance of the book lies in the fact that the applied solutions it contains can contribute to the creation and activation of agricultural "growth points" in rural areas, reducing their vulnerability to climate change and triggering a process of de-urbanization—the movement of population toward rural areas as green spaces characterized by rapid economic growth and a high quality of life.

The social significance of the book is explained by the fact that its scientific and practical recommendations for improving the effectiveness of climate-smart agricultural management in rural areas support the sustainable development of these territories. In doing so, they accelerate progress and ensure the integrated and synergistic implementation of the 17 UN Sustainable Development Goals (SDGs), particularly SDG 1, SDG 4, SDG 16, and SDG 17 (in terms of improving the socio-economic well-being of rural residents); SDG 2 and SDG 6 (in terms of strengthening food security, including agricultural irrigation); SDG 3 and SDGs 12–15 (in terms of "green" development and combating climate change in rural areas); and SDGs 8–11 (in terms of stimulating high-productivity and knowledge-intensive employment for digital agricultural professionals, high-tech and green economic growth, and the sustainable development of rural territories).

The materials of this book are consistently presented across five parts. The first part summarizes the pressing challenges of sustainable rural development under conditions of climate change. The importance of this part for the overall topic of the book lies in its explanation of the specific features of climate-related agricultural management from the perspective of spatial economics.

The second part is devoted to highlighting the advanced potential of climate-smart technologies in agriculture. The value of this part in disclosing the book's topic lies in its explanation of the benefits of AI-driven automation of agricultural production for improving climate adaptation in rural areas.

The third part explains the contribution of climate-smart technologies to the sustainable development of rural areas. Its role in developing the book's topic is determined by the detailed examination of mechanisms for initiating socio-economic development in rural territories based on climate-smart agricultural management.

The fourth part focuses on international experience in the sustainable development of rural areas through the application of climate-smart technologies in agriculture. The usefulness of this part lies in its inclusion of numerous case studies and examples from global practice in public and corporate climate-related agricultural management in rural territories.

The concluding fifth part presents ready-to-implement applied solutions for the sustainable development of rural areas based on climate-smart technologies. The necessity of this part is reflected in the fact that it offers recommendations for optimizing climate-smart agricultural management that are highly relevant to contemporary rural territories.

The primary target audience of this book consists of scholars studying issues related to combating climate change in agriculture. They will find in it an original scientific approach to climate-smart agricultural management in support of the sustainable development of rural areas.

The secondary target audience includes, first, representatives of agribusiness, for whom the book provides applied solutions aimed at improving the effectiveness of climate-smart management of agricultural production and venture investment in agriculture in rural areas.

Second, it addresses representatives of public authorities responsible for rural governance, offering scientific and practical recommendations for improving state regulation of rural economies through better coordination of socio-economic and environmental policies, as well as policies aimed at stimulating the digital development of the spatial economy.

Aziza B. Karbekova

About This Book

This book is devoted to the development of a scientific concept of climate change management in rural areas, grounded in smart agricultural technologies, the systematization of international experience in this field, and the formulation of recommendations for the organizational and economic optimization of such management in support of the sustainable development of rural territories. The uniqueness of the book lies in its consideration of the territorial context of climate-smart modernization of agriculture, which allows it to shape a comprehensive understanding of climate-smart management of agricultural entrepreneurship as an actor in the spatial economy and a growth point for rural areas.

The first of the book's five parts addresses the pressing challenges of sustainable rural development under conditions of climate change. The second part explores the advanced potential of climate-smart technologies in agriculture. The third part details the contribution of climate-smart technologies to the sustainable development of rural areas. The fourth part examines international experience. The final, fifth part formulates recommendations for optimizing the practical application of these technologies.

The book is primarily intended for scholars, offering an original scientific approach to climate-smart agricultural management in support of the sustainable development of rural territories. Additionally, it is aimed at representatives of agribusiness and public authorities responsible for rural governance, who will find practical recommendations for improving public and corporate climate management.

Contents

Current Challenges to the Sustainable Development of Rural Areas Under Climate Change

Bridging the Gap Between Green Finance and Digital Transformation: Do These Synergies Impact the Global Debt Burden and Sustainable Economic Development?

Hayk G. Mnatsakanyan and Liana H. Karapetyan

Abstract In the current era of digital innovation, the global transition toward climate neutrality has ushered in a dual transformation: a shift toward green energy and the rise of digital technologies. However, both processes require substantial investments, often financed through public debt—raising concerns about long-term fiscal sustainability, especially in emerging economies. Against the backdrop of the fourth wave of global debt, considering climate deadlines, energy security, and the digital carbon paradox, this research aims to identify whether and how investments in renewable energy and the expansion of digital technologies affect the debt burden. As global public debt soared to a record $100 trillion in 2024, equivalent to 93% of global GDP, and climate impacts continued to reduce GDP growth by 1–2 percentage points each year, particularly in developing countries, the challenge of financing investments has become increasingly urgent. To achieve net zero emissions by 2050, the world needs to invest $9.2 trillion annually in green infrastructure, with developing countries receiving only 9% of global sustainable finance flows in 2023, pushing countries into a new green debt trap. The development of digital transformation in this context is an opportunity and a challenge. The ICT sector is projected to have the potential to reduce (5–10% by 2030) global emissions. However, the growing energy demand associated with ICT will exacerbate existing energy crises, with the sector's energy consumption estimated to reach approximately 1000 TWh by 2026. This paradox requires closer coordination between financing mechanisms for ICT growth and renewable energy sources.

H. G. Mnatsakanyan (✉) · L. H. Karapetyan
Armenian State University of Economics, Yerevan, Armenia
e-mail: hayk.mnatsakanyan@asue.am

L. H. Karapetyan
e-mail: liana.karapetyan@ysu.am

L. H. Karapetyan
Yerevan State University (YSU), Yerevan, Armenia

A. B. Karbekova (ed.), *Sustainable Development of Rural Territories Based on Climate-Smart Technologies*, Climate Change Management,
https://doi.org/10.1007/978-3-032-20106-5_1

Keywords Green finance · Digital transformation · Climate debt trap · Renewable energy · Debt burden

JEL Codes E62 · G18 · G28 · O33 · Q01 · Q55

1 Introduction

The global challenge of achieving climate neutrality by 2050 involves two transformative processes: a shift to renewable energy and the rapid expansion of digital technologies, which require unprecedented public and private investment, often financed by public debt. Emerging economies are struggling to raise capital. Current challenges threaten to throw countries into a "green debt trap." By examining the interplay between green finance and digital transformation in the context of debt sustainability, this research aims to analyze how investments in renewable energy and ICT infrastructure affect the debt-to-GDP ratio, presenting strategic priorities that will make it possible to harmonize the development of these sectors to ensure sustainable growth.

The analysis addresses a broad range of macroeconomic challenges, including the impact of ICT on the financial system, the macroeconomic links between green and digital investments and public debt dynamics, the role of digital transformation in driving the green transition and managing the energy burden, and the potential of energy markets and cross-sector synergies to inform policy frameworks, illustrated through relevant case studies.

2 Literature Review

The synergistic interactions between green finance and digital transformation are at the heart of academic research, all driven by the growing importance of green agendas and the rapid development of digital technologies. This review examines the issue in the context of deepening debt tensions among countries, presenting the threat of a "green debt trap." Various studies show that effective investment financing mechanisms can significantly improve countries' debt positions while investing in key growth sectors leading to climate neutrality, such as renewable energy and digital technologies (United Nations Environment Program 2022).

The global digital transformation has significantly impacted the financing system (Afjal 2023; Carè et al. 2025; Harsono and Suprapti 2024; Tao and Chao 2024). The digital economy, as an emerging economic paradigm, is a key driver of the green transformation of the global economic framework (Guo et al. 2024; Li et al. 2024; Yu et al. 2024). The analysis, which maps the evolution and scope of sustainable finance development, also documents the central role of technological innovations in advancing sustainable finance practices (Judijanto et al. 2024).

Financing the green transformation through advanced financial instruments often defers the debt burden to the future, potentially intensifying the ongoing fourth wave of debt accumulation. Achieving climate neutrality by 2050 will require large-scale investments in green infrastructure. Given the nature of many of these investments in public goods, government (co-)financing will be essential in bridging the funding gap (Kantorowicz et al. 2023). Additional public investment of 0.5–4.5% of GDP will be required to finance green investments over the next decade (Jaumotte and Schwerhoff 2021). Policies that place the burden of financing green investments on the private sector could lead to capital flight from economies (International Monetary Fund 2022).

Debt-financed, ICT-driven green investment strategies face the challenge of the digital–carbon paradox, where efforts to achieve environmental sustainability through digitalization may unintentionally lead to fiscal strain and higher carbon emissions. Striking a balance between innovation and green transformation remains a key challenge in advancing sustainable and environmentally responsible development in the fintech sector (Mohammed et al. 2024). On the one hand, according to forecasts, the development of the ICT sector can reduce global greenhouse gas emissions by 5–10% by 2030 (World Economic Forum 2024). On the other hand, the growth of the ICT sector and digitalization contribute to an increase in electricity consumption and greenhouse gas emissions. In 2020, ICT-related emissions made up just 1.4% of global emissions, and electricity consumption accounted for 4% (Malmodin et al. 2024). However, projections raise concern, with estimates predicting a 14–24% increase in emissions by 2030–2040. Various forecasts highlight several scenarios for the development of the IT sector, the most unfavorable of which assumes that by 2030 the electricity consumed by the sector could reach about half (51%) of global production. At the same time, within the framework of the most favorable scenario, this indicator is estimated at a level close to 8 percent of global electricity. Even under the conditions of the projected rates of expansion of renewable energy production, some studies show that in the case of an unfavorable development scenario, by 2030 about a quarter (23%) of global greenhouse gas emissions could reach and be due to the activities of the IT sector (Godlovitch et al. 2021).

The reviewed literature highlights the urgency of harmonizing green finance and digital transformation to prevent fiscal and environmental crises. While ICTs facilitate new green instruments, their energy requirements need systemic reforms in energy markets, exposing countries to new debt crises. Future research should focus on identifying strategies to help countries avoid the green debt trap, conducting empirical analyses of integrated financing mechanisms, and exploring policy innovations to address the digital-carbon paradox.

3 Methodology

The research uses a mixed methods approach to assess the interaction between the transition to green energy and digital technologies in achieving climate neutrality, focusing on macroeconomic resilience, debt dynamics, and energy market developments. The methodology includes data analysis, comparative case studies, and policy evaluation to achieve the research objectives. The methodological framework includes the following:

- Literature review and bibliographic analysis: The analysis of relevant academic literature, research, and policy documents was conducted to map the evolution of green finance and digital innovation;
- Quantitative data analysis: The research analyzed macroeconomic databases to provide empirical support for the theoretical propositions. The proposed solutions include scaling issues and the dynamics of green finance, investment, digital transformation, and debt. To highlight the connections between the energy market and its structures, the authors studied the reports and working documents of the World Bank, OECD, IMF, IEA, and other international organizations.

4 Results

4.1 The Growing Volume of Data: How ICT Can Sustainably Manage This Growth

With the ongoing digital transformation, the global economy is driving an explosive growth in data volumes. According to forecasts (Taylor 2024), 181 zettabytes of data will be created, stored, and used worldwide by 2025, nearly tripling the amount of 2020. According to the International Energy Agency (IEA), electricity demand from the ICT sector could reach 1,000 terawatt hours (TWh) by 2026—approximately equal to the total electricity consumption of Japan (IEA 2024). This growth trend is certainly driving a sharp increase in energy consumption, which threatens the sustainability of energy systems and could lead to a sharp increase in emissions.

In a conscious effort to reduce its environmental impact, the ICT sector has become the largest buyer of renewable energy in the world, signing about 60% of renewable energy purchase agreements in 2021 (World Bank 2024b). According to estimates, tech giants (e.g., Google, Amazon, Meta, and Microsoft) are the largest corporate buyers of renewable energy, accounting for over 40% of the contracted capacity during 2017–2022 (Chauhan 2023).

The liberalization of energy markets presents an opportunity to advance the digital economy and the development of the energy sector. The ICT sector plays a crucial role as an alternative source of green financing for developing the renewable energy sector. It enables governments to shift part of the investment burden to the private sector, helping to avoid the so-called "green debt trap." At the same time, ICT serves as a catalyst for renewable energy growth by driving technological innovation, optimizing energy systems, and supporting the creation of decentralized, efficient energy ecosystems. In this way, ICT fosters the advancement of one of the most vital sectors of the economy—energy—reducing the dependence of many countries on traditional energy prices and diminishing the influence of OPEC countries in the global economy. However, it is essential to be able to design effective transition mechanisms that will allow the development of digital technologies to curb further growth of global debt. The key step here is to target capital correctly.

Countries with more competitive electricity market structures have higher levels of capital attraction from the private sector. The Single Buyer Model (SBM), often viewed as a transitional stage toward full market liberalization, has emerged as the most widely adopted approach, currently implemented in 89 countries (International Finance Corporation 2024). Private capital flows to renewable energy projects have increased significantly since 2010. The volume of financing reached $1.48 trillion between 2010 and 2023. However, similar to the trend in total private investment flows, emerging markets received just 16% of all sustainable financing flows during this timeframe. This gap has widened recently, with emerging markets receiving just 9% of total financing in 2023 (International Finance Corporation 2024).

Liberalizing the energy sector will create the opportunity for a freer flow of capital. As a result, capital will begin to flow in, creating a self-reinforcing cycle in which the expansion of renewable energy drives digital growth, increasing the demand for cleaner energy. For developing countries, promoting renewable energy can serve as a strategy to stimulate economic growth and accelerate ICT development while advancing the shift toward sustainable energy. Armenia presents a compelling example where the growth of the ICT sector has the potential to catalyze renewable energy projects, reducing reliance on imported fuels and enhancing economic flexibility. Integrating the development of digital technologies with renewable energy initiatives offers a dual engine for economic modernization and sustainability. By aligning ICT expansion with green energy transitions, countries can strengthen economic resilience, attract green investment, and reshape their position within the global economic landscape.

4.2 The Global Economy is Shifting Towards Electricity Generation

The year 2023 set new records for both electricity production and consumption. Global electricity generation grew by 2.5%, reaching an all-time high. This growth rate, 25% faster than that of total primary energy consumption, highlights the accelerating electrification of the global energy system. Renewable energy consumption expanded six times faster than total primary energy consumption, with renewables accounting for 14.6% of total primary energy use. Combined with nuclear energy, low-carbon sources represented over 18% of global primary energy consumption.

Despite this progress, greenhouse gas emissions from energy use and industrial processes rose by 2.1%, surpassing the previous record set in 2022. For the first time, global energy-related emissions exceeded 40 gigatonnes (Gt). Meanwhile, renewables' share of total power generation increased slightly, from 29 to 30% (Energy Institute (EI) 2024). The change in the structure of energy production and consumption is a result of the development of digital technologies and countries' green agendas (Shabalov et al. 2021).

4.3 Energy-Related Factors Leading to the Debt Crisis: Historical Parallel

The origins of the debt crisis of developing countries can be traced back to the chain created by the oil price shock of 1973–1974. The formation of OPEC in the 1960s was a turning point in terms of national resource sovereignty; OPEC decisions began to play an important role in the global energy market, especially oil, and international relations (Adelman 1990). The flow of "petrodollars" accumulated in banks to developing countries formed a debt trap in which the main actors were the energy giants. This flow of funds from OPEC member countries to banks, and then to developing countries, began the process of recycling petrodollars (Baumgartner et al. 1981; Watson et al. 1992). Thus, the approach that servicing the economy and structural transformation at the expense of debt is justified is a trap in which petrodollars circulate.

The share of global debt in global GDP has risen significantly, from approximately 100% in 1970 to over 326% by the third quarter of 2024. This growth is substantial and accelerating. Notably, public debt continues to expand rapidly, playing an increasingly prominent role in the overall structure of global debt (Tiftik et al. 2024). Global public debt is at a historically high level, representing approximately 93% of global GDP (as of 2024). It is projected to approach 100% by 2030. This represents a 10% increase in the debt-to-GDP ratio compared to the pre-pandemic level in 2019 (International Monetary Fund 2024). Such developments create a situation where developing countries struggle to combine the actions necessary for sustainable economic growth with climate change mitigation and adaptation (NGFS 2022; OECD 2024).

4.4 Climate Debt Trap: Direct Investment Versus Debt

The global financial system is being gripped by a fourth wave of global debt, which will directly impact the ability of countries to combat climate change and have far-reaching consequences. Many economies are already caught in a green debt trap (International Monetary Fund 2023). Statements from groups such as V20,

G24, UNECA, CELAC, and ASEAN have highlighted that rising debt burdens and tightening global financing conditions are hampering the ability of countries to invest in climate resilience (Alayza et al. 2023).

A sustainable development vision will significantly depend on developing sustainable infrastructure, which is mainly provided by public investment (Weber et al. 2016). There is an increasing gap worldwide between the need for financing infrastructure and the capacity of national budgets to meet this demand. This gap becomes even more pronounced when considering the additional investment required to achieve the Sustainable Development Goals (SDGs), with the gap tripling in size. National budgets alone are insufficient to cover the extensive financing needs for infrastructure development and SDG-related investments (Woetzel et al. 2016). Many studies have investigated the sustainability of public debt in different countries from various perspectives, employing diverse methodologies. Individual debt ceilings may vary by country. However, they usually range from 40 to 70% of GDP (Davoodi et al. 2022).

Over the past decade, a major shift has occurred in the structure of energy sector financing. Until 2013, the public sector provided more than two-thirds (67%) of development finance, which has declined to 50% (2022) over the past decade. Simultaneously, private sector investment has grown rapidly: from almost zero in 2013 to 14% in 2017 and 30% in 2022 (REN21 2024). The structural shift toward a greater role for the private sector in developing and financing the energy sector reflects a deliberate reorientation of global investment strategies within the framework of the green transformation. This shift is driven by the limited fiscal capacity of governments to fund large-scale investments and the ideological commitment to liberalizing the energy sector. This trend can significantly contribute to debt reduction, foster economic development, and enhance overall economic stability.

4.5 Prioritizing ICT–Finance Synergies to Break the Climate-Debt Deadlock

Public investment and incentives play a crucial role in creating and shaping markets that serve the public interest. However, governments face the complex challenge of using technological products as tools and ensuring they are applied in ways that do not exacerbate the problems they aim to solve. A key question arises: "Can countries govern at the pace at which technologies evolve?" For instance, at least 50 countries are already employing weather-modifying technologies developed in China—highlighting the urgent need for regulatory frameworks that can keep up with rapid technological advancements (Hamilton and Stephens 2022). Renewable energy climate projects are triply effective due to their high return on investment (ROI), potential for debt restructuring, and positive environmental impact. A striking example is Rwanda, which has made remarkable progress over the past 15 years—boosting household access to electricity from just 6% in 2009 to 75% by March 2024 (World

Bank 2024). Countries with high debt burdens and climate risks are caught in a dangerous cycle (Krogstrup and Oman 2020). Climate disasters reduce GDP growth by 1–2 percentage points in affected countries and increase public debt by 2.5% of GDP annually due to reconstruction costs and lost revenues (Aligishiev 2022, Fuje et al. 2023). Despite territorial, economic, demographic, and socio-political differences, the macroeconomic assessments of countries regarding public investment and sovereign debt levels are almost identical. According to the assessments, public debt becomes increasingly risky when managing public investment if the debt-to-GDP ratio rises towards the unsustainable debt zone. In this context, the IMF's estimates are clear: a 1% increase in the debt-to-GDP ratio beyond a sustainable level is linked to a 0.3% decline in public infrastructure investment, as rising debt servicing costs divert resources away from capital spending (World Bank 2024c). If public debt crosses risk thresholds, such financing becomes detrimental to the financing of public infrastructure. If a country has initiated the green transformation process with an already high level of accumulated public debt, pursuing sustainable development through unsustainable financing models will inevitably lead to unsustainable economic outcomes (Ari and Koc 2018). The key to success lies in strategically combining climate projects with tailored debt instruments grounded in local conditions and informed by global best practices. Countries such as Norway and Chile demonstrate how integrating liberalized markets and ICT partnerships can drive sustainable growth, diversify economies, and replace fossil fuel–driven volatility with long-term stability (Bolton 2023; Verdejo-Fredes 2022).

5 Conclusion

The combination of renewable energy and digital infrastructure can rewrite the history of debt by developing climate-resilient, technology-driven economies. The synthesis of green energy and digital technologies is a strategic solution to overcome the oil crisis and debt trap originating from the last century, focusing on the following aspects:

- Climate-friendly and ICT-driven financial instruments;
- Diversification of energy sources and reduce oil dependency;
- Leverage financial and technological innovation;
- Breaking dependence on the petrodollar by weakening OPEC's influence;
- Turn debt into a source of sustainable growth by reducing energy dependence through green financing;
- Achieve economic sovereignty through digital infrastructure and renewable energy.

Thus, the gap between green finance and digital transformation will be bridged, and overcoming the debt trap is possible:

(a) Renewable energy as an investment magnet for ICT sector: Breaking the petrodollar cycle;
(b) Liberalized energy markets as dual catalysts: Countering OPEC's legacy;
(c) Green digital infrastructure and economic resilience: Rewriting the debt narrative;
(d) Synthesis: From oil shocks to green stability.

This development strategy replaces oil volatility with green digital stability, creating a society free from fossil fuel price shocks and debt traps. As a result, sustainability (Kibert et al. 2020) redefines its meaning, acting as a guarantee of economic security, where the synergy of green finance, debt, climate, and technology serves as a strategy of liberation. The research concludes that the proposed development strategy replaces oil volatility with green-digital stability, creating the economic system free from fossil fuel price shocks and energy and green debt traps.

References

Adelman MA (1990) The first oil price explosion 1971–1974. Working Paper No. MIT-CEPR 90–013WP. Cambridge, MA, MIT

Afjal M (2023) Bridging the financial divide: a bibliometric analysis on the role of digital financial services within FinTech in enhancing financial inclusion and economic development. Humanit Soc Sci Commun 10:645. https://doi.org/10.1057/s41599-023-02086-y

Alayza N, Laxton V, Neunuebel C (2023) Developing countries won't beat the climate crisis without tackling rising debt. World Resources Institute. Retrieved from https://tinyurl.com/3zvx6bba (Accessed 20 May 2025)

Aligishiev Z, Massetti E, Bellon M (2022) Macro-fiscal implications of adaptation to climate change. Staff Climate Notes 2022/002. https://doi.org/10.5089/9798400201608.066

Ari I, Koc M (2018) Sustainable financing for sustainable development: understanding the interrelations between public investment and sovereign debt. Sustainability 10(11):3901. https://doi.org/10.3390/su10113901

Baumgartner U, Johnson GG, Dillon KB, Williams RC, Keller PM, Tyler M, Reichmann T (1981) External indebtedness of developing countries. IMF, Washington, DC. https://doi.org/10.5089/9781451935875.084

Bolton R (2023) A history of electricity liberalisation: origins and evolution of the Nordic model. Oil, Gas and Energy Law 21(2)

Carè R, Boitan IA, Stoian AM, Fatima R (2025) Exploring the landscape of financial inclusion through the lens of financial technologies: a review. Financ Res Lett 72:106500. https://doi.org/10.1016/j.frl.2024.106500

Chauhan A (2023) Global corporate clean energy procurement crosses 50 GW with Asia as the largest region in 2022. S&P Global—Commodity Insights. Retrieved from https://tinyurl.com/bdhsxhew (Accessed 20 May 2025)

Davoodi HR, Elger P, Fotiou A, Garcia-Macia D, Han X, Lagerborg A, Medas PA (2022) Fiscal rules and fiscal councils: recent trends and performance during the COVID-19 pandemic. IMF Working Paper No. 2022/011. Washington, DC, IMF

Energy Institute (EI) (2024) In: Energy institute statistical review of world energy 2024. 73rd ed. London, UK, EI. Retrieved from https://tinyurl.com/4pp44xsc (Accessed 20 May 2025)

Fuje Y, Yao L, Choi J, Mighri N (2023) Disasters in emerging market and developing economies (IMF Working Paper No. WP/23/261)

Godlovitch L, Louguet A, Baischew D, Wissner M, Pirlot A (2021) In: Environmental impact of electronic communications: final study report. Bad Honnef, Germany, WIK-Consult and Ramboll

Guo B, Hu P, Lin J (2024) The effect of digital infrastructure development on enterprise green transformation. Int Rev Financ Anal 92:103085. https://doi.org/10.1016/j.irfa.2024.103085

Hamilton C, Stephens H (2022) In: Geoengineering the climate: reining in the "weather warriors." London, UK, Chatham House

Harsono I, Suprapti IAP (2024) The role of fintech in transforming traditional financial services. Account Stud Tax J (COUNT) 1(1):81–91. https://doi.org/10.62207/gfzvtd24

IEA (2024) Electricity 2024. Paris, France, IEA. Retrieved from https://tinyurl.com/yyrysd59 (Accessed 20 May 2025)

International Finance Corporation (2024) In: Repurposing power markets: the path to sustainable and Affordable Energy for all. Washington, DC, IFC. Retrieved from https://tinyurl.com/2s6x2rh7 (Accessed 20 May 2025)

International Monetary Fund (IMF) (2022) Chapter 2—scaling up private climate finance in emerging market and developing economies: challenges and opportunities. In: Global financial stability report: navigating the high-inflation environment. Washington, DC, International Monetary Fund, pp 45–64

International Monetary Fund (IMF) (2023) Climate crossroads: fiscal policies in a warming world [Fiscal monitor]. Washington, DC, IMF. Retrieved from https://tinyurl.com/bdhyb69v (Accessed 20 May 2025)

International Monetary Fund (IMF) (2024) Putting a lid on public debt [Fiscal monitor]. Washington, DC, IMF. Retrieved from https://tinyurl.com/4ycbkada (Accessed 20 May 2025)

Jaumotte F, Schwerhoff G (2021) Reaching net zero emissions. IMF Blog. Retrieved from https://tinyurl.com/3frfjbj9 (Accessed 20 May 2025)

Judijanto L, Qosim N, Syamsulbahri S, Utami EY, Afandy C (2024) Sustainable finance: a bibliometric analysis of green finance and its role in global markets. West Sci Business Managem 2(04):1206–1215. https://doi.org/10.58812/wsbm.v2i04.1500

Kantorowicz J, Collewet M, DiGiuseppe M, Vrijburg H (2023) How to finance green investments? The role of public debt. Energy Policy 184:113899. https://doi.org/10.1016/j.enpol.2023.113899

Kibert CJ, Thiele L, Peterson A, Monroe M (2020) In: Ethics of sustainability. Gainesville, FL, Powell Center, University of Florida. Retrieved from http://rio20.net/wp-content/uploads/2012/01/Ethics-of-Sustainability-Textbook.pdf (Accessed 20 May 2025)

Krogstrup S, Oman W (2020) Macroeconomic and financial policies for climate change mitigation. Working Paper No. 2019/185. Washington, DC, IMF

Li J, Zhang T, Du X, Li Y (2024) How can the digital economy drive low-carbon city performance in China to achieve sustainable development goals? A multiple-output perspective. J Clean Prod 454:142316. https://doi.org/10.1016/j.jclepro.2024.142316

Malmodin J, Lövehagen N, Bergmark P, Lundén D (2024) ICT sector electricity consumption and greenhouse gas emissions—2020 outcome. Telecommun Policy 48(3):102701. https://doi.org/10.1016/j.telpol.2023.102701

Mohammed KS, Serret V, Jabeur SB, Nobanee H (2024) The role of artificial intelligence and fintech in promoting eco-friendly investments and non-greenwashing practices in the US market. J Environ Manage 359:120977. https://doi.org/10.1016/j.jenvman.2024.120977

NGFS (2022) In: Climate scenarios for central banks and supervisors. Paris, France, NGFS Secretariat. Retrieved from https://tinyurl.com/y3z9xs76 (Accessed 20 May 2025)

OECD (2024) Climate finance provided and mobilized by developed countries in 2013–2022, climate finance and the USD 100 billion goal. OECD Publishing, Paris, France. https://doi.org/10.1787/19150727-en

REN21 (2024) In: Renewables 2024 global status report collection, global overview. Paris, France, REN21 Secretariat

Shabalov MY, Zhukovskiy YL, Buldysko AD, Gil B, Starshaya VV (2021) The influence of technological changes in energy efficiency on the infrastructure deterioration in the energy sector. Energy Rep 7:2664–2680. https://doi.org/10.1016/j.egyr.2021.05.001

Tao Z, Chao J (2024) Investing in green, sustaining the planet: the role of fintech in promoting corporate green investment in the Chinese energy industry. J Environ Manage 370:122990. https://doi.org/10.1016/j.jenvman.2024.122990

Taylor, P. (2024). Amount of data created, consumed, and stored 2010–2023, with forecasts to 2028. Retrieved from https://tinyurl.com/bdhvabkr (Accessed 10 March 2025)

Tiftik E, Mahmood K, Aycock R (2024) Winds of change—prospects for debt markets in 2025. IIF Global Debt Monitor. Retrieved from https://tinyurl.com/ynxy49vf (Accessed 20 May 2025)

United Nations Environment Program (UNEP) (2022) In: Emissions gap report 2022: the closing window—climate crisis calls for rapid transformation of societies. Nairobi, Kenya, UNEP. Retrieved from https://tinyurl.com/298c79rt (Accessed 20 May 2025)

Verdejo-Fredes H, García-Muñoz F, Tobar F, Becker C, Olivares M, Zolezzi J, Guzmán G (2022) Retail electricity market liberalization: an overview of international experience and effects on the Chilean regulated tariff. Sustainability 14(21):13996. https://doi.org/10.3390/su142113996

Watson CM, Regling KP (1992) Chapter 4—history of the debt crisis. In: Effros RC (ed) Current legal issues affecting central banks (Volume I). IMF, Washington, DC, pp 67–78

Weber B, Staub-Bisang M, Alfen HW (2016) In: Infrastructure as an asset class: investment strategy, sustainability, project finance and PPP. 2nd ed. Hoboken, NJ, Wiley. https://doi.org/10.1002/978 1119226574

Woetzel J, Garemo N, Mischke J, Hjerpe M, Palter R (2016) In: Bridging global infrastructure gaps. New York, NY, McKinsey Global Institute. Retrieved from https://tinyurl.com/2murzj42 (Accessed 20 May 2025)

World Bank (2024) Ingredients for accelerating universal electricity access: lessons from Rwanda's inspirational approach. Retrieved from https://tinyurl.com/5b84utmt (Accessed 20 May 2025)

World Bank (2024b) Digital progress and trends report 2023. World Bank, Washington, DC. https://doi.org/10.1596/978-1-4648-2049-6

World Bank (2024c) International debt report 2024. World Bank, Washington, DC

World Economic Forum (2024) AI and energy: will AI help reduce emissions or increase power demand? Here's what to know. Retrieved from https://tinyurl.com/e2rcsfyn (Accessed 20 May 2025)

Yu Z, Liu S, Li S (2024) Research on the spatial effect of digital economy development on urban carbon reduction. J Environ Manage 357:120764. https://doi.org/10.1016/j.jenvman.2024.120764

The Pension Fund Resources as a Source for Mortgage Loan Repayments: New Opportunities for Citizens

Karapet E. Movsisyan

Abstract The utilization of accumulated pension funds for mortgage loan payments serves as a crucial tool in enhancing housing affordability. This mechanism can reduce the financial burden on citizens and stimulate the development of the mortgage market, which is particularly important for individuals from lower and middle-income groups who face difficulties in acquiring their own homes. This research aims to examine international experiences, particularly the cases of Sweden and Singapore, and, based on these examples, propose effective models for the use of accumulated pension funds in Armenia. To achieve this goal, it is essential to address potential legislative changes that would support the implementation of such a system. The research will also analyze the possible risks associated with this policy and explore ways to mitigate any negative impact on financial stability. As a result, the research will identify key steps to enhance public trust in the pension accumulation system while aligning with citizens' needs, thereby creating new opportunities for the housing market. Moreover, introducing such a mechanism can contribute to the overall stability of the financial system and strengthen state involvement in social protection. Simultaneously, the increased demand for housing in Armenia's business environment may also stimulate the construction sector, positively impacting the country's economic growth.

Keywords Pension · Fund · Mortgage · Housing · Stability · Opportunity · Development

JEL Codes G21 · G23 · H55 · R21 · R31

K. E. Movsisyan (✉)
Armenian State University of Economics, Yerevan, Armenia
e-mail: karapetmovsisyan@gmail.com

A. B. Karbekova (ed.), *Sustainable Development of Rural Territories Based on Climate-Smart Technologies*, Climate Change Management,
https://doi.org/10.1007/978-3-032-20106-5_2

1 Introduction

In contemporary economic systems, the effective management of accumulated pension funds has become a crucial factor in ensuring the stability of financial markets. In Armenia, the development of mortgage lending and the issue of housing affordability require innovative approaches that can reduce the financial burden on citizens and facilitate homeownership.

Using accumulated pension funds as a source for mortgage loan payments is a novel approach that can contribute to the development of financial markets and the enhancement of citizens' social security. This mechanism is widely implemented in several countries; its potential introduction in Armenia could stimulate investment activity, ensure long-term financial stability, and increase public interest in the pension accumulation system.

The research objective is to assess the feasibility of integrating accumulated pension funds into the mortgage lending system, analyze its benefits and risks, and propose possible policy directions based on international experience.

2 Methodology

This research employs qualitative and quantitative methods to assess the interrelation between accumulated pension funds and mortgage loans. The primary methods used are as follows:

- **Comparative analysis** (Simister and Scholz 2017): The research examines the experiences of Sweden and Singapore, where accumulated pension funds are integrated into mortgage loan payment systems through unique models;
- **Economic modeling** (Christensen and Kiefer 2009): Calculations are conducted to estimate the extent to which the involvement of pension funds in Armenia could reduce borrowers' financial burden.
- **Data analysis** (Eldridge 2025): The authors analyzed statistical data on Armenia's minimum salary, accumulated pension funds, and the mortgage market, comparing trends over different years. This analysis helps evaluate the correlation between salaries and mortgage interest rates, as well as the impact of pension funds on the housing loan market.

3 Results

The research objective is to analyze how the utilization of the accumulated pension system can be applied to make the housing market more accessible for individuals from middle- and low-income groups. For this purpose, the authors examined the experiences of Sweden and Singapore.

Table 1 presents the volume and structure of accumulated pension funds in the countries considered.

According to Table 1, in Sweden, the volume of pension funds constitutes 30% of GDP, indicating the country's long-term financial stability. This figure reflects Sweden's well-established and developed pension system, where public and private components contribute to the strong financial backing of the social security system. Additionally, the fund management model minimizes systemic risks through the joint operation of public and private sectors, enabling efficient resource management.

In the case of Singapore, where pension funds make up 40% of GDP, the system is managed under a centralized model, fully controlled by the Central Provident Fund (CPF). The Singaporean pension system is characterized by government oversight, which allows authorities to direct fund flows, ensuring economic security for the population. Moreover, the centralized model ensures broad coverage, as all employed citizens are included in the system.

In both countries, citizen participation in the pension system is extensive, which is essential for maintaining social stability. In Sweden, after ten years, 80% of accumulated funds can be used for mortgage loan repayment. In Singapore, this figure stands at 70%. However, while Sweden has a more diversified management model, Singapore's centralized approach allows the government to have more direct influence over fund expenditures and investment allocations.

Figure 1 presents the structure of pension investments in Armenia.

In Armenia, most accumulated pension fund resources are directed toward purchasing government bonds (including Eurobonds), which comprise 40.43% of the total investments. Equities also occupy a significant share, accounting for 30.98%, indicating the funds' willingness to invest in market-risk assets. Deposits make up

Table 1 Volume and structure of accumulated pension funds in Sweden and Singapore, 2024

Indicators	Sweden	Singapore
Type of accumulated pension system	Mandatory public and private components	Central provident fund (CPF)
Volume of accumulated pension funds (as % of GDP)	30%	40%
Number of participants	All employed citizens and permanent residents	All employed citizens and permanent residents
Fund management model	Combination of public and private management companies	Centrally managed public fund
Contribution rate from monthly salary	16–18%	20–37%
Percentage of accumulated funds Available for mortgage loan repayment over ten years	80%	70%

Source Compiled by the authors based on data from the Singapore Department of Statistics (2023) and Statistics Sweden (2023)

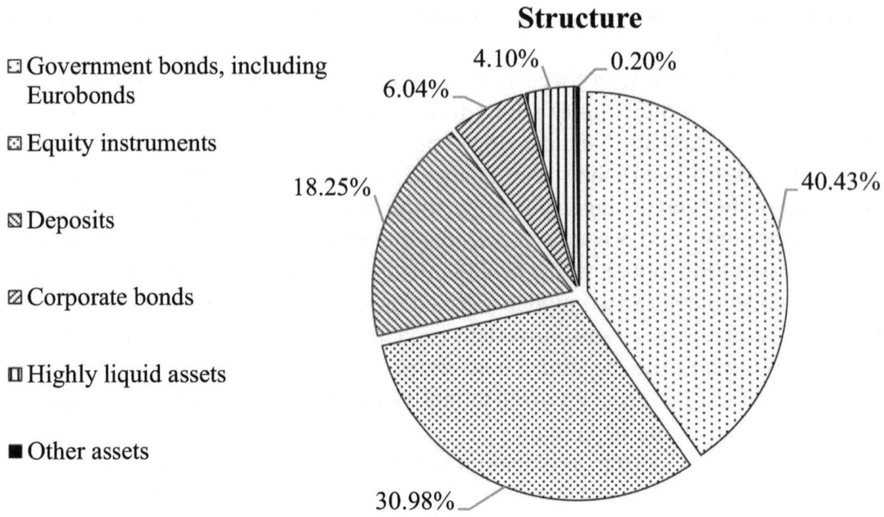

Fig. 1 Structure of pension investments in Armenia, 2023–2024 (average % allocation). *Source* Compiled by the authors based on data from AMX Pension Service (2023)

18.25%, ensuring a degree of stability and guaranteed returns. Corporate bonds account for 6.04%. Highly liquid assets make up 4.10%, providing diversification of investments.

Unlike Sweden and Singapore, where accumulated pension funds can be used for mortgage loan repayments, in Armenia, they are not used for such purposes, which limits the potential for improving accessibility to the housing market.

A significant portion of the accumulated pension fund assets, 67%, is invested in Armenian dram, reducing currency exchange rate risks. Investments in US dollars account for 23%, establishing a connection with international markets. The share in euros is 10%, allowing for some diversification. This structure indicates that the funds are primarily focused on the local economy while maintaining a degree of international investment exposure.

The Swedish pension system is highly developed and has several features, some of which are also related to housing acquisition opportunities:

1. **The accumulation process**: In Sweden, every citizen contributes to their pension by setting aside 16% to 18% of their monthly salary. This amount is directed into their pension account;
2. **Using accumulated funds for a mortgage**: Sweden has a system that allows individuals to use their accumulated pension funds as a down payment for a mortgage loan. Local banks can accept these funds as part of the mortgage payment, covering 10–20% of the housing price, making the home-buying process easier. After ten years, 80% of the total accumulated amount is refunded to the citizen.

If a person works for a monthly salary of 600,000 SEK (Swedish Krona) and saves 108,000 SEK into their pension fund (18%), then over a decade, the accumulated amount would be as follows (Kotok et al. 2010):

$$600000 \times 18/100 = 108000\,SEK \tag{1}$$

$$108000 \times 120 = 12960000\,SEK \tag{2}$$

After ten years, the citizen is refunded 80% of the total accumulated amount, which can be used as a down payment to purchase a house:

$$12960000 \times 80/100 = 10368000\,SEK \tag{3}$$

Singapore also has a well-organized pension system called the Central Provident Fund (CPF). In Singapore, pension funds are more actively used to solve housing issues:

1. The accumulation process: In Singapore, an employed individual is required to contribute between 20 and 37% of their income to the pension fund. These contributions are accumulated in various accounts designed for purchasing a home;
2. Using accumulated funds for a mortgage: In Singapore, citizens can use 70% of the funds accumulated in their CPF over ten years to purchase a home or pay off an existing mortgage.

If a citizen earns a monthly salary of 500 SGD (Singapore Dollars) and contributes 37% of it (185 SGD) to the pension fund, the accumulated amount over a decade would be as follows (Lim and (Stanley) and Cheong 2017):

$$500 \times 37\% = 185\,SGD \tag{4}$$

$$185 \times 120 = 22200\,SGD \tag{5}$$

$$(22200 \times 70)/100 = 15540\,SGD \tag{6}$$

This amount (i.e., 15,540 SGD) can be used for the mortgage payment.

Thus, in Sweden and Singapore, accumulated funds are used as down payments, providing a beneficial opportunity for individuals to purchase a home or pay off their mortgage.

Figure 2 presents the number of registered employees in Armenia.

From 2020 to 2024, the number of registered employees in Armenia has shown steady growth. In 2020, the number was approximately 606,000. By 2024, it had reached around 750,000, indicating an expansion of the labor market. The growth dynamics were especially noticeable in 2022–2023 when the number of registered employees increased by more than 64,000. This trend may be attributed to the economic recovery, stricter tax discipline, and the impact of the mandatory pension

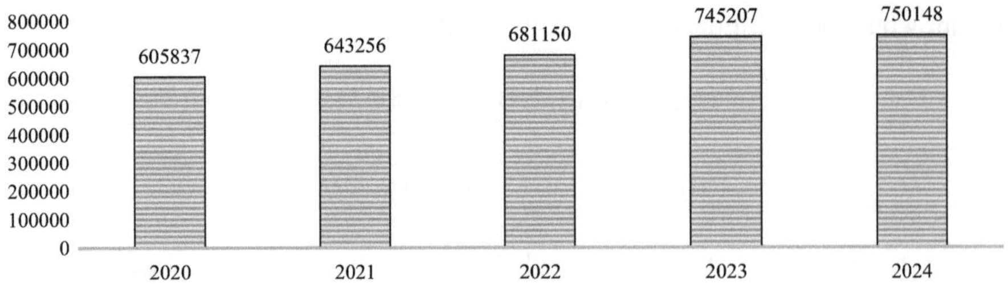

Fig. 2 Number of registered employees in Armenia, 2020–2024. *Source* Compiled by the authors based on the Legal Information System of Armenia (2023)

system. Although the growth rate slowed down in 2024, the overall increase in the number of registered employees is a positive signal for the Armenian economy.

Table 2 shows how much money has been transferred to the pension fund from the minimum salary over the past five years.

Now, using the experience of Singapore and Sweden, the authors will present the possible amount that can be given to a registered working citizen from their accumulated funds. It is possible to use a simple approach by considering the annual contribution and annual rate. The calculation formula will have the following form.

Let the minimum salary for ten years be fixed at 75,000 AMD. We assume the combined contribution from the citizen and the state for one month to be fixed at 10%.

$$(75000 \times 10\%)/100 = 7500\,AMD \tag{7}$$

The monthly contribution is 7500 AMD (citizen's + state's contribution). After a decade, the return rate will be 80% based on the Swedish model. The duration is ten years or 120 months.

$$7500 \times 120 = 900000\,AMD \tag{8}$$

Table 2 Amount transferred from minimum salary to the accumulated pension fund

Year	Minimum salary (AMD)	Employee monthly contribution (%)	State monthly contribution (%)	Employee contribution (AMD)	State contribution (AMD)	Total amount (AMD)
2020	68,000	2.5	7.5	1700	5100	6800
2021	68,000	3.5	6.5	2380	4420	6800
2022	68,000	4.5	5.5	3060	3740	6800
2023	75,000	5	5	3750	3750	7500
2024	75,000	5%	5	3750	3750	7500

Source Compiled by the authors based on data from the Legal Information System of Armenia (2023)

After ten years, the citizen will receive 720,000 AMD.

$$(900000 \times 80)/100 = 720000\,AMD \tag{9}$$

With the above amount, the citizen can pay for the mortgage loan. Considering that salaries vary, it can be noted that the amount could be significantly higher for citizens with higher salaries.

The authors propose that this rule be applicable for citizens earning up to 500,000 AMD (inclusive). If we assume that a citizen earns 500,000 AMD, the calculation will be as follows.

The combined contribution of the citizens and the state is 10%:

$$500000 \times 10\% = 50000 \tag{10}$$

This means the monthly contribution will be 50,000 AMD (citizen's and state's contribution). The duration is ten years or 120 months.

$$50000 \times 120 = 6000000 \tag{11}$$

Then, based on Sweden's example, the return rate will be 80%:

$$6000000 \times 80\% = 4800000\,AMD \tag{12}$$

In this case, if the salary is 500000 AMD, after ten years, the citizen will receive 4,800,000 AMD, which can also be used for mortgage payments or other investment purposes. Thus, as we can see, a higher salary results in a larger accumulated amount, significantly increasing the citizen's financial independence.

Therefore, the effective implementation of the pension system can play a significant role in improving the accessibility of the housing market in Armenia, especially for middle and lower social groups. The experiences of Sweden and Singapore demonstrate that a well-structured policy and mechanisms allow pension fund resources to be directed towards long-term investments, including housing provision.

For Armenia, it is important to consider the successful models of these countries by creating clear and reliable mechanisms that will ensure citizen participation in the pension system and encourage investments in the housing market. Implementing systemic steps in this direction can improve the level of housing security and contribute to stable economic development.

4 Conclusion

Thus, the research showed that the experiences of Sweden's and Singapore's pension systems could serve as valuable foundations for improving the affordability of the housing market in Armenia. Both countries have successfully implemented different approaches, where pension savings are used as a tool to promote higher homeownership rates in the housing market. In Sweden and Singapore, systems that combine public and private components allow citizens to use their accumulated funds to purchase homes or pay off mortgage debts.

First and foremost, it must be considered that the population of Armenia faces significant housing challenges, especially for individuals with middle and low incomes. A developed pension system, which allows citizens to use their pension savings to purchase a home or pay off a mortgage, could become a key tool in making the housing market more accessible. Implementing such a system could provide long-term financial stability, which would reduce social polarization, contributing to broader engagement in the real estate market.

In the context of Armenia, the authors propose introducing a similar pension system model, where citizens can accumulate funds for purchasing a home, with corresponding state subsidies. Within the framework of the proposed system, people would be able to receive updated information about the amount accumulated in their accounts.

Given that Armenia's market faces economic uncertainty and the loans provided by banks often have higher interest rates, the introduction of a pension system could make mortgages more affordable, especially for individuals from lower social backgrounds.

In Armenia, the mandatory pension funds are managed not by the state, as many citizens believe, but by two private companies: the French-rooted "Amundi-ACBA Asset Management" LLC and the Austro-German "C-Quadrat Ampega Asset Management Armenia" LLC. These companies receive fees for their management services, which are charged to the funds. It is also suggested that the management of the pension funds, currently overseen by private companies, be transformed into state ownership, allowing for stricter control over the selection of financial directions.

Based on the experience of Singapore and Sweden, where pension funds are successfully managed by state systems, the authors can say that the nationalization of the funds in Armenia would lead to improvements in several key areas. First and foremost, a state-run system could better manage financial risks, reducing the likelihood of financial mistakes made by private companies that were not carefully considered. Second, state ownership would enhance the transparency of investment strategies and limit the impact of political and economic risks on fund management. Third, the state could directly manage the funds, using them to improve the economic and social well-being of the population.

These steps could lead to a more stable and predictable economic environment and increase opportunities for the lower and middle social classes to acquire housing.

References

AMX Pension Service (2023) Pension system reports [Used for data compilation by the authors]. Retrieved from https://theamx.com/solutions/amxconnect-overview (Accessed 5 Apr 2025)

Christensen JB, Kiefer NM (2009) Economic modeling and inference. Cambridge University Press, Cambridge, UK
Eldridge S (2025) Data analysis. In: Encyclopedia britannica. Retrieved from https://www.britannica.com/science/data-analysis (Accessed 5 Apr 2025)
Kotok DR, Sciarretta V, Stephansen K (2010) Invest in Europe now!: Why Europe's markets will outperform the US in the coming years. Wiley, Hoboken, NJ
Legal Information System of Armenia (2023) National legal framework [Used for data compilation by the authors]
Lim PS (Stanley), Cheong MH (2017) In: Value investing in Asia: the definitive guide to investing in Asia. Chichester, UK, Wiley Finance
Simister N, Scholz V (2017) Qualitative comparative analysis (QCA). INTRAC. Retrieved from https://www.intrac.org/app/uploads/2017/01/Qualitative-comparative-analysis.pdf (Accessed 5 Apr 2025)
Singapore Department of Statistics (2023) Statistical database [Used for data compilation by the authors]. Retrieved from https://tablebuilder.singstat.gov.sg (Accessed 5 Apr 2025)
Statistics Sweden (2023) Macroeconomic indicators [Used for data compilation by the authors]. Retrieved from https://www.scb.se/en/ (Accessed 5 Apr 2025)

Christensen JB, Kiefer NM (2009) Economic modeling and inference. Cambridge University Press, Cambridge, UK

[illegible] (2023) Data analysis [illegible]. Retrieved from https://www.[illegible] 2025

[illegible] [illegible], Seamon [illegible] (2012) [illegible] markets [illegible] the U.S. [illegible]

Legal Information System of Armenia [illegible] (Compiled by the authors)

Lin [illegible], Chang MH (2011) [illegible] investing in Asia [illegible]. Chichester, UK: Wiley Finance

[illegible] (2023) [illegible]

[illegible] (2025)

Statistical Committee of the Republic of Armenia [illegible] (2023) [illegible]

[illegible] (2023) [illegible]
[illegible]

Digital Innovations in the Hotel Industry: Ecological Benefits and Risks

Gayane R. Tovmasyan and Tigran E. Stepanyan

Abstract The hotel industry is increasingly incorporating digital innovations to enhance operational efficiency, guest experiences, and sustainability. Technologies such as smart energy management systems, AI-driven customer service, mobile check-ins, and automated resource monitoring contribute to reduced environmental impact, optimized resource management, and improved sustainability practices. However, these advancements also pose ecological risks, including increased electronic waste, higher energy consumption, and resource-intensive manufacturing. This research explores the main digital innovations in the hotel industry, examining their ecological benefits and potential risks. The research also provides recommendations for mitigating the risks, including the adoption of renewable energy sources, responsible e-waste management practices, smart building technologies, etc. By balancing technological advancement with sustainability goals, the hotel industry can reduce its ecological footprint and contribute to long-term environmental preservation. This research could be valuable to hotel industry professionals, researchers, policymakers, technology providers, environmental organizations, and consumers by providing insights into the ecological benefits and potential risks associated with digital innovations in the hospitality sector.

Keywords Digital innovations · Hotel industry · Sustainability · Ecological benefits · Ecological risks · Resource management

JEL codes L83 · Q01 · Q55

G. R. Tovmasyan (✉)
Armenian State University of Economics, Yerevan, Armenia
e-mail: tovmasyangayane@yahoo.com

G. R. Tovmasyan · T. E. Stepanyan
Public Administration Academy of the Republic of Armenia, Yerevan, Armenia
e-mail: stepanyan84@icloud.com

G. R. Tovmasyan
Yerevan State University, Yerevan, Armenia

Brusov State University, Yerevan, Armenia

A. B. Karbekova (ed.), *Sustainable Development of Rural Territories Based on Climate-Smart Technologies*, Climate Change Management,
https://doi.org/10.1007/978-3-032-20106-5_3

1 Introduction

Digital innovations shape all areas of the economy nowadays (Popkova 2024). The tourism sector is also changing due to new technologies. Nowadays, it is easier for tourists to book hotels online, buy air tickets online, search for tourist information, and plan their travel ahead using travel websites, etc. Digital technologies affect tourists' decision-making (Bilan et al. 2024; Costa Liberato et al. 2018; Pencarelli 2020) and provide new opportunities for developing virtual and smart tourism (Tovmasyan 2021, 2022).

Hotel industries also implement various digital technologies that enable them to increase operational efficiency, decrease environmental impact, better guest experiences, and optimize resource management. Various technologies, including smart energy management systems, mobile check-ins, AI-driven customer service, and automated resource monitoring, help optimize energy and water usage, minimize waste, and promote sustainability. However, these advancements also pose ecological risks, including increased electronic waste and higher energy consumption from data centers.

The research aims to explore the main digital innovations in the hotel industry and analyze their ecological benefits and risks.

2 Materials and Methods

Many authors analyze different aspects of digital innovations in different areas of hotel management (Alrawadieh et al. 2021; Elkhwesky et al. 2024; Varelas et al. 2021; Wynn and Jones 2022).

As hotels try to provide high-quality services (Tovmasyan 2020), they always try to introduce innovations in their performance.

The competitiveness and operational efficiency of hotels rely heavily on digitalization (Bole 2024).

A study by Zaragoza-Sáez et al. (2024) highlights that innovation activities alone do not enhance hotel competitiveness. To fully leverage innovation for efficient operations and differentiated services, hotels must integrate digitalization into their processes.

According to Kolobkova et al. (2021), the application of digital technologies in the hotel industry enhances service quality, reduces service time, boosts hotel competitiveness, and lowers consumer service costs.

According to Suder et al. (2024), digitalization has a significantly positive impact on hotel performance. Moreover, it mediates the relationship between entrepreneurial behaviors and performance. Specifically, digitalization fully mediates the effect of proactiveness on firm growth and the impact of innovation on market performance.

Ben Youssef and Zeqiri (2022) outline five conditions under which Industry 4.0 can contribute to combating climate change in the hospitality sector. First, integrating

Industry 4.0 technologies boosts energy efficiency and lowers greenhouse gas emissions. Second, it optimizes water usage by reducing consumption and enhancing efficiency. Third, it aids in cutting down on food waste. Fourth, it facilitates the advancement of Circular Hospitality 4.0. Finally, it helps mitigate the environmental impact of transportation and travel. The research applies quantitative and qualitative methods, case studies, comparative analysis, and a survey. The authors conducted a survey in 200 hotels in Yerevan and ten regions to learn what kind of digital innovations they implement.

3 Results

3.1 Introducing Hotel Industry in Armenia

In the Republic of Armenia, the Law on Tourism defines hotel services as those related to accommodation, overnight stays, and the organization and provision of food within hotel facilities.

In the Republic of Armenia, the following types of hotel facilities are operating: hotel, aparthotel, motel, sanatorium, hostel, specialized camp, holiday home (boarding house), camping, bed and breakfast (B&B), and guesthouse (Republic of Armenia 2023).

Hotel facilities can be classified based on various qualification standards, which vary by country. In the Republic of Armenia, hotel qualification is voluntary and not mandated by national regulations.

In 2023, the Republic of Armenia had 1140 hotel facilities with a total of 12,507 rooms and 8205 employees. However, out of 1140 facilities, only three had an official star qualification: Multi Grand Hotel (five-star deluxe), Ani Plaza, and Ararat Hotels (four-star) (Ministry of Economy of the Republic of Armenia 2023).

Most hotel facilities in Armenia are concentrated in Yerevan. The other ten regions of the country collectively account for only about 46% of all hotel establishments. Additionally, Yerevan generates 79% of the total volume of accommodation services, highlighting the uneven territorial development of the hospitality sector (Table 1).

Hotels offer numerous economic and social benefits, including the following:

- They generate significant revenue for local businesses and contribute to the state budget;
- The hotel sector creates extensive employment opportunities, including direct jobs (hotel staff) and indirect roles (suppliers and service providers);
- By attracting tourists and investments, hotels play a crucial role in the development of rural and remote areas;
- Beyond accommodation, hotels enhance the visitor experience by offering additional services such as restaurants, spas, and sports centers;
- Many hotels showcase local culture, traditions, and cuisine, enriching the overall tourism experience;

Table 1 Volume of accommodation services and the number of hotel facilities in Yerevan and regions of the Republic of Armenia, 2022–2023

Regions of the republic of Armenia/Yerevan	2022		2023	
	Quantity, unit	Volume, million AMD	Quantity, unit	Volume, million AMD
Yerevan	598	44,986.2	618	45,996.1
Aragatsotn	12	316.3	16	333.1
Ararat	12	121.4	13	122.3
Armavir	21	104.8	23	109.8
Gegharqunik	50	936.4	58	948.5
Lori	37	701.2	44	757.6
Kotayk	118	4469.3	130	4588.3
Shirak	41	546.3	46	755.8
Syunik	50	680.1	58	888.7
Vayots Dzor	31	989.6	36	986.4
Tavush	87	2097.3	98	2719.6
Total for the country	1057	55,948.9	1140	58,206.2

Source Developed by the authors based on the "Tourism" data provided by the Statistical Committee of the Republic of Armenia (2024)

- They collaborate with transport companies, tour operators, and event organizers, strengthening the tourism ecosystem;
- Hotels often provide tourist information, guides, and other assistance to visitors;
- Many hotels implement sustainable initiatives, minimizing environmental impact, supporting local producers, and promoting responsible tourism.

Therefore, the hotel industry is essential for tourism development and can also support regionally balanced socio-economic growth within the country.

Surveys conducted in Armenian hotels reveal that some establishments are adopting various digital innovations. These include online booking and management systems, digital payments, and improved Wi-Fi connectivity. Additionally, a few hotels have started using AI-powered chatbots and virtual assistants, smart room technologies, customer data analytics, VR and AR experiences, as well as sustainable technologies. However, such innovations are not yet widespread and still require further development.

3.2 *Types of Digital Innovations in the Hotel Industry*

In recent years, the following digital innovations have shaped the development of the hotel industry (EHL Graduate School 2024; Lacalle 2023, Swiss Education Group n.d).

- AI and machine learning enhance the ability to deliver personalized experiences for each guest. By analyzing past behaviors and preferences, AI systems can recommend services and amenities that align with individual tastes, ensuring that every stay feels unique. Additionally, these technologies streamline operations, enabling guests to check in, request assistance, or receive quick responses via AI-powered chatbots and voice assistants;
- Contactless check-in and check-out allow guests to skip traditional front desk interactions, reducing wait times and minimizing physical contact. Via mobile apps or kiosks, guests can complete the check-in process, access their rooms using digital keys, and settle their bills seamlessly without the need for direct interaction with reception staff;
- The Internet of Things (IoT) and smart room technology are transforming the hospitality industry by automating key in-room functions like lighting, temperature control, and entertainment systems. Guests can easily adjust these settings using their smartphones or voice commands, offering a more personalized and convenient experience. Beyond enhancing comfort, IoT also plays a crucial role in maintaining high service standards. For instance, IoT systems can detect issues with heating, ventilation, or air conditioning and alert staff to address them before they affect the guest;
- Mobile apps are enhancing the guest experience by integrating advanced features such as virtual concierge services. These allow guests to easily request amenities, book activities, or order room service with just a few taps. Additionally, the apps include feedback options, enabling guests to communicate directly with hotel staff, voice concerns, offer suggestions, share positive feedback, or make special requests;
- Virtual reality (VR) and augmented reality (AR) are revolutionizing the hospitality industry. VR enables potential guests to take virtual tours of hotel rooms, amenities, and nearby attractions before making a booking, helping them make more informed decisions. On-site, AR enhances the guest experience by overlaying digital content onto the physical environment. Guests can use their smartphones to access interactive maps, view detailed information about hotel features, or explore local points of interest. AR also facilitates real-time translations;
- Robotics and automation are streamlining operations in hotels by taking over tasks such as delivering luggage, providing room service, and offering basic concierge services. These innovations improve efficiency by handling routine tasks and add a modern touch that enhances the guest experience. Hotels that incorporate robots typically see an average increase of 15% in operational efficiency, making these technologies a valuable asset;

- Facial recognition technology offers quick and secure check-ins by verifying guests' identities without the need for physical documents like passports or ID cards. This not only speeds up the check-in process but also minimizes human error and ensures that only registered guests can access rooms and other secure areas, enhancing both efficiency and security;
- Sustainable technologies are increasingly important as travelers prioritize eco-conscious choices in their accommodations. In fact, around 81% of travelers plan to select sustainable lodging options. Many hotels are increasingly relying on renewable energy sources, such as solar panels and wind turbines, to significantly lower their carbon footprint. Additionally, water conservation has become a priority, with the adoption of technologies like low-flow fixtures and advanced irrigation systems, enabling hotels to minimize water usage while maintaining high service standards.

Thus, digital innovations offer several advantages to the hotel industry, including increased efficiency, improved communication, and an enhanced guest experience. However, there are also notable disadvantages, such as overreliance on technology, security concerns, job displacement, high implementation costs, ethical concerns, and others.

3.3 Ecological Benefits and Risks of Digital Innovations in the Hotel Industry: Steps Towards Minimizing the Risks

Digital innovations in the hotel industry introduce several ecological benefits, which contribute to sustainability and environmental conservation:

- Energy efficiency: IoT-based systems automatically adjust lighting, heating, and cooling based on occupancy, reducing energy consumption;
- Reduction of paper waste due to digital check-ins and check-outs, e-billing and digital receipts, QR code menus and guides;
- Water conservation is improved through IoT sensors that track water usage in hotel rooms, identifying overuse and leaks in real-time. Smart technology can also be integrated with devices such as low-flow showerheads to reduce water consumption further. Moreover, some hotels implement automated irrigation systems that leverage local weather data to optimize garden watering schedules, effectively minimizing water waste;
- Food waste reduction is supported by IoT sensors that monitor refrigerators, flagging any inefficiencies that could lead to spoilage. These sensors also track expiration dates and notify kitchen staff when food items need to be used. Automated inventory systems use big data to monitor usage, regulate stock levels, and prevent over-ordering, effectively reducing food waste;
- Sustainable transportation and mobility based on electric vehicles;

- Carbon footprint reduction: AI-powered chatbots replace printed directories and reduce human resource energy consumption. Additionally, remote work and virtual meetings help lower business travel emissions by enabling online conferences, minimizing the need for travel and its associated environmental impact;
- Eco-friendly housekeeping: Guests can request services digitally, reducing unnecessary laundry and cleaning.

While digital innovations in the hotel industry offer many ecological benefits, they also come with certain ecological risks that need to be addressed, including the following:

- E-waste generation: Frequent upgrades of digital devices (e.g., smart TVs, tablets, kiosks, and keyless entry systems) contribute to e-waste;
- High energy consumption of digital systems: Smart automation, AI-driven management systems, and cloud computing require significant energy to operate. Data centers supporting hotel management software and reservation platforms consume vast amounts of electricity;
- Resource-intensive manufacturing: The production of digital devices relies on rare earth metals and non-renewable resources, which contribute to environmental degradation;
- Cybersecurity and digital infrastructure maintenance: The need for constant software updates, cybersecurity measures, and system maintenance increases electricity usage. Physical infrastructure like servers and routers requires cooling, adding to energy consumption;
- The carbon footprint of cloud computing and AI: Cloud-based hotel management, AI-driven chatbots, and big data analytics rely on energy-intensive data centers. Large-scale data processing contributes to greenhouse gas emissionsж
- Short lifespan of digital technologies: Rapid technological advancements make digital systems obsolete quickly, leading to frequent replacements and increased e-waste. Many devices are difficult to recycle due to complex material compositions;
- Digital divide and social inequality: High costs of implementing green digital solutions may lead to only high-end hotels adopting them, creating inequality in sustainable practices. Over-reliance on technology may lead to job displacement, affecting social sustainability.

To minimize these risks, hotels should do the following:

- Use renewable energy sources to power digital systems;
- Implement AI-driven energy optimization to reduce unnecessary consumption;
- Use energy-efficient algorithms, opting for software solutions that require less processing power and energy;
- Implement responsible e-waste recycling programs, choose high-quality, durable digital equipment with upgradable software instead of frequent replacements;
- Choose eco-friendly digital suppliers with sustainable practices;
- Optimize cloud usage and data processing efficiency; regularly clean up and optimize hotel management software and databases to reduce energy use;

- Extend the lifecycle of digital devices, focus on software enhancements rather than full device replacements;
- Promote sustainable digital literacy among staff and guests; educate staff on using digital innovations responsibly to reduce environmental impact; encourage guest participation by offering incentives for guests who choose digital check-ins, e-receipts, or eco-friendly digital services;
- Adopt smart building technologies: automated climate control systems, smart lighting systems, intelligent window tinting and insulation;
- Monitor the carbon footprint of digital operations by tracking and publishing reports on energy consumption, e-waste management, and sustainability efforts.

By implementing these strategies, hotels can leverage digital innovations while ensuring minimal environmental harm.

4 Conclusions

Digital innovations have significantly transformed the hotel industry, offering numerous ecological benefits while also introducing potential environmental risks. Technologies such as AI-driven personalization, IoT-enabled smart rooms, contactless services, and sustainable energy solutions contribute to enhanced guest experiences, increased efficiency, and reduced resource consumption. These advancements support sustainability initiatives by minimizing energy waste, optimizing water usage, and promoting eco-friendly practices. However, the increased reliance on digitalization also raises concerns regarding electronic waste, high energy consumption from data centers, potential cybersecurity risks, etc.

In Armenia, the hotel industry plays a prominent role in tourism development, driving economic benefits and job creation while supporting regional development. The adoption of digital technologies in Armenian hotels has the potential to further enhance service quality, competitiveness, and sustainability. However, addressing the ecological risks associated with digitalization requires a balanced approach that includes responsible e-waste management, energy-efficient infrastructure, and regulatory frameworks supporting green innovations.

As the hospitality industry continues to evolve, the integration of digital technologies must align with long-term sustainability goals, ensuring that innovation not only enhances operational efficiency and guest satisfaction but also minimizes the environmental footprint of hotels.

Further studies should explore what kind of digital innovation policies the hotels in Armenia implement and how they minimize the ecological risks of these innovations.

The research may be useful for hotel industry professionals, researchers, policymakers, technology providers, environmental organizations, and consumers by offering insights into the ecological benefits and risks of digital innovations in the hospitality sector.

References

Alrawadieh Z, Alrawadieh Z, Cetin G (2021) Digital transformation and revenue management: evidence from the hotel industry. Tour Econ 27(2):328–345. https://doi.org/10.1177/1354816620901928

Ben Youssef A, Zeqiri A (2022) Hospitality industry 4.0 and climate change. Circular Econ Sustain 2(3):1043–1063. https://doi.org/10.1007/s43615-021-00141-x

Bilan Y, Tovmasyan G, Dallakyan S (2024) The impact of digital technologies on tourists' travel choices and overall experience. J Tourism Serv 15(29):153–175. https://doi.org/10.29036/jots.v15i29.805

Bole DK (2024) Hotel digitalization: exploring the benefits, challenges, and potential solutions. In: Youssef AB, Dutta PK, Doshi R, Sajnani M (eds) AI, blockchain, and metaverse in hospitality and tourism industry 4.0. 1st ed. New York, NY, Chapman and Hall/CRC, pp 12–26

da Costa Liberato PM, Alén-González E, de Azevedo Liberato DFV (2018) Digital technology in a smart tourist destination: the case of Porto. J Urban Technol 25(1):75–97. https://doi.org/10.1080/10630732.2017.1413228

EHL Graduate School (2024) Key hospitality technology trends to watch in 2025. Retrieved from https://hospitalityinsights.ehl.edu/technology-trends-hospitality-industry (Accessed 5 Feb 2025)

Elkhwesky Z, El Manzani Y, Elbayoumi Salem I (2024) Driving hospitality and tourism to foster sustainable innovation: a systematic review of COVID-19-related studies and practical implications in the digital era. Tour Hosp Res 24(1):115–133. https://doi.org/10.1177/14673584221126792

Kolobkova VA, Romanov AA, Frolova EA (2021) Digital technologies in the hotel industry: new prospects for sustainable development. In: Popkova EG, Ostrovskaya VN, Bogoviz AV (eds) Socio-economic systems: paradigms for the future. Cham, Switzerland, Springer, pp 387–394. https://doi.org/10.1007/978-3-030-56433-9_41

Lacalle E (2023) Innovative digital transformation ideas for hotels [Blog]. Mews. Retrieved from https://www.mews.com/en/blog/hotel-digital-transformation (Accessed 5 Feb 2025)

Ministry of Economy of the Republic of Armenia (2023) Information on qualified hotel facilities in the Republic of Armenia. Retrieved from https://mineconomy.am/media/22944/qualified_hotels_03%E2%80%A404%E2%80%A42023.pdf (Accessed 5 Feb 2025)

Pencarelli T (2020) The digital revolution in the travel and tourism industry. Inform Technol Tourism 22(3):455–476. https://doi.org/10.1007/s40558-019-00160-3

Popkova EG (Ed.) (2024) Smart technologies and innovations in E-business. Hershey, PA, IGI Global. https://doi.org/10.4018/978-1-6684-7840-0

Republic of Armenia (2023) Law of the Republic of Armenia "On tourism," Article 12 (adopted on December 22, 2023). Yerevan Armenia. Retrieved from https://www.arlis.am/DocumentView.aspx?DocID=188696 (Accessed 5 Feb 2025)

Statistical Committee of the Republic of Armenia (2024) Tourism. Yerevan Armenia. Retrieved from https://armstat.am/file/article/marzer_2024_18.pdf (Accessed 16 Apr 2025)

Suder M, Duda J, Kusa R, Mora-Cruz A (2024) At the crossroad of digital and tourism entrepreneurship: mediating effect of digitalization in hospitality industry. Eur J Innov Manag 27(4):1057–1081. https://doi.org/10.1108/EJIM-08-2022-0422

Swiss Education Group (n.d.) 8 innovative technologies in the hospitality industry. Retrieved from https://www.cesarritzcolleges.edu/en/news/technology-in-hospitality-industry/ (Accessed 5 Feb 2025)

Tovmasyan G (2020) Evaluating the quality of hotel services based on tourists' perceptions and expectations: the case study of Armenia. J Int Stud 13(1):93–107. https://doi.org/10.14254/2071-8330.2020/13-1/6

Tovmasyan G (2021) Organization of "smart" tourism using technological solutions and innovations. AMBERD Bulletin 3:50–56. https://doi.org/10.52174/2579-2989_2021_3_50

Tovmasyan G (2022) Virtual tourism: travel from home. AMBERD Bulletin 3:136–143. https://doi.org/10.52174/2579-2989_2022.3-136

Varelas S, Karvela P, Georgopoulos N (2021) The impact of information technology and sustainable strategies in hotel branding, evidence from the Greek environment. Sustainability 13(15):8543. https://doi.org/10.3390/su13158543

Wynn M, Jones P (2022) IT strategy in the hotel industry in the digital era. Sustainability 14(17):10705. https://doi.org/10.3390/su141710705

Zaragoza-Sáez P, Marco-Lajara B, Úbeda-García M, Manresa-Marhuenda E (2024) Exploratory and co-exploratory innovation. The mediating role of digitalization on competitiveness in the hotel industry. Technol Forecast Soc Change 199:123069. https://doi.org/10.1016/j.techfore.2023.123069

Challenges of Digital Innovations in the Green Education Ecosystem

Armen G. Tshughuryan, Liana H. Grigoryan, Tigran R. Manukyan, Qnarik H. Khachatryan, and Hayk A. Melkumyan

Abstract The use of digital innovations in the green education system is currently considered a necessity. Given the unique nature of organizing green education, digital tools are being integrated into the learning process, with the potential to evolve into applications powered by artificial intelligence. Therefore, it is considered relevant to analyze the effectiveness of the transition from traditional organization methods to digital transformations in the green education ecosystem. The research examines the advantages of using digital hubs and the possibilities of their formation in the green education ecosystem. Particularly, the authors present recommendations in assessing the effectiveness of information communication and decision-making using digital hubs in the green education organization ecosystem based on artificial intelligence and an aggregate index of green edification results. Based on this approach, indicators are proposed in a numerical format, expressed as the ratio of "actual achievement to intended result." The collection and evaluation of data on actual performance and expected outcomes in the areas of environment and sustainable development are conducted through a digital hub. Consequently, it is appropriate to calculate an aggregate index of green education outcomes that combines behavioral indicators related to business activity, academic performance, and environmental protection.

A. G. Tshughuryan (✉) · L. H. Grigoryan · T. R. Manukyan · Q. H. Khachatryan · H. A. Melkumyan
Armenian State University of Economics, Yerevan, Armenia
e-mail: Jarmen2005@gmail.com

L. H. Grigoryan
e-mail: lianagrigoryan11@gmail.com

T. R. Manukyan
e-mail: tigran-19@mail.ru

Q. H. Khachatryan
e-mail: qnarik91khachatryan@mail.ru

H. A. Melkumyan
e-mail: hayk.melkumyan12@gmail.com

A. B. Karbekova (ed.), *Sustainable Development of Rural Territories Based on Climate-Smart Technologies*, Climate Change Management,
https://doi.org/10.1007/978-3-032-20106-5_4

Keywords Environmental economics · Beneficiaries networking · Green digitization · Digital innovation · Digital hub

JEL Codes O3 · Q5

1 Introduction

The effective organization of green education is a pressing necessity of our time and is closely linked to the goals of sustainable development. However, green education has distinct characteristics that set it apart from traditional educational approaches. (Chassignol et al. 2018). In particular, green education is a dynamic process that accompanies individuals throughout their lives. Therefore, it can be formal, non-formal, and informal. In this case, using digital innovations can significantly affect the organization of green education and the expected learning outcomes. Additionally, the importance of networked communication among green education beneficiaries, collaboration through digital technologies, and ensuring accountability within an eco-social context is strongly emphasized.

The research aims to identify the possibilities of applying digital innovations in the green education ecosystem. Digital innovations in this area can act as tools for ensuring the processing and transmission of information. They provide opportunities for applying artificial intelligence (AI), with which managerial decisions are made (Khosravi et al. 2022). Therefore, the current problem is to observe and analyze how digital innovations contribute to the effective functioning of the green education ecosystem (Cope et al. 2021). In this context, the research object is the scope of digital transformation within the green education ecosystem. The research subject is the evaluation of the effectiveness of digital innovations applied in this domain.

2 Methods

The research uses the results of professional analyses that researchers have obtained while studying digital innovations in business and education. Particularly, the authors analyzed the opportunities for applying digital innovations in the ecosystem of organizing green education and the risks associated with them. The authors paid special attention to the advantages of using digital hubs in terms of coordinating digital flows and the possibilities of drawing qualitative conclusions based on them. A toolkit for measuring the effectiveness of the application of digital innovations was used with performance indicators that provide qualitative assessments of the ratio "actual achievement/intended result" from digital innovations.

3 Results

Green education, organizationally different from traditional education (Andersson and Öhman 2016), operates in a special ecosystem, where informational interconnected manifestations arise (Fig. 1). In turn, informational interconnections require digital innovations that enhance the effectiveness of green education. These innovations help avoid redundancy in educational services, facilitate the coordination and processing of interdisciplinary subject matter, and support data-driven management decisions to improve educational efficiency by using AI.

Thus, when a student transitions from high school to university, chatbots can be used to identify which areas of green education will be offered at the university, and, importantly, to what depth and level of specialization they will be covered (Southworth et al. 2023). After graduating from university, when there is a need to improve environmental awareness through non-formal education, AI is already trying to build optimized non-formal learning programs (Yang et al. 2022). Green education also continues in an informal format, when an individual acquires information on ecological balancing from various sources during daily activities. Similar alternative information is also obtained as a result of community engagement on environmental protection issues. Therefore, it is necessary to constantly interconnect this multi-layered and multi-sectoral information and apply digital innovations in this regard (Zhang and Aslan 2021).

The imperative to introduce digital innovations into the green education ecosystem is shaped by the global environmental crisis (global warming, relentless exploitation of non-renewable resources, and emissions of harmful waste) and the failure to integrate environmental approaches into current education systems (Benavot and McKenzie 2021).

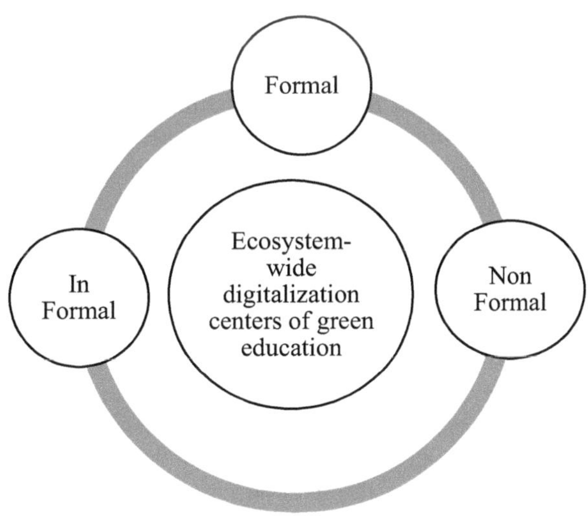

Fig. 1 Areas of digital transformation in the green education ecosystem. *Source* Developed by the authors

The analysis of 100 national curricula shows that almost half of them (47%) do not include awareness of the challenges of climate change and the interpretation of opportunities to prevent the corresponding risks. Only a third of teachers can explain the impact of climate change in their regions. According to UNESCO studies, 70% of young people surveyed in 2021 were unable to describe the risks of climate change (UNESCO 2024).

Green education integrates environmental knowledge with information from disciplines such as geography, biology, and chemistry, as well as essential self-management skills, without which an individual's capacity to protect the environment would be incomplete. In essence, green education requires a digitally transformed, interdisciplinary learning approach. This enables systemic information processing, monitoring of green behavioral norms, and the application of AI to address ecological challenges (Evans et al. 2017).

Green education is not limited to transferring environmental knowledge to students. It forms environmentally friendly behavior in learners. Unlike other forms of education, green education accompanies individuals throughout their lives and often extends beyond the academic setting. It provides environmental knowledge and skills not only in classrooms or university lecture halls but also through activities such as excursions, hikes, field research, and other experiential, extracurricular environments.

Thus, green education is a lifelong learning process aimed at the environmental education of individuals, the manifestation of environmentally friendly behavior, and the realization of opportunities for supporting sustainable development through interdisciplinary learning (Fig. 2).

On the other hand, not being considered discontinuous and accompanying an individual throughout life, green education needs the perception and coordination of multi-layered information flows, which are formed in the academic, business, or public sectors. In other words, the effective delivery of green education currently faces the challenge of establishing a digital hub (Interreg Europe 2020), where diverse information from multiple sources is consolidated into a single node. This creates a large volume of digital data, the processing and coordination of which enables artificial intelligence to function effectively in delivering educational services.

In the digital hub for green education, information inflows are collected for various purposes. For example, data from the public sector often reflects the population's indifferent attitudes toward the environment and the degree to which environmental norms are consistently observed (Arbidane and Khachatryan 2024). Data on the results of green learning and methods of implementing educational programs are transferred from the academic sector to the information hub. The business sector presents data on corporate social responsibility (CSR), sustainable development of economic management, and a green education information hub. Naturally, such a multi-layered database undergoes systemic processing in an information hub, offering innovative digital solutions in the green education ecosystem (Fig. 2).

By analyzing the information aggregated in the digital hub, AI can identify priority areas for reform in green education. For instance, if signals from the business sector indicate shortcomings in social responsibility or sustainable economic development,

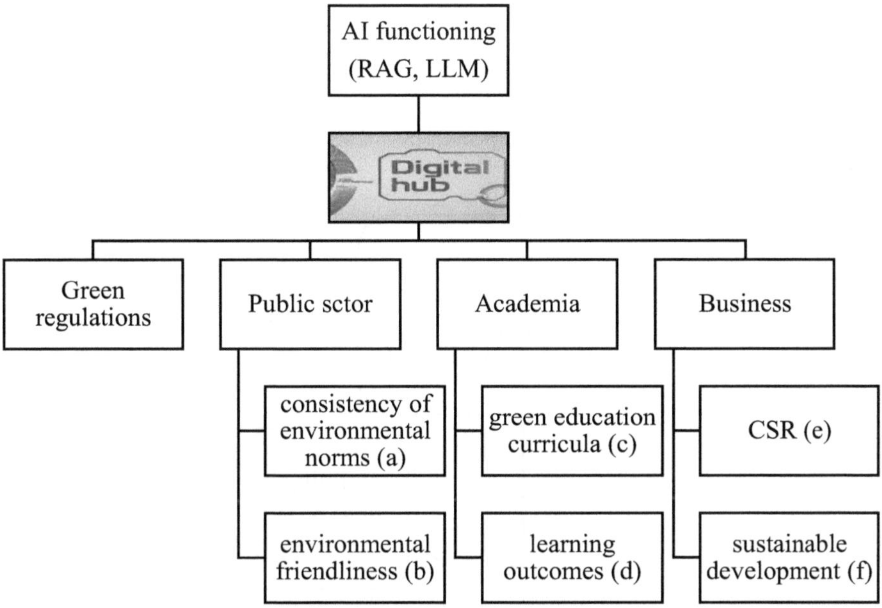

Fig. 2 Design arrangement of green education digital hub. *Source* Developed by the authors

the integration of creative and problem-solving skills becomes a key focus in green education to help prevent environmental harm. Similarly, when data from the public sector reveals inconsistent adherence to environmental protection norms by individuals, AI can generate recommendations for systemic reforms in informal education, targeting behavioral change and awareness (Hopcan et al. 2022).

Therefore, the question arises as to what algorithmic approaches will be used to implement the AI model developed in the digital hub (Ouyang and Jiao 2021). In this regard, the authors propose considering the following priorities: determining the scope of providing the digital tablet with information sources, specifying information communication channels and feedback, and defining indicators for monitoring and improving the existing green education system (Fig. 3).

Digital hubs function most effectively when built within a clearly defined framework of predetermined network communication. In countries with relatively small territories, such as Armenia, it is advisable to establish a centralized digital hub at the national level. This hub would facilitate continuous environmental information exchange among public institutions, the business community, and the academic sector.

For example, local self-government bodies could regularly submit data on public attitudes toward climate change prevention, household waste reduction, and the use of alternative energy sources. Business organizations would contribute information on their achievements and untapped potential in sustainable development. Meanwhile, academic institutions would provide insights related to the structure and delivery

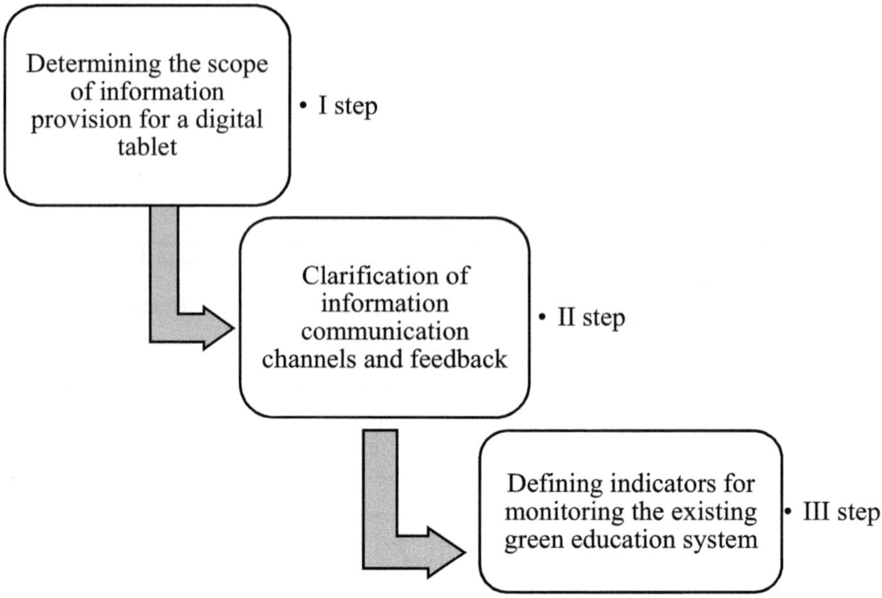

Fig. 3 The process of defining algorithmic priorities for an AI activity model, aimed at improving green education in a digital hub. *Source* Developed by the authors

of green education during program evaluations. To ensure coherence and relevance, the information collected in the digital hub should primarily pertain to intra-national contexts.

A large language model (LLM) is a type of AI that excels at processing, understanding, and generating human language. LLMs are useful for analyzing, summarizing, and creating content across many industries. Retrieval-Augmented Generation (RAG) is an advanced AI architecture that enhances LLMs by integrating them with real-time data retrieval capabilities. Rather than relying solely on static knowledge from its training data, a RAG system retrieves relevant, up-to-date information from authoritative internal knowledge bases and operational systems, using this data as augmented context during generation. This approach enables the LLM to produce more informed, accurate, and reliable responses.

Equally important is that the communication and processing of information within the digital hub be built on the principle of feedback. This ensures that the hub facilitates interconnection among reporting systems across various sectors and enables the formulation of specific qualitative insights related to ecosystem protection. Ultimately, these insights will serve as a foundation for AI-driven decision-making aimed at enhancing green education within the republic.

It is crucial to define the basis on which such decisions will be made. In the authors' view, this can be effectively achieved using performance indicators, expressed numerically as the ratio of "actual achievement to intended result." The digital hub is responsible for collecting and evaluating data on these actual achievements and intended outcomes in the fields of environmental protection and sustainable development. In this context, the authors propose calculating an aggregate index of green education outcomes—one that integrates behavioral indicators related to business activity, academic performance, and environmental stewardship (Fig. 3).

$$\mathbf{Ie} = \mathbf{Ia} \times \mathbf{Ib} \times \mathbf{Ic} \times \mathbf{Id} \times \mathbf{If}$$

Moreover, the closer the proposed aggregate index approaches 1.0 (or exceeds it), the more positively the performance of AI within the digital hub is evaluated in relation to the organization of green education. Conversely, if the index falls short, the AI system generates systemic recommendations for improving green education practices.

4 Conclusions

The ecosystem of organizing green education contains multifaceted information flows used by various stakeholders. Moreover, effective management of information flows requires the use of digital innovations. Digital transformations of organizing green education also require a systemic approach, when information flows are collected in a digital pill from business, academic, and community platforms. The inclusion of "green regulations" defined by legislative and sub-legislative normative acts is also emphasized (Fig. 2). This enables informational and analytical comparisons between strategic goals and actual achievements in green behavior, particularly in relation to the activities of both the private and public sectors.

In this regard, the authors propose to eventually include a framework of AI in the digital transformations of the ecosystem to organize green education, where decisions are made on green education improvements using RAG data processing and LLM tools. Moreover, AI interacts with stakeholders through feedback mechanisms by presenting behavioral indicators related to business activity, academia, and environmental protection. These indicators clearly highlight the strengths and weaknesses of the existing green education ecosystem, serving as the basis for informed decisions to improve the delivery of educational services.

References

Andersson E, Öhman J (2016) Young people's conversations about environmental and sustainability issues in social media. Environ Educ Res 23(4):465–485. https://doi.org/10.1080/13504622.2016.1149551

Arbidane I, Khachatryan N (2024) Challenges of green school quality assessment. Econ Finan Account 2(14):23–33. https://doi.org/10.59503/29538009-2024.2.14-23

Benavot A, McKenzie M (2021) Learn for our planet: a global review of how environmental issues are integrated in education. Paris, France, UNESCO. Retrieved from https://unesdoc.unesco.org/ark:/48223/pf0000377362 (Accessed 19 Apr 2025)

Chassignol M, Khoroshavin A, Klimova A, Bilyatdinova A (2018) Artificial Intelligence trends in education: a narrative overview. Proc Comput Sci 136:16–24. https://doi.org/10.1016/j.procs.2018.08.233

Cope B, Kalantzis M, Searsmith D (2021) Artificial intelligence for education: knowledge and its assessment in AI-enabled learning ecologies. Educ Philos Theory 53(12):1229–1245. https://doi.org/10.1080/00131857.2020.1728732

Evans NS, Stevenson RB, Lasen M, Ferreira J-A, Davis J (2017) Approaches to embedding sustainability in teacher education: a synthesis of the literature. Teach Teach Educ 63:405–417. https://doi.org/10.1016/j.tate.2017.01.013

Hopcan S, Polat E, Ozturk ME, Ozturk L (2022) Artificial intelligence in special education: a systematic review. Interact Learn Environ 31(10):7335–7353. https://doi.org/10.1080/10494820.2022.2067186

Interreg Europe (2020) Digital innovation hubs. Retrieved from https://www.interregeurope.eu/good-practices/digital-innovation-hubs (Accessed 19 Apr 2025)

Khosravi H, Shum SB, Chen G, Conati C, Tsai Y-S, Kay J, Gašević D (2022) Explainable artificial intelligence in education. Comput Educ: Artif Intell 3:100074. https://doi.org/10.1016/j.caeai.2022.100074

Ouyang F, Jiao P (2021) Artificial intelligence in education: the three paradigms. Comput Educ: Artif Intell 2:100020. https://doi.org/10.1016/j.caeai.2021.100020

Southworth J, Migliaccio K, Glover J, Glover J, Reed D, McCarty C, Thomas A (2023) Developing a model for AI Across the curriculum: transforming the higher education landscape via innovation in AI literacy. Comput Educ: Artif Intell 4:100127. https://doi.org/10.1016/j.caeai.2023.100127

UNESCO (2024) Green school quality standard: greening every learning environment. Paris, France: UNESCO. Retrieved from https://www.unesco.org/en/articles/green-school-quality-standard-greening-every-learning-environment (Accessed 19 Apr 2025)

Yang D, Zhao W, Du J, Yang Y (2022) Approaching artificial intelligence in business and economics research: a bibliometric panorama (1966–2020). Technol Anal Strategic Managem 36(3):563–578. https://doi.org/10.1080/09537325.2022.2043268

Zhang K, Aslan AB (2021) AI technologies for education: recent research and future directions. Comput Educ: Artif Intell 2:100025. https://doi.org/10.1016/j.caeai.2021.100025

Smart Entrepreneurship in Industry: Development Opportunities in the Conditions of Geo-Economic Fragmentation

Sergey A. Makovetsky, Aleksey V. Matyushin, Svetlana S. Bukreeva, and Anna S. Savostyanova

Abstract The research aims to identify the potential for the growth of smart entrepreneurship in the context of the relationship between trade openness and GDP growth amid the emerging reality of geoeconomic fragmentation. The global economy's vulnerability to financial and economic crises compels industrial enterprises to prioritize adopting new technologies. Analyzing successful international practices supports the effective application of achievements from various national programs and strategies while helping avoid past mistakes in designing and implementing smart industry development strategies. This approach contributes to cost reduction and risk mitigation. Using regression analysis, the authors examined how the openness index impacts GDP growth rates in the two major global blocs: BRICS and the G7. The authors provided a five-year forecast for the potential development of smart entrepreneurship in the Russian Federation within the BRICS framework. A horizontal analysis was conducted to examine the dynamics of growth rate changes. The research concludes that international organizations significantly involved in global trade have been largely ineffective, with protectionist policies contributing to the deterioration of international trade. The authors found that G7 countries are less dependent on trade openness than BRICS nations. As a result, given the lower economic baseline of the BRICS countries, they possess greater potential for advancing smart entrepreneurship.

Keywords Geoeconomic fragmentation · Outdoor index · BRICS · G7 · Smart entrepreneurship · Industry · Trade

JEL Codes F02 · L51 · L52 · O11 · O40

S. A. Makovetsky (✉) · A. V. Matyushin · S. S. Bukreeva · A. S. Savostyanova
Free Economic Society, Moscow, Russia
e-mail: ups.dn@mail.ru

A. B. Karbekova (ed.), *Sustainable Development of Rural Territories Based on Climate-Smart Technologies*, Climate Change Management,
https://doi.org/10.1007/978-3-032-20106-5_5

1 Introduction

Against the backdrop of current international political and economic events, countries are increasingly becoming participants in one of the two main power blocs—either BRICS or the G7. Vishnevsky et al. (2025) justified that 70% of global production and trade falls on two world blocks. That is, more than two-thirds of the world economy is concentrated in 12 camps. Geo-economic fragmentation gives rise to conflicting processes in international economic relations due to political factors. On the one hand, it reduces the openness in interaction with unfriendly countries, limiting global cooperation. On the other hand, this promotes the development of "macro-regional globalization," which strengthens economic ties between partners and stimulates inclusive economic growth within such associations on "their own" and "strangers." As a result, the key characteristics of international economic relations amid geoeconomic fragmentation include the leading powers' retreat from unconditional support for globalization, the increasing use of force or threats to resolve economic disputes, and the growing division of the world into "allies" and "outsiders." Consequently, the authors selected these two blocs of countries for comparative analysis.

Geo-economic fragmentation and related changes in international economic relations create new challenges and opportunities for business. In conditions when globalization is no longer an unconditional priority, and trade and investment ties are increasingly formed based on geopolitical preferences, entrepreneurs are forced to adapt to a new reality. This requires rethinking the strategies for entering the markets, building the supply chains, and choosing partners, considering political risks and sanctions regimes.

In this dynamic environment, the concept of smart entrepreneurship becomes especially relevant. It involves flexibility, adaptability, and the ability to work in conditions of uncertainty, which has recently increased due to fragmentation using advanced technologies, big data analytics, and digital platforms for effective business management. Smart entrepreneurship focuses on developing innovative solutions that help minimize the risks associated with the fragmentation of the global economy and identify new opportunities for growth.

Companies that adopt digitalization and automation gain a substantial competitive edge by swiftly adapting their logistics processes, reducing reliance on volatile markets, and forming partnerships with aligned economic blocs. Additionally, innovative technologies like artificial intelligence and blockchain enhance the transparency and security of financial transactions, which is increasingly crucial in today's fragmented global landscape (Kurkarevich and Turban 2022).

Consequently, intellectual entrepreneurship is becoming a way of survival and a strategy for stable development in the context of geo-economic fragmentation.

The most important feature of smart entrepreneurship in the conditions of geo-economic fragmentation is the ability to quickly respond to changes in the external environment. This includes flexibility in resource management and strategic forecasting, which allows for minimizing potential threats and using opening opportunities. Industrial enterprises that leverage advanced technologies and analytical tools can rapidly adapt to changes in international trade, taxation, sanctions policies, and regulatory requirements.

Additionally, the digital transformation of the business opens up opportunities for more effective management of finance, marketing, and operating activities (Gataullin et al. 2021). Thus, in conditions of geo-economic fragmentation, smart entrepreneurship becomes not just a competitive advantage but a necessary tool for survival and development. It allows companies to maintain stability in a changing global environment, find new markets and partners, and build long-term growth strategies.

Intelligent entrepreneurship and global trade openness are closely interconnected, as enterprising business leaders actively leverage international trade opportunities to grow their businesses. They utilize advanced technologies (e.g., artificial intelligence, blockchain, and the Internet of Things) to enhance product and service quality and optimize management and logistics processes.

2 Literature Review

The G7 economies are historically more open to trade and investment than the BRICS economy. As several researchers note, this is attributed to their more developed economies, membership in the WTO and other international organizations, and a strong commitment to the principles of free trade (Grabner et al. 2021; Ha and Ngoc Theng 2023; Haas 2019; Kavlin 2024; Larionova and Shelepov 2019).

In recent decades, the BRICS economy, especially China and India, has demonstrated growing openness (Hölscher et al. 2010). A significant factor in economic integration into the world economy was the entry of China into the WTO in 2001. However, some BRICS countries (e.g., Russia) pursue a more protectionist policy.

Many studies (e.g., the work of the IMF and the World Bank) show that the openness coefficient in the G7 countries is, on average, higher than in the BRICS countries (Becker et al. n.d.). However, in this regard, China often reaches levels comparable to those of some G7 countries. For example, V. Portyakov (2019) demonstrated that China's openness coefficient was approximately 35%, compared to about 27% in the United States.

Despite the significant amount of research, it is necessary to highlight the following factors that change the openness of the economy:

1. The higher the level of a country's economic development, the greater its openness, making the level of development a key determinant;
2. Membership in the WTO and other international organizations helps reduce trade barriers and attract foreign investment;
3. Political stability and the rule of law typically foster a more favorable investment climate;
4. Geographic location, particularly proximity to major trade routes, supports the development of industry and trade;

5. Domestic government policy on production, trade, investment, and currency regulation is one of the most significant factors influencing overall economic openness.

Thus, the openness of the BRICS and G7 economies varies significantly. The G7 countries have great resources and experience, while the BRICS countries actively build their capabilities and seek to take leading positions in the global economy.

Global openness of international trade allows smart entrepreneurs to expand their client base and increase income. They can enter new markets, cooperate with foreign partners, and invest in foreign projects. This stimulates the economic growth and development of technology.

3 Materials and Methods

To obtain the most reliable results, the authors used a quantitative-quality methodology in accordance with the systemic approach.

For this purpose, the authors calculated the Trade Openness Index for BRICS and G7 countries from 2001 to 2023, using the ratio of total exports and imports to GDP based on data from the World Development Indicators (WDI). They then computed the growth rate of the openness index. Additionally, for the same period, they determined the population growth rate and the GDP growth rate (in 2015 constant prices). The results are presented in Table 1.

Source Developed by the authors based on (World Bank 2025).

The research model looks as follows:

$$y = a + b^{*} x \tag{1}$$

where:

y—GDP growth rate per 1 person;

a, b—coefficients;

x—opening index growth rate.

The reliability of the regression model is assessed using Fisher's F-criterion and the correlation coefficient. In the second stage, a five-year forecast is developed to evaluate the impact of the openness coefficient on GDP growth within the respective country groups.

4 Results

To determine the influence of potential factors on the economic development of the BRICS and G7 countries, the authors carried out a factor analysis of the openness index in 2023. As a result, the following economy metric model for BRICS was

Table 1 Dynamics of the growth rate of openness index and per capita GDP in the BRICS and G7 countries

Year	BRICS		G7	
	Openness index growth rate	GDP growth rate, 1 person, %	Openness index growth rate	GDP growth rate, 1 person, %
2002	3.512	5.210	−1.159	0.584
2003	28.752	6.259	2.658	1.314
2004	9.488	7.347	7.880	2.329
2005	4.964	7.524	6.863	1.949
2006	2.812	8.740	9.416	1.935
2007	−3.027	9.995	5.813	1.520
2008	2.235	6.379	2.595	−0.683
2009	−18.979	4.415	−17.772	−4.270
2010	8.465	8.001	9.917	2.285
2011	2.481	6.510	8.033	1.013
2012	−2.198	5.272	−1.782	0.858
2013	−0.942	5.239	0.921	1.030
2014	−3.412	4.834	0.790	1.455
2015	−7.732	4.062	−5.206	1.645
2016	−6.431	4.380	−3.324	0.992
2017	1.451	4.968	4.723	1.773
2018	3.488	4.983	3.249	1.612
2019	−3.862	4.133	−2.767	1.238
2020	−3.460	−0.405	−11.100	−4.545
2021	9.345	7.489	9.651	5.467
2022	2.735	2.811	10.492	2.076
2023	−4.487	4.742	−3.699	1.348

obtained:

$$y = 5.4844 + 0.0886^{*}x \quad (2)$$

According to the resulting model (2) for the BRICS countries, a 1% annual increase in the openness index leads to a 0.0886-point increase in GDP per capita.

The following economy metric model was obtained for the G7 countries:

$$y = 0.668 + 0.2274^{*}x \quad (3)$$

According to model (3) for the G7 countries, a 1% annual increase in the openness index results in a 0.2274-point increase in GDP per capita.

Let us move on to the detailed results of regression analysis (Table 2).

Table 2 Detailed analysis for BRICS and G7 BRICS

Regression statistics					
Multiple r	0.359428				
R^2	0.129188				
Normalized R^2	0.085648				
Standard error	2.095484				
Observations	22				
Dispersion analysis					
	Df	*SS*	*MS*	*F*	*Significance F*
Regression	1	13.02862	13.02862	2.967081	0.100404
Remainder	20	87.82109	4.391054		
Total	21	100.8497			

G7

Regression statistics					
Multiple r	0.791252				
R^2	0.626079				
Normalized R^2	0.607383				
Standard error	1.296789				
Observations	22				
Dispersion analysis					
	Df	*SS*	*MS*	*F*	*Significance F*
Regression	1	56.31422786	56.31423	33.48727	1.16E-05
Remainder	20	33.6332195	1.681661		
Total	21	89.94744737			

Source Calculated and compiled by the authors

The coefficient of determination for the G7 model indicates a stronger and more reliable impact of openness on economic growth. For BRICS, this influence is relatively weak. This suggests that openness plays a less significant role in GDP growth for BRICS countries. The economies of the G7 are largely driven by international trade, while BRICS economies rely more heavily on domestic consumption. As a result, BRICS countries are less dependent on the international exchange of goods and services, making them more self-sufficient. However, this self-sufficiency also points to a greater untapped potential for expanding international trade and, consequently, increasing production. For countries like Russia, this implies that economic potential should be redirected from raw material extraction to producing goods with high value-added. This shift further supports the potential for the growth of smart entrepreneurship in both industry and trade.

Smart entrepreneurship positively affects global trade and production as follows:

1. It accelerates export-import transactions. That is, the mechanisms of e-commerce reduce the time necessary for the production, export, and import of goods and services;
2. It expands the circle of participants in the cross -border turnover, involving new resources in production and trade and expanding the circle of potential buyers and manufacturers;
3. It facilitates international production and cross-border trade. Smart contracts automate customs clearance, calculation of tariffs, and verification of conformity;
4. It facilitates efficient supply chain management. Smart contracts enable real-time tracking of goods, automate inventory control, and streamline coordination among all involved parties.

According to WTO forecasts, by reducing costs and increasing performance, digital technologies can provide additional growth in international trade in the amount of up to 34% by 2030.

According to the statistics, 81% of online buyers conduct research before making a purchase, with 60% beginning their search on a search engine before visiting a website with product descriptions. Moreover, 66% of consumers conduct this research on a PC or laptop at home, while 15% use a mobile device. This highlights the growing importance of optimizing digital resources for all screen types, making responsive design a critical priority for marketers.

Smartphones have become the primary device for online shopping worldwide. In the USA, 73% of respondents use their smartphones for purchases. In China, 92% of respondents shop online via their phones. In India, the figure stands at 88% (Andersen 2025).

The current state of the global economy is marked by economic instability and a profound political crisis of trust. The rise of populism, nationalism, and xenophobia, coupled with the rejection of globalization, open markets, and immigration, has created an atmosphere of distrust that undermines effective international cooperation. As Richard Haass aptly observed, the term "new world mess" more accurately captures the current reality, characterized by unresolved conflicts in regions such as the Middle East and Ukraine, along with escalating tensions in areas ranging from cyberspace to the space race. The absence of trust and a functional system of global governance lays the groundwork for the worsening of global challenges. Over the long term, this environment may contribute to slower economic growth and heightened geopolitical instability.

5 Conclusion

Thus, the transition from the traditional production model based exclusively on in-house production capacities to a hybrid smart model of distributed automated production has several undeniable advantages and makes it possible to achieve the following effects:

1. Thanks to distributed production, industrial enterprises can leverage the resources of various suppliers and partners, enabling them to optimize the use of production capacities and reduce costs;
2. The smart model enables industrial production to quickly adapt to market changes and evolving consumer demands, significantly shortening the time between product development and market release;
3. The system of production adaptation and cooperation accelerates the exchange of technological expertise, leading to improved product quality and, consequently, higher customer satisfaction;
4. Automation of design and production optimize design and production processes, reducing the time and costs of developing new products;
5. Thanks to a quick response to changes in the market situation and improving the quality of manufactured products, industrial enterprises better meet customer needs.

Thus, transitioning from a traditional model to a hybrid model enables industrial enterprises to enhance production efficiency by reducing costs, accelerating the development and launch of new products, improving product quality, and increasing customer satisfaction.

To successfully shift to smart manufacturing, an industrial enterprise must initially ensure real-time access to critical information throughout the production process. It is also essential to eliminate manual verification steps, where errors are most likely to occur and establish a cyber-physical feedback loop between production and planning.

Once these foundational improvements are implemented and tested, the next step is to simulate the physical environment by creating a virtual replica of the production floor. This digital twin enables automation and ensures that products meet quality standards through continuous real-time monitoring and measurement.

References

Andersen, D. (2025). 37 statistics retail marketers need to know in 2025. Invoca [Blog]. Retrieved from https://www.invoca.com/blog/retail-marketing-statistics. Accessed 15 Mar 2025

Gideon Becker, Claudius Heimberger, Philipp Kapeller, Jakob Springholz, Florian (n.d.). Measuring economic openness: A review of existing measures and empirical practices. Retrieved from https://www.econstor.eu/bitstream/10419/193623/1/WP84.pdf. Accessed 15 March 2025

Gataullin TM, Gataullin ST, Ivanova KV (2021) Modeling an electronic auction. In: Popkova EG, Sergi BS (eds) Smart technologies for society, state and economy. Springer, Cham, Switzerland, pp 1108–1117. https://doi.org/10.1007/978-3-030-59126-7_122

Grabner C, Heimberger P, Kapeller J, Springholz F (2021) Understanding economic openness: a review of existing measures. Rev World Econ 157:87–120. https://doi.org/10.1007/s10290-020-00391-1

Ha TM, Ngoc Theng D (2023) Economic sanction and global sourcing complexity: a cross-country analysis. World Econ 46(4):1017–1050. https://doi.org/10.1111/twec.13350

Haas, R. (2019). A world in disarray: American foreign policy and the crisis of the old order (V. Zhelninova, Transl. from English). AST Publishing House:Moscow, Russia, (Original work published 2017)

Hölscher J, Marelli E, Signorelli M (2010) China and India in the global economy. Econ Syst 34(3):212–217. https://doi.org/10.1016/j.ecosys.2010.02.002

Kavlin, D. B. (2024) Big game: Where BRICS and G7 lead the global economy *Kommersant*. Retrieved from https://www.kommersant.ru/doc/7219509. Accessed 15 Mar 2025

Kurkarevich TS, Turban GV (2022) Digital transformation of international business. Minsk, Belarus: Rivsh

Larionova MV, Shelepov AV (2019) Group twenty, BRICS and "Semey group" in the global economic administration. Bulletin of International Organizations 14(4):48–71

Portyakov V (2019) Policy of foreign economic openness in the People's Republic of China: Main results and prospects. Outlines of Global Transformations: Politics, Economics, Law 12(1):44–68. https://doi.org/10.23932/2542-0240-2019-12-1-44-68

Vishnevsky VP, Makovetsky SA, Vishnevskaya EN (2025) Effects of geoeconomic fragmentation: BRICS vs G7. Journal of Applied Economic Research 24(1):6–33. https://doi.org/10.15826/vestnik.2025.24.1.001

World Bank (2025) World Development Indicators (WDI): World economic demography database. Retrieved from https://data.worldbank.org/. Accessed 15 Mar 2025

Sustainability Reporting as a Key Approach to Promoting Sustainable Innovation in Entrepreneurship for Addressing Global Economic Challenges

Tatyana U. Druzhilovskaya, Emilia S. Druzhilovskaya, Yulia V. Prikazchikova, Yuliia V. Zakharova, and Tatyana V. Stozharova

Abstract This research aims to identify current challenges in sustainability reporting (SR) that hinder the promotion of sustainable innovation in entrepreneurship as a means of addressing global economic issues (SIEGE) and propose solutions to overcome these obstacles. During the research, the authors analyzed the SR of organizations, which is publicly available on their official websites. The authors also studied the SR of organizations from different countries worldwide in various fields for 2022–2023. This approach was used to obtain results on the problems of the formation of SR related to improving the ensuring SIEGE, which occurs in various existing organizations. The research employed analysis, synthesis, comparison, grouping, and logical approach. The authors studied the existing Russian and international practices of forming sustainability reporting and identified the most relevant problems in this area related to improving and ensuring SIEGE. The authors then specified problems and combined them into several large groups. As a result of the conducted research, the authors present recommendations for solving the identified problems, contributing to ensuring SIEGE. The original recommendations presented in this research can be used to improve regulatory frameworks to form SR at the national and international levels.

Keywords Sustainable development · Sustainability reporting (SR) · Sustainable innovation · Global economy · Entrepreneurship

JEL Code M41

T. U. Druzhilovskaya · E. S. Druzhilovskaya (✉) · Y. V. Prikazchikova · Y. V. Zakharova · T. V. Stozharova
Lobachevsky State University of Nizhni Novgorod, Nizhny Novgorod, Russia
e-mail: druzhilovskaya@yandex.ru

A. B. Karbekova (ed.), *Sustainable Development of Rural Territories Based on Climate-Smart Technologies*, Climate Change Management,
https://doi.org/10.1007/978-3-032-20106-5_6

1 Introduction

Sustainability reporting (SR) is a key component in promoting sustainable innovation in entrepreneurship as a means of addressing global economic challenges (hereinafter referred to as ensuring SIEGE). The development and continuous improvement of such reporting enable stakeholders to access vital information needed for informed decision-making in today's dynamic environment.

The authors also revealed that significant challenges persist in developing sustainability reporting (SR). This research is dedicated to analyzing these issues in detail.

This research examines the main problems arising in preparing SR by Russian and international organizations, which hinder improving the quality of ensuring SIEGE. The problems identified in this area have been grouped into several major categories. Additionally, this research offers recommendations for addressing these challenges, which can be applied to develop sustainability reporting (SR) for Russian and international organizations.

2 Methodology

During the research, the authors analyzed SR published on the official websites of various organizations. The research focused on the SRs of well-known companies prepared for the years 2022 and 2023. The authors examined Russian and international organizations from diverse sectors. This approach was used to ensure that the findings are relevant and applicable across a wide range of existing organizations.

In recent years, scientific literature has paid much attention to issues related to the formation of financial information (Abu Alia et al. 2024; Druzhilovskaya et al. 2020; Kulikova et al. 2018; Kulikova et al. 2020) and data on the sustainable development of organizations (Al-Shaer and Hussainey 2022; Debnath et al. 2024; Di Tullio and La Torre, 2022; Jamil et al. 2021; Kılıçarslan 2024; Muhammed 2024; Roy and Tanna 2021; Stănciulescu 2024; Ya et al. 2024; Tilt et al. 2021). Different authors consider various aspects of this area. Some scientists analyze the specifics of SR in organizations in different fields of activity. Such publications include the following:

- The works of Debnath et al. (2024), who examine the specifics of the SR for the banking sector;
- The works of Di Tullio and La Torre (2022) and Ya et al. (2024) focus on universities;
- The study by (Muhammed 2024) that consider non-financial organizations;
- Jamil et al. (2021), who examine the specifics of the SR for corporate government structures.

The countries whose experience scientists consider in relation to the development of the SR are also not the same. For example, (Muhammed, 2024) analyzed the specifics of such reporting in Nigeria, (Roy and Tanna, 2021)—India, (Kılıçarslan, 2024); (Stănciulescu 2024); (Jamil et al. 2021; (Tilt et al. 2021); (Al-Shaer and Hussainey, 2022); (Di Tullio and La Torre 2022); (Ya et al. 2024)—Ireland.

The works of the authors mentioned, along with those of many others, offer in-depth studies on issues related to sustainability reporting (SR). However, significant challenges in developing such reporting still persist, posing a major obstacle to enhancing the effectiveness of ensuring SIEGE. This research is dedicated to analyzing these challenges and proposing solutions to address them.

The research employed analysis, synthesis, comparison, grouping, and logical approach.

3 Results

The authors analyzed the practice of forming sustainability reporting of well-known Russian and international organizations in various fields and identified several significant challenges that complicate the use of SR to ensure sustainable innovation in entrepreneurship to solve global economic problems.

The research has shown that problems begin with the titles that organizations use to designate their SR. Table 1 shows examples of various titles for this reporting used by Russian and international organizations. It is important to note that Table 1 is based on the analysis of sustainability reports (SR) prepared in 2023 by 60 organizations.

Source Developed by the authors

Table 1 Examples of inconsistencies in the titles of SR of organizations

Titles of SR used by different organizations	Number of organizations using such titles	The percentage of these organizations (%)
Sustainability report	33	55
Annual report	18	30
Social report	3	5
Report about activities in the field of sustainable development	1	1,(6)
Public report	1	1,(6)
ESG report	1	1,(6)
Impact report	1	1,(6)
Corporate social responsibility and sustainable development	1	1,(6)
Consolidated report	1	1,(6)
Total	60	100

Table 1 clearly illustrates significant variations in the titles of sustainability reports (SR). As a result, users may struggle to identify which reports contain the information they need, negatively affecting the effectiveness of ensuring SIEGE.

The research also revealed varying approaches among organizations regarding the number of sustainability reports (SR) issued within a single reporting period. Some organizations consolidate all necessary information into a single comprehensive report, while others distribute the information across multiple reports, such as a "Social Report" and an "Environmental Report." In some cases, organizations publish both a sustainability report and separate social and environmental reports on their websites. Since sustainability reports typically include both social and environmental aspects, this practice can create confusion for users about whether the sustainability report alone contains all the required information or if additional reports must be consulted. Such inconsistencies hinder the overall quality and effectiveness of ensuring SIEGE.

The next group of problems is the discrepancy between the sections within which organizations systematize information in their SR. Figure 1 presents examples of such discrepancies. The names of the organizations examined have been replaced with conditional ones (i.e., "A", "B," and "C").

As shown in Fig. 1, the structure and sections of sustainability reports (SR) vary significantly in practice. This inconsistency complicates users' ability to analyze the reports effectively and make informed decisions based on them. Consequently, this undermines the overall quality and effectiveness of ensuring SIEGE.

Another group of problems is the discrepancy between the information provided by organizations in the same sections of their SR. The examples of such differences

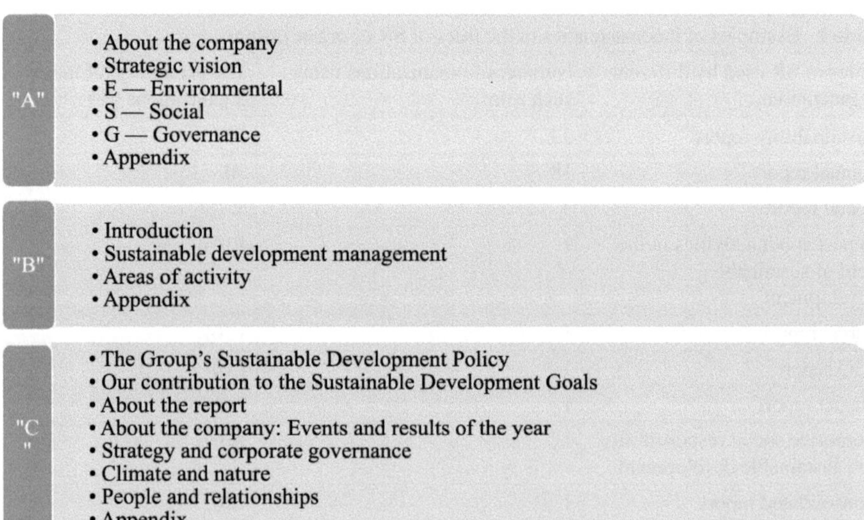

Fig. 1 Examples of differences in the sections of the SR of organizations. *Source* Developed by the authors

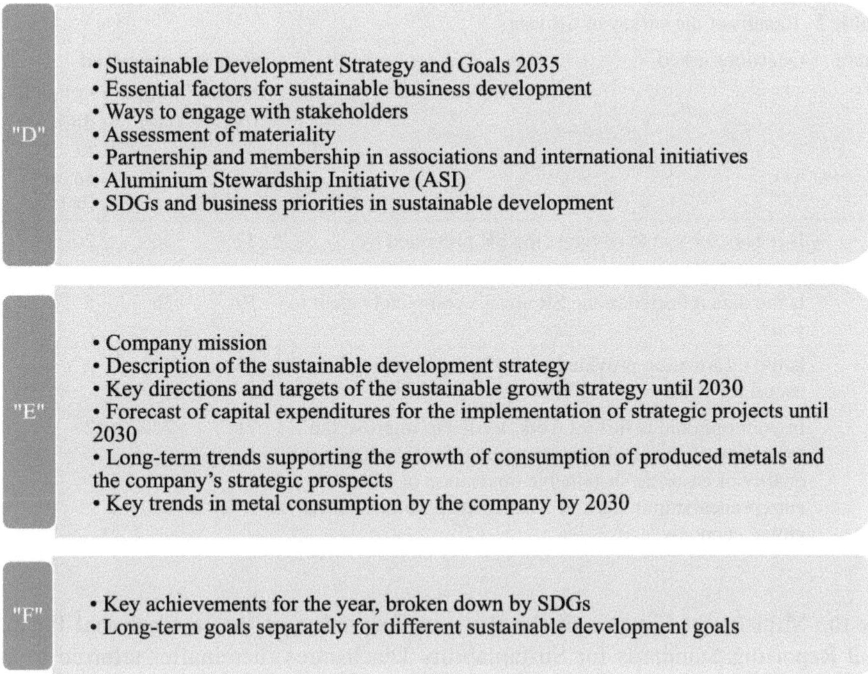

Fig. 2 Examples of the inconsistency of information presented in the section "Sustainable Development Strategy" of the SR by organizations (if they have such a section). *Source* Developed by the authors

are presented in Fig. 2. The authors would like to note that the names of organizations have been replaced by conventional ones (i.e., "D," "E," and "F").

Figure 2 clearly demonstrates significant discrepancies in how different organizations present information within the same sections of their sustainability reports (SR). As a result, users may find it difficult to interpret the relevant data, and the comparability of information across companies is significantly reduced. This, in turn, impairs the overall quality and effectiveness of ensuring SIEGE.

The identified problems are relevant not only within individual countries but also across different countries. This undoubtedly has a negative impact on the effectiveness of ensuring SIEGE. The findings presented are further supported by a survey of sustainability report (SR) users conducted as part of our research. The results of this survey are summarized in Table 2.

Source Developed by the authors.

As can be seen from Table 2, many users of the SR still consider such reporting to be insufficiently informative and user-friendly. Additionally, a fairly large proportion of respondents consider it advisable to further improve approaches to the preparation of the SR to improve the quality ensuring SIEGE.

Thus, it is necessary to carry out work in the above-mentioned area. A significant step forward in this area for Russian organizations is the planned implementation

Table 2 Results of the survey of SR users

Item No.	Questions asked	Responses received		
		"Yes" (%)	"No" (%)	"I find it difficult to answer" (%)
1.	Is it easy for you to navigate the SR presented by organizations?	52	46	2
2.	Is the data reflected in the SR always completely clear to you?	59	36	5
3.	Is the information provided in the SR by organizations useful for you?	54	43	3
4.	In your opinion, is further work needed to improve the formation of sustainability reporting (SR) to enhance the quality of ensuring sustainable innovation in entrepreneurship in light of solving the problems of the global economy?	71	27	2

by the Ministry of Finance of the Russian Federation of the International Financial Reporting Standards for Sustainability Disclosures (hereinafter referred to as IFRS Ss). These standards are designed to enhance the approaches to sustainability reporting and improve the comparability of information across different organizations. However, no system is without flaws. As the research has shown, the currently approved IFRS Ss still contain several open and debatable issues related to the presentation of sustainability information.

Considering the above, the authors consider it appropriate to introduce the following improvements to IFRS S1 to address the existing problems:

1. Specify that if an entity presents sustainability information separately from its financial statements, this information should be included in a distinct report titled "Sustainability Report";
2. Establish an indicative list of sections to be included in the Sustainability Report (SR) while allowing organizations to add additional sections if doing so enhances the report's informational value for users;
3. Provide a recommended set of indicators for each of the aforementioned sections while allowing organizations to supplement these with additional indicators if necessary to enhance the informational value of the Sustainability Report (SR);
4. Include in the appendices to IFRS S1 examples of Sustainability Reports (SRs) for organizations operating in various sectors.

The authors hope that the recommendations presented will be utilized to enhance approaches to preparing sustainability reports (SR), thereby improving the overall quality of ensuring SIEGE.

4 Conclusion

The research has shown several problems associated with the formation of sustainability reporting (SR) by organizations, which hinder its effective use in promoting sustainable innovation in entrepreneurship as a means of addressing global economic challenges. In the authors' view, these problems can be categorized into the following major groups:

1. Lack of unity in practice in the titles used for SR;
2. The practice by some organizations of presenting sustainability information across multiple separate reports rather than in a single comprehensive report, which complicates users' ability to access and obtain the necessary information;
3. Inconsistency between the sections of SR in the practice of different organizations;
4. Significant discrepancies in the information provided within the same sections of SR by different organizations (if such sections exist).

The identified problems pose a significant obstacle to enhancing the quality of sustainable innovation in entrepreneurship in the context of addressing global economic challenges.

To help resolve these issues, this paper presents a set of recommendations for amending the first International Standard for the disclosure of sustainability information. These proposed changes aim to improve the formation of sustainability reporting (SR), thereby contributing to more effective SIEGE assurance.

References

Abu Alia M, Amarneh KH, Abdeljawad I (2024) The relevance of IFRS accounting information: evidence from a pure IFRS environment. An-Najah Univ J Research—B (Humlties) 38(5):987–1018. Retrieved from https://staff.najah.edu/media/published_research/2024/05/21/Value_relevance.pdf. Accessed 15 May 2025

Al-Shaer H, Hussainey K (2022) Sustainability reporting beyond the business case and its impact on sustainability performance: UK evidence. J Environ Manag 311:114883. https://doi.org/10.1016/j.jenvman.2022.114883

Debnath P, Das K, Bhuyan AK, Saikia B, Das S, Kanoo R, Saha A, Debi H (2024) Corporate sustainability reporting practices in the banking sector: a governance implication. J GovAnce Regul 13(4):167–177. https://doi.org/10.22495/jgrv13i4art16

Di Tullio P, La Torre M (2022) Sustainability reporting at a crossroads in Italian universities: is web-based media adoption deinstitutionalising sustainability reporting? Adm Sci 12(1):34. https://doi.org/10.3390/admsci12010034

Druzhilovskaya TY, Druzhilovskaya ES, Korshunova TN, Denisova IP, Denisov PV (2020) Accounting policies for financial instruments of organizations in the modern conditions of the innovative economy. In: Popkova E, Sergi B (Eds.), Scientific and technical revolution: yesterday, today and tomorrow, Springer, Cham, Switzerland, pp 1588–1595. https://doi.org/10.1007/978-3-030-47945-9_169

Jamil A, Mohd Ghazali NA, Puat Nelson S (2021) The influence of corporate governance structure on sustainability reporting in Malaysia. Soc Responsib J 17(8):1251–1278. https://doi.org/10.1108/srj-08-2020-0310

Kılıçarslan A (2024) Sustainability reporting in local governance: an analysis of the Istanbul environment management company (IEMC) sustainability report. J Corp GovAnce, Insur, Risk Manag 11(1):75–84. https://doi.org/10.56578/jcgirm110105

Kulikova LI, Druzhilovskaya TY, Druzhilovskaya ES (2018) Challenges of the practical application of international accounting standard on property, plant and equipment by Russian companies. J Eng Appl Sci 13(3):569–574

Kulikova LI, Gubaidullina AR, Mukhametzyanov RZ, Druzhilovskaya TU, Druzhilovskaya ES (2020) International financial reporting standard (IFRS) regulations for supply chain management. Int J Supply Chain Manag 9(4):570–575. Retrieved from https://ojs.excelingtech.co.uk/index.php/IJSCM/article/view/5183/2620. Accessed 15 May 2025

Muhammed KU (2024) Sustainability committee, audit firm type and corporate sustainability reporting of non-financial firms in Nigeria. Copernic J Financ & Account 13(1):63–80. https://doi.org/10.12775/cjfa.2024.004

Roy P, Tanna H (2021) Sustainability reporting in India—a short content analysis of sustainability reports of selected companies of NSE. J Dev Res 14(4):1–9

Stănciulescu E (2024) Sustainability reporting. The Romanian sustainability code. CECCAR Bus Rev 5(5):32–41. https://doi.org/10.37945/cbr.2024.05.04

Tilt CA, Qian W, Kuruppu S, Dissanayake D (2021) The state of business sustainability reporting in sub-Saharan Africa: an agenda for policy and practice. Sustain Account, Manag Policy J 12(2):267–296. https://doi.org/10.1108/sampj-06-2019-0248

Ya S, Zhao D, Cao Y, Xu C (2024) New trends in sustainability reporting: exploring the online sustainability reporting practices by Irish universities. J Innov Knowl 9(3):100529. https://doi.org/10.1016/j.jik.2024.100529

Assessment of the Current State and Prospects for the Development of Engineering Literacy in Single-Industry Towns

Zhanna V. Smirnova, Mariia V. Mochalina, Olga T. Cherney, Elena G. Skachkova, and Evgeny Y. Denisov

Abstract The research aims to investigate the issue of insufficient engineering literacy development in the single-industry towns of the Nizhny Novgorod Region and its restraining influence on the region's socio-economic dynamics. The purpose of this research is to identify the key factors that determine the level of engineering literacy and to develop methodological approaches for its improvement under the specific conditions of monoprofile cities. To achieve this goal, the authors conducted a sociological study to identify the causes of the low level of engineering training among schoolchildren. Based on the data obtained, the authors propose recommendations to enhance educational programs, foster interaction between educational institutions and industrial enterprises, and promote engineering specialties among young people. During the research, the authors developed methodological support for several measures aimed at developing engineering literacy. The proposed measures aim to mitigate the negative consequences of the lack of qualified engineering personnel and enhance the region's competitiveness.

Keywords Industrial potential · Single-industry towns · Engineering literacy · Qualified personnel · Economic development · Questioning of students

JEL Codes L5 · N6 · O1

Z. V. Smirnova (✉) · M. V. Mochalina · O. T. Cherney · E. G. Skachkova · E. Y. Denisov
Minin Nizhny Novgorod State Pedagogical University, Nizhny Novgorod, Russia
e-mail: z.v.smirnova@mininuniver.ru

A. B. Karbekova (ed.), *Sustainable Development of Rural Territories Based on Climate-Smart Technologies*, Climate Change Management,
https://doi.org/10.1007/978-3-032-20106-5_7

1 Introduction

In the context of the country's economic development, the importance of single-industry towns with significant potential is steadily increasing. Their growth plays a crucial role in attracting investment, enhancing living standards, modernizing infrastructure, and generating new jobs. However, due to their specific characteristics, single-industry towns face several unique challenges. One of the most pressing issues is the insufficient level of engineering literacy among the population, which hinders innovation and reduces the competitiveness of these regions. It is essential to recognize that engineering literacy encompasses theoretical knowledge in the fields of engineering and technology, practical skills, and a practical understanding.

2 Literature Review

In economic science, the development of single-industry towns is a widely discussed topic, with particular emphasis on improving regional production structures and expanding training opportunities.

The works examined analyze research on the socio-economic development of single-industry towns. For example, Sfrerartseva, Manaeva, Lampón, Pérez-Moure, Carballo-Cruz, and Velando-Rodríguez highlight the role of single-industry towns in the system of regional productive force deployment in different countries (Lampon et al. 2024; Li et al. 2025; Rastvortseva and Manaeva 2016; Sun et al. 2024; Voronina et al. 2021a). The role of these territories in diversifying regional development has been studied (Bukhvald 2017; Tan et al. 2024; Voronina et al. 2021b). The development strategies of single-industry towns are described by Faikov, Baidarov, Bezverbny, Mikryukova, and Miryazova (Faikov and Baidarov 2022; Bezverbny et al. 2020). The issues of population mobility in single-industry towns, along with the characteristics and development prospects of these territories, are analyzed in the works of Bobkov, Klagova, Garnova, Anikieva, Lu, Yin, Wen, and Xu (Anikieva 2020; Bobkov et al. 2020; Lu et al. 2023).

Significant contributions to studying key issues related to harnessing the potential of single-industry towns for the country's economic development have been made by Boyko, Kozodubov, Shabelnikova, and others (Boyko 2017; Kozodubov 2017; Shabelnikova 2019).

Despite the attention given to the problems of single-industry towns, many important issues remain insufficiently explored from a scientific perspective. Particularly, the absence of a comprehensive concept for strengthening engineering potential and the limited development of specific aspects in this field highlight the need for further research and innovative solutions. Within this research, the authors focus on the problem of improving the level of engineering literacy in the single-industry towns of the Nizhny Novgorod Region.

The Nizhny Novgorod Region is the largest industrial region of Russia (Avezova and Mansurova, 2021; Guo et al. 2025; Jianan et al. 2024; Mukhina et al. 2020). In industry, the priority sectors of the Nizhny Novgorod Region include the automotive, mechanical engineering, and metallurgical industries, as well as the petrochemical, defense-industrial, and electric power sectors. These industries have traditionally shaped the region's industrial landscape and continue to define its future development trajectory.

The industrial diversity of the Nizhny Novgorod Region is supported by the large-scale industrial production of Nizhny Novgorod and the enterprises of single-industry towns such as Zavolzhie, Gorodets, Balakhna, Proks, Pavlovo, Kulebaki, and others. However, the low level of engineering literacy among residents of these towns leads to a shortage of qualified personnel, which complicates the implementation of innovative projects and startups in the region. Many residents possess limited skills in working with advanced technologies, which reduces their ability to adapt to rapidly changing working conditions. This hinders the introduction of new technologies into production and, as a result, negatively impacts the competitiveness of local enterprises.

Therefore, the problem of an insufficient level of engineering literacy among the population of single-industry towns in the Nizhny Novgorod Region causes serious consequences for their economic and technological development.

3 Materials and Methods

The relevance of developing engineering literacy in single-industry towns lies in the need to increase their competitiveness and adapt to new technologies. The key condition for this is to increase the population's engineering literacy.

This research aims to identify factors that affect the formation of engineering literacy and develop practical methods to enhance it in monofunctional cities (Bagryanskaya and Mochalina 2023; Gao et al. 2024; Huang et al. 2025; Rathee and Sadhukhan 2025).

To achieve the goal, the following tasks were set:

- To assess the current level of engineering literacy in the single-industry towns of the Nizhny Novgorod Region;
- To identify factors affecting the formation of engineering literacy in students;
- To develop recommendations for increasing the engineering literacy of the population of single-industry towns;
- To create a set of measures to increase engineering literacy in single-industry towns of the Nizhny Novgorod Region;
- To test developed measures through the implementation of projects based on the Minin Nizhny Novgorod State Pedagogical University.

The research methodology included a sociological survey of students from single-industry towns to assess their knowledge and interest in the field of engineering. The analysis of the questionnaire results revealed the main causes of the insufficient level of engineering literacy and highlighted the need to improve methods for teaching engineering disciplines.

4 Results

The industrial potential of single-industry towns plays an important role in the economy of the Nizhny Novgorod Region. Large and medium-sized enterprises located in their territories ensure the functioning of the region's key industries. The significance of single-industry towns for the Nizhny Novgorod Region's industry is confirmed by the example of the Volga area, where sectors such as energy, petrochemicals, mechanical engineering, and the food industry are well-developed.

PJSC "Zavolzhsky Motor Plant" is a large enterprise of the engineering industry. It is part of the "Sollers" PJSC. The main areas of activity of PJSC "Zavolzhsky Motor Plant" are the production of internal combustion engines corresponding to contemporary environmental standards for cars (e.g., SUV and LCV) and buses (e.g., Pavlovsky Automobile Plant), and spare parts for them, as well as the production of auto components for orders of world car companies (Smirnova et al. 2021; Yu et al. 2025).

LLC "Rosalith Foundry Plant" is the largest manufacturer of aluminum castings and complex parts, including their mechanical processing and the production of specialized equipment. The plant is equipped with a full range of advanced technological systems: automated IDRA pressure machines, Kurtz robotic complexes for gravity casting, the Fill Construction finishing complex, and high-precision machining centers such as Heckert, Comau, and Grob. Its facilities also include the Dea Global Advantage coordinate measuring machine, the YXLON Multiplex universal X-ray inspection and tomography system, and the Columbus spectrometer (Smirnova et al. 2022).

The shipyard-building corporation (SBC) in Gorodets plays a significant role in the industrial development of the Nizhny Novgorod Region. Since its founding, the corporation has been building and repairing various ships, making it a significant player in the shipbuilding market. The main activities of the SBC include fleet modernization, restoration of vessel technical condition, and the development of new projects. The SSK serves as an important economic center, contributing to the dynamic industrial growth of Gorodets.

The Kulebaksky Metallurgical Plant "Rupolim" plays a vital role in the industrial development of the entire region. The plant specializes in the production of high-quality metal products and alloys, strengthening Russia's position in the global metal industry market. Rupolim actively implements new technologies and modernizes its equipment, which increases production efficiency and reduces costs. Its products serve a wide range of industries, including construction, engineering, and energy.

The JSC "Kulebaksky Plant of Metal Structures" (JSC "KZMK") is a major enterprise in the industry, producing metal structures used in the construction of industrial and technological facilities, warehouse terminals, office buildings, sports and cultural centers, bridges, and other specialized structures.

Thus, production enterprises in single-industry towns actively contribute to the development of their territory's industry and the regional economy.

The Nizhny Novgorod Region, a major industrial center in Russia with strong traditions and an attractive investment climate, has demonstrated steady growth in high-tech production, particularly in electronics and IT. Enterprises actively implement innovations and modernize their equipment. However, the operation of these high-tech industries requires specialists with advanced engineering literacy, which poses a serious challenge for single-industry towns. The low level of engineering literacy among the younger generation, combined with insufficient motivation to pursue engineering and a weak interest in developing their hometowns, leads to an outflow of youth to larger cities. As a result, ensuring a supply of engineering personnel for enterprises in single-industry towns becomes a crucial factor in their future economic development and the overall prosperity of the region.

Educational institutions play a crucial role in training personnel for industrial enterprises in single-industry towns. With its many years of experience in training specialists across various fields, Minin Nizhny Novgorod State Pedagogical University recognizes the importance of addressing this problem. In response to today's social demand for engineering personnel in single-industry towns, the university has launched research activities to tackle this challenge (Mukhina et al. 2022). This initiative aims to develop a model for training engineering personnel with a high level of engineering literacy who are motivated to work in their hometowns. As part of this effort, Minin Nizhny Novgorod State Pedagogical University conducted a large-scale sociological study among schoolchildren in the single-industry towns of the Nizhny Novgorod Region (Balakhna, Gorodets, and Volga) to assess the initial level of engineering literacy and identify the factors influencing its formation. More than 1100 schoolchildren and secondary vocational education students participated in the research. The number of respondents is presented in Table 1.

Source Developed by the authors

For the study, the authors developed a questionnaire that included open- and closed-ended questions. Designed using Google Forms, the questionnaire consisted of 18 items aimed at assessing the level of engineering literacy among schoolchildren. Particular attention was given to questions that allow for drawing well-founded conclusions about the formation of engineering literacy, analyzing the results,

Table 1 The number of respondents surveyed

Total in the Nizhny Novgorod region		Gorodets	Volga	Kulebaki
Schoolchildren	755	170	377	93
Secondary vocational education	356	58	66	231

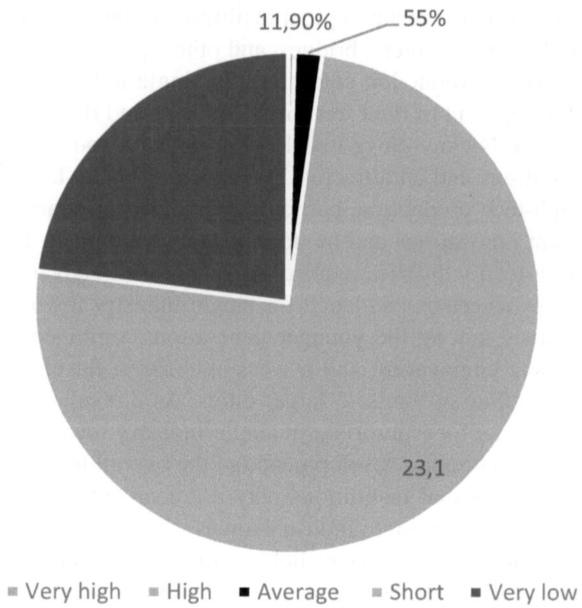

Fig. 1 Students' interest in engineering. *Source* Developed by the authors

and developing effective methods for its enhancement in the context of single-industry towns. One of the most significant questions focused on clarifying adolescents' personal attitudes toward engineering: "How do you assess your interest in engineering?" The distribution of responses to this question is presented in Fig. 1.

The insignificant share of the respondents (14.8%) expressed a high or very high interest in engineering, which indicates that only a few teenagers consider this area as a future profession. The most common answer was the average interest (55%), indicating the presence of a certain interest without a clear desire for an engineering career. Simultaneously, a high percentage of respondents with an average level of interest opens up opportunities to increase involvement in engineering through educational programs, popularize engineering professions, practical projects, and interaction with local enterprises. A significant proportion of respondents (30.2%) demonstrated low or very low interest, indicating a need to improve the image of engineering and create more attractive career prospects to attract a larger number of young people.

The responses of schoolchildren to the question "How do you evaluate your knowledge in the field of engineering?" show several key points. Most respondents (40.2%) rated their knowledge as being at an average level (3 points), indicating an awareness of the need for further training and practical experience. Only a small proportion (3.4%) considered their knowledge excellent, which underscores the high complexity of the engineering field. Simultaneously, a significant share of students (43.8%) rated their knowledge as low (1 or 2 points), indicating a lack of confidence in their skills and the need to improve teaching quality and provide more practical training. Additionally, 12.7% of respondents rated their knowledge at four points, showing the presence of a motivated group of students who could be engaged in more advanced study and projects through networked collaboration with industrial enterprises in single-industry towns. To gain a deeper understanding of the challenges faced in studying engineering disciplines, students were also asked what difficulties they encounter when studying engineering subjects. The answers to this question made it possible to identify the main problem areas (Chaykina et al. 2023).

The main difficulty for most students (34.7%) is understanding theoretical concepts, which requires a revision of the teaching methodology. Teachers are recommended to use visual materials and practical examples more actively to facilitate the perception of complex topics. Another significant factor is the lack of interest in the subject (33.7%), which may be due to insufficient motivation or the weak connection between theory and practice. To increase student engagement, it is recommended to introduce projects that demonstrate the practical application of concepts and invite specialists from local industrial enterprises to conduct master classes. A considerable share of students (29.3%) also noted a lack of practice, which highlights the need to integrate more hands-on activities, laboratory work, and design tasks into the educational process. A smaller proportion of respondents (2.3%) mentioned a lack of time, which should nonetheless be considered when planning the curriculum to avoid student overload.

A study conducted among the youth of single-industry towns revealed the main causes of the low level of engineering literacy. These include insufficient educational resources (e.g., the lack of modern institutions and equipment), low motivation to study engineering disciplines (influenced by social stereotypes and limited information about career prospects), and restricted access to modern technologies at local enterprises. The results highlight the need for comprehensive improvements in the quality of engineering education and adjustments in teaching methods. To address this issue, Minin Nizhny Novgorod State Pedagogical University is implementing several projects to increase engineering literacy in the single-industry towns of the Nizhny Novgorod Region. These projects include the development and implementation of practice-oriented educational programs, the updating of methodological support for educational institutions, the establishment of networked cooperation between schools and enterprises, and initiatives to promote engineering professions. The university relies on state support to carry out these initiatives, which aim to advance engineering education and professional training.

5 Conclusion

The insufficient level of engineering literacy among students in single-industry towns leads to serious negative consequences for these territories, including slower innovative development, reduced competitiveness of local enterprises due to a shortage of qualified personnel, and increased youth migration to more developed regions. Enhancing the level of engineering education among the population is therefore a key factor not only for the development of individual single-industry towns but also for strengthening the region's overall competitiveness. Achieving sustainable socio-economic development in these areas requires the combined efforts of educational institutions, businesses, and government structures.

Acknowledgments The publication was prepared as part of the state assignment of the Ministry of Education of the Russian Federation No. 073-00056-25-00 for the implementation in 2025 of research work on the topic "Closter model for the training of a graduate of the Secondary Professional Education in Partnership with a Technopark of Professional Education."

References

Anikieva AV (2020) Labor mobility of people from the monotowns in Sverdlovsk oblast. Monit Public Opin: Econ Soc Chang 5:449–465. https://doi.org/10.14515/monitoring.2020.5.536

Avezova MM, Mansurova MG (2021) Priority directions for increasing diversification of the regional economy. Management 9(1):5–15. https://doi.org/10.26425/2309-3633-2021-9-1-5-15

Bagryanskaya ES, Mochalina MV (2023) Designing spaces for social projects of the Nizhny Novgorod region. In: *Actual issues of sustainable development of modern society and economics: a collection of scientific articles of the 2nd all-Russian scientific and practical conference* (volume 3). CJSC University Book, Kursk, Russia, pp 120–122

Bezverbny VA, Mikryukov NY, Miryazov TR (2020) Model of the Ural Economic District: features and prospects of development. Sci Rev Ser 1 Econ Law 5:91–106. https://doi.org/10.26653/2076-4650-2020-5-08

Bobkov VN, Claudov YG, Garnov AP (2020) Regulation of precarious employment in single-industry towns. Econ RegNs 16(2):522–534. https://doi.org/10.17059/2020-2-14

Boyko NN (2017) Tax and legal regulation of the territories of ahead of socio-economic development in single-industry towns. Taxes 5:3–6

Bukhvald EM (2017) Mono-company towns within the strategic planning system in Russia. Theory Pract Soc Dev 12:75–78. https://doi.org/10.24158/tipor.2017.12.16

Chaykina ZV, Mochalina MV, Uskova AA, Alexandrova VY (2023) Educational robotics as a means of developing the technical thinking of students in institutions of additional education. Mod Sci 4:152–157

Faikov DY, Baidarov DY (2022) "Big Sarov": some of the not-so-obvious strategies for the development of single-industry towns. Probl Dev Territ 26(1):10–26. https://doi.org/10.15838/ptd.2022.1.117.2

Gao Y, Zhao P, Zhang M (2024) Discovering temporal-spatial features of jobs-housing relationship from a regional perspective: a nationwide study. J Transp Geogr 121:104020. https://doi.org/10.1016/j.jtrangeo.2024.104020

Guo L, Tang M, Wu Y, Bao S, Wu Q (2025) Government-LED regional integration and economic growth: evidence from a quasi-natural experiment of urban agglomeration development planning policies in China. Cities 156:105482. https://doi.org/10.1016/j.cities.2024.105482

Huang Z, Shi T, Tong D (2025) Understanding the role of government in regional industrial evolution: evidence from land supply in China. Pap Reg Sci 104(1):100070. https://doi.org/10.1016/j.pirs.2024.100070

Jianan Z, Tiejian H, Qun Y (2024) Based on gated recurrent network analysis of advanced manufacturing cluster and unified large market to promote regional economic development. Comput Ind Eng 197:110575. https://doi.org/10.1016/j.cie.2024.110575

Kozodubov AA (2017) Legal regulation of local self -government in single -industry towns. State Power Local Self-GovMent 10:34–39

Lampon JF, Perez-Moure H, Carballo-Cruz F, Velando-Rodríguez ME (2024) New mobility technologies and regional status in the automotive industry value chain: the case of Spain and Portugal. Technol Soc 78(1):102624. https://doi.org/10.1016/j.techsoc.2024.102624
Li H, Hu T, Ma X, Cheng B (2025) The impact of transportation infrastructure on the regional economic integration in China: a CGE analysis. Int Rev Econ & Financ 99:104045. https://doi.org/10.1016/j.iref.2025.104045
Lu L, Yin S, Wen F, Xu Q (2023) The spatial structure of labour force employment in China's industries: measurement and extraction. Econ Anal Policy 77:472–486. https://doi.org/10.1016/j.eap.2022.12.001
Mukhina MV, Chaykina ZV, Tsapina TN, Kutepova LI, Smirnova ZV (2022) Research of the psychological structure of technical thinking. In: Bogoviz AV, Suglobov AE, Maloletko AN, Kaurova OV (eds) Cooperation and sustainable development. Springer, Cham, Switzerland, pp 487–493. https://doi.org/10.1007/978-3-030-77000-6_58
Mukhina MV, Smirnova ZV, Yargin MA, Galtsina AS (2020) Evaluation of competitiveness of the Nizhny Novgorod region. Sci Krasn 9(3–3):73–77
Rastvortseva SN, Manaeva IV (2016) Single-industry towns in the system of distribution of productive forces of Russian regions. Territ Dev Issues 5(35):3
Rathee C, Sadhukhan S (2025) Regional impact assessment of highways and policy interventions: lessons from Haryana. India Case Stud Transp Policy 20:101421. https://doi.org/10.1016/j.cstp.2025.101421
Shabelnikova SI (2019) Specific features of single-industry towns' development in the Russian Far East. Federalism 1:172–182. https://doi.org/10.21686/2073-1051-2019-1-172-182
Smirnova ZV, Chelnokova EA, Mukhina MV, Cherney OT, Kozlova EP (2022) Digital technologies in the teacher's professional activities. In: Ostrovskaya VN, Bogoviz AV (eds) Big data in the Govtech system. Springer, Cham, Switzerland, pp 47–52. https://doi.org/10.1007/978-3-031-04903-3_6
Smirnova ZV, Kutepova LI, Kutepov MM, Razorenov VA, Mukhina MV (2021) Differentiation of education as a means of improving the quality of technological education. In: Popkova EG, Ostrovskaya VN, Bogoviz AV (eds) Socio-economic systems: paradigms for the future. Springer, Cham, Switzerland, pp 1155–1162. https://doi.org/10.1007/978-3-030-56433-9_121
Sun Y, Tang S, Dou Z, Wang T (2024) How environment and technology affect the regional manufacturing industry development. Heliyon 10(15):e35321. https://doi.org/10.1016/j.heliyon.2024.e35321
Tan Y, Wang Y, Duan L, Gu Z (2024) Analysis of spatial-temporal evolution and investment model of industrial cross-cross-cross network-taking the Pearl River Delta urban agglomeration as an example. Heliyon 10(17):e37623. https://doi.org/10.1016/j.heliyon.2024.e37623
Voronina EV, Milchakova NN, Sergeeva IV (2021a) Monogorod: diversification trajectory. Surgut State Univ J 4(34):28–38. https://doi.org/10.34822/2312-3419-2021-4-28-38
Voronina EV, Milchakova NN, Sergeeva IV (2021b) Monogorod: diversification trajectory. Surgut State Univ J 4(34):28–38. https://doi.org/10.34822/2312-3419-2021-4-28-38
Yu H, Lyu Q, Jiang Y (2025) Environmental regulation, regional finance development and enterprises' cross-regional capital flow. Int Rev Econ & Financ 99:103984. https://doi.org/10.1016/j.iref.2025.103984

The Ecological Footprint of Tourism: Challenges and Opportunities for Sustainable Rural Development

Nelli I. Akylbekova, Elvira Sh. Satylganova, Irina B. Mukambaeva, Duishekan O. Kerimbekova, and Saltanat K. Omurova

Abstract This research examines the development of rural tourism in the Kyrgyz Republic between 2019 and 2024, highlighting its economic, social, cultural, and environmental impacts. Utilizing official statistics from the National Bank of the Kyrgyz Republic and a comprehensive SWOT analysis, the research explores the dynamics of formal and informal tourism sectors, including the effects of the COVID-19 pandemic on visitor numbers. The findings reveal that while rural tourism significantly contributes to local economic growth and cultural preservation, it also presents challenges such as seasonal income fluctuations, infrastructure strain, and environmental degradation. The research emphasizes the critical role of local communities and sustainable practices in balancing tourism growth with ecological conservation. A coherent strategy for advancing sustainable rural tourism in Kyrgyzstan should be predicated upon the modernization of infrastructural systems, the cultivation of ecological consciousness among visitors and local communities, and the institutionalization of robust regulatory mechanisms. Such a multidimensional approach is intended to safeguard the singular ecological and cultural patrimony of the country over the long term and engender inclusive socio-economic dividends that reinforce local livelihoods, stimulate endogenous development trajectories, and integrate peripheral regions into broader circuits of sustainable growth.

Keywords Ecological footprint · Sustainable tourism · Rural development · Environmental impact · Ecotourism · Resource management · Biodiversity · Local communities · Tourism policy · Environmental sustainability

JEL Classification H41 · L83 · O13 · Q01 · Q26 · Q57 · R11

N. I. Akylbekova (✉) · E. Sh. Satylganova · I. B. Mukambaeva · D. O. Kerimbekova · S. K. Omurova
Kyrgyz National University named after Jusup Balasagyn, Bishkek, Kyrgyzstan
e-mail: nelly_a@bk.ru

A. B. Karbekova (ed.), *Sustainable Development of Rural Territories Based on Climate-Smart Technologies*, Climate Change Management,
https://doi.org/10.1007/978-3-032-20106-5_8

1 Introduction

Tourism has emerged as a powerful engine for economic growth and social transformation in rural areas across the globe. It offers unique opportunities to diversify rural economies, generate employment, and promote cultural exchange.

The methodological approaches employed enabled the calculation of an integrated index that assesses the relative development of agriculture in each EAEU country. This index provides a foundation for designing targeted strategies to improve the efficiency and productivity of national agricultural sectors while facilitating coordinated, synergistic growth across the EAEU (Mukambaeva et al. 2024a).

In terms of sustainable development, agriculture plays a crucial role because it drives economic growth and food security and supports environmental conservation and rural livelihoods, thereby contributing to the social and ecological well-being of the region (Mukambaeva et al. 2024b).

Tourism serves as a key driver of economic growth by creating jobs, attracting investment, and generating foreign exchange, which is particularly important for the macroeconomic stability of developing economies (Li et al. 2024). The increasing emphasis on sustainable tourism reflects a global commitment to balancing economic benefits with environmental protection and cultural heritage preservation (Jia et al. 2021). Achieving this balance requires the adoption of integrated development strategies that address environmental, economic, and social goals (Raihan 2024). Sustainable tourism ensures long-term viability and plays a key role in the protection and transmission of intangible cultural heritage (Zhao et al. 2024). In this context, rural tourism has emerged as a viable development strategy, offering multiple socio-economic benefits for rural communities (Yarimoglu and Binboga 2019).

International experiences in rural tourism highlight a range of effective practices and models that emphasize community involvement, environmental conservation, and cultural authenticity. These insights have been thoroughly examined in the works of contemporary researchers (Ashraf et al. 2020), providing valuable guidance for policymakers and tourism planners.

Numerous empirical studies (Chu et al. 2022; Croes et al. 2021; Sun et al. 2020) have demonstrated that tourism serves not only as a catalyst for economic growth but also as a stabilizing factor in the broader socio-economic landscape. These factors collectively stimulate long-term economic advancement and social transformation.

Nonetheless, tourism is intricately linked to the exploitation of natural ecosystems, given its frequent reliance on landscapes, biodiversity, and climate as central attractions. The expansion of tourism activities often coincides with increased transportation demands, heightened energy consumption, and greater pressure on natural habitats, manifested through practices such as overhunting or resource extraction (Adebayo et al. 2022; Akadiri et al. 2020; Khoi et al. 2021; Wang et al. 2021). Consequently, the environmental sustainability of tourism is increasingly questioned because its growth may come at the cost of ecological degradation. One of the most pressing contemporary challenges lies in reconciling the economic benefits of tourism with the imperative to preserve environmental quality (Akadiri et al. 2019; Debbage and Debbage 2019; Dolnicar 2020; Sun et al. 2020).

This has led to a growing scholarly consensus on the necessity of designing and implementing tourism policies that minimize environmental harm while sustaining economic momentum. Researchers argue that understanding the nuanced relationship between tourism and environmental degradation is crucial for developing effective governance strategies and promoting sustainable development (Akadiri et al. 2020; Bramwell et al. 2016; Brauer et al. 2019; Smith and Diekmann 2017).

Tourism is not alone in shaping environmental dynamics; economic growth and globalization also exert significant influence. These variables interact with the environment through three principal mechanisms: scale, composition, and technique effects. According to Ahmed et al. (2021), while the composition and technique effects of GDP and globalization may encourage cleaner production methods and the adoption of environmentally sound technologies, the scale effect associated with economic expansion tends to elevate environmental pressure by increasing production volumes and resource extraction.

Traditionally, the environmental impact of such macroeconomic variables has been evaluated through carbon dioxide (CO_2) emissions. However, scholars such as Dogan et al. (2020) contend that CO_2 emissions alone fail to capture the full extent of anthropogenic pressure on ecosystems.

Despite this methodological advancement, the empirical literature remains inconclusive regarding the precise effects of tourism, economic growth, and globalization on EF. For instance, Mikayilov et al. (2019) and Godil et al. (2020) reveal that tourism intensifies environmental degradation in countries such as Azerbaijan and Turkey. In contrast, studies conducted in Thailand and Singapore (Khoi et al. 2021; Kongbuamai et al. 2020) suggest that tourism can play a positive role in enhancing environmental quality, possibly due to stricter regulations and sustainable practices. Adding further complexity, Akadiri et al. (2020) highlight a nonlinear relationship between GDP and carbon emissions in small island developing states, where tourism and the squared term of GDP exhibit a mitigating effect, while GDP and globalization, when considered linearly, aggravate emissions.

Given the heterogeneity of findings across different regional and developmental contexts, many researchers (Alola et al. 2019) emphasize the necessity of conducting region-specific investigations.

2 Materials and Methods

To analyze the development of rural tourism in the Kyrgyz Republic and its impact on the environment and rural development, the analysis draws upon statistical materials compiled by the National Statistical Committee of the Kyrgyz Republic. A comparative analysis of the dynamics of tourist flow was conducted considering the impact of global events such as the COVID-19 pandemic. The socio-economic and environmental impacts of tourism growth in rural areas were assessed based on a summary of data from scientific publications, government reports, and field observations. The methodological approach includes a SWOT analysis to identify the strengths, weaknesses, opportunities, and threats to the development of sustainable rural tourism. Additionally, a system analysis was used to develop strategic recommendations aimed at balancing economic development, cultural heritage preservation, and environmental protection.

3 Results

In recent years, the Kyrgyz Republic has increasingly asserted itself on the global tourism map. Its unique natural landscapes, rich cultural heritage, and the enduring image of nomadic civilization have drawn growing interest from international travelers. Particularly attractive are those areas that have retained their pristine character: remote regions situated far from urban centers, often in close proximity to rural communities, mountain ranges, and alpine lakes.

From 2019 to 2024, tourism in the Kyrgyz Republic underwent significant fluctuations due to global disruptions, followed by a strong recovery driven by the appeal of its natural landscapes and cultural heritage (Fig. 1).

Between 2019 and 2024, water consumption in the Kyrgyz Republic grew by 26%, from 5211.1 to 6565.6 million cubic meters, outpacing the rate of water abstraction and indicating more intensive use of available resources, likely due to improved delivery systems or the expansion of water-intensive sectors. Over the same period, wastewater volumes, although comparatively small, increased by 38%, rising from

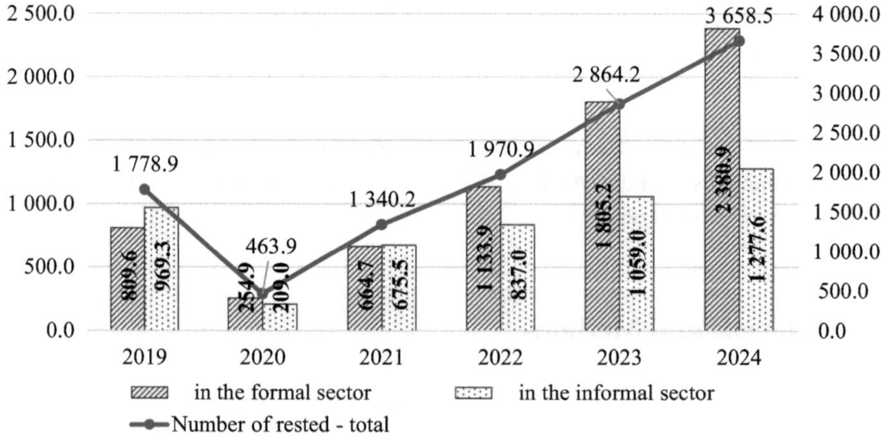

Fig. 1 The number of visitors to the Kyrgyz Republic, thousand people. *Source* Compiled by the authors based on the National Statistical Committee of the Kyrgyz Republic (National Statistical Committee of the Kyrgyz Republic n.d.-b)

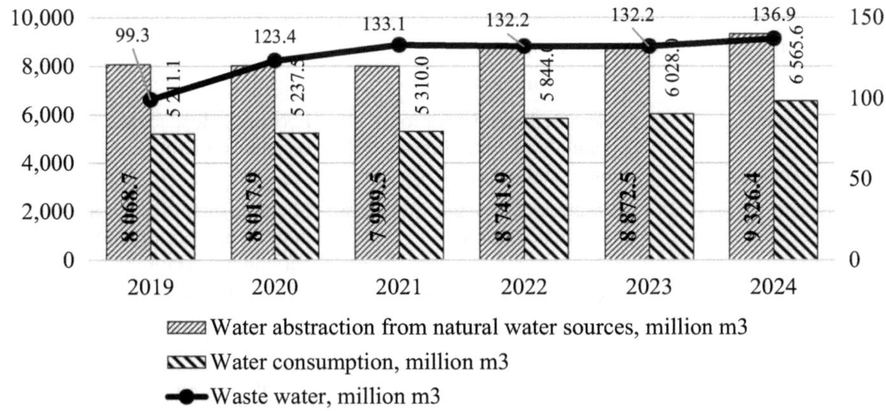

Fig. 2 Water consumption and wastewater in the Kyrgyz Republic, million m^3. *Source* Compiled by the authors based on the National Statistical Committee of the Kyrgyz Republic (National Statistical Committee of the Kyrgyz Republic n.d.-a)

99.3 to 136.9 million cubic meters. This suggests a gradual growth in return flows accompanying the overall increase in water use (Fig. 2).

The relatively low volume of wastewater compared to total water consumption likely indicates the prevalence of non-return uses, such as irrigation, where most water is absorbed by soil or lost through evaporation rather than discharged. At the same time, land use patterns are undergoing transformation. Opportunities are emerging for diversification through activities such as yurt camps, ethnoparks, and agritourism farms. However, in the long term, there is a risk that agriculture could be displaced, with fertile land being lost to tourism-related construction (Table 1).

Safeguarding rights and ensuring equitable access to resources are crucial to prevent marginalization by large external investors. The sector's development is constrained by several factors: poor infrastructure in remote areas, limited environmental monitoring, and low awareness of sustainable practices among tourists and local residents. Additionally, the lack of mandatory environmental certification reduces accountability. Overcoming these challenges requires comprehensive, integrated measures. These include investing in renewable energy and eco-friendly transportation, implementing educational programs to raise environmental awareness, and developing national strategies that guide tourism growth while maintaining ecological sustainability.

Unregulated tourism growth, compounded by climate change and excessive commercialization, threatens fragile ecosystems, depletes natural resources, and erodes cultural identity, particularly when protected areas become oversaturated without visitor limits (Table 2).

Timely realization of opportunities can neutralize weaknesses and prevent negative outcomes, ensuring harmonious coexistence between tourism and nature.

Table 1 The impact of the increase in the number of tourists in the Kyrgyz Republic on the development of rural areas and the environment

Aspect	Positive impact	Negative impact
Rural economy	Increased income for local populations (guesthouses, tours, and handicrafts)	Economic dependence on seasonal tourism
Employment	Creation of new jobs in tourism and related services	Shortage of skilled labor and outflow of workforce from the agricultural sector
Infrastructure	Development of roads, communication, sanitation, and transport	Overload of infrastructure due to rapid growth in tourist numbers
Social sector	Improved living standards and increased investment in education and healthcare	Social inequality between communities involved in tourism and those left out
Cultural preservation	Support for traditional crafts and promotion of cultural heritage	Commercialization and distortion of cultural traditions
Ecology and environment	Growing interest in ecotourism and environmental initiatives (clean-ups, reserves)	Increased pollution, waste generation, and ecosystem degradation
Land use	Opportunities for land use diversification (yurt camps, ethnoparks, and agrotourism)	Loss of agricultural land to tourism facilities, imbalance in land use
Local communities	Greater participation and self-governance	Dispossession of residents through large-scale investment and the clash of priorities

Source Developed by the authors

4 Discussion

A non-cooperative evolutionary game theory approach involving three players is employed to uncover behavioral patterns and pathways toward sustainable governance. Based on the analysis of this model, the adoption of an enhanced public participation mechanism is proposed, capable of achieving a cooperative evolutionary equilibrium among the parties, minimizing environmental risks, and strengthening the institutional foundations for sustainable development (Chu et al. 2022).

To support the low-carbon transition of rural tourism in the study area and accelerate progress toward the "dual carbon" targets, it is recommended to improve public transport infrastructure, promote green mobility options, increase the energy efficiency of rural dining and accommodation facilities, and strengthen awareness of low-carbon tourism practices (Wan et al. 2025).

Selecting an appropriate climate change adaptation strategy presents a complex challenge for any ski resort because its successful implementation may be hindered by divergent perspectives among stakeholders (Cholakova and Dogramadjieva 2023).

Table 2 SWOT analysis of the impact of tourism on the environment

SWOT element	Description
Strengths	– Abundant natural and cultural resources, including mountainous terrains, freshwater lakes, and diverse ecosystems; – Intensifying demand for eco-tourism experiences and environmentally sustainable approaches; – Active assistance and backing from international institutions committed to sustainable development; – Increasing involvement of local populations in fostering and advancing eco-tourism initiatives
Weaknesses	– Lack of infrastructure and environmental standards in tourist areas; – Limited resources for monitoring and controlling tourism's impact; – Low awareness among tourists and locals about sustainable consumption principles; – Lack of a compulsory environmental accreditation framework for tour operators
Opportunities	– Advancement of environmentally friendly technologies and the integration of renewable energy sources within the tourism sector; – Implementation of educational initiatives and training programs aimed at fostering principles of sustainable tourism; – Mobilization of financial resources and stimulation of investment flows into eco-oriented infrastructure projects; – Formulation of national regulatory frameworks, standards, and strategic guidelines to promote sustainable tourism development
Threats	– Unregulated expansion of tourist flows can trigger ecological degradation and undermine ecosystem resilience; – Adverse impacts of climate change are increasingly manifest in the depletion and destabilization of natural resource bases; – Intensified commodification of tourism often occurs at the cost of ecological integrity and broader societal values; – Risk of overloading protected natural areas

Source Developed by the authors

Ecotourism holds considerable promise as a driver of sustainable development. However, there remains a lack of universally accepted standards and methodologies for effectively managing and monitoring the specific impacts of individual tourism packages on natural ecosystems and local communities (Mancini et al. 2022).

Ecological footprint analysis is one of the most useful models for the environmental impact assessment of human activities. The tourism industry in Hosainabad relies on an area beyond this village to meet its biological needs and environmental sustainability. Findings suggest that decision-makers must pay enough attention to tourists' activities in small areas in order to prevent further environmental disruption (Habibi et al. 2023).

The ecological footprint of tourism in the Beijing Jiufeng National Forest Park is calculated based on the theory and method of the component method for the ecological footprint. This research suggests that the flow of visitors in forest parks with large ecological deficits should be directed to those with ecological surpluses (Zhang et al. 2022).

In China, the integration of rural tourism with agricultural production led to reduced water, labor, and fertilizer use, as well as increased environmental sustainability (Zhang et al. 2022). A separate study in China showed that catering, transportation, and leisure activities within the tourism sector significantly contribute to air pollution (GHG and PM2.5), with catering services having the most negative impact. Transportation and leisure show mixed effects (Li et al. 2024).

According to recent research grounded in the Stimulus-Organism-Response framework, residents' environmentally responsible behavior positively influences tourists' green consumption through the mediating effects of environmental concern and destination identification, thereby supporting sustainable rural tourism development (Hu et al. 2021).

Public participation benefits cooperative equilibrium by reducing governance costs and improving policy flexibility, especially when long-term aspirations of reputational effects and environmental tax reform are embraced. Strong support and ensured independence are crucial for unleashing the full power of public participation in environmental governance (Chu et al. 2022).

5 Conclusions

The unique combination of majestic mountain landscapes, pristine lakes, and rich biodiversity in the Kyrgyz Republic provides a solid foundation for promoting eco-oriented tourism models that are in line with global trends in responsible tourism. However, realizing this potential requires overcoming a range of ecological and organizational challenges, including the risks of environmental degradation, insufficient infrastructure, and limited local capacity.

Addressing these issues demands a comprehensive and integrated approach that combines the introduction of eco-certification systems, the promotion of low-impact rural tourism, investment in green infrastructure, and the empowerment of local communities as key custodians of natural and cultural heritage.

International cooperation, knowledge exchange, and the adoption of innovative technologies can significantly enhance the sustainability of rural tourism initiatives. Under the aegis of well-articulated national policy frameworks and through the sustained involvement of all relevant stakeholders, rural tourism has the capacity to mature into a pivotal engine for inclusive economic growth, the stewardship of ecological systems, and the safeguarding of cultural patrimony, thereby harmonizing development imperatives with environmental and socio-cultural sustainability. In this way, Kyrgyzstan can position itself as a competitive and responsible tourism destination while ensuring the long-term protection of its unique natural wealth for future generations.

References

Adebayo TS, Akadiri SS, Asuzu OC, Pennap NH, Sadiq-Bamgbopa Y (2022) Impact of tourist arrivals on environmental quality: a way towards environmental sustainability targets. Curr Issues Tour 26(6):958–976. https://doi.org/10.1080/13683500.2022.2045914

Ahmed Z, Zhang B, Cary M (2021) Linking economic globalization, economic growth, financial development, and ecological footprint: evidence from symmetric and asymmetric ARDL. Ecol Indic 121:107060. https://doi.org/10.1016/j.ecolind.2020.107060

Akadiri SS, Alkawfi MM, Uğural S, Akadiri AC (2019) Towards achieving environmental sustainability target in Italy. The role of energy, real income and globalization. Sci Total Environ 671:1293–1301. https://doi.org/10.1016/j.scitotenv.2019.03.448

Akadiri SS, Uzuner G, Akadiri AC, Lasisi TT (2020) Environmental Kuznets curve hypothesis in the case of tourism island states: the moderating role of globalization. Int J Financ Econ 26(2):2846–2858. https://doi.org/10.1002/ijfe.1938

Alola AA, Bekun FV, Sarkodie SA (2019) Dynamic impact of trade policy, economic growth, fertility rate, renewable and non-renewable energy consumption on ecological footprint in Europe. Sci Total Environ *685*:702–709. https://doi.org/10.1016/j.scitotenv.2019.05.139

Ashraf MS, Hou F, Kim WG, Ahmad W, Ashraf RU (2020) Modeling tourists' visiting intentions toward ecofriendly destinations: Implications for sustainable tourism operators. Bus Strat Environ 29(1):54–71. https://doi.org/10.1002/bse.2350

Bramwell B, Higham J, Lane B, Miller G (2016) Twenty-five years of sustainable tourism and the journal of sustainable tourism: looking back and moving forward. J Sustain Tour 25(1):1–9. https://doi.org/10.1080/09669582.2017.1251689

Brauer R, Dymitrow M, Tribe J (2019) The impact of tourism research. Ann Tour Res 77:64–78. https://doi.org/10.1016/j.annals.2019.05.006

Cholakova S, Dogramadjieva E (2023) Climate change adaptation in the ski industry: Stakeholders' perceptions regarding a mountain resort in Southeastern Europe. J Outdoor Recreat Tour 42:100611. https://doi.org/10.1016/j.jort.2023.100611

Chu Z, Bian C, Yang J (2022) How can public participation improve environmental governance in China? A policy simulation approach with multi-player evolutionary game. Environ Impact Assess Rev *95*:106782. https://doi.org/10.1016/j.eiar.2022.106782

Croes R, Ridderstaat J, Bąk M, Zientara P (2021) Tourism specialization, economic growth, human development and transition economies: the case of Poland. Tour Manag 82:104181. https://doi.org/10.1016/j.tourman.2020.104181

Debbage KG, Debbage N (2019) Aviation carbon emissions, route choice and tourist destinations: are non-stop routes a remedy? Ann Tour Res 79:102765. https://doi.org/10.1016/j.annals.2019.102765

Dogan E, Ulucak R, Kocak E, Isik C (2020) The use of ecological footprint in estimating the environmental Kuznets curve hypothesis for BRICST by considering cross-section dependence and heterogeneity. Sci Total Environ 723:138063. https://doi.org/10.1016/j.scitotenv.2020.138063

Dolnicar S (2020) Designing for more environmentally friendly tourism. Ann Tour Res 84:102933. https://doi.org/10.1016/j.annals.2020.102933

Godil DI, Sharif A, Rafique S, Jermsittiparsert K (2020) The asymmetric effect of tourism, financial development, and globalization on ecological footprint in Turkey. Environ Sci Pollut Res 27(32):40109–40120. https://doi.org/10.1007/s11356-020-09937-0

Habibi K, Pira M, Rahimi A, Hemmati G, Alizadeh H (2023) Ecological footprint analysis of tourism management in rural areas. J Environ Plann Manag 67(6):1169–1181. https://doi.org/10.1080/09640568.2022.2156852

Hu J, Xiong L, Lv X, Pu B (2021) Sustainable rural tourism: linking residents' environmentally responsible behaviour to tourists' green consumption. Asia Pacific J Tour Res 26(8):879–893. https://doi.org/10.1080/10941665.2021.1925316

Jia J, Chen D, Ju M, Lu C, Zhong C (2021) China's CO_2 emissions: an innovative framework for analyzing carbon reduction in sustainable tourism under the guidance of the United Nations' sustainable development goals. J Clean Prod 430:139752. https://doi.org/10.1016/j.jclepro.2023.139752

Khoi NH, Le NH, Ngoc BH (2021) The effect of tourism development on the ecological footprint in Singapore: evidence from asymmetric ARDL method. Curr Issue Tour 25(15):2500–2517. https://doi.org/10.1080/13683500.2021.1971165

Kongbuamai N, Viet BQ, Ullah YMA, Yun L (2020) The impact of tourism and natural resources on the ecological footprint: a case study of ASEAN countries. Environ Sci Pollut Res 27(16):19251–19264. https://doi.org/10.1007/s11356-020-08582-x

Li Y, Liu Y, Solangi YA (2024) Analysis of factors and strategies for the implementation of sustainable tourism in a green economic structure in China. J Clean Prod 434:140011. https://doi.org/10.1016/j.jclepro.2023.140011

Mancini MS, Barioni D, Danelutti C, Barnias A, Bračanov V, Capanna Piscè G et al (2022) Ecological footprint and tourism: development and sustainability monitoring of ecotourism packages in Mediterranean protected areas. J Outdoor Recreat Tour 38:100513. https://doi.org/10.1016/j.jort.2022.100513

Mikayilov JI, Mukhtarov S, Mammadov J, Azizov M (2019) Re-evaluating the environmental impacts of tourism: does EKC exist? Environ Sci Pollut Res 26(19):19389–19402. https://doi.org/10.1007/s11356-019-05269-w

Mukambaeva I, Sayakbaeva A, Akylbekova N, Mukambaev N, Ermekova A, Shambetova E (2024b) Influence of the state investment budget on the Kyrgyz agricultural sector through data analysis. In: 2024 IEEE 4th international conference on smart information systems and technologies (SIST). IEEE, Astana, Kazakhstan, pp 409–414. https://doi.org/10.1109/SIST61555.2024.10629277

Mukambaeva IB, Akylbekova NI, Mukambaev NJ, Lailieva EJ, Nam IE (2024a) Comparing the agricultural sectors of the EAEU countries through the sustainability index. In: Sergi BS, Popkova EG, Ostrovskaya AA, Chursin AA, Ragulina YV (eds) Ecological footprint of the modern economy and the ways to reduce it. Springer, Cham, Switzerland, pp 431–435. https://doi.org/10.1007/978-3-031-49711-7_71

National Statistical Committee of the Kyrgyz Republic (n.d.-a). Environment. Retrieved from https://stat.gov.kg/ru/statistics/turizm-otdyh-ohrana-okruzhayushej-sredy/. Accessed 27 June 2025

National Statistical Committee of the Kyrgyz Republic (n.d.-b) Tourism. Retrieved from https://stat.gov.kg/ru/statistics/turizm/. Accessed 27 June 2025

Raihan A (2024) The influence of tourism on the road to achieving carbon neutrality and environmental sustainability in Malaysia: the role of renewable energy. Sustain Anal Model 4:100028. https://doi.org/10.1016/j.samod.2023.100028

Smith MK, Diekmann A (2017) Tourism and well-being. Ann Tour Res 66:1–13. https://doi.org/10.1016/j.annals.2017.05.006

Sun Y-Y, Cadarso MA, Driml S (2020) Tourism carbon footprint inventories: a review of the environmentally extended input-output approach. Ann Tour Res 82:102928. https://doi.org/10.1016/j.annals.2020.102928

Wan J, Wang P, Wang M, Huang Y, Luo J (2025) Research on the carbon footprint of rural tourism based on life cycle assessment: a case study of a village in Guangdong, China. Sustainability 17(14):6495. https://doi.org/10.3390/su17146495

Wang Y, Wang L, Liu H, Wang Y (2021) The robust causal relationships among domestic tourism demand, carbon emissions, and economic growth in China. SAGE Open 11(4):1–14. https://doi.org/10.1177/21582440211054478

Yarimoglu E, Binboga G (2019) Understanding sustainable consumption in an emerging country: the antecedents and consequences of the ecologically conscious consumer behavior model. Bus Strat Environ 28(4):642–651. https://doi.org/10.1002/bse.2270

Zhang Y, Pan J, Chen K (2022) Study on the ecological footprint of tourism in the Beijing Jiufeng National Forest Park based on the component method. In: Zhang W, Yu F (eds) Global ecological governance and ecological economy. Springer, Singapore, pp 141–156. https://doi.org/10.1007/978-981-16-7025-1_9

Zhao X, Elahi E, Wang F, Xing H, Khalid Z (2024) Sustainable tourism development for traditional Chinese drama's intangible cultural heritage. Heliyon 10(3):e25483. https://doi.org/10.1016/j.heliyon.2024.e25483

The Impact of Climate Change on Agriculture in Kyrgyzstan and its Role in Ensuring Sustainable Economic Growth

Nelli I. Akylbekova, Liu Yujie, Gulmira D. Tokoeva, Batima B. Nasyrova, and Saltanat K. Omurova

Abstract This research examines the development of rural tourism in the Kyrgyz Republic between 2019 and 2024, highlighting its economic, social, cultural, and environmental impacts. Utilizing official statistics from the National Bank of the Kyrgyz Republic and a comprehensive SWOT analysis, the research explores the dynamics of formal and informal tourism sectors, including the effects of the COVID-19 pandemic on visitor numbers. The findings reveal that while rural tourism significantly contributes to local economic growth and cultural preservation, it also presents challenges such as seasonal income fluctuations, infrastructure strain, and environmental degradation. The research highlights the crucial role of local communities and sustainable practices in striking a balance between tourism growth and ecological conservation. A coherent strategy for advancing sustainable rural tourism in Kyrgyzstan should be predicated upon the modernization of infrastructural systems, the cultivation of ecological consciousness among visitors and local communities, and the institutionalization of robust regulatory mechanisms. Such a multidimensional approach is intended to safeguard the singular ecological and cultural patrimony of the country over the long term and engender inclusive socioeconomic dividends that reinforce local livelihoods, stimulate endogenous development trajectories, and integrate peripheral regions into broader circuits of sustainable growth.

Keywords Agricultural production · Crop and livestock dynamics · Climate change · Sustainability · SWOT analysis · Food security · Agroecosystem resilience · Digital agriculture

JEL Classification Q10 · Q12 · Q15 · Q18 · Q54 · O13

N. I. Akylbekova (✉) · G. D. Tokoeva · B. B. Nasyrova · S. K. Omurova
Kyrgyz National University named after Jusup Balasagyn, Bishkek, Kyrgyzstan
e-mail: nelly_a@bk.ru

L. Yujie
Kyrgyz State University named after Ishenaly Arabaev, Bishkek, Kyrgyzstan

A. B. Karbekova (ed.), *Sustainable Development of Rural Territories Based on Climate-Smart Technologies*, Climate Change Management,
https://doi.org/10.1007/978-3-032-20106-5_9

1 Introduction

Kyrgyzstan is traditionally considered an agrarian country due to the significant role that agriculture plays in its economy, employment, and food security. More than half of the country's population lives in rural areas. A considerable share of them is directly engaged in farming, livestock breeding, and related agricultural activities. Agriculture ensures the livelihood of rural households and contributes to national economic stability through the supply of food products and exports of certain goods, such as meat, dairy products, wool, and crops.

The agrarian character of Kyrgyzstan is primarily determined by its geographical and natural conditions. Located in the heart of Central Asia, the country is predominantly mountainous: about 90% of its territory is covered by ranges of the Tien Shan and Pamir-Alai. Although this reduces the share of land suitable for cultivation (arable land accounts for less than 7% of the territory), it provides vast areas for pastures, which serve as a foundation for livestock breeding—the leading branch of agriculture.

The climatic conditions of Kyrgyzstan also strongly influence its agrarian profile. The country has a sharply continental climate characterized by hot summers, cold winters, and significant temperature fluctuations between day and night. Annual precipitation varies by region. The Fergana Valley in the south has relatively favorable conditions for crop farming, but many other regions depend heavily on irrigation due to low rainfall. Glaciers and mountain rivers act as important sources of freshwater, ensuring irrigation systems and hydropower, but they are vulnerable to climate change and melting processes.

Kyrgyzstan's agrarian nature is shaped by the combination of its mountainous geography, pasture resources, and continental climate, which determine the predominance of livestock breeding alongside crop farming in fertile valleys.

The multifaceted impact of climate change on the agricultural sector of Kyrgyzstan constitutes a decisive determinant in shaping the trajectory of the nation's sustainable economic development.

2 Materials and Methods

This research adopts an integrative mixed-methods design, combining quantitative analysis, comparative evaluation, and sustainability diagnostics to explore the transformational dynamics of Kyrgyzstan's agricultural sector amid climatic perturbations. Utilizing official datasets from the National Statistical Committee (2018–2024) and corroborated by an extensive review of scientific and institutional sources, the research applies descriptive statistics and SWOT analytics to elucidate structural shifts, resilience potential, and systemic vulnerabilities. By correlating production indicators with climatic variables (i.e., temperature anomalies, precipitation variability, and glacial retreat), the research advances a multidimensional understanding of agroecological adaptation and resource efficiency. This sustainability-oriented framework enables a holistic interpretation of the interlinkages between agricultural productivity, environmental constraints, and economic growth, thereby providing a robust empirical foundation for climate-resilient policy formulation.

3 Results

In recent years, the Kyrgyz Republic has increasingly asserted itself on the global tourism map. Its unique natural landscapes, rich cultural heritage, and the enduring image of nomadic civilization have drawn growing interest from international travelers. Particularly attractive are those areas that have retained their pristine character: remote regions situated far from urban centers, often in close proximity to rural communities, mountain ranges, and alpine lakes.

Kyrgyzstan is experiencing noticeable climate change driven by a combination of global and domestic factors. On the global scale, the enhanced greenhouse effect and the overall trend of global warming are the dominant forces. Rising concentrations of greenhouse gases (e.g., carbon dioxide and methane in the atmosphere have led to a steady increase in average global temperatures). Due to its mountainous geography and continental climate, Kyrgyzstan is particularly vulnerable to these shifts.

Although the country contributes a relatively small share to global emissions, its specific geographical configuration and fragile ecological balance make it one of the most climate-vulnerable territories in Central Asia. This paradox (i.e., low emissions yet high exposure) requires an in-depth examination of exogenous and endogenous determinants of climate dynamics.

A central driver is the global warming trajectory shaped by the greenhouse effect. The cumulative rise in carbon dioxide (CO_2), methane (CH_4), and other greenhouse gases, largely produced by industrialized nations, has triggered a systemic alteration of atmospheric and thermal regimes. While Kyrgyzstan imports a significant portion of industrial goods and hydrocarbons, its domestic reliance on coal-based energy further exacerbates its carbon footprint, even if on a smaller scale compared to major emitters (Table 1).

The intensification of fossil fuel consumption presents another vector of climatic stress. Despite the dominance of hydropower in electricity generation, coal and petroleum derivatives remain the principal sources of energy for heating and transportation. This dual structure not only amplifies local air pollution but also increases cumulative CO_2 emissions, thereby reinforcing global warming feedback loops.

Equally significant is the degradation of ecosystems and land resources, particularly the overexploitation of pasturelands, which leads to erosion, desertification, and biodiversity loss. The progressive decline in vegetative cover diminishes the ecological capacity for carbon sequestration and disrupts the natural regulatory mechanisms of the regional climate.

The SWOT analysis of Kyrgyz agriculture (Table 2) highlights its strategic strengths, notably vast pasturelands exceeding nine million hectares, and abundant

Table 1 Key drivers of climate change in the Kyrgyz Republic and their implications

Factor	Description	Implications for Kyrgyzstan
Global warming and greenhouse effect	Rising global emissions of CO_2, CH_4, and other greenhouse gases drive systemic changes in atmospheric and thermal regimes. Kyrgyzstan, though a minor emitter, is affected through imported goods and reliance on coal energy	Increased average temperatures, amplified climate vulnerability despite low national emissions
Fossil fuel consumption	Hydropower dominates electricity generation, yet coal and petroleum remain the primary sources for heating and transport	Higher CO_2 emissions, local air pollution, and reinforcement of global warming feedback loops
Ecosystem and land degradation	Overuse of pastures leads to erosion, desertification, and biodiversity decline; reduced vegetation limits carbon sequestration	Loss of ecological resilience, weakened natural climate regulation, and worsening rural livelihoods.
Glacier retreat	Accelerated melting of Tien Shan glaciers, which serve as strategic freshwater reserves	Altered hydrological cycle, water shortages, frequent floods and droughts, risks to agriculture and energy security
Regional climatic specificity	A sharply continental climate with extreme temperature fluctuations makes Kyrgyzstan highly sensitive to global variations	Even minor global shifts cause severe local impacts, intensifying socio-economic instability
Urbanization and anthropogenic pressure	Population concentration in cities, deforestation, poor waste management, and low energy efficiency of infrastructure	Intensified local climate stress, degradation feedback loops, and increased ecological vulnerability

Source Developed by the authors

freshwater resources from glaciers and rivers, which underpin livestock production, support irrigation-based cropping, and enhance the sector's resilience to water variability.

The country's geographical positioning along emerging trade corridors, including the EAEU and proximity to China, further potentiates export opportunities and integration into regional value chains.

Table 2 SWOT analysis of agriculture in the Kyrgyz Republic

Strengths	*Weaknesses*
– Favorable natural conditions for livestock breeding due to vast pasturelands (over 9 million hectares)	– Low level of technological modernization and mechanization in farming
– Availability of freshwater resources from glaciers and mountain rivers for irrigation	– High dependence on irrigation with outdated water infrastructure and losses in water distribution
– Traditional knowledge and experience in livestock and crop production	– Small-scale, fragmented farms limiting economies of scale
– Growing demand for organic and eco-friendly products, with potential for niche markets	– Low productivity in crop and livestock sectors compared to global averages
– Strategic geographic location along trade routes (EAEU, China, and Central Asia)	– Insufficient storage, logistics, and processing facilities leading to post-harvest losses
Opportunities	*Threats*
– Expansion of exports of meat, dairy, fruits, and vegetables to neighboring countries and beyond	– Climate change: Glacier retreat, water scarcity, increased droughts and floods
– Development of agro-tourism and organic farming as sources of rural income	– Land degradation and overgrazing reducing the long-term productivity of pastures
– Integration of digital technologies and precision farming for efficiency	– Strong dependence on external markets and vulnerability to global price fluctuations
– International cooperation and investments in sustainable agriculture and infrastructure	– Outmigration of the rural labor force, reducing human capital in agriculture
– Access to the EAEU market and potential for deeper trade integration	– Limited resilience to external shocks (pandemics, geopolitical instability)

Source Developed by the authors

Conversely, the weaknesses reveal structural and technological constraints. The predominance of small, fragmented farms impedes economies of scale and limits capital accumulation for mechanization and innovation adoption.

The opportunities highlighted in the analysis suggest avenues for transformative growth. Expansion of exports in high-demand commodities such as meat, dairy, fruits, and vegetables can generate foreign exchange earnings, while agro-tourism and organic farming present alternative rural income streams that align with the global "experience economy" and sustainable consumption trends. The integration of digital agronomy, including remote sensing,

However, the threat landscape underscores the sector's susceptibility to multiple systemic shocks. Climate change constitutes a principal exogenous stressor, with glacier retreat, hydrological variability, and increasing incidence of droughts and floods threatening both crop yields and pasture productivity.

The analysis of Kyrgyzstan's agricultural production from 2018 to 2024 reveals heterogeneous growth across crop and livestock sectors, with total grain output rising from 1741.5 to 2085.5 thousand tons. Wheat exhibited steady gains (615.9 → 664.8

thousand tons), barley expanded markedly (429.3 → 611.1 thousand tons), and corn increased to 805.3 thousand tons, reflecting strategic land allocation and adoption of climate-resilient agronomic technologies, whereas rice showed modest growth and legumes, particularly beans, contracted, indicating adjustments in crop rotation and market-driven cultivation patterns.

Industrial and cash crops in Kyrgyzstan between 2018 and 2024 exhibit divergent trajectories: sugar beet output expanded markedly (773.0 → 957.4 thousand tons), reflecting intensification of high-value root crops. Cotton, oilseeds, and tobacco declined due to market rationalization and policy realignment. Horticultural and vegetable production consistently grew: vegetables (1094.9 → 1227.5 thousand tons), melons (249.1 → 280.9 thousand tons), and fruits and berries (251.4 → 346.2 thousand tons). Grape yields decreased (8.8 → 6.0 thousand tons), likely reflecting climatic sensitivity or land-use competition. Livestock outputs also expanded, with meat (221.3 → 252.9 thousand tons), milk (1589.7 → 1823.8 thousand tons), and eggs (533.2 → 790.3 million pieces), indicating improved productivity and feed management efficiencies (National Statistical Committee of the Kyrgyz Republic n.d.-a).

Wool output remained relatively stable, indicating steady maintenance of small ruminant herds (Fig. 1).

These production patterns collectively suggest that Kyrgyzstan's agricultural sector is undergoing qualitative and quantitative evolution, characterized by the optimization of crop portfolios, adoption of high-efficiency livestock systems, and strategic alignment with domestic and international market demands. The observed trends underscore the interplay between agronomic innovation, resource allocation, and policy interventions, which are critical for enhancing agroecosystem resilience and promoting sustainable food security within the context of climatic variability and economic globalization.

Between 2018 and 2024, exports from Kyrgyzstan's agriculture, forestry, and fishing sector exhibited a generally upward yet fluctuating trajectory, rising from \$128.0 million in 2018 to a peak of \$162.8 million in 2019 before experiencing a slight pandemic-induced decline to \$160.8 million in 2020. Nevertheless, the sector demonstrated resilience, recovering in 2021 with exports reaching \$169.9 million.

Export revenues in Kyrgyzstan's agricultural sector peaked in 2022 at \$294.1 million, driven by enhanced production efficiency, commodity diversification, and strategic regional market engagement, but declined to \$254.1 and \$242.0 million in 2023–2024, reflecting global price volatility, climatic perturbations, and trade disruptions (National Statistical Committee of the Kyrgyz Republic n.d.-b; National Statistical Committee of the Kyrgyz Republic n.d.-c) (Fig. 2).

In 2020–2021, agriculture's GDP share temporarily rose to 12.2–12.4% due to pandemic-driven emphasis on domestic food production. However, from 2022 to 2024, it declined sharply to 8.6%, reflecting structural economic diversification, expansion of higher value-added sectors, and persistent agrarian challenges (e.g., land degradation, water scarcity, limited technological adoption, and rural labor outmigration), signaling a process of relative deagrarianization despite stable or growing absolute production volumes.

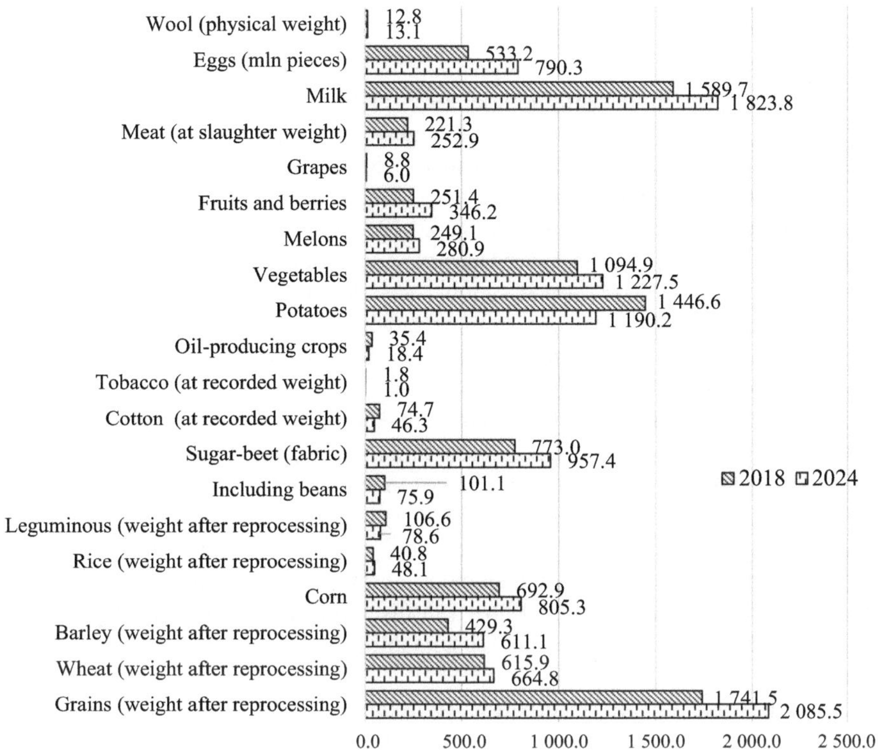

Fig. 1 Production of main agricultural products by types of farms of the Kyrgyz Republic, thousand tons. *Source* Compiled by the authors based on the National Statistical Committee of the Kyrgyz Republic (n.d.-a)

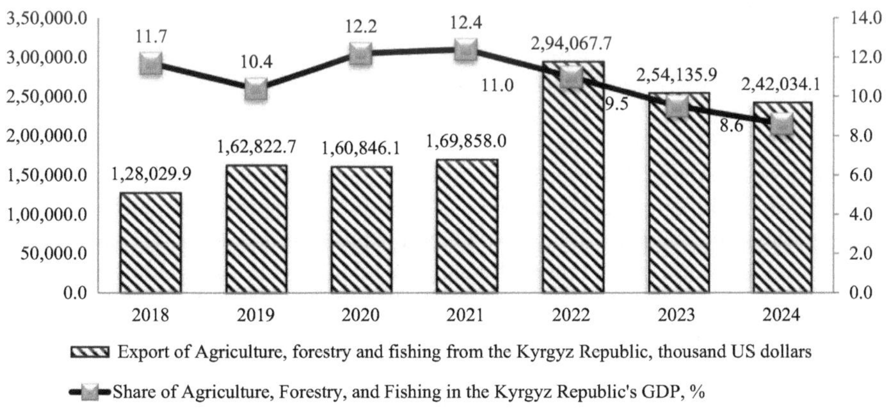

Fig. 2 Export of agriculture, forestry, and fishing from the Kyrgyz Republic, $ thousand. *Source* Compiled by the authors based on the National Statistical Committee of the Kyrgyz Republic (n.d.-b, n.d.-c)

4 Discussion

Climate change, with its rising temperatures and unpredictable weather, seriously threatens agriculture by reducing crop and livestock yields and creating major economic challenges, especially in warmer regions (Singh et al. 2021).

Agriculture's vulnerability arises from its climatic dependency on temperature and precipitation dynamics (Fazal and Wahab 2013), necessitating proactive adaptation and mitigation strategies to safeguard agrarian capital and curtail emission intensities (Ioan and Rădulescu 2015). Food supply faces dual threats: direct climate change impacts and potential alterations from greenhouse gas emission reduction efforts (McCarl et al. 2001). Addressing these challenges is essential for maintaining global food security and sustainable economic growth.

Anthropogenic climate perturbations increasingly destabilize agronomic systems through extreme weather events, threatening yields, food security, and socio-ecological resilience. Key emission sources include livestock enteric fermentation, rice methane, fertilizer-derived N_2O, residue burning, and land-use change (Majeed et al. 2023). Addressing these challenges requires climate-smart, regenerative practices (e.g., precision nutrient management, agroforestry, conservation tillage, and crop diversification) integrated within a transdisciplinary framework to align agricultural productivity with environmental sustainability and long-term resilience (Majeed et al. 2023).

Anthropogenic climate perturbations profoundly affect agroecosystems by intensifying droughts, floods, erratic precipitation, and thermal stress, undermining crop phenology, yield stability, and food system resilience, while key emission sources include enteric fermentation, rice methanogenesis, fertilizer-derived N_2O, and biomass burning, with land-use changes further amplifying atmospheric forcing. Mitigation and adaptation require climate-smart and regenerative practices (e.g., conservation agriculture, precision nutrient management, agroforestry, and crop diversification), integrated within a transdisciplinary framework to enhance soil carbon sequestration, hydrological resilience, and adaptive capacity (Malhi et al. 2021).

Economic globalization and financial development exert complex, non-linear effects on environmental sustainability through trade, capital flows, and technological diffusion, with empirical evidence demonstrating that growth-finance dynamics shape ecological footprints in context-specific ways (Ahmed et al. 2021). Effective mitigation thus requires integrating financial regulation with sustainability-oriented economic policies. Achieving environmental resilience in developed economies necessitates targeted decarbonization strategies and optimized resource allocation within globalized markets (Akadiri et al. 2019).

Alola, Bekun, and Sarkodie demonstrate that Europe's ecological footprint is shaped not only by energy intensity and economic output but also by demographic trends, technological adoption, and energy mix composition, underscoring the need to integrate energy transition strategies within broader socio-economic and demographic frameworks to ensure long-term environmental resilience (Alola et al. 2019).

Dogan, Ulucak, Kocak, and Isik empirically validated the Environmental Kuznets Curve for BRICST nations using ecological footprint metrics, demonstrating that economic expansion's environmental outcomes are contingent upon industrialization thresholds and technological diffusion (Dogan et al. 2020). Collectively, prior research confirms that sustainable development necessitates the integration of macroeconomic governance, energy transition, and ecosystem-based policy frameworks (Ahmed et al. 2021; Akadiri et al. 2019). Complementarily, Mukambaeva, Akylbekova, Mukambaev, Lailieva, and Nam employed a multidimensional Sustainability Index to expose inter-country asymmetries in agricultural efficiency and resilience within the EAEU (Mukambaeva et al. 2024a). Mukambaeva, Sayakbaeva, Akylbekova, Mukambaev, Ermekova, and Shambetova demonstrated the catalytic role of state investment budgets in advancing technological modernization and sustainable agrarian transformation in Kyrgyzstan (Mukambaeva et al. 2024b).

Yarimoglu and Binboga highlight that fostering environmentally conscious consumer behavior through awareness, ethical values, and perceived efficacy can substantially reduce societal ecological footprints, emphasizing the integration of behavioral insights into policy and market strategies to advance sustainability in emerging economies (Yarimoglu and Binboga 2019).

5 Conclusions

The comprehensive analysis of the agricultural sector in the Kyrgyz Republic demonstrates a multidimensional landscape characterized by significant endogenous strengths and pronounced systemic vulnerabilities.

Transformative growth in agriculture can be driven by export expansion, niche markets for organic and eco-certified products, and adoption of digital technologies such as precision farming and blockchain-based supply chains. International cooperation, sustainable infrastructure investment, and climate-smart policy frameworks are essential for resilience, competitiveness, and sustainable value creation.

The sustainable development of Kyrgyz agriculture requires a comprehensive, synergistic approach that integrates climate-smart practices, sustainable water and land management, ecosystem restoration, and digital agronomic technologies to enhance resilience against climatic variability. Strengthening institutional capacity, policy frameworks, and financial mechanisms, alongside capacity building, knowledge transfer, and participatory research, is essential for effective climate adaptation. Promoting rural economic diversification through agro-tourism, organic production, and value-added enterprises, combined with robust monitoring, evaluation, and data-driven decision-making, will enhance adaptive capacity and foster long-term sustainability. Ultimately, aligning policy, investment, and scientific interventions ensures that agriculture remains a driver of economic growth, a stabilizing force for rural communities, and a central pillar of Kyrgyzstan's resilience to environmental and economic challenges.

References

Ahmed Z, Zhang B, Cary M (2021) Linking economic globalization, economic growth, financial development, and ecological footprint: evidence from symmetric and asymmetric ARDL. Ecol Indic 121:107060. https://doi.org/10.1016/j.ecolind.2020.107060

Akadiri SS, Alkawfi MM, Uğural S, Akadiri AC (2019) Towards achieving environmental sustainability target in Italy. The role of energy, real income and globalization. Sci Total Environ 671:1293–1301. https://doi.org/10.1016/j.scitotenv.2019.03.448

Alola AA, Bekun FV, Sarkodie SA (2019) Dynamic impact of trade policy, economic growth, fertility rate, renewable and non-renewable energy consumption on ecological footprint in Europe. Sci Total Environ *685*:702–709. https://doi.org/10.1016/j.scitotenv.2019.05.139

Dogan E, Ulucak R, Kocak E, Isik C (2020) The use of ecological footprint in estimating the environmental Kuznets curve hypothesis for BRICST by considering cross-section dependence and heterogeneity. Sci Total Environ 723:138063. https://doi.org/10.1016/j.scitotenv.2020.138063

Fazal SA, Wahab SA (2013) Economic impact of climate change on agricultural sector: a review. J Transform Entrep 1(1):39–49. Retrieved from https://tinyurl.com/2ybhxpw7. Accessed 9 Oct 2025

Ioan I, Rădulescu CV (2015) New challenges for agriculture within the context of climate change. Theor Appl Econ 22:253–262. Retrieved from https://store.ectap.ro/articole/1150.pdf. Accessed 9 Oct 2025

Majeed N, Sharif R, Sheeraz MN, Sajjad S, Maqsood H, Imran M et al (2023) Impact of climate change on agriculture & vice versa: a review. J Qual Assur Agric Sci *3*(01):1–15. https://doi.org/10.38211/jqaas.2023.3.31

Malhi GS, Kaur M, Kaushik P (2021) Impact of climate change on agriculture and its mitigation strategies: a review. Sustainability 13:1318. https://doi.org/10.3390/su13031318

McCarl BA, Adams RM, Hurd BH (2001) Global climate change and its impact on agriculture. College Station, TX: Texas A&M College of Agriculture & Life Sciences. Retrieved from https://agecoresearch.tamu.edu/mccarl/wp-content/uploads/sites/4/2023/12/879.pdf. Accessed 9 Oct 2025

Mukambaeva IB, Akylbekova NI, Mukambaev NJ, Lailieva EJ, Nam IE (2024a) Comparing the agricultural sectors of the EAEU countries through the sustainability index. In: Sergi BS, Popkova EG, Ostrovskaya AA, Chursin AA, Ragulina YV (eds) Ecological footprint of the modern economy and the ways to reduce it. Springer, Cham, Switzerland, pp 431–435. https://doi.org/10.1007/978-3-031-49711-7_71

Mukambaeva I, Sayakbaeva A, Akylbekova N, Mukambaev N, Ermekova A, Shambetova E (2024b) Influence of the state investment budget on the Kyrgyz agricultural sector through data analysis. In: In: 2024 IEEE 4th international conference on smart information systems and technologies (SIST). IEEE, Astana, Kazakhstan, pp 409–414. https://doi.org/10.1109/SIST61555.2024.10629277

National Statistical Committee of the Kyrgyz Republic (n.d.-a) Agriculture. Retrieved from https://stat.gov.kg/ru/statistics/selskoe-hozyajstvo/. Accessed 27 Sept 2025

National Statistical Committee of the Kyrgyz Republic (n.d.-b) Foreign trade. Retrieved from https://stat.gov.kg/ru/statistics/vneshneekonomicheskaya-deyatelnost/. Accessed 27 Sept 2025

National Statistical Committee of the Kyrgyz Republic (n.d.-c) National accounts. Retrieved from https://stat.gov.kg/ru/statistics/nacionalnye-scheta. Accessed 27 Sept 2025

Singh P, Vaidya MK, Pathania K (2021) Economic impact of climate change on agriculture: present, past and future. J Pharmacognosy Phytochem 10(SI1):19–27

Yarimoglu E, Binboga G (2019) Understanding sustainable consumption in an emerging country: the antecedents and consequences of the ecologically conscious consumer behavior model. Bus Strat Environ 28(4):642–651. https://doi.org/10.1002/bse.2270

Advanced Potential of Climate-Smart Technologies in Agriculture

Best Practices in Climate-Smart Transitions of Agrarian Economies to Support Sustainable Rural Development in BRICS+

Aida S. Makeeva, Firuza N. Bayboboeva, Marina V. Safronchuk, and Yuliya V. Chutcheva

Abstract The research is devoted to a theoretical explanation of the essence of the climate-smart transition of the BRICS+ agrarian economy and the development of scientific, methodological, and practical recommendations for accelerating this transition and ensuring that it contributes to the sustainable development of rural areas in BRICS+. Drawing on statistics from 2023–2025, the research examines advanced practices in the climate-smart transition of the agrarian economy in support of sustainable rural development within BRICS+, analyzed using trend and regression methods. As a result, the authors constructed a model of sustainable rural development in BRICS+ under the influence of land-related factors and a set of organizational and technological factors associated with the climate-smart transition of the agrarian economy in 2023–2025. The theoretical significance of the developed model lies in its identification of the patterns of sustainable rural development specific to BRICS+. Forecasts of the spatial economy of BRICS+ up to 2028 demonstrate that the climate-smart transition of the agrarian economy can reverse the trend of population outflow from rural areas. A methodological framework for crisis management of the socio-economic development of rural territories in BRICS+ is proposed, centered on increasing the flexibility of business processes in climate-smart agriculture. The research also offers original recommendations for improving the organizational and technological foundations of agricultural digitalization, the implementation of which will allow the climate-smart agrarian economy to become

A. S. Makeeva
Jalal-Abad State University Named After B. Osmonov, Manas, Kyrgyzstan

F. N. Bayboboeva
Tashkent State University of Economics, Tashkent, Uzbekistan
e-mail: f.bayboboeva@tsue.uz

M. V. Safronchuk (✉)
MGIMO University, Moscow, Russia
e-mail: msafronchouk@gmail.com

Y. V. Chutcheva
Russian State Agrarian University—Moscow Timiryazev Agricultural Academy, Moscow, Russia

A. B. Karbekova (ed.), *Sustainable Development of Rural Territories Based on Climate-Smart Technologies*, Climate Change Management,
https://doi.org/10.1007/978-3-032-20106-5_10

a growth driver for rural areas in BRICS+ . In turn, this will enable more sustainable development of BRICS+ rural territories through increases in agricultural labor productivity and a rising share of rural residents within the population structure, thus initiating a process of de-urbanization in BRICS+ .

Keywords Climate-smart technologies · Climate transition · Agrarian economy · Sustainable territorial development · Rural areas · BRICS+

JEL Classification P25 · Q18 · R58

1 Introduction

BRICS+ is an association of countries whose members are geographically dispersed, as most lack shared borders and are also located far from one another, representing different parts of the world. Consequently, unlike many other country alliances, cooperation within BRICS+ is based not on geographical proximity but on shared economic interests. Simultaneously, the spatial organization of the BRICS+ economy is shaped to a much greater extent by natural and climatic factors than in most other international groupings due to several characteristics of the BRICS+ economic system.

One such characteristic is that BRICS+ countries possess vast territories and, as a result, a complex spatial structure in which rural areas are numerous and significant in scale. Population outflow from these rural areas significantly exacerbates internal imbalances in the BRICS+ economic system, creating extensive sparsely populated territories that lag critically behind urban areas in terms of the level and pace of socio-economic development. Thus, unlike in most other countries, where urbanization merely restructures the spatial economy, in BRICS+ nations, urbanization generates a spatial-economic crisis requiring state intervention.

Another characteristic is the remarkable success BRICS+ countries have achieved in addressing climate change. In recent years, these economies have managed to combine dynamic and productive greening of economic activity with rapid economic growth, resulting in solid progress in green development. Additionally, BRICS+ is one of the frontrunners of the Fifth Industrial Revolution. Smart technologies that have become widespread in the BRICS+ digital economy are increasingly used in environmental practices, including climate-change mitigation and adaptation.

Taken together, these factors underscore the relevance of utilizing the potential of climate-smart modernization of rural territories in BRICS+ to accelerate the development of agricultural production in these areas. Such modernization would transform rural territories into rapidly growing agricultural hubs with broad employment opportunities, high investment attractiveness, and a favorable environment, ultimately triggering a process of de-urbanization in BRICS+ and promoting the sustainable development of its rural regions.

However, this potential is hindered by the insufficient scientific and methodological support for the organizational and technical mechanisms that integrate smart technologies into climate adaptation practices in agriculture. As a result, these mechanisms remain underdeveloped and only partially effective. This issue must be addressed because it slows the climate-smart transition of the agrarian economy needed to support sustainable rural development in BRICS+ . In light of this, the research objective is to formulate a theoretical understanding of the climate-smart transition of the BRICS+ agrarian economy and develop scientific, methodological, and practical recommendations for accelerating this transition and ensuring its contribution to the sustainable development of rural territories in BRICS+.

2 Literature Review

This research is grounded in the fundamental principles of the climatic concept of spatial economics, which established the theoretical basis for examining urbanization as a factor in the sustainable development of rural areas (Torosyan et al. 2011). A content analysis of the existing scientific literature relevant to the topic of this study has identified the following organizational and technological factors of the climate-smart transition of the agrarian economy:

- ICT infrastructure for climate-smart agriculture (Litvinova et al. 2019),
- Availability of digital human resources for the climate-smart agrarian economy (Bratukhina et al. 2025),
- Pace of innovation-driven development in the agrarian economy (Bogoviz et al. 2018),
- Flexibility of business processes in climate-smart agriculture (Civelek et al. 2025),
- Adequacy of venture financing for agrarian modernization (Morhachov et al. 2024).

In the literature pertaining to the aforementioned concept, the following indicators are cited as manifestations of the sustainable development of rural territories:

- Growth in agricultural labor productivity (Civelek et al. 2025),
- Population movement from urban to rural areas (Sahakyan and Antamoshkina 2022).

The literature review conducted for this research suggests that, despite a substantial body of research on related issues, the core problem remains unresolved due to the lack of understanding of the patterns governing sustainable rural development, specifically within BRICS+ as a unique economic system.

This gap in the literature gives rise to a research question: "What drives the sustainable development of rural areas in BRICS+?" Some scholars Bezrukova et al. (2020) argue that rural population outflow is caused by declining soil fertility and shrinking agricultural land due to climate change.

Other researchers Gunjal et al. (2024) emphasize the importance of climate-smart technologies in boosting agricultural productivity, reducing dependency on land resources, and minimizing vulnerability to climate-related risks. Building on this latter perspective, the research puts forward the hypothesis that in the digital economy of BRICS+, the sustainable development of rural territories is determined primarily by the organizational and technological factors of the climate-smart transition of the agrarian economy, while the land factor plays a relatively minor role.

3 Materials and Methods

The research sample includes all ten current BRICS+ member countries, with statistical data for 2023–2025 consolidated into a single empirical dataset. The methodology applied in this research involves a factor analysis of the sustainable development of rural territories in BRICS+ , carried out using regression analysis.

The research examines the dependence of agricultural labor productivity (Agr) (based on World Bank data (2025a) and the share of rural residents in the population structure (Rur) (based on World Bank data (2025c) on the share of regularly cultivated agricultural land within the territorial composition (Fl) (based on World Bank data (2025b), as well as on a set of organizational and technological factors associated with the climate-smart transition of the agrarian economy, using data from UNCTAD (2025):

- ICT infrastructure (Fot_1);
- Availability of digital personnel (Fot_2);
- Pace of innovation-driven development (Fot_3);
- Flexibility of business processes (Fot_4);
- Adequacy of venture financing (Fot_5).

The compiled statistical data for these indicators are systematized in Table 1.

Based on the results of the regression analysis, those organizational and technological factors of the climate-smart transition of the agrarian economy (among the variables $Fot_{1–5}$), are selected for which the regression coefficients are positive in Agr анв Rur. The proposed hypothesis is considered confirmed if, in the resulting regression equations, the coefficients for Fl are smaller than the coefficients for the selected variables from the $Fot_{1–5}$ group.

Trend analysis is also applied to identify three-year trends (calculated as the ratio of 2025 values to 2023 values) in the changes of the dependent variables (Agr and Rur) and the selected factor variables (from the $Fot_{1–5}$ set). A forecast of the spatial economy of BRICS+ is then constructed under the assumption that these trends continue into the next three-year period (2026–2028).

As an alternative, a forecast of the sustainable development of rural territories in BRICS+ under the climate-smart transition of their agrarian economy up to 2028 is produced. For this forecast, the maximum attainable values of the selected factor

Table 1 Statistics on climate-smart agriculture and the sustainable development of rural territories in BRICS+ in 2023–2025

Year	BRICS+ country	ICT (shares of 1)	Skills (shares of 1)	R&D (shares of 1)	Industry activity (shares of 1)	Access to finance (shares of 1)	Permanent cropland (% of land area)	Cereal yield (tons per hectare)	Rural population (% of total population)
		Fot_1	Fot_2	Fot_3	Fot_4	Fot_5	Fl	Agr	Rur
2023	Brazil	0.8	0.6	0.7	0.6	0.9	0.93	4.48	12.45
	China	0.5	0.5	1.0	0.8	1.0	1.98	6.32	36.44
	Egypt	0.6	0.5	0.4	0.5	0.8	0.96	7.34	57.03
	Ethiopia	0.1	0.2	0.4	0.3	0.7	2.07	2.90	77.34
	India	0.6	0.4	0.8	0.8	0.8	4.70	3.50	64.13
	Indonesia	0.6	0.4	0.4	0.7	0.8	13.74	5.70	42.07
	Iran	0.6	0.5	0.5	0.4	0.9	1.17	2.30	23.19
	Russia	0.8	0.8	0.7	0.5	0.9	0.12	2.71	24.87
	South Africa	0.7	0.5	0.5	0.6	0.9	0.34	5.50	31.67
	UAE	0.8	0.8	0.5	0.7	0.9	0.58	19.81	12.46
2024	Brazil	0.8	0.6	0.7	0.6	0.9	0.93	4.91	12.21
	China	0.5	0.5	1.0	0.8	1.0	2.03	6.38	35.43
	Egypt	0.6	0.4	0.5	0.5	0.8	0.96	7.08	56.90
	Ethiopia	0.1	0.2	0.5	0.3	0.7	1.82	2.82	76.84
	India	0.6	0.3	0.9	0.8	0.8	4.78	3.59	63.64
	Indonesia	0.6	0.4	0.5	0.6	0.8	13.80	5.83	41.43
	Iran	0.6	0.5	0.5	0.4	0.9	1.17	2.32	22.74
	Russia	0.8	0.7	0.7	0.5	0.9	0.12	3.43	24.67
	South Africa	0.7	0.5	0.5	0.6	0.9	0.34	5.33	31.18

(continued)

Table 1 (continued)

Year	BRICS+ country	ICT (shares of 1)	Skills (shares of 1)	R&D (shares of 1)	Industry activity (shares of 1)	Access to finance (shares of 1)	Permanent cropland (% of land area)	Cereal yield (tons per hectare)	Rural population (% of total population)
		Fot_1	Fot_2	Fot_3	Fot_4	Fot_5	Fl	Agr	Rur
	UAE	0.8	0.7	0.5	0.6	0.9	0.58	20.37	12.22
2025	Brazil	0.8	0.5	0.6	0.6	0.9	0.93	5.34	11.98
	China	0.6	0.5	1.0	0.9	1.0	2.04	6.42	34.46
	Egypt	0.6	0.4	0.5	0.4	0.8	0.96	7.40	56.74
	Ethiopia	0.1	0.2	0.4	0.3	0.8	1.82	2.86	76.34
	India	0.6	0.4	0.9	0.8	0.9	4.86	3.63	63.13
	Indonesia	0.6	0.4	0.5	0.6	0.8	13.92	5.83	40.80
	Iran	0.6	0.5	0.5	0.4	0.9	1.17	2.42	22.30
	Russia	0.8	0.7	0.7	0.5	0.9	0.12	3.17	24.45
	South Africa	0.7	0.5	0.5	0.6	0.7	0.34	5.73	30.70
	UAE	0.8	0.7	0.6	0.6	0.9	0.58	23.31	11.99

Source Developed by the authors based on the materials from the UNCTAD (2025) and World Bank (2025a, b, c)

variables (from the $Fot_{1\text{-}5}$ group) are substituted into the regression equations for Agr and Rur.

4 Results

The regression-based analysis of the statistical data presented in Table 1 has resulted in the development of the following economic–mathematical model of the sustainable development of rural territories in BRICS+ under the influence of the land factor and a set of organizational and technological factors associated with the climate-smart transition of the agrarian economy during 2023–2025:

$$\begin{cases} \text{Agr} = 2.5135 - 16.2511\text{Fot}_1 + 29.1450\text{Fot}_2 - 23.1218\text{Fot}_3 + 31.3246\text{Fot}_4 \\ -4.3670\text{Fot}_5 - 0.2859\text{Fl},\ \text{R}^2 = 0.7229,\ \text{significance F} = 4.1960; \\ \text{Rur} = 160.4657 - 47.0817\text{Fot}_1 - 38.7182\text{Fot}_2 + 25.9219\text{Fot}_3 + 14.5757\text{Fot}_4 \\ -114.3951\text{Fot}_5 - 0.2576\text{Fl},\ \text{R}^2 = 0.9099,\ \text{significance F} = 18.4364 \end{cases} \tag{1}$$

Model (1) indicates that in BRICS+, a one-unit increase in the development level of ICT infrastructure for climate-smart agriculture leads to a decrease in agricultural labor productivity by 16.2511 tons per hectare and a reduction in the share of rural residents in the population structure by 47.0817%. An increase of one unit in the availability of digital personnel in the climate-smart agrarian economy results in a rise in agricultural productivity by 29.1450 tons per hectare, but a decline in the share of rural residents by 38.7182%.

A one-unit acceleration in the pace of innovation-driven development of the agrarian economy corresponds to a decline in agricultural productivity by 23.1218 tons per hectare but an increase in the share of rural residents by 25.9219%. A one-unit increase in the flexibility of business processes in climate-smart agriculture triggers an increase in agricultural productivity by 31.3246 tons per hectare and a rise in the share of rural residents by 14.5757%.

A one-unit increase in venture financing for agrarian modernization causes a decrease in agricultural productivity by 4.3670 tons per hectare and a reduction in the share of rural residents by 114.3951%. At the same time, a 0.2859% increase in the share of regularly cultivated agricultural land in the territorial structure results in a decrease in agricultural productivity by 16.2511 tons per hectare and a reduction in the share of rural residents by 0.2576%.

Thus, both regression coefficients in the model's equations are positive only for a single organizational-technological factor of the climate-smart transition of the agrarian economy: Fot_4 (flexibility of business processes). In both regression equations, the coefficients for Fl are negative and, therefore, smaller than the coefficients for the selected variable (Fot_4), confirming the initial hypothesis.

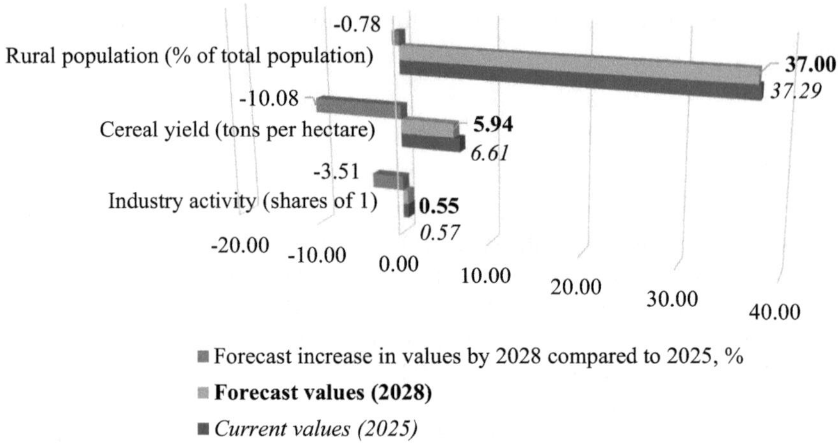

Fig. 1 Crisis forecast of the spatial economy of BRICS + assuming the continuation of the rural population outflow trend through 2028. *Source* Developed by the authors

Trend analysis of the data in Table 1 has identified the following three-year (2022–2025) trends in climate-change mitigation within the digital economy of BRICS+ rural territories:

- A decline in the flexibility of business processes in climate-smart agriculture by a factor of 0.97;
- A rise in agricultural labor productivity by a factor of 1.09;
- A reduction in the share of rural residents in the population structure by a factor of 0.98.

A forecast of the spatial economy of BRICS+ , assuming these trends continue into the next three-year period (2026–2028), is shown in Fig. 1. This forecast is referred to as a crisis scenario because it anticipates a continued outflow of population from rural territories, meaning a deepening of the socio-economic crisis in these areas.

As shown in Fig. 1, a further decline of 3.51% in the flexibility of business processes in climate-smart agriculture (from 0.57 in 2025 to 0.55) in BRICS + countries by 2028 is projected to lead to the following:

- A 10.08% decrease in agricultural labor productivity (from 6.61 tons per hectare in 2025 to 5.94 tons per hectare);
- A 0.78% reduction in the share of rural residents in the population structure (from 37.29% in 2025 to 37.00%).

If the forecast in Fig. 1 materializes, it will slow the socio-economic development of rural territories in BRICS+ countries. As an alternative to this crisis scenario, a forecast of the sustainable development of rural territorial systems in BRICS+ under the climate-smart transition of their agrarian economy through 2028 is constructed

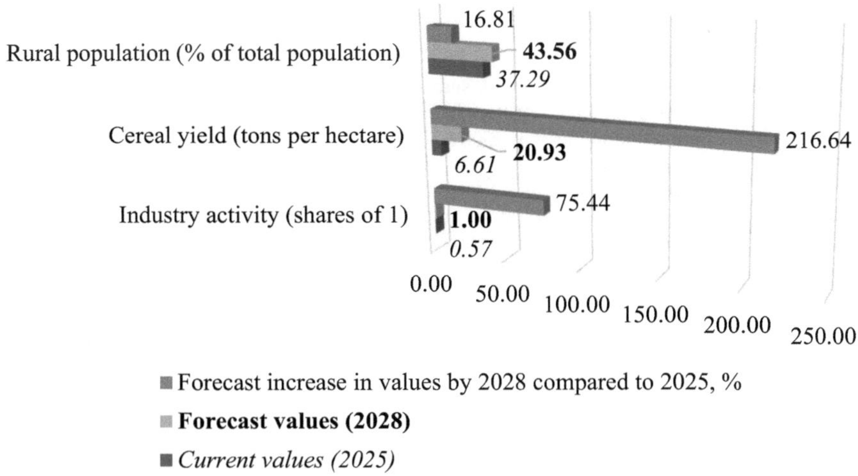

Fig. 2 Forecast of the sustainable development of rural territories in BRICS+ under the climate-smart transition of their agrarian economy through 2028. *Source* Developed by the authors

(Fig. 2), using model (1) with the maximum attainable value (1.00) substituted for the selected factor variable (Fot_4).

As shown in Fig. 2, an increase of 75.44% in the flexibility of business processes in climate-smart agriculture (from 0.57 in 2025 to 1.00) in BRICS+ countries by 2028 is projected to lead to the following:

- A 216.64% increase in agricultural labor productivity (from 6.61 tons per hectare in 2025 to 20.93 tons per hectare);
- A 16.81% rise in the share of rural residents in the population structure (from 37.29% in 2025 to 43.56%).

Practical implementation of the forecast shown in Fig. 2 will ensure the sustainable development of rural territories in BRICS+ countries.

5 Discussion

The theoretical contribution of the new results obtained in this study lies in extending the line of research conducted by Civelek et al. (2025), Sahakyan and Antamoshkina (2022), and Torosyan et al. (2011), and in enriching the climatic concept of spatial economics by revealing previously unknown patterns of sustainable rural development specifically within BRICS+ , summarized in Table 2.

Table 2 Patterns of the sustainable development of rural territories specifically within BRICS+ in comparison with earlier literature

Type of factors	Factors of a climate-smart transition in the agricultural economy	Disclosure of factors in existing literature	The influence of factors in BRICS+	
			The productivity of agricultural labor	The share of rural residents in the population structure
Land factor	Share of regularly cultivated agricultural land in the territorial structure	Bezrukova et al. (2020)	Negative	Negative
Organizational and technological factors	ICT infrastructure	Litvinova et al. (2019)	Negative	Negative
	Availability of digital personnel	Bratukhina et al. (2025)	Positive	Negative
	The pace of innovative development	Bogoviz et al. (2018)	Negative	Positive
	Business process flexibility	Civelek et al. (2025)	Positive	Positive
	Sufficiency of venture financing	Morhachov et al. (2024)	Negative	Negative

Source Developed by the authors

As reflected in Table 2, consistent with Civelek et al. (2025), the flexibility of business processes in climate-smart agriculture is the key organizational and technological factor of the climate-smart transition of the agrarian economy that supports the sustainable development of rural territories in BRICS+.

Some organizational and technological factors of the climate-smart transition proved contradictory, only partially supporting the sustainable development of rural territories in BRICS+:

- The availability of digital personnel for the climate-smart agrarian economy: unlike the findings of Bratukhina et al. (2025), it does not increase the share of rural residents in the population structure,
- The pace of innovation-driven development of the agrarian economy: unlike the findings of Bogoviz et al. (2018), it does not contribute to increasing agricultural labor productivity.

Negative (i.e., inhibiting rather than promoting the sustainable development of rural territories in BRICS+) were the following organizational and technological factors of the climate-smart transition:

- ICT infrastructure for climate-smart agriculture (in contrast to the findings of Litvinova et al. (2019)),
- Adequacy of venture financing for agrarian modernization (in contrast to the findings of Morhachov et al. (2024).

Contrary to the findings of Bezrukova et al. (2020), this research has established that in BRICS + countries, expanding the share of regularly cultivated agricultural land within the territorial structure does not support the sustainable development of rural territories: it increases neither agricultural labor productivity nor the movement of population from urban to rural areas.

In line with Gunjal et al. (2024), the hypothesis has been confirmed that in the digital economy of BRICS +, the sustainable development of rural territories is determined primarily by the organizational and technological factors of the climate-smart transition of the agrarian economy, while the land factor is of minor significance.

6 Conclusion

Thus, the principal outcome of this research is the model developed to explain the sustainable development of rural territories in BRICS+ under the influence of the land factor and a set of organizational and technological factors associated with the climate-smart transition of the agrarian economy in 2023–2025. The model possesses fundamental value because it reveals the underlying patterns of sustainable rural development specific to BRICS+ .

The forecasts of the spatial economy of BRICS+ up to 2028 show that the climate-smart transition of the agrarian economy will make it possible to reverse the trend of population outflow from rural territories. The practical significance of this research lies in the methodological framework created for crisis management of the socio-economic development of rural areas in BRICS+ , whose central element is the enhancement of business-process flexibility in climate-smart agriculture.

Implementing the proposed recommendations for improving the organizational and technological foundations of agricultural digitalization will allow the climate-smart agrarian economy to become a growth point for rural territories in BRICS+ . The advantages of applying these recommendations for sustainable rural development include increased agricultural labor productivity and a growing share of rural residents in the population structure, thus initiating a process of de-urbanization in BRICS+ .

References

Bezrukova TL, Gyiazov AT, Kuksova IV (2020) Socio-ecological and economic efficiency of the territories of forestry. IOP Conf Ser: Earth Environ Sci 595(1):012050. https://doi.org/10.1088/1755-1315/595/1/012050

Bogoviz AV, Taranov PM, Shuvaev AV (2018) Innovational tools for the provision of food security through state support for the agro-industrial complex in the conditions of digital economy. In: Popkova E (ed) The impact of information on modern humans. Springer, Cham, Switzerland, pp 659–665. https://doi.org/10.1007/978-3-319-75383-6_84
Bratukhina EA, Lapteva IP, Denisov IV, Zhailauov YB, Bencic S (2025) Globalisation as a factor of training of digital personnel and intelligent machines in emerging economies: international cooperation vs. global competition. Int J Econ Policy Emerg Econ 21(3):287–300. https://doi.org/10.1504/IJEPEE.2025.146583
Civelek M, Brezina I, Maňák R, Hajduk Ľ, Vrátný T (2025) The importance of risk management and marketing innovation activities of firms during health, political, energy, and economic crises. Acta Montanistica Slovaca 30(2):332–346. https://doi.org/10.46544/AMS.v30i2.05
Gunjal A, Pednekar R, Kanere S, Ruke S (2024) Dynamic resource allocation in edge computing environments: a machine learning approach. J Trends Challenges Artif Intell 1(4):133–138. https://doi.org/10.61552/JAI.2024.04.003
Litvinova TN, Tolmachev AV, Saenko II, Iskandaryan GO (2019) Role and meaning of the ICT infrastructure for development of entrepreneurial activities in the Russian agricultural machinery market. In: Popkova E, Ostrovskaya V (eds) Perspectives on the use of new information and communication technology (ICT) in the modern economy. Springer, Cham, Switzerland, pp 793–799. https://doi.org/10.1007/978-3-319-90835-9_91
Morhachov I, Antamoshkina E, Snapkouskaya S, Sahakyan M, Gavrineva N (2024) Limit allowable value of the Bank of Russia key rate: based on the example of agricultural holdings. In: Samoylenko I, Rajabov T (eds) Innovations in sustainable agricultural systems, vol 1. Springer, Cham, Switzerland, pp 490–500. https://doi.org/10.1007/978-3-031-70673-8_51
Sahakyan M, Antamoshkina E (2022) Managerial decision-making in the sphere of tourism under the conditions of risk and uncertainty. In: Gibadullin A (ed) Digital and information technologies in economics and management. Springer, Cham, Switzerland, pp 117–129. https://doi.org/10.1007/978-3-030-97730-6_11
Torosyan G, Matevosyan G, Harutyunyan S (2011) Use of natural materials for wastewater treatment & improvement of water quality. In: Hlavinek P, Winkler I, Marsalek J, Mahrikova I (eds) Advanced water supply and wastewater treatment: a road to safer society and environment. Springer, Dordrecht, Netherlands, pp 171–180. https://doi.org/10.1007/978-94-007-0280-6_15
UNCTAD (2025) Frontier technology readiness index, annual (analytical). Retrieved from https://unctadstat.unctad.org/datacentre/dataviewer/US.FTRI. Accessed 2 Dec 2025
World Bank (2025a) Cereal yield (kg per hectare). Retrieved from https://data.worldbank.org/indicator/AG.YLD.CREL.KG. Accessed 2 Dec 2025
World Bank (2025b) Permanent cropland (% of land area). Retrieved from https://data.worldbank.org/indicator/AG.LND.CROP.ZS. Accessed 2 Dec 2025
World Bank (2025c) Rural population (% of total population). Retrieved from https://data.worldbank.org/indicator/SP.RUR.TOTL.ZS. Accessed 2 Dec 2025

The Eurasian Model for Sustainable Rural Development Based on Climate-Smart Agricultural Technologies: Practical Solutions for the EAEU

Ira K. Panosyan, Marina V. Safronchuk, Angelina I. Seliankina, and Yuliya V. Chutcheva

Abstract The research aims to develop a Eurasian model for the sustainable development of rural territories based on climate-smart agricultural technologies, using the EAEU as a case study. To achieve this, the authors carried out a systematic analysis of EAEU statistics for 2024–2025 and 2016 using regression and trend analysis methods. The resulting model comprises several components. First, it includes a system of equations that reveals the economic and mathematical relationships between the diffusion of climate-smart agricultural technologies and the sustainable development of rural territories in the EAEU in 2024–2025. Second, it incorporates a roadmap for the sustainable development of EAEU rural territories based on climate-smart agricultural technologies through 2029. Third, it presents ten-year trends in the share of rural residents in the total population of EAEU countries (as of 2025 compared with 2016). Fourth, it offers a set of applied solutions for improving government regulation of the spread of climate-smart agricultural technologies in support of sustainable rural development in each EAEU country. The proposed recommendations focus on strengthening investment support for telecommunications services, which form the core of the authors' approach to state regulation of climate-smart agricultural technologies dissemination. Fifth, the model provides alternative forecasts of the share of rural residents in the total population of EAEU countries through 2029: one under the current policy approach and another under the newly recommended approach. Implementing the model will make it possible to

I. K. Panosyan
Armenian State University of Economics, Yerevan, Armenia

M. V. Safronchuk (✉)
MGIMO University, Moscow, Russia
e-mail: msafronchouk@gmail.com

A. I. Seliankina
RUDN University, Moscow, Russia

Y. V. Chutcheva
Russian State Agrarian University—Moscow Timiryazev Agricultural Academy, Moscow, Russia

A. B. Karbekova (ed.), *Sustainable Development of Rural Territories Based on Climate-Smart Technologies*, Climate Change Management,
https://doi.org/10.1007/978-3-032-20106-5_11

reverse the decline in the rural population share observed over the past decade and to initiate a process of sustainable development in rural territories across all EAEU countries.

Keywords Eurasian model · Sustainable development · Rural territories · Climate-smart agricultural technologies · EAEU · Agro-climatic economy

JEL Classification N54 · N55 · O18 · Q54

1 Introduction

When examining international experience in combating climate change, it is advisable to study Eurasian practice separately. Special attention should be paid to it because, unlike other global examples of transnational economic cooperation, Eurasian economic integration is not confined to a single part of the world but spans countries across two continents: Europe and Asia. The unique geography of the Eurasian economic space results in a heterogeneous composition of national-scale green economic systems.

Alongside this, the establishment of the EAEU (in 2015) has, over the past decade (up to 2025), institutionalized strong economic ties among the member states. This creates the prerequisites and opportunities for the coordinated implementation of joint environmental initiatives by the EAEU countries. Among the key initiatives on the EAEU climate agenda is the strengthening of climate resilience in agriculture to stimulate the socio-economic development of rural territories and reduce their developmental lag behind urban areas. This agro-climatic initiative is enshrined in Article 95, "Main Directions of Coordinated (Harmonized) Agro-Industrial Policy and Measures of State Support for Agriculture," of the Treaty on the Eurasian Economic Union of May 29, 2014 (Republic of Belarus, the Republic of Kazakhstan and the Russian Federation 2014).

Under contemporary digital-economy conditions, this agro-climatic initiative requires technological support in the form of climate-smart agricultural technologies, whose implementation in other countries has accelerated agricultural progress. Although climate-smart agricultural technologies are being adopted within the EAEU, their deployment remains fragmented and small-scale, lacking a coherent program and receiving insufficient scientific attention.

This gives rise to a scientific and practical problem: the uncertainty surrounding the conditions under which digital ICT-driven modernization of the Eurasian economy can promote the widespread adoption of climate-smart technologies, ensure their systemic use in agriculture, and support the sustainable development of rural areas. Aiming to address this problem through scholarly inquiry, this research seeks to develop a Eurasian model of sustainable rural development based on climate-smart agricultural technologies, using the EAEU as a case study.

2 Literature Review

This research is conducted on the theoretical foundation of the agro-climatic economy concept as described in the literature by Antamoshkina et al. (2022), expressed through the following four core scholarly theses. The first thesis: the sustainability of rural development is interpreted in terms of the ratio between rural and urban population sizes within national socio-economic systems (Isayan and Mayilyan 2022).

The second thesis: the main determinants of sustainable rural development are the expansion of value added in agriculture and the growth of agricultural productivity (Altukhov et al. 2019). The third thesis: under current climate-change conditions, the sustainable development of rural territories is constrained by the negative impact of the economy's ecological footprint and by the risks of natural disasters affecting agricultural value added and productivity (Bogoviz et al. 2023).

The fourth thesis: climate-smart agricultural technologies possess the potential to increase agricultural value added and productivity, and their adoption in an agro-climatic economy is shaped by the following set of general conditions:

- Coverage of rural territories by mobile 3G/4G Internet (Hazarika et al. 2024);
- The level of cybersecurity of Internet communications (Petrenko et al. 2019);
- Adequacy of investment support for telecommunications services (Gyiazov et al. 2023);
- Prevalence of ICT-based employment among agricultural producers in rural territories (Bannikova et al. 2019);
- Patent activity in the field of climate-smart agricultural technologies (Litvinova 2020).

The literature review related to this concept shows that most existing sources focus on international experience, while publications on Eurasian practices of economic activity within an agro-climatic economy remain scarce. This creates a gap in the literature: a lack of clarity regarding the causal relationships behind the sustainable development of rural territories based on climate-smart agricultural technologies in the EAEU.

Addressing this gap, the research poses a research question: "Which specific conditions (from the generalized list above) determine the spread of climate-smart agricultural technologies in Eurasian agriculture, to what extent these technologies reduce agriculture's vulnerability to climate change, and how they support the sustainable development of rural territories in the EAEU?" To answer this question, the research undertakes a detailed examination of the empirical experience of sustainable rural development based on climate-smart agricultural technologies within the EAEU.

3 Materials and Methods

The research is conducted using a sample of five EAEU countries (Armenia, Belarus, Kazakhstan, Kyrgyzstan, and Russia), drawing on statistical data for 2024–2025 and for 2016 (for selected indicators). At the first stage of the research, the authors determined three economic–mathematical relationships in the Eurasian agroclimatic economy using regression analysis, illustrated through the example of the EAEU:

1. The dependence of the share of rural residents in the total population of EAEU countries (RT), based on World Bank data (2025c), on the share of agricultural value added in GDP (AC_1), according to World Bank data (2025a), and on crop productivity (AC_2), according to World Bank data (2025b), expressed as the function RT = F(AC_1, (AC_2);
2. The dependence of agricultural value added in GDP (AC_1) and crop productivity (AC_2) on climate-change factors (the ecological footprint of the economy (EV_1) and the risk of natural disasters (EV_2)), as well as the general set of conditions for the dissemination of climate-smart agricultural technologies:

   - Mobile 3G/4G Internet coverage of rural areas (TC_1);
   - The level of cybersecurity of Internet communications (TC_2);
   - Adequacy of investment support for telecommunication services (TC_3);
   - Prevalence of ICT employment among agricultural producers in rural areas (TC_4);
   - Patent activity in the field of climate-smart agricultural technologies (TC_5).

For the quantitative measurement of these factor variables, statistics from UNDP and MBRF (2024) are used. The following functions are constructed: AC_1 = F($EV_{1–2}$, $TC_{1–5}$) и AC_2 = F($EV_{1–2}$, $TC_{1–2}$). The conditions that effectively promote the dissemination of climate-smart agricultural technologies are then identified by selecting those factor variables $TC_{1–5}$, for which the regression coefficients are positive in the AC_1 and AC_2 functions. The empirical information base for the research consists of the statistical data presented in Table 1.

At the second stage of the research, the authors developed a roadmap for the sustainable development of rural areas in the EAEU based on climate-smart agricultural technologies through 2029. To this end, the functions for AC_1 and AC_2 are first supplied with the average and then with the maximum values of the selected factor variables TC_{1-5} taken from the sample in Table 1, allowing the projected values of the resulting variables AC_1, AC_2, and RT to be identified.

At the third stage, applied solutions are developed to improve state regulation of the dissemination of climate-smart agricultural technologies in support of the sustainable development of rural areas in the EAEU. For this purpose, the authors carried out a trend analysis on the share of rural residents in the total population of the EAEU member states (RT), comparing data for 2025 with data from a decade earlier (2016).

Target conditions for the spread of climate-smart technologies (control values of the selected factor variables $TC_{1–5}$) are determined by year (for 2026–2029); the

Table 1 Statistics on rural development and the spread of climate-smart agricultural technologies in the EAEU in 2024–2025

Year	EAEU countries	3G/4G mobile network coverage (% population)	Secure Internet services per 1 million population	Investment in telecommunication services (% GDP)	ICT employment (%)	ICT PCT patent applications (per 100 billion GDP)	Ecological footprint per capita (global hectares)	Natural hazard exposure (score 0–100)	Agriculture, forestry, and fishing, value added (% of GDP)	Crop production index (2014–2016 = 100)	Rural population (% of total population)
		TC_1	TC_2	TC_3	TC_4	TC_5	EV_1	EV_2	AC_1	AC_2	RT
2024	Armenia	100.0	51.3	23.3	22.2	2.4	87.2	56.0	8.48	71.89	36.26
	Belarus	98.7	71.3	23.7	28.0	2.1	8.7	65.3	7.25	89.63	19.27
	Kazakhstan	89.4	64.7	8.5	0.0	0.7	73.2	58.0	3.83	108.65	41.82
	Kyrgyzstan	85.1	48.2	28.6	8.7	6.4	90.8	50.0	9.49	97.94	62.20
	Russia	93.3	75.8	24.2	20.3	4.4	55.3	43.0	2.97	111.37	24.67
2025	Armenia	100.0	50.4	15.4	31.5	36.8	87.2	65.0	7.92	73.65	36.07
	Belarus	98.7	70.7	11.3	30.2	35.6	65.3	83.0	6.87	93.16	18.88
	Kazakhstan	88.3	64.0	3.9	0.0	35.5	73.2	60.0	3.94	129.35	41.61
	Kyrgyzstan	99.1	47.2	27.4	10.8	50.6	90.8	61.0	8.61	106.82	61.83
	Russia	91.6	75.3	13.1	24.9	41.8	55.3	49.0	2.74	133.67	24.45

Source Developed by the authors based on the materials from the UNDP and MBRF (2024), World Bank (2025a, b, c)

projected values of the resulting variables AC_1, AC_2, and RT under these conditions are identified. The expected forecast values of the share of rural residents in the total population of the EAEU countries are then compared under the current approach and under a new approach to state regulation of the dissemination of climate-smart agricultural technologies, one that follows the authors' recommendations.

4 Results

As part of the first stage of this research, the authors conducted a regression analysis of the statistical data in Table 1, resulting in the development of the following system of equations, which reveals the economic and mathematical relationships between the spread of climate-smart agricultural technologies and the sustainable development of rural areas in the EAEU in 2024–2025:

$$\begin{cases} \mathrm{RT} = 69.4183 + 6.6669\mathrm{AC}_1 + 0.6370\mathrm{AC}_2,\ \mathrm{R}^2 = 0.7277, \\ \mathrm{AC}_1 = 19.8266 - 0.1230\mathrm{TC}_1 - 0.1831\mathrm{TC}_2 + 0.1291\mathrm{TC}_3 + 0.0556\mathrm{TC}_4 \\ -0.0125\mathrm{TC}_5 - 0.0156\mathrm{EV}_1 + 0.1248\mathrm{EV}_2,\ \mathrm{R}^2 = 0.9941, \\ \mathrm{AC}_2 = 149.5805 - 0.8988\mathrm{TC}_1 + 1.0962\mathrm{TC}_2 + 0.2663\mathrm{TC}_3 - 0.9016\mathrm{TC}_4 \\ +0.6798\mathrm{TC}_5 - 0.0929\mathrm{EV}_1 - 0.4765\mathrm{EV}_2,\ \mathrm{R}^2 = 0.9954 \end{cases} \tag{1}$$

System of Eqs. (1) indicates that a 1% increase in the share of agricultural value added in GDP contributes to a 6.6669% rise in the share of rural residents in the total population of the EAEU countries. A one-point increase in crop production productivity leads to a 0.6370% rise in the share of rural residents in the total population. Together, agricultural value added and agricultural productivity determine 72.77% of the sustainability of rural development in the EAEU (the equation for RT is significant at the 0.10 level, with F-significance = 0.0714).

An increase in the ecological footprint of the economy by one global hectare reduces the share of agricultural value added in GDP by 0.0156% and decreases crop productivity by 0.0929. A one-point increase in disaster risk raises the share of agricultural value added in GDP by 0.1248% but lowers crop productivity by 0.4765.

Expanding mobile 3G/4G Internet coverage in rural areas by 1% leads to a 0.1230% decline in the share of agricultural value added in GDP and a 0.8988% drop in crop productivity. A one-unit increase in the cybersecurity level of Internet communications results in a 1.0962 rise in crop productivity but a 0.1831% decrease in the share of agricultural value added in GDP.

Strengthening investment support for telecommunication services by 1% of GDP ensures a 0.1291% increase in the share of agricultural value added in GDP and a 0.2663% rise in crop productivity. A 1% increase in the prevalence of ICT employment among agricultural producers in rural areas leads to a 0.0556% rise in the share of agricultural value added in GDP. However, it results in a 0.9016 decrease in crop productivity.

An increase of 1 per 100 billion GDP in patent activity in the field of climate-smart agrotechnologies is accompanied by a 0.6798 rise in crop productivity but a 0.0125% decline in the share of agricultural value added in GDP. Taken together, climate change factors and the conditions for the dissemination of climate-smart agricultural technologies determine 99.41% of agricultural value added in the EAEU (the AC_1 equation is significant at the 0.05 level, F-significance = 0.0409) and 99.54% of agricultural productivity in the EAEU (the AC_2 equation is significant at the 0.05 level, F-significance = 0.0316).

System of Eq. (1) thus allows us to conclude that the only consistent condition that genuinely contributes to the spread of climate-smart agricultural technologies in the EAEU is investment support for telecommunication services (the regression coefficients for factor variable TC_3 are positive in the functions for AC_1 and AC_2.

Within the second stage of this research, the authors prepared a roadmap for the sustainable development of rural areas in the EAEU based on climate-smart agricultural technologies through 2029 was prepared (Table 2).

As shown in Table 2, in 2026, it is expected that investment support for telecommunication services will be increased to the EAEU average for 2024–2025, that is, to 17.94% of GDP. This will lead to a rise in the share of agricultural value added in GDP to 6.60% and an increase in crop production productivity to 108.65. As a result, the share of rural residents in the total population will increase to an average of 41.39% across the EAEU.

In 2027, it is advisable to raise investment support for telecommunication services to the EAEU average for 2024, that is, to 21.66% of GDP. This is expected to increase the share of agricultural value added in GDP to 7.19% and raise crop production productivity to 109.97. Consequently, the share of rural residents in the total population will reach an average of 46.26% across the EAEU.

Table 2 Roadmap for sustainable development of rural areas in the EAEU based on climate-smart agricultural technologies until 2029

Indicator	Base values in 2025	Recommended values				Growth in 2029 compared to 2025
		2026	2027	2028	2029	
Investment in telecommunication services (% GDP)	14.22	17.94	21.66	27.40	28.60	2.01
Agriculture, forestry, and fishing, value added (% of GDP)	6.02	6.60	7.19	8.10	8.28	1.38
Crop production index (2014 – 2016 = 100)	107.33	108.65	109.97	112.01	112.43	1.05
Rural population (% of total population)	36.57	41.39	46.26	53.64	55.19	1.51

Source Developed by the authors

In 2028, investment support for telecommunication services should be increased to the maximum recorded in the EAEU for 2025, that is, to 27.40% of GDP. This is projected to raise the share of agricultural value added in GDP to 8.10% and increase crop production productivity to 112.01. This will bring the share of rural residents in the total population to an average of 53.64% across the EAEU.

In 2029, investment support for telecommunication services needs to be elevated to the maximum level recorded in the EAEU for 2024 (i.e., to 28.60% of GDP (2.01 times higher than in 2025)). This will result in an increase in the share of agricultural value added in GDP to 8.28% (1.38 times higher than in 2025) and a rise in crop production productivity to 112.43 (1.05 times higher than in 2025). Consequently, the share of rural residents in the total population will reach an average of 55.19% across the EAEU (1.51 times higher that in 2025).

As part of the third stage of this research, Fig. 1 presents a trend analysis of the share of rural residents in the total population of the EAEU countries (RT), comparing data for 2025 with data from a decade earlier (2016).

As shown in Fig. 1, the greatest decline in the share of rural residents in the total population in 2025 compared with 2016 among the EAEU countries occurred in Belarus (0.83). The decrease in Russia amounted to 0.94, in Kyrgyzstan to 0.96, in Kazakhstan to 0.97, and in Armenia to 0.98. Applied solutions have been developed to improve state regulation of the dissemination of climate-smart agricultural technologies in support of sustainable rural development in the EAEU. These include increasing investment in telecommunication services (relative to 2025):

- Armenia: by 16.49% in 2026, 40.65% in 2027, 77.92% in 2028, and 85.71% in 2029;
- Belarus: by 58.76% in 2026, 91.68% in 2027, 142.48% in 2028, and 153.10% in 2029;
- Kazakhstan: by 360.00% in 2026, 455.38% in 2027, 602.56% in 2028, and 633.33% in 2029;

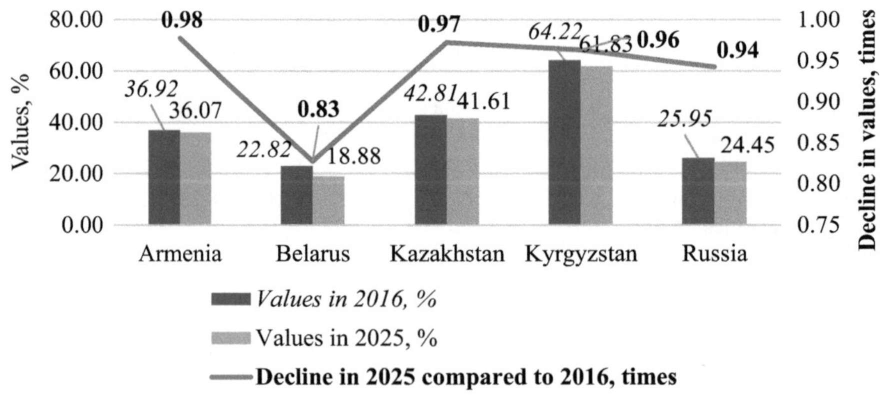

Fig. 1 Trend analysis of the share of rural residents in the total population by EAEU countries in 2025 compared to 2016. *Source* Developed by the authors

- Kyrgyzstan: by 4.38% in 2029;
- Russia: by 36.95% in 2026, 65.31% in 2027, 109.16% in 2028, and 118.32% in 2029.

Using System of Eq. (1), the authors determined the target conditions for the dissemination of climate-smart technologies (control values of the selected factor variables TC_{1-5}) for 2026–2029; the corresponding projected values of the resulting variables AC_1, AC_2, and RT were also identified. The expected forecast values for the share of rural residents in the total population of the EAEU countries, under the current approach and under the new approach to state regulation of the dissemination of climate-smart agricultural technologies, which follows the authors' recommendations, are compared in Fig. 2.

As shown in Fig. 2, the implementation of the recommendations will make it possible to achieve an increase in the share of rural residents in the total population by 2029:

- In Armenia to 54.47%, instead of a decline to 32.88%;
- In Belarus to 28.50%, instead of a decline to 8.84%;
- In Kazakhstan to 62.84%, instead of a decline to 37.15%;
- In Kyrgyzstan to 93.37%, instead of a decline to 53.13%;
- In Russia to 36.92%, instead of a decline to 19.28%.

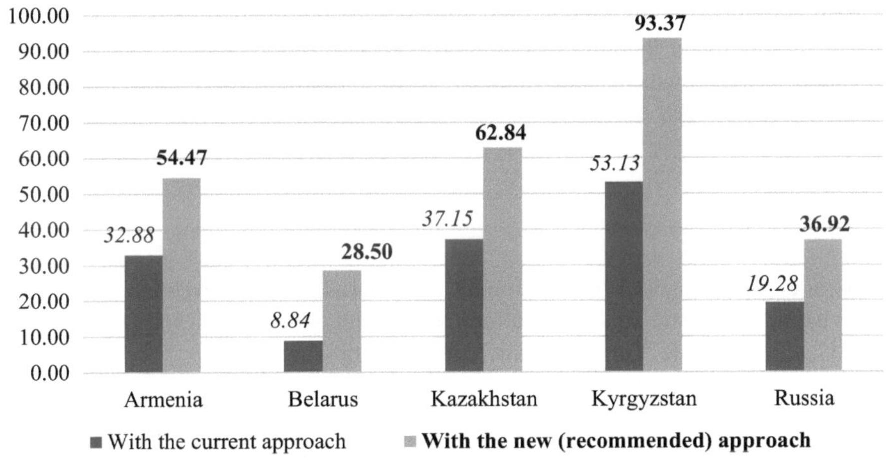

Fig. 2 Alternative projections of the share of rural residents in the total population by EAEU country until 2029, %. *Source* Developed by the authors

5 Discussion

The innovative results obtained in this research contribute to the works of Altukhov et al. (2019), Antamoshkina et al. (2022), Bogoviz et al. (2023), and Isayan and Mayilyan (2022), advancing the theoretical concept of agro-climatic economics by clarifying the causal relationships underlying the sustainable development of rural areas based on climate-smart agricultural technologies in the EAEU.

Confirming the findings of Gyiazov et al. (2023) it is demonstrated that sufficient investment support for telecommunication services systematically enhances the sustainability of rural development in the EAEU, simultaneously increasing both value added and labor productivity in agriculture.

Contrary to Hazarika et al. (2024), it is found that mobile 3G/4G Internet coverage in rural areas does not contribute to the sustainable development of rural territories in the EAEU. It is further revealed that only partial support for sustainable rural development in the EAEU is provided by the following:

- The level of cybersecurity in Internet communications, which fosters productivity growth but not value added in agriculture (unlike the findings of Petrenko et al. (2019));
- The prevalence of ICT employment among agricultural producers in rural areas, which increases value added but not productivity in agriculture (unlike the findings of Bannikova et al. (2019));
- Patent activity in the field of climate-smart agrotechnologies, which promotes productivity growth but not value added in agriculture (unlike the findings of Litvinova (2020)).

6 Conclusion

Thus, a comprehensive Eurasian model for the sustainable development of rural areas based on climate-smart agricultural technologies has been developed using the EAEU as an example. This model includes the following:

- A system of equations reflecting the economic–mathematical relationships between the dissemination of climate-smart agricultural technologies and the sustainable development of rural areas in the EAEU in 2024–2025;
- A roadmap for the sustainable development of rural territories in the EAEU based on climate-smart agricultural technologies through 2029;
- Ten-year trends in changes in the share of rural residents in the total population of EAEU member states (2025 compared with 2016);
- Applied solutions for improving state regulation of the dissemination of climate-smart agricultural technologies to support sustainable rural development in each EAEU country, unified into an original regulatory approach;

- Alternative forecasts of the share of rural residents in the total population of EAEU member states through 2029 under the current approach and under the new (recommended) approach.

The practical significance of the proposed recommendations lies in the fact that their implementation will make it possible to reverse the decline in the rural population share observed over the past decade and to initiate a process of sustainable development in these territories in every EAEU country, achieving significant progress (raising this share to 55.19% compared with 36.57% in 2025) by 2029.

References

Altukhov AI, Bogoviz AV, Kuznetsov IM (2019) Creation of an information system—A necessary condition of rational organization of agricultural production. In: Popkova E, Ostrovskaya V (eds) Perspectives on the use of new information and communication technology (ICT) in the modern economy. Springer, Cham, Switzerland, pp 800–809. https://doi.org/10.1007/978-3-319-90835-9_92

Antamoshkina EN, Lutsenko JV, Daeva TV, Radionova EA, Sahakyan MA (2022) Regional food systems: assessment of the level of self-sufficiency in food. IOP Conf Ser: Earth Environ Sci 1069:012032. https://doi.org/10.1088/1755-1315/1069/1/012032

Bannikova NV, Astrakhantseva EY, Morozova IA, Litvinova TN (2019) Mastering of the information and communication technologies by labor resources of rural territories as a new vector of their development in modern Russia. In: Popkova E, Ostrovskaya V (eds) Perspectives on the use of new information and communication technology (ICT) in the modern economy. Springer, Cham, Switzerland, pp 701–708. https://doi.org/10.1007/978-3-319-90835-9_81

Bogoviz AV, Tolmachev AV, Sozinova AA, Gyiazov AT (2023) The leading experience of ESG management of business in support of decarbonisation and development of the green economy in the Eurasian Economic Union (EAEU). In: Popkova EG, Sergi BS (eds) ESG management of the development of the green economy in Central Asia. Springer, Cham, Switzerland, pp 145–153. https://doi.org/10.1007/978-3-031-46525-3_16

Gyiazov AT, Isakov NZ, Kyzy ZA, Darkenbaev TS (2023) Practical experience of climate-responsible business in the digital economy markets in the Eurasian Economic Union (EAEU). In: Popkova EG (ed) Smart green innovations in Industry 4.0. Springer, Cham, Switzerland, pp 77–83. https://doi.org/10.1007/978-3-031-45830-9_9

Hazarika A, Madhukullya S, Hazarika A (2024) An introspection into the efficient data analysis with artificial intelligence. J Trends Challenges Artif Intell 1(3):81–84. https://doi.org/10.61552/JAI.2024.03.001

Isayan L, Mayilyan F (2022) Causations or payoffs? The interaction between countries' economic results and competing proxies of social capital. Econ Sociol 15(2):60–73. https://doi.org/10.14254/2071-789X.2022/15-2/4

Litvinova TN (2020) Managing the development of infrastructural provision of AIC 4.0 on the basis of artificial intelligence: case study in the agricultural machinery market. In: Popkova E, Sergi B (eds) Digital economy: Complexity and variety vs. rationality. Springer, Cham, Switzerland, pp 317–323. https://doi.org/10.1007/978-3-030-29586-8_37

Petrenko ES, Benčič S, Koroleva AA (2019) Digital business in the cyber economy: the organization of production and distribution based on the breakthrough technologies of Industry 4.0. In: Filippov VM, Chursin AA, Ragulina JV, Popkova EG (eds) The cyber economy. Springer, Cham, Switzerland, pp 11–17. https://doi.org/10.1007/978-3-030-31566-5_2

Republic of Belarus, the Republic of Kazakhstan and the Russian Federation (2014) Treaty on the Eurasian Economic Union (adoption on May 29, 2014). Astana, Kazakhstan. Retrieved from https://docs.eaeunion.org/upload/iblock/3b2/0aq0urjrfygepanpcxgllxg9x68ce6bd/itia_05062014_doc.pdf. Accessed 3 Dec 2025

UNDP and MBRF (2024) Global Knowledge Index 2023–2024. UNDP and MBRF, New York, NY; Dubai, UAE. Retrieved from https://www.undp.org/arab-states/publications/global-knowledge-index-2024. Accessed 3 Dec 2025

World Bank (2025a) Agriculture, forestry, and fishing, value added (% of GDP). Retrieved from https://data.worldbank.org/indicator/NV.AGR.TOTL.ZS. Accessed 3 Dec 2025

World Bank (2025b) Crop production index (2014–2016 = 100). Retrieved from https://data.worldbank.org/indicator/AG.PRD.CROP.XD. Accessed 3 Dec 2025

World Bank (2025c) Rural population (% of total population). Retrieved from https://data.worldbank.org/indicator/SP.RUR.TOTL.ZS. Accessed 3 Dec 2025

Green Infrastructure for Agriculture: Adapting to Climate Change Through Sustainable Construction

Aiganysh A. Saiakbaeva, Nelli I. Akylbekova, Temirlan Taalaibek, Ainura J. Kenzhekaraeva, and Kanyshai A. Toktalieva

Abstract The research examines the potential, implementation strategies, and challenges associated with the development of green infrastructure in agricultural construction across the diverse regions of the Kyrgyz Republic. Employing a mixed-methods approach, the authors integrate quantitative analysis of official statistical data with qualitative evaluation of policy frameworks, expert opinions, and field observations. A comparative regional assessment encompassing nine administrative areas of the Kyrgyz Republic, including Bishkek and Osh, reveals significant disparities in investment levels, construction intensity, and the adoption of sustainable practices. The proposed model of sustainable agricultural construction incorporates environmental technologies, water resource management, renewable energy integration, social inclusion, economic viability, and climate adaptation measures. The results of this research underscore the importance of coherent policy coordination, capacity-building programs, and the diffusion of innovative, resource-efficient technologies. The research concludes that strengthening institutional support and promoting cross-sectoral cooperation are critical for enhancing the resilience, productivity, and environmental sustainability of Kyrgyzstan's agricultural sector in the context of climate change.

Keywords Sustainable construction · Green infrastructure · Agriculture · Kyrgyz Republic · Climate adaptation · Energy efficiency · Water conservation · Economic sustainability · Regional development · Rural areas · Eco-friendly technologies

JEL Codes Q15 · Q18 · Q24 · Q54 · R11

A. A. Saiakbaeva · N. I. Akylbekova (✉)
Kyrgyz National University Named After Jusup Balasagyn, Bishkek, Kyrgyzstan
e-mail: nelly_a@bk.ru

T. Taalaibek
National Agency for Investments Under the President of the Kyrgyz Republic, Bishkek, Kyrgyzstan

A. J. Kenzhekaraeva · K. A. Toktalieva
Kyrgyz State Technical University Named After I. Razzakov, Bishkek, Kyrgyzstan

A. B. Karbekova (ed.), *Sustainable Development of Rural Territories Based on Climate-Smart Technologies*, Climate Change Management,
https://doi.org/10.1007/978-3-032-20106-5_12

1 Introduction

Green infrastructure (GI) represents a vital tool for enhancing the adaptive capacity of agricultural systems to climate change while safeguarding ecosystem services. Through practices such as agroforestry, conservation buffers, and wetlands restoration, GI strengthens soil health, improves water retention, and fosters biodiversity, thereby increasing resilience to extreme weather events. However, its large-scale adoption in resource-limited settings requires targeted institutional support, tailored investment strategies, and integration of ecological objectives with agricultural productivity.

Recent advancements reveal an increasing reliance on GI-based technological interventions to bolster agricultural efficiency, enrich biodiversity, and stabilize microclimatic conditions. Applications of GI are now being actively explored and implemented through agroforestry systems, conservation buffer zones, and precision agriculture techniques. Moreover, the integration of nature-based approaches, such as regenerative soil management and soil health monitoring technologies, further exemplifies the role of GI in fostering environmentally sustainable and climate-resilient farming practices (Ogwu and Kosoe 2025).

In urban settings, green infrastructure (e.g., green roofs, vegetated swales, urban forests, and permeable pavements) serves as a nature-based solution to mitigate the urban heat island effect and manage surface runoff, thereby reducing the risk of urban flooding (Gill et al. 2007). The integration of green infrastructure with conventional grey systems creates multifunctional landscapes that simultaneously meet engineering needs, deliver essential ecosystem services, and improve urban environmental quality (Nakamura et al. 2020).

The benefits of GI extend well beyond environmental advantages. It also positively influences human health by reducing stress, promoting physical activity, and lowering heat-related illness (Pitman et al. 2015). Improved community livability, aesthetic enhancement, and recreational opportunities further underscore the socio-economic value of green infrastructure. Moreover, effective water management through GI systems reduces the burden on municipal water systems, cuts treatment costs, and contributes to increased economic resilience.

Despite its wide-ranging benefits, the implementation of green infrastructure (GI), particularly in low-resource and developing contexts, faces significant challenges. The most pressing obstacles include high upfront investment costs, limited technical expertise, and weak institutional frameworks that fail to prioritize or incentivize nature-based solutions (Ogwu and Kosoe 2025). Additionally, the absence of coherent and integrated policy support hampers large-scale adoption and mainstreaming of GI in development planning.

To overcome these obstacles, the establishment of supportive frameworks is essential. This includes the provision of financial incentives such as subsidies and tax breaks, capacity-building programs, public–private partnerships, and the promotion of knowledge-sharing platforms that disseminate best practices and success stories

(Nakamura et al. 2020; Ogwu and Kosoe Ogwu and Kosoe 2025). Furthermore, integrating GI into national climate adaptation strategies and urban/rural development plans can institutionalize its application and unlock funding through climate finance mechanisms.

Green infrastructure offers a wide range of significant and multifaceted benefits that contribute to sustainable development across urban and rural landscapes. Furthermore, in the context of rapid urbanization, green infrastructure presents new opportunities for expanding urban food production, thereby reinforcing food security and enhancing the resilience of cities to environmental and socio-economic challenges (Pitman et al. 2015).

Nature-based solutions offer a promising pathway to enhance the resilience of agricultural systems to climate change by addressing mitigation and supply-side adaptation. The integration of productive green infrastructure into agricultural landscapes represents a viable and forward-looking approach to confronting the multifaceted challenges posed by a changing climate. Such practices support ecological stability and contribute to the sustainability and long-term productivity of farming systems under increasing environmental stress (Anderson and Gough 2021). Long-term environmental water management is essential for maintaining river basin health and ensuring ecological flow sustainability, as demonstrated in large systems like the Murray-Darling Basin (Gawne et al. 2019). The integration of green infrastructure into construction practices serves as a key driver for achieving sustainability targets and mitigating environmental impacts through innovative, nature-based approaches (Umar et al. 2024).

Advancing such initiatives demands coordinated efforts, adequate investment, and active engagement from governments, communities, and the private sector to embed green infrastructure as a fundamental element of climate adaptation strategies.

2 Materials and Methods

This research employed a mixed-methods approach to assess the potential, implementation, and challenges of developing GI in the construction sector of agricultural regions in Kyrgyzstan. The research combined quantitative statistical analysis with qualitative policy review and expert evaluation. Quantitative data were collected from publicly available sources, including the National Statistical Committee of the Kyrgyz Republic, particularly indicators on the volume of self-performed construction work and investments in fixed capital by region (2019–2024). A comparative analysis was conducted by region to identify differences and growth trends. The analysis focused on nine administrative territories, including the cities of Bishkek and Osh.

The data were synthesized into a model of sustainable construction in agriculture of Kyrgyzstan, which describes key components of sustainable agricultural construction, such as environmental technologies, water conservation, energy efficiency, social inclusiveness, economic sustainability, and adaptation to climate change.

3 Results

Green infrastructure is increasingly recognized as a vital approach to sustainable development, climate adaptation, and biodiversity conservation worldwide. In the Kyrgyz Republic, with its unique mountainous landscapes, diverse ecosystems, and growing environmental challenges, GI presents significant opportunities to enhance ecological resilience and support socio-economic development. The Kyrgyz Republic's natural environment is characterized by vast forests, alpine meadows, rivers, and wetlands, which form the basis for many ecosystem services essential for agriculture, water supply, and biodiversity. Although the concept of green infrastructure is relatively new in the national development agenda, several initiatives already reflect GI principles. Protected areas and natural reserves maintain biodiversity and regulate hydrological cycles. Afforestation and reforestation projects combat soil erosion and desertification. Wetland restoration helps regulate floods and purify water. Urban green spaces and parks in cities like Bishkek contribute to microclimate regulation and public well-being. However, these efforts are often fragmented and not formally integrated into broader infrastructure or land-use planning strategies.

The volume of construction work self-performed by the territory in the Kyrgyz Republic serves as an important indicator of regional development dynamics and the capacity of local economies to independently implement infrastructure projects. Analyzing the volume of construction work self-performed across the territories of the Kyrgyz Republic in 2019 and 2024, a significant increase is observed in all regions.

In the Batken Region, the volume nearly doubled, rising from 636.9 to 1233.6 million soms. The Jalal-Abad Region shows especially rapid growth, with construction volumes increasing from 6261.4 to 19,624.0 million soms, indicating a dynamic expansion of the construction sector in this area.

The Issyk-Kul Region also demonstrates substantial growth, with volumes increasing from 5957.3 to 10,658.4 million soms. The Naryn Region more than doubled its construction volume, rising from 1387.4 to 2499.1 million soms. The Osh Region saw its volume of self-performed construction work grow from 4531.9 to 9590.4 million soms. The Talas Region experienced a more moderate increase from 2977.5 to 3604.2 million soms. The Chui Region showed significant growth, increasing from 13,745.9 to 23,784.7 million soms. The capital city, Bishkek, leads in absolute terms with construction volumes nearly doubling from 47,155.4 to 89,060.5 million soms. The city of Osh also experienced considerable growth, with volumes rising from 6791.1 to 20,066.1 million soms (Fig. 1).

Overall, the data reflects a dynamic development of the construction industry throughout the country, indicating increased investment activity, economic growth, and rising demand for construction services and infrastructure projects.

Investments in fixed capital by territory in the Kyrgyz Republic serve as a key indicator of regional economic development, reflecting the scale and direction of financial resources allocated for infrastructure, industry, and social projects across the country.

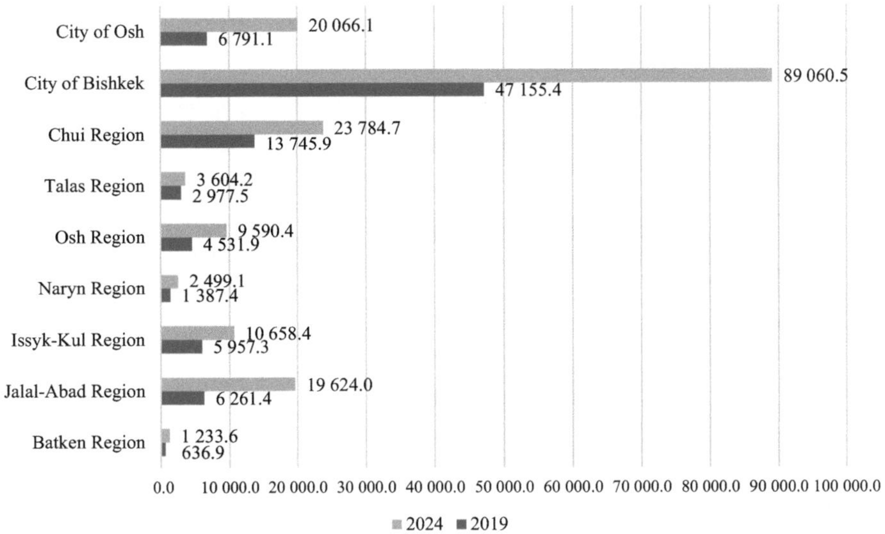

Fig. 1 Volume of construction work, self-made by the territory of the Kyrgyz Republic, million soms. *Source* Compiled by the authors based on the National Statistical Committee of the Kyrgyz Republic (2025)

Bishkek retained its leading position in terms of total investment volume, growing from 59,556.0 to 72,320.2 million soms. As the country's political and economic center, the capital continues to attract the largest share of fixed capital investment. Finally, Osh demonstrated a sharp increase in investments, from 5683.6 to 15,953.2 million soms, signaling strong urban development and greater investor interest in the country's second-largest city (Fig. 2).

Overall, the data indicate an upward trend in fixed capital investment across most regions, reflecting broader economic growth, improved infrastructure development, and increased investor confidence in the Kyrgyz Republic.

Despite its potential, several challenges hinder the widespread adoption of green infrastructure in Kyrgyzstan. The absence of clear policies and legal frameworks explicitly supporting GI in infrastructure development creates significant regulatory and institutional gaps. Financial constraints, including limited funding and a lack of incentives, further discourage investment in green projects, particularly in rural and underserved areas. A shortage of trained professionals and inadequate knowledge-sharing limit effective planning, design, and maintenance of GI systems. Weak monitoring mechanisms and insufficient environmental data impede evidence-based decision-making. Additionally, fragmentation among government agencies responsible for agriculture, environment, urban planning, and water resources undermines the integrated development of green infrastructure (Table 1).

The effective implementation of green infrastructure is limited by weak regulatory frameworks, high upfront costs with minimal financial support, and low stakeholder awareness, all of which constrain its broader adoption and potential

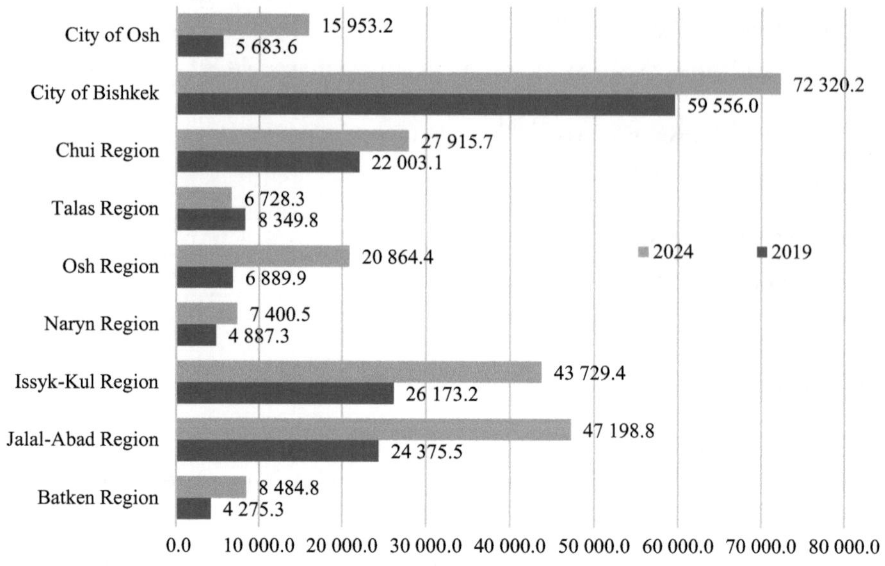

Fig. 2 Investments in the main capital by territory of the Kyrgyz Republic, million soms. *Source* Compiled by the authors based on the National Statistical Committee of the Kyrgyz Republic (2025)

Table 1 Problems of implementing green infrastructure in the Kyrgyz Republic

Problem category	Description of the problem
Insufficient regulatory framework	Lack of clear legal mechanisms and standards for the implementation of GI at the national level
Financial constraints	High initial costs, absence of government subsidies, and lack of incentive mechanisms
Limited awareness	Low level of awareness among local authorities, farmers, and developers about the benefits of GI
Lack of technical expertise	Shortage of specialists and knowledge required for the design, implementation, and maintenance of green infrastructure
Climatic and geographic challenges	Extreme weather conditions and terrain features complicate the implementation of certain GI components
Low level of cross-sector collaboration	Lack of coordination among the agriculture, ecology, construction, and water resource sectors
Limited access to data	Insufficient monitoring infrastructure and scientific research related to GI and ecosystem services

Source Developed by the authors

benefits. Climatic and geographic factors, including extreme weather conditions and challenging terrain, further restrict the feasibility of certain GI solutions in specific regions. Additionally, limited cross-sector collaboration hampers integrated planning and resource sharing among the agriculture, ecology, construction, and water management sectors. Finally, insufficient access to relevant data and scientific research undermines evidence-based decision-making and the effective monitoring of GI outcomes. Collectively, these challenges highlight the complex, multifaceted barriers that must be addressed to successfully mainstream green infrastructure initiatives.

4 Discussion

Siddiqui, Khan, and Akhtar (Siddiqui et al. 2025) emphasize that integrating sustainable concrete solutions (e.g., permeable pavements, green roofs, and bioswales) into green infrastructure plays a crucial role in enhancing environmental resilience and urban sustainability.

According to Siddiqui et al. (Siddiqui et al. 2025), the use of optimized concrete mixtures with recycled aggregates and supplementary cementitious materials significantly reduces carbon emissions and supports eco-friendly urban development.

Jonnalagadda et al. (Jonnalagadda et al. 2024) propose a comprehensive framework that addresses construction, maintenance, and policy adaptations to enhance infrastructure resilience in the face of climate change impacts. There is an urgent need to promptly adapt our engineering approaches, design methods, and construction materials to ensure effective infrastructure performance and extend the intended service life, reduce the carbon footprint of infrastructure projects, and improve the resilience and long-term durability of built structures.

According to Rosa (Rosa 2022), sustainable irrigation expansion on underperforming rainfed croplands presents a significant opportunity to enhance global food security and climate resilience, provided that such practices avoid depleting freshwater resources and damaging aquatic ecosystems.

Liu et al. (Liu et al. 2020) emphasize that the successful implementation of green infrastructure practices plays a critical role in promoting healthy watershed management, despite facing technical, institutional, and financial challenges.

Sustainable construction technologies, such as green buildings and eco-friendly materials, are essential for reducing environmental impact and improving energy efficiency (Perez Cuadros et al. 2023). The integration of GI and sustainability initiatives in the construction industry offers numerous benefits, including reduced environmental impact, improved energy efficiency, and long-term cost savings (Umar et al. 2024). Research in sustainable infrastructure is growing exponentially, with main streams focusing on GI, sustainable buildings, and assessment methods (Thomé et al. 2016). Emerging topics include cost-effectiveness, project management, and the integration of human and economic values in environmental sustainability (Thomé et al.

2016). However, challenges remain in implementing these practices, particularly in low-resource settings (Ogwu and Kosoe 2025).

Sustainable development in agriculture is crucial for national food security and economic stability (Ilyukhin and Ilyukhina 2025). According to A. A. Ilyukhin and S. A. Ilyukhina (Ilyukhin and Ilyukhina 2025), effective modeling of influencing factors is crucial for ensuring the sustainable development of agriculture in contemporary economic conditions.

Key factors include improving rural infrastructure and social environments to enhance the attractiveness of rural living and working conditions (Ilyukhin and Ilyukhina 2025). Constructing social infrastructure in rural areas is essential to address population decline and aging, as well as to improve the quality of life (Kuzmich 2021). This requires prioritizing agriculture in state economic policy and providing substantial government support for rural development (Kuzmich 2021). Additionally, implementing resource-saving technologies in agriculture is becoming increasingly important due to limited natural resources and climate change (Dronova 2024). These technologies optimize production processes and reduce negative environmental impacts, contributing to the sustainable development of municipal districts (Dronova 2024). Overall, a comprehensive approach involving infrastructure development, social improvements, and resource-efficient practices is necessary for sustainable agricultural growth.

Sustainable construction in the Kyrgyz Republic faces challenges and opportunities. In rural areas, traditional earthen buildings dominate, contributing to high energy consumption for heating (Mehta et al. 2022). Implementing energy-efficient measures could potentially reduce domestic space heating demand by 50%–70% (Mehta et al. 2022). Urban housing development requires consideration of scientific, technical, and practical issues, with emphasis on proper urban planning and architectural principles (Sasykeev 2020). Additionally, prosecutorial supervision plays an important role in enforcing legislation related to construction safety and sustainable development (Zhumashev et al. 2024). These efforts collectively contribute to the ongoing improvement of sustainable construction practices in the Kyrgyz Republic.

The methodological approaches applied have made it possible to calculate an integrated index that reflects the relative sustainability of agricultural development in each EAEU member state. This index serves as a strategic tool for formulating targeted policies aimed at enhancing the ecological efficiency, productivity, and resilience of national agricultural systems, while promoting coordinated and balanced growth throughout the EAEU (Mukambaeva et al. 2024a).

Within the sustainable development framework, agriculture serves as a cornerstone for economic growth and food security, while also playing a vital role in environmental stewardship and the empowerment of rural communities. Promoting sustainable practices, it contributes to long-term ecological balance and supports inclusive social well-being across the region (Mukambaeva et al. 2024b).

5 Conclusions

To maximize the benefits of green infrastructure, it is essential to incorporate GI principles into national and regional development plans, environmental regulations, and climate adaptation strategies. Strengthening Kyrgyzstan's capacity for green infrastructure requires coordinated actions that combine technical training, financial incentives, and inclusive stakeholder engagement, supported by pilot projects adapted to the country's climatic and geographic realities. An integrated model for sustainable agricultural construction, encompassing environmental technologies, water conservation, energy efficiency, social inclusion, and economic viability, offers a strategic pathway to enhance resilience, promote ecological stewardship, and stimulate long-term economic growth (Table 2).

Integrating climate adaptation strategies (e.g., drought-resistant crops, soil conservation techniques, and nature-based construction solutions) into Kyrgyzstan's agricultural sector is essential for enhancing resilience, safeguarding food security, and mitigating the impacts of climate variability. This holistic approach to sustainable construction balances environmental preservation with economic viability and

Table 2 Model of sustainable construction in agriculture of Kyrgyzstan

Component	Description	Examples/methods	Expected effect
Environmental technologies	Use of resource-saving and eco-friendly materials and technologies	Biomaterials, waste recycling, and energy saving	Reduction of environmental impact
Water conservation	Optimization of water use in agriculture	Drip irrigation and rainwater harvesting	Efficient use of limited water resources
Energy efficiency	Implementation of energy-saving systems in agricultural construction	Use of solar panels and passive heating	Reduction of energy costs and carbon emissions
Social inclusion	Involvement of local communities and training in sustainable methods	Training programs and farmer engagement	Increased social sustainability and quality of life
Economic sustainability	Increasing profitability and long-term efficiency of agricultural facilities	Reducing operational costs and process optimization	Growth of income and sustainable farm development
Climate adaptation	Construction of facilities resilient to climate change and extreme weather conditions	Use of drought-resistant plant varieties and soil protection	Minimization of production risks and losses

Source Developed by the authors

social well-being, positioning green infrastructure as a cornerstone of the country's long-term climate adaptation and rural development strategy.

References

Anderson V, Gough WA (2021) Nature-based resilience: a multi-type evaluation of productive green infrastructure in agricultural settings in Ontario Canada. Atmosphere 12(9):1183. https://doi.org/10.3390/atmos12091183

Dronova MV (2024) The use of resource-saving technologies in agriculture as a factor in the sustainable development of the municipal district. J Econ Entrep 18(4):632–637

Gawne B, Hale J, Stewardson MJ, Webb JA, Ryder DS, Brooks SS, Campbell CJ, Capon SJ, Everingham P, Grace MR, Stoffels RJ (2019) Monitoring of environmental flow outcomes in a large river basin: the commonwealth environmental water holder's long-term intervention in the Murray-Darling Basin Australia. Rivers Res Appl 36(SI4):630–644. https://doi.org/10.1002/rra.3504

Gill SE, Handley J, Ennos AR, Pauleit S (2007) Adapting cities for climate change: the role of the green infrastructure. Built Environ 33(1):115–133. https://doi.org/10.2148/benv.33.1.115

Ilyukhin AA, Ilyukhina SV (2025) Sustainable development of agriculture: modeling of factors. J Econ Entrep 18(12):590–592

Jonnalagadda S, Jonnalagadda V, Jonnalagadda N (2024) A framework for construction, maintenance, function and policy adaptations to build resilience to infrastructure against the impacts of the climate change. Eur Mod Stud J 7(6):185–193. https://doi.org/10.59573/emsj.7(6).2023.19

Kuzmich N (2021) Construction of socially significant facilities to improve the living conditions of the rural population. J Econ Entrep 1(126):401–405. https://doi.org/10.34925/EIP.2021.126.01.078

Liu B, Wang M, Chen T, Tseng P, Sun Y, Chiang AC, Chiang P-C (2020) Establishment and implementation of green infrastructure practice for healthy watershed management: challenges and perspectives. Water-Energy Nexus 3:186–197. https://doi.org/10.1016/j.wen.2020.05.003

Mehta K, Zörner W, Greenough RT (2022) Residential building construction techniques and the potential for energy efficiency in Central Asia: example from high-altitude rural settlement in Kyrgyzstan. Energies 15(23):8869. https://doi.org/10.3390/en15238869

Mukambaeva IB, Akylbekova NI, Mukambaev NJ, Lailieva EJ, Nam IE (2024a) Comparing the agricultural sectors of the EAEU countries through the sustainability index. In: Sergi BS, Popkova EG, Ostrovskaya AA, Chursin AA, Ragulina YV (eds) Ecological footprint of the modern economy and the ways to reduce it, pp 431–435. Springer, Cham, Switzerland. https://doi.org/10.1007/978-3-031-49711-7_71

Mukambaeva I, Sayakbaeva A, Akylbekova N, Mukambaev N, Ermekova A, Shambetova E (2024b) Influence of the state investment budget on the Kyrgyz agricultural sector through data analysis. In: 2024 IEEE 4th international conference on smart information systems and technologies (SIST), pp 409–414. IEEE, Astana, Kazakhstan. https://doi.org/10.1109/SIST61555.2024.10629277

Nakamura F, Ishiyama N, Yamanaka S, Higa M, Akasaka T, Kobayashi Y, Ono S, Fuke N, Kitazawa M, Morimoto J, Shoji Y (2020) Adaptation to climate change and conservation of biodiversity using green infrastructure. Rivers Res Appl 36(6):921–933. https://doi.org/10.1002/rra.3576

National Statistical Committee of the Kyrgyz Republic. (n.d.). Construction. Retrieved from https://stat.gov.kg/en/statistics/stroitelstvo/. Accessed 27 June 2025

Ogwu MC, Kosoe EA (2025) Integrating green infrastructure into sustainable agriculture to enhance soil health, biodiversity, and microclimate resilience. Sustainability 7(9):3838. https://doi.org/10.3390/su17093838

Perez Cuadros JJ, Rivero Landeo AA, Neyra Torres JL (2023) Sustainable construction technologies for green infrastructure (eco-friendly). In: Proceedings of LEIRD 2023: the 3rd LACCEI international multiconference on entrepreneurship, innovation and regional development: "igniting the spark of innovation: emerging trends, disruptive technologies, and innovative models for business success". Latin American and Caribbean Consortium of Engineering Institutions, Buenos Aires, Argentina. https://doi.org/10.18687/leird2023.1.1.506

Pitman SD, Daniels CB, Ely ME (2015) Green infrastructure as life support: urban nature and climate change. Trans R Soc S Aust 139:97–112. https://doi.org/10.1080/03721426.2015.1035219

Rosa L (2022) Adapting agriculture to climate change via sustainable irrigation: biophysical potentials and feedbacks. Environ Res Lett 17(6):063008. https://doi.org/10.1088/1748-9326/ac7408

Sasykeev UT (2020) City planning and architectural principles and regulations of housing construction of the Kyrgyz Republic. The Herald of KSUCTA named after N. A. Isanov 70(4):45–52. https://doi.org/10.35803/1694-5298.2020.4.500-504

Siddiqui AR, Khan RA, Akhtar MN (2025) Sustainable concrete solutions for green infrastructure development: a review. J Sustain Constr Mater Technol 10(1):8. https://doi.org/10.47481/jscmt.1667793

Thomé AM, Ceryno PS, Scavarda AJ, Remmen A (2016) Sustainable infrastructure: a review and a research agenda. J Environ Manage 184(2):143–156. https://doi.org/10.1016/j.jenvman.2016.09.080

Umar MO, Umana AU, Matthew B, Garba P, Omowole BM, Queen A (2024) Green infrastructure and sustainability initiatives in the construction industry: a review of current practices and future directions. World J Adv Res Rev 24(02):1061–1081. https://doi.org/10.30574/wjarr.2024.24.2.3415

Zhumashev AB, Lustig S, Rysmendeev BD, Mazitov R, Mamatazizova EK, Mamatazizova NK (2024) Prosecutor's supervision as a means of ensuring safety in the construction sector. E3S Web Conf 533:02033. https://doi.org/10.1051/e3sconf/202453302033

Sustainable Cherry Production in Kyrgyzstan: Field Analysis of Soil, Climate, and Technology Solutions

Elmira A. Midinova, Zhyrgal T. Samieva, and Begimai M. Atakulova

Abstract Sweet cherry (*Prunus avium*) production is a key agricultural sector in southern Kyrgyzstan. However, its long-term sustainability is challenged by soil degradation, climate variability, and outdated orchard management. This research evaluated cherry productivity in seven experimental plots across southern Kyrgyzstan and neighboring Uzbekistan by examining soil properties (humus content, pH, and macronutrient levels), climatic conditions, tree age, and applied production technologies. The results showed that orchards with higher soil organic matter and near-neutral pH, particularly those in Dosmat (4.83% SOM, pH 7.0) and Mady (2.57% SOM, pH 7.4), produced 16–18 kg of fruit per tree. Productivity declined in older orchards (>19 years) and at high-altitude sites. In contrast, the Vodil plot in Uzbekistan achieved the highest yield (32.25 kg/tree) despite low soil organic matter (0.88%), owing to intensive fertilization, drip irrigation, and modern pest control, demonstrating how advanced agrotechnologies can compensate for suboptimal soil conditions. Conversely, orchards with minimal fertilization and irrigation, such as Vishenka-Osh (2 kg/tree), experienced substantial yield losses, underscoring the importance of adequate nutrient and water inputs. Satellite imagery and remote sensing further enhanced the assessment by providing early detection of plant stress and pest outbreaks. Overall, the findings indicate that sustainable cherry production in mountainous regions depends on an integrated strategy that combines soil fertility management, altitude-appropriate climate adaptation, and precision agriculture. Broader adoption of remote sensing, GIS-based planning, and decision-support tools can significantly strengthen resilience to environmental stress and align local practices with global sustainability goals. This research offers practical guidance for farmers and a strategic foundation for scaling climate-smart cherry production across Central Asia.

E. A. Midinova (✉) · B. M. Atakulova
Osh State University, Osh, Kyrgyzstan

Z. T. Samieva
Kyrgyz-Uzbek International University Named After B. Sydykov, Osh, Kyrgyzstan

A. B. Karbekova (ed.), *Sustainable Development of Rural Territories Based on Climate-Smart Technologies*, Climate Change Management,
https://doi.org/10.1007/978-3-032-20106-5_13

Keywords Southern Kyrgyzstan · Prunus avium · Precision agriculture · Sustainable farming · Orchard management

JEL Codes Q12 · Q15 · Q16 · Q54 · Q56 · O13

1 Introduction

As global demand for sustainable agriculture intensifies, ensuring high crop yields while preserving natural resources has become a pressing priority. Mountainous regions, such as southern Kyrgyzstan, present opportunities and unique challenges for fruit production, particularly the cultivation of sweet cherry (*Prunus avium*). The foothill plains (1000–1300 m), encompassing the lower elevations of the Fergana Valley, experience a continental climate with mean annual precipitation of 98–502 mm, scorching summer temperatures (up to 47 °C), and winter frosts, factors contributing to ecological stress and yield variability.

Research indicates that stone fruits require specific temperature thresholds for successful pollination and ripening (Alburquerque et al. 2008). Late spring frosts, sudden temperature drops, and erratic precipitation patterns pose critical threats during vulnerable phenological phases such as flowering and bud break (Kaya and Kose 2021, 2022; Kaya et al. 2020; Matzneller et al. 2016; Rojas et al. 2021; Salazar-Gutiérrez et al. 2014). Late spring frosts coincide with flowering, which might cause more losses of cherry fruit than any other biological and environmental hazards (Chmielewski et al. 2018; Kamilova et al. 2022; Khasanov et al. 2021). Exposure to – 2 to – 5 °C during bud opening can cause significant damage or crop failure as buds lose frost resistance as they develop (Dilishatov et al. 2021).

Traditional gardening methods face sustainability issues due to soil erosion, declining humus, and water inefficiency (Duulatov et al. 2021). These environmental stresses underscore the urgent need for adaptive strategies in high-altitude horticulture. Soil characteristics such as pH (approximately 6–6.8), good drainage, adequate depth ($\geq$ 1 m), and appropriate macro- and micronutrients are essential for tree health and fruit quality (Gonçalves et al. 2024; Karpushova et al. 2022). Targeted application of nitrogen, potassium, boron, selenium, and calcium fertilizers can improve fruit yield and quality under specific conditions (Ekinci et al. 2016; Roeva et al. 2024; Wójcik 2024; Wójcik and Wójcik 2006). However, such interventions must be adapted to local soil properties and orchard management practices.

At the orchard level, knowledge transfer is uneven: clustered production zones benefit from information exchange, while scattered orchards often continue traditional practices that limit productivity.

Precision and climate-smart agriculture (drip irrigation, sensor networks, remote sensing, and decision-support tools) have shown promise in similar mountain horticultural systems by improving water use efficiency and early stress detection (Anand et al. 2023; Fountas et al. 2020; Muangprathub et al. 2019).

However, these technologies remain under-adopted locally, representing a research gap and an opportunity for scalable interventions.

This research analyses soil properties, tree age, climatic variability, and technology interventions across seven experimental plots in southern Kyrgyzstan and neighboring Uzbekistan. By linking plot-level soil and management data with yield outcomes, we identify context-specific, sustainable strategies for enhancing cherry productivity and resilience, contributing to regional adaptation efforts and the UN SDGs (2, 13, 15).

2 Materials and Methods

Nutrient composition was evaluated in seven orchard plots: six located in the Osh and Batken provinces of southern Kyrgyzstan and one in Fergana, Uzbekistan (Tables 1–2 presented in section Data Availability). Laboratory analyses were performed at the Southern Regional Soil and Agrochemical Station. All soil samples were collected in spring 2023 from a depth of 30 cm. Annual precipitation across the study plots ranged from 280 mm in Kadamjay (Batken) to 350 mm in Uzgen (Osh). Mean temperatures varied from − 4 °C in January to + 32 °C in July, depending on elevation and geographic location. Plot sizes ranged from 0.8 to 6 ha, encompassing foothill and low-mountain agroecosystems characterized by contrasting soils and management practices.

A descriptive analysis was applied to evaluate soil fertility and nutrient availability. Key macronutrients (N, P, K, Ca, Mg, and S) and micronutrients (Fe, B, and Zn) with environmental factors were considered due to their established importance for cherry growth and fruit quality. Nitrogen promotes vegetative and trunk development; phosphorus supports root formation and flowering; potassium is essential for fruit quality and ripening. Micronutrients, such as iron (Fe), boron (B), and zinc (Zn), play vital roles in enzymatic activity and plant disease resistance. These factors are central to photosynthesis, cellular metabolism, and overall fruit production (Chawla and Kumar Sharma 2024; Roeva et al. 2024).

Soil organic matter (SOM) was measured using Tyurin's method, which estimates organic carbon (C_org) through potassium dichromate oxidation and applies a 1.724 conversion coefficient, based on the assumption that carbon accounts for roughly 58% of SOM (Pribyl 2010).

Total nitrogen content was determined using the Moscow CIRAO method, a classical technique involving the oxidation of nitrogen-containing compounds, with subsequent quantification of liberated ammonia. Nitrogen concentration is calculated based on the titration of released ammonia, incorporating the reagent volume and sample parameters into standard equations.

Plant-available forms of phosphorus and potassium were determined using the Machigin method, a widely adopted agrochemical technique used to assess the bioavailable nutrient pool in soils (Kim et al. 2021; Shamrikova et al. 2022).

3 Results and Discussion

The analysis of seven experimental plots showed significant variation in cherry yields driven by soil quality, fertilizer application, altitude, and climatic conditions.

Soil types across the seven plots ranged from light to heavy loams, with humus content varying from 0.88% to 4.83%. Plots with higher humus levels and balanced pH, such as Dosmat (4.83% SOM, pH 7.0, 18 kg/tree), consistently produced higher yields. These results align with previous studies showing that elevated humus content and a neutral to slightly alkaline pH (7.0–7.6) are strongly associated with improved sweet cherry productivity. Recent research has also emphasized the importance of SOM and total nitrogen (TN) in enhancing soil aggregate stability and overall soil quality (Chen et al. 2025; Palacios-Peralta et al. 2022). Studies by Wójcik and Wójcik (2006) demonstrated that boron, zinc, and selenium fertilization can significantly increase yield and fruit quality. Ekinci et al. (2016) highlighted the benefits of foliar calcium applications for improving storage quality and fruit firmness. Roeva et al. (2016) emphasized the role of foliar calcium applications in improving storage quality and firmness. Roeva et al. (2024) confirmed that nitrogen and potassium fertilization significantly enhance soil fertility and yield potential in cherry orchards. Particularly, the Mady plot (1.2 mg/kg N, 51 mg/kg P_2O_5, 210 mg/kg K_2O) demonstrated a high yield (16 kg/tree), confirming that a balanced macronutrient supply is critically important for tree growth and high-quality fruits.

These findings collectively show that precision agriculture, including irrigation, fertilization, and orchard management, can offset the limitations of poor soil conditions. The Vodil site in Uzbekistan (Plot 7) recorded the highest yield (32.25 kg/tree) despite having very low SOM (0.88%), largely due to intensive fertilization and advanced agricultural technologies. This indicates that advanced management practices can compensate for poor soil conditions in the short term but may also increase the risk of long-term soil degradation if not balanced with sustainable approaches (Anand et al. 2023).

Optimal levels of nitrogen and phosphorus improved harvest (16 kg) in Plot 6 (Mady), where consistent use of over-composted manure and nutrients was shown. In contrast, plots without fertilizer application, such as Plot 2 (Vishenka-Osh) and Plot 5 (Kyzyl-Bulak), showed lower yields (2 kg and 13 kg, respectively). Whereas, excess nitrogen can negatively affect cherry productivity and soil health. Field studies in sour cherry orchards have shown that increasing nitrogen fertilization beyond recommended levels (e.g., from 60 to 120 kg N/ha) did not improve yield or fruit quality and even led to reductions in total soluble solids (TSS), fruit color, and soil enzymatic activity (Rutkowski and Łysiak 2023; Rutkowski et al. 2022).

In southern Kyrgyzstan, commercial sweet cherry orchards situated between approximately 580 m and 1267 m above sea level already lie within a favorable thermal and altitudinal range for meeting dormancy and flowering requirements. Many cultivars grown in Mediterranean climates achieve sufficient chilling above ~ 650 m, and similar elevations ($\approx$ 700–1400 m) are commonly used in neighboring highland regions to ensure adequate chilling accumulation and high fruit quality

(Alburquerque et al. 2008). Altitudinal gradients also tend to enhance fruit development, with improvements in fresh weight and size frequently observed at higher elevations. Although annual rainfall in these areas (280–450 mm) is lower than global rain-fed sweet cherry standards (500–2700 mm), current orchard yields remain commercially viable. Moreover, they can be substantially increased through precision irrigation, fertigation, and targeted soil-fertility management, complemented by improved canopy management and pest control techniques to reduce harvest-time losses.

Weather forecasting and pest control technologies can improve tree health and reduce crop loss. Plot 5 and Plot 3, located at higher altitudes, faced frost and temperature fluctuations that reduced yields. Drought conditions in Plot 2 (Vishenka-Osh) and Plot 3 (Dosmat) decreased soil moisture availability, further stressing the trees. This highlights the need for improved irrigation systems in regions prone to water shortages. Moderate irrigation before harvesting reduced fruit size and yield, as also demonstrated under regulated deficit irrigation regimes in sweet cherry (Blanco et al. 2019). Plot 4 (Dostuk) and Plot 5 (Kyzyl-Bulak) showed reduced yields due to poor water management, inadequate pest control, and insufficient climate adaptation.

The age of trees also influences yield variability. Young trees exhibited fluctuations in yield, while mature trees (ages 12–14) in Dosmat, Mady, and Vodil outperformed older trees with declining immunity and poor pest and disease management, such as those in Airport-Osh (19 years). Decline in productivity is attributed to tree age, declining immunity, and the transformation of natural soil into "low-fertility soil" due to long-term agronomic practices. These soils, referred to as "depleted soils," cannot support crop or yield growth without additional fertilizers because of their low nutrient capacity and poor condition.

Using one of our plots as an example, the authors demonstrate the importance of digital technologies in analyzing and optimizing agricultural processes. The authors used satellite imagery and aerial photography from Google Earth Pro (GPS Logger 20230613-095702) (date of receiving: September 17, 2024) to demonstrate how helpful it is to detect early signs of stress, diseases, or pest infestations, enabling farmers to implement timely management strategies and optimize their harvesting and marketing efforts.

The authors selected Plot 1 (40.60833° N, 72.79333° E), which contains 19-year-old cherry trees, because the farmer reported substantial losses in both tree numbers (declining from 478 to 274 trees) and yield (reduced to 80–90% of the 2022 level). To investigate these changes, the plot's geographical and environmental characteristics were visualized for the period 2010–2024.

Through Google Earth's visualization, the authors examined changes in color, texture, and size over time, revealing key insights into the orchard's health (Tsiaras And Domakinis 2023; Zhao et al. 2023).

By comparing satellite images across different seasons, we identified vegetation color patterns that reflect tree health, soil conditions, and potential issues such as pest infestations or diseases. Shifts in coloration, from green to brown or yellow, indicate early signs of stress or pest activity, enabling timely interventions and more precise pest management. Textural changes in the imagery also provide valuable information on soil moisture status, canopy density, and overall tree vigor, helping to optimize irrigation scheduling and fertilizer applications according to crop needs. Our imagery analysis revealed notable changes during 2017–2018, including a gradual decline in NDVI and tree vigor. These trends intensified, culminating in a major outbreak in 2022, when an invasive population of the cherry fruit fly (Rhagoletis cerasi) caused severe damage and an 80–90% reduction in yield.

Furthermore, observation of variations in bloom timing, intensity, and canopy health is evident through changes in vegetation index profiles of Plot 1.

Overall, the research underscores the importance of a holistic strategy that integrates soil nutrient management, altitude-specific climate adaptation, and technological innovation. These conclusions are consistent with global sustainability frameworks and echo findings from Central Chile, where Rojas et al. (2021) highlighted the value of decision-analysis tools and precision technologies for enhancing cherry orchard resilience under extreme weather conditions.

4 Conclusions

This research confirms that cherry productivity in southern Kyrgyzstan and Fergana, Uzbekistan, is driven by multiple interacting factors, including soil organic matter, pH balance, macro-nutrient levels, irrigation systems, technological interventions, and tree age.

Plots with higher humus content and optimal pH, such as Dosmat and Mady, demonstrated enhanced yields, whereas Plot 7 (Vodil) achieved the highest productivity through intensive nutrient management and modern technologies despite its low soil quality.

In contrast, Plots 2 (Vishenka-Osh) and 4 (Dostuk), which lacked fertilization and irrigation systems, produced lower yields, highlighting the need for modernization.

The integration of remote sensing and satellite imagery proved effective in detecting orchard health issues and yield-related stress. Our findings support the adoption of precision agriculture practices in cherry orchards, particularly in mountainous regions.

Future research should explore the combined use of GIS, remote sensing, and data-driven decision-making to enhance productivity and resilience in cherry production.

Authors' Contribution This work is derived from the dissertation of E. A. Midinova, completed under the supervision of Prof. Dr. Zh. T. Samieva. E. A. Midinova was responsible for the conceptualization, methodology, data collection, statistical analysis, interpretation, and writing of the manuscript. Prof. Dr. Zh. T. Samieva contributed to the conceptualization and provided a critical review of the manuscript. B. M. Atakulova contributed to data collection and assisted with manuscript preparation.

Data Availability Table 1 shows the specific plots and soil and fertilizer parameters. It is available at https://figshare.com with the identifier https://doi.org/https://doi.org/10.6084/m9.figshare.27141951.

Table 2 shows the relationship between environmental factors (altitude, temperature, and precipitation), cultivation methods, irrigation, and technologies addressing yield challenges. It is available at https://figshare.com with the identifier https://doi.org/https://doi.org/10.6084/m9.figshare.30028171.

Figure 1, "Multi-year imagery of Plot No. 1 (Nariman Village, Kara Suu District, Kyrgyzstan)," is available at https://figshare.com with the identifier https://doi.org/https://doi.org/10.6084/m9.figshare.30030232.

Figure 2, "Multi-year blooming variations of Plot No. 1 highlighting variations in canopy health and bloom phenology," is available at https://figshare.com with the identifier https://doi.org/https://doi.org/10.6084/m9.figshare.30030295.

References

Alburquerque N, García-Montiel F, Carrillo A, Burgos L (2008) Chilling and heat requirements of sweet cherry cultivars and the relationship between altitude and the probability of satisfying the chill requirements. Environ Exp Bot 64(2):162–170. https://doi.org/10.1016/j.envexpbot.2008.01.003

Anand S, Kumar P, Alok A, Kumar R (2023) Chapter-4. Precision agriculture: technology and implementation. Integrated Publications TM, New Delhi, India, pp 65–92

Blanco V, Torres-Sánchez R, Blaya-Ros PJ, Pérez-Pastor A, Domingo R (2019) Vegetative and reproductive response of 'Prime Giant' sweet cherry trees to regulated deficit irrigation. Sci Hortic 249:478–489. https://doi.org/10.1016/j.scienta.2019.02.016

Chawla R, Kumar Sharma S (2024) Nitrogen fertilization of stone fruits: a comprehensive review. J Plant Nutr 48(3):445–485. https://doi.org/10.1080/01904167.2024.2405990

Chen M, Feng S, Wang J, Gao M, Liu M, Wang K, Shangguan ZP, Zhang Y (2025) The stability of soil aggregates in sweet cherry (*Prunus avium* L.) orchards of different ages and varieties. Heliyon 11(3):e42189. https://doi.org/10.1016/j.heliyon.2025.e42189

Chmielewski F-M, Götz K-P, Weber KC, Moryson S (2018) Climate change and spring frost damages for sweet cherries in Germany. Int J Biometeorol 62(2):217–228. https://doi.org/10.1007/s00484-017-1443-9

Dilishatov OU, Omurbekova GK, Khasanov BU (2021) Analysis of agroclimatic conditions of the Kadamzhay district of the Batken region. Bull Sci Pract 7(11):112–117. https://doi.org/10.33619/2414-2948/72/13

Duulatov E, Pham QB, Alamanov S, Orozbaev R, Issanova G, Asankulov T (2021) Assessing the potential of soil erosion in Kyrgyzstan based on RUSLE, integrated with remote sensing. Environ Earth Sci 80:658. https://doi.org/10.1007/s12665-021-09943-6

Ekinci N, Özdüven F, Gür E (2016) Effects of preharvest foliar calcium applications on the storage quality of '0900 Ziraat' sweet cherry cultivar. Erwerbs-Obstbau 58:227–231. https://doi.org/10.1007/s10341-016-0281-y

Fountas S, Espejo-García B, Kasimati A, Mylonas N, Darra N (2020) The future of digital agriculture: technologies and opportunities. IT Professional 22(1):24–28. https://doi.org/10.1109/MITP.2019.2963412

Gonçalves B, Aires A, Oliveira I, Baltazar M, Cosme F, Afonso S, Pinto T, Anjos MR, Inês A, Morais MC, Vilela A, Silva AP (2024) From orchard to wellness: unveiling the health effects of sweet cherry nutrients. Nutrients 16(21):3660. https://doi.org/10.3390/nu16213660

Kamilova LT, Samieva ZhT, Khasanov BU, Midinova EA (2022) Cultivation of the cerasus avium cherry variety in the Kadamjay District of the Batken region under the conditions of climate change. Sci Educ Eng 2(74):66–73. https://doi.org/10.54834/16945220_2022_2_66

Karpushova SE, Bazieva AM, Fomenko NM, Akopova ES (2022) Benefits of circular agriculture for the environment: international experience of using digitalization and higher education development. In: Popkova, EG, Sergi BS (eds) Sustainable agriculture, pp 139–148. Springer, Singapore. https://doi.org/10.1007/978-981-16-8731-0_14

Kaya O, Kose C (2021) How sensitive are the flower parts of the sweet cherry in sub-zero temperatures? Use of differential thermal analysis and critical temperatures assessment. N Z J Crop Hortic Sci 50(1):17–31. https://doi.org/10.1080/01140671.2021.1890156

Kaya O, Kose C (2022) Sensitivity of some sweet cherry (*Prunus avium* L.) cultivars to late spring frosts during different phenological stages following bud burst. Theor Appl Climatol 148(3):1713–1725. https://doi.org/10.1007/s00704-022-04030-7

Kaya O, Kose C, Donderalp V, Gecim T, Taskın S (2020) Last updates on cell death point, bud death time and exothermic characteristics of flower buds for deciduous fruit species by using differential thermal analysis. Sci Hortic 270:109403. https://doi.org/10.1016/j.scienta.2020.109403

Khasanov BU, Karimov EM, Erkali UU (2021) Analysis of the climatic conditions of Kadamzhay district of the Batken region. Herald of KRSU 21(8):164–166. https://doi.org/10.36979/1694-500X-2021-21-8-164-166

Kim M-S, Jeon S-H, Lee T-G, Jung H-I, Kim C-W, Kim Y-K (2021) Comparison of wet oxidation and dry combustion methods for organic matter analysis of soils derived from granite, limestone, and volcanic ash. Korean J Soil Sci Fert 54(4):674–683. https://doi.org/10.7745/KJSSF.2021.54.4.674

Matzneller P, Götz K-P, Chmielewski F-M (2016) Spring frost vulnerability of sweet cherries under controlled conditions. Int J Biometeorol 60:123–130. https://doi.org/10.1007/s00484-015-1010-1

Muangprathub J, Boonnam N, Kajornkasirat S, Lekbangpong N, Wanichsombat A, Nillaor P (2019) IoT and agriculture data analysis for smart farm. Comput Electron Agric 156:467–474. https://doi.org/10.1016/j.compag.2018.12.011

Palacios-Peralta C, Reyes-Díaz M, González-Villagra J, Ribera-Fonseca A (2022) The potential roles of the N and P supplies on the internal browning incidence in sweet cherries in the Southern Chile. Horticulturae 8(12):1209. https://doi.org/10.3390/horticulturae8121209

Pribyl DW (2010) A critical review of the conventional SOC to SOM conversion factor. Geoderma 156(3–4):75–83. https://doi.org/10.1016/j.geoderma.2010.02.003

Roeva TA, Leonicheva EV, Leonteva LI (2024) Fertility indicators of Haplic Luvisol of a sour cherry orchard associated with application of nitrogen and potassium fertilizers. Mosc Univ Soil Sci Bull 78(S1):S28–S36. https://doi.org/10.3103/S0147687423060054

Rojas G, Fernandez E, Whitney C, Luedeling E, Cuneo IF (2021) Adapting sweet cherry orchards to extreme weather events—decision analysis in support of farmers' investments in Central Chile. Agric Syst 187:103031. https://doi.org/10.1016/j.agsy.2020.103031

Rutkowski K, Łysiak GP (2023) Effect of nitrogen fertilization on tree growth and nutrient content in soil and cherry leaves (*Prunus cerasus* L.). Agriculture 13(3):578. https://doi.org/10.3390/agriculture13030578

Rutkowski K, Łysiak GP, Zydlik Z (2022) Effect of nitrogen fertilization in the sour cherry orchard on soil enzymatic activities, microbial population, and fruit quality. Agriculture 12(12):2069. https://doi.org/10.3390/agriculture12122069

Salazar-Gutiérrez MR, Chaves B, Anothai J, Whiting M, Hoogenboom G (2014) Variation in cold hardiness of sweet cherry flower buds through different phenological stages. Sci Hortic 172:161–167. https://doi.org/10.1016/j.scienta.2014.04.002

Shamrikova EV, Kondratenok BM, Tumanova EA, Vanchikova EV, Lapteva EM, Zonova TV, Lu-Lyan-Min EI, Davydova AP, Libohova Z, Suvannang N (2022) Transferability between soil organic matter measurement methods for database harmonization. Geoderma 412:115547. https://doi.org/10.1016/j.geoderma.2021.115547

Tsiaras S, Domakinis C (2023) Use of GIS in selecting suitable tree crop cultivation sites in mountainous less favoured areas: an example from Greece. Forests 14(6):1210. https://doi.org/10.3390/f14061210

Wójcik P (2024) Responses of 'burlat' sweet cherry trees to selenium fertilisation under low soil selenium conditions. J Soil Sci Plant Nutr 24:8343–8352. https://doi.org/10.1007/s42729-024-02119-8

Wójcik P, Wójcik M (2006) Effect of boron fertilization on sweet cherry tree yield and fruit quality. J Plant Nutr 29(10):1755–1766. https://doi.org/10.1080/01904160600897471

Zhao W, Wang M, Pham VT (2023) Unmanned aerial vehicle and geospatial analysis in smart irrigation and crop monitoring on IoT platform. Mob Inf Syst 2023(1):1–12. https://doi.org/10.1155/2023/4213645

Smart Farming in Armenia: The Role of Innovative Livestock Buildings in Sustainable Agriculture

Gayane R. Tovmasyan, Samvel S. Avetisyan, Vergine L. Kirakosyan, and Tereza G. Shahrimanyan

Abstract The growing global population and increasing demand for food have accelerated the adoption of smart technologies in agriculture. This research examines the role of smart livestock buildings in modernizing and developing animal farming practices in the Republic of Armenia. Through the integration of sensors, automation systems, and data analytics, these smart facilities enhance productivity, animal welfare, and environmental sustainability. The research reviews Armenia's agricultural context, the implementation of state-supported programs for the construction or reconstruction of smart livestock buildings, and the expected outcomes, including increased milk yield, reduced disease incidence, and improved competitiveness of livestock products. The findings highlight the potential of smart livestock technologies to transform traditional farming practices, promote sustainable development, and provide a model for future agricultural innovation. The results may be useful for a wide range of stakeholders, including government institutions and policymakers seeking to improve agricultural support programs, farmers, potential investors, and agribusiness stakeholders. They can also serve as a valuable resource for researchers and scholars working in the fields of agriculture, sustainable development, and smart technologies.

Keywords Smart farming · Smart livestock buildings · Internet of Things (IoT) · Sustainable agriculture · Milk production · Armenia

JEL codes Q16 · Q55

G. R. Tovmasyan (✉) · S. S. Avetisyan · V. L. Kirakosyan · T. G. Shahrimanyan
Armenian State University of Economics, Yerevan, Armenia
e-mail: tovmasyangayane@yahoo.com

T. G. Shahrimanyan
e-mail: shahrimanyan@list.ru

G. R. Tovmasyan
Yerevan State University, Yerevan, Armenia

Brusov State University, Yerevan, Armenia

Public Administration Academy of the Republic of Armenia, Yerevan, Armenia

A. B. Karbekova (ed.), *Sustainable Development of Rural Territories Based on Climate-Smart Technologies*, Climate Change Management,
https://doi.org/10.1007/978-3-032-20106-5_14

1 Introduction

The world's population is constantly growing, which increases the demand for food. This forces agriculture to work more efficiently and productively.

In recent years, the rapid development of smart technologies has radically changed almost all aspects of the economy. Agriculture has also begun to actively apply these technologies. Nowadays, agriculture must ensure high results and long-term sustainable development, reducing environmental damage. Smart agriculture has become a necessity in the face of population growth, increasing food demand, climate change, and resource limitations. Technological innovations make it possible to increase productivity, save water and land, and ensure sustainable development of agriculture.

The creation of smart livestock buildings has become a significant part of the progressive developments in the agricultural sector. They are of revolutionary importance in agriculture because they improve animal welfare and the sustainable development of the environment. As we know, traditional livestock facilities face several problems related to resource shortages, disease management, and unsuitable conditions for animals. In this regard, smart livestock facilities enable the overcoming of the mentioned problems. These facilities are equipped with automated systems, data analysis tools, and the necessary sensors that enable continuous monitoring, climate control, proper feeding, and efficient waste management practices.

The integration of smart technologies into livestock facility management enables increased efficiency, promotes sustainable farming practices, and supports the implementation of food safety and environmental protection goals.

The main purpose of this research is to explore the design, functionality, and benefits of smart livestock buildings, highlighting how they are reshaping the landscape of today's agriculture and supporting the development of a more resilient, efficient, and animal-friendly farming sector in Armenia.

2 Methods

Different studies explore the importance and mechanisms of smart agriculture (Akkem et al. 2023; Anubhav et al. 2023; Giua et al. 2022; Idoje et al. 2021; Pivoto et al. 2018; Sharma et al. 2022).

Agriculture 4.0 marks the fourth agricultural revolution, characterized by the integration of digital technologies aimed at creating a more intelligent, efficient, and environmentally sustainable farming sector. Innovative agricultural technologies are being developed to promote sustainability and optimize farming practices. This transformation includes the widespread adoption of digitalization and automation in agriculture and everyday life, leveraging tools such as big data, artificial intelligence (AI), robotics, the Internet of Things (IoT), and virtual and augmented reality (Javaid et al. 2022).

Smart farming has emerged as a contemporary concept in the agricultural sector, aiming to revolutionize traditional farming methods through the integration of Information and Communication Technologies (ICT). Specifically, advancements such as Uncrewed Aerial Vehicles (UAVs), Uncrewed Ground Vehicles (UGVs), image processing, machine learning, big data, cloud computing, and wireless sensor networks (WSNs) are expected to drive significant transformation. These technologies give many benefits. They increase the productivity of agricultural activities, reduce operational costs by optimizing the use of inputs such as fuel, fertilizers, and pesticides, minimize labor requirements, and enhance the quality of agricultural products (Moysiadis et al. 2021).

Smart farming uses ICT during the whole agricultural cycle (starting from planting, ending with harvesting) to optimize resource management (Virk et al. 2020).

Emerging technologies like the Internet of Things (IoT) and cloud computing promote smart farming. They make it possible to integrate robotics and AI in agriculture. This is possible by the concept of big data, which makes it possible to conduct a comprehensive analysis of the whole farming cycle and make informed decisions (Wolfert et al. 2017).

As noted by Mohamed et al. (2021), the IoT may be used in different agricultural activities, like farm monitoring, irrigation management, pest control, and harvesting. The government support, especially in developing states, is very crucial for increasing the benefits of such smart tools.

While the growing integration of IoT-based smart farming (SF) has revolutionized agriculture by boosting productivity and efficiency, its complete realization is hampered by obstacles such as cybersecurity threats, limited rural connectivity, and device incompatibility (Prashanth et al. 2025).

In the Robs4Crops project, Papadopoulos et al. (2025) explored the views of agricultural stakeholders on robotic farming technologies using a mixed-method approach of quantitative and qualitative questionnaires. Feedback was collected from 104 participants, including farmers, researchers, tech providers, policymakers, and others, during demonstrations held in France, Greece, Spain, and the Netherlands, as well as from 54 agricultural students at the Agricultural University of Athens. The research revealed that while stakeholders recognized the potential of agricultural robots, noting benefits such as reduced labor needs, cost efficiency, and improved productivity, they also identified several challenges. These included high initial costs, the lack of necessary skills, and a degree of skepticism, particularly among older generations.

Advanced AI tools in agriculture enable higher quality standards and impact the price-to-quality ratio, thereby enhancing the competitiveness of agricultural products in national and international markets. Furthermore, leveraging ICT to improve agricultural product quality supports the UN's goals for ensuring global food security (Litvinova et al. 2025).

IoT technologies are used in livestock farming to monitor animal behavior and health, automate daily farm operations, enhance farm security, track environmental conditions, and predict future events by analyzing the data collected (Terence et al. 2024). These technologies enable the monitoring of feed, health, and well-being of cattle, sheep, goats, and cows, while supporting the traceability of wild animals and promoting sustainability across various livestock processes (Farooq et al. 2022). Within livestock facilities, tools such as camera systems, radio-frequency identification (RFID), beacon sensors, and environmental sensors gather data that is stored in databases, forming the foundation for automated livestock management systems (Kim et al. 2021).

The main methods applied are data collection, analysis, and synthesis.

3 Results

3.1 Agriculture Sphere in Armenia

Agriculture is an important area of the Armenian economy (Avagyan et al. 2022; Avetisyan et al. 2022). Due to its heavy reliance on agriculture for economic stability and food security, Armenia is particularly vulnerable to the substantial challenges posed by evolving climatic patterns (Manucharyan 2025).

Armenia possesses limited land resources, with a total area of 29,743 km^2. Of this, agricultural land accounts for approximately 2041.4 thousand hectares. This agricultural land is distributed as follows:

- Arable land (441.6 thousand ha or 21.6%);
- Perennial grass (40.2 thousand ha or 2.0%);
- Plough-land (121.3 thousand ha or 5.9%);
- Pastures (1049.2 thousand ha or 51.4%);
- Other categories (389.1 thousand ha or 19.1%) (Statistical Committee Republic of Armenia 2024, p. 370).

Table 1 shows the gross agricultural output by farms from 2019 to 2023. In 2023, it was 948.3 billion Armenian drams. The percentage of the agricultural output of commercial organizations was only 7.3%. Household plots accounted for 92.7% of the total. Following the rapid privatization of land (excluding pastures), livestock, buildings, agricultural machinery, and other fixed assets between 1991 and 1993, approximately 340,000 small peasant farms were established, replacing 380 large state and collective farms. On average, each farm had 1.4 hectares of usable agricultural land, of which 0.9 hectares was arable. Despite the inevitable challenges of intensive land use, the strong commitment of land users helped prevent a decline in production volumes and fostered the development of effective management practices under the new conditions. The characteristics of Armenia's agrarian sector, particularly the diversity of small family farms, highlight the importance of advancing smart agriculture. Key areas for development include intensive gardening, greenhouse cultivation of vegetables and flowers, and the growth of dairy cattle breeding and poultry farming using smart livestock buildings.

Table 2 shows the production of main animal husbandry products for 2019–2023.

Table 1 Gross agricultural output by farms, (at current prices, billion drams)

	2019	2020	2021	2022	2023
Total	853.3	833.3	934.4	1021.7	948.3
Including					
Plant growing	410.9	399.5	469.1	519.2	447.8
Animal husbandry	442.4	433.8	465.3	502.5	500.5
Commercial organizations	52.7	49.5	54.3	64.4	69.1
Including					
Plant growing	13.9	11.2	12.1	12.1	12.8
Animal husbandry	38.8	38.3	42.2	52.3	56.3
Household's plots	800.6	783.8	880.1	957.3	879.2
Including					
Plant growing	397.0	388.3	457.0	507.1	435.0
Animal husbandry	403.6	395.5	423.1	450.2	444.2

Source Composed by the authors based on (Statistical Committee Republic of Armenia 2024, p. 371)

Table 2 Production of main animal husbandry products

Total	2019	2020	2021	2022	2023
Meat (slaughter weight), 1000 tons	107.3	107.7	110.6	103.7	100.3
Milk, 1000 tons	667.9	654.3	670.7	623.1	591.6
Eggs, million pieces	720.6	754.6	710.4	749.1	732.2
Wool (physical weight), tons	981	1048	1083	1074	1101
Commercial organizations					
Meat (slaughter weight), 1000 tons	11.4	11.8	12.5	14	14.2
Milk, 1000 tons	7.1	7.9	7.5	7.1	9.4
Eggs, million pieces	264.1	276	249.1	320	347.8
Wool (physical weight), tons	0	0	3	3	0
Household's plots					
Meat (slaughter weight), 1000 tons	95.9	95.9	98.1	89.7	86.1
Milk, 1000 tons	660.8	646.4	663.2	616.0	582.2
Eggs, million pieces	456.5	478.6	461.3	429.1	384.4
Wool (physical weight), tons	981	1048	1080	1071	1101

Source Composed by the authors based on (Statistical Committee Republic of Armenia 2024, p. 382)

In recent years, the development of agriculture has slowed, with noticeable declines in the production of several key food products, including grain, sugar, vegetable oil, butter, and poultry meat. Simultaneously, more than half of the arable land and pastures remain unused for their intended purpose due to various factors. This situation highlights the need to place greater emphasis on the development of intensive forms of land use and livestock farming. In plant breeding, there has been rapid growth in indoor vegetable cultivation and floriculture. In livestock farming, the focus has been on increasing milk, meat, and egg production through the use of smart livestock buildings. However, beef and poultry meat production remain underdeveloped, primarily due to a shortage of feed resources.

3.2 Smart Technologies Used in the Armenian Livestock Sector

One of the key sectors of Armenia's economy is agriculture, which contributes about 9% of the country's gross domestic product (GDP). In recent years, livestock production has accounted for over 50% of the total agricultural output. However, production and economic indicators in livestock breeding remain insufficient, primarily due to the continued reliance on extensive breeding methods. The vast majority of farms involved in cattle and buffalo breeding are traditional subsistence farms.

According to the Statistical Committee of the Republic of Armenia, as of January 1, 2024, there were 491,594 heads of cattle, of which 211,190 (or 43%) were cows. Cattle breeding remains the leading branch of Armenia's livestock sector, with 95% of the country's milk production coming from cattle.

At the current stage of development, it is necessary to apply the latest technologies to form commodity farms that will allow increasing the milk yield of cows (Government of the Republic of Armenia 2024). In this case, smart livestock buildings are the best solution.

Since 2019, a total of 49 smart livestock buildings have been constructed under a state support program aimed at modernizing livestock facilities. These barns are designed to ensure proper drinking water supply, ventilation, lighting, and milking equipment, all critical for animal health and productivity. For instance, a cow requires 3–4 L of water per day to produce one liter of milk, with this demand occurring at various times throughout the day. Therefore, cows must have unrestricted access to water. Lighting is also a vital factor: maintaining 16 h of stable light per day can significantly boost milk yields. Ventilation is equally important, as inadequate airflow can raise barn temperatures above 25 °C, potentially reducing milk production by 20–30% (Manukyan 2023).

Looking ahead, the 2025–2028 program for the construction and modernization of smart livestock buildings aims to further improve animal welfare conditions through state support for both building infrastructure and technological upgrades. As a result, these efforts are expected to increase productivity indicators across Armenia's livestock sector.

To achieve the specified goal, it is necessary to solve the following problems:

1. Offering affordable options for smart livestock buildings and their technological support;

2. Availability of necessary investments for the construction of livestock buildings and partial reimbursement of costs;
3. Assessment of possible risks related to the implementation of the program and development of measures aimed at their mitigation;
4. Stocking smart livestock buildings exclusively with pedigree animals.

Smart livestock buildings are a comprehensive solution to increase the efficiency of the milk production chain. The following three models are proposed for construction or reconstruction and technological support:

1. Model 1 is designed for 20–25 animals, the costs of construction and technological support of which are set at a maximum of 30 million Armenian drams;
2. Model 2 is designed for 40–45 animals, the costs of construction and technological support of which are set at a maximum of 46 million Armenian drams;
3. Model 3 is designed for 95–100 animals, the costs of construction and technological support of which are set at a maximum of 114 million Armenian drams.

Within the project framework, it is planned to build or reconstruct 50 livestock buildings. The amount of compensation will be about 1,163,290,000 drams (Government of the Republic of Armenia 2024).

Within the program framework, the government compensates 50% of the cost. There are also preferential conditions for individuals in border areas and members of cooperatives. For people with disabilities obtained during military service, approximately 70% of costs are covered through compensation.

The following results are expected from the program implementation:

Due to the implementation of the program, the average milk yield of cows in livestock buildings in accordance with zoohygienic standards will increase by 50–60%, from 2550 kg to almost 4000 kg. That of buffaloes is to increase from 1200 to 2000 kg;
The incidence of animal diseases will decrease by 30%. Milk will be produced with exceptionally high quality indicators;
The presence of about 50 (baseline: 58 livestock buildings) cattle and buffalo farms operating in accordance with modern technological standards in the Republic of Armenia;
Introduction of advanced technologies for cattle and buffalo farming, industry intensification;
Creation of favorable conditions for animal behavior (ensuring zoohygienic conditions, effective use of genetic potential);
Increase in livestock by 0.5%;
Decrease in the cost of milk by at least 20% and increase in production volumes by 40–50%;
Increase in income of economic entities by almost 30%, increase in competitiveness;

Increase in professional capabilities of economic entities;
The created farms can serve as an example for a wide range of economic entities in terms of introducing them to modern approaches to cattle and buffalo breeding, as well as an educational base for students (Government of the Republic of Armenia 2024).

4 Conclusion

The integration of smart technologies into livestock farming represents a transformative step for Armenian agriculture, addressing productivity and sustainability challenges. Traditional farming methods (e.g., extensive livestock practices) do not have enough efficiency and very often fail to ensure optimal animal welfare. In this regard, the development of smart livestock buildings will provide a comprehensive solution to these challenges based on their technological solutions like sensors, automated feeding systems, climate control, and monitoring tools. The current experience of Armenia shows that the existing state-supported program of developing smart livestock buildings contributes to increasing milk yield, improving animal health, reducing disease incidence, and decreasing production costs. Such innovative solutions increase the competitiveness of Armenian livestock products, contribute to the sustainable use of land and water resources, and create educational opportunities for farmers and students alike. As Armenia continues adopting and expanding these technologies, the livestock sector is expected to experience long-term improvements in efficiency, sustainability, and quality, aligning with global standards and advancing the country's agricultural development goals.

References

Akkem Y, Biswas SK, Varanasi A (2023) Smart farming using artificial intelligence: a review. Eng Appl Artif Intell 120:105899. https://doi.org/10.1016/j.engappai.2023.105899

Anubhav K, Agarwal M, Aashish K (2023) Smart farming for future: a structural relation analysis of attitude, facilitating condition, economic benefit and government support. Technol Anal Strat Manag 37(2):187–201. https://doi.org/10.1080/09537325.2023.2283550

Avagyan G, Petrosyan G, Mkhitaryan K, Vardanyan K, Mirumyan A (2022) Evaluation of the effectiveness of state assistance programs implemented in the RA agricultural sector through a survey among farmers and experts. Messenger ASUE 6:22–42. https://doi.org/10.52174/1829-0280_2022.6-22

Avetisyan S, Avagyan G, Mkhitaryan K, Mirumyan A, Karapetyan S (2022) Development trends and analysis of structural shifts of RA agricultural sector. Messenger ASUE 5:64–80. https://doi.org/10.52174/1829-0280_2022_5_64

Farooq MS, Sohail OO, Abid A, Rasheed S (2022) A survey on the role of IoT in agriculture for the implementation of smart livestock environment. IEEE Access 10:9483–9505. https://doi.org/10.1109/ACCESS.2022.3142848

Giua C, Materia VC, Camanzi L (2022) Smart farming technologies adoption: which factors play a role in the digital transition? Technol Soc 68:101869. https://doi.org/10.1016/j.techsoc.2022.101869

Government of the Republic of Armenia (2024) Program for support for the construction or reconstruction of "smart" livestock buildings and their technological support for 2025–2028 (appendix no. 4 to the RA government resolution no. 1392-L of 22 August 2024). Yerevan, Armenia. Retrieved from https://www.arlis.am/hy/acts/207595/print/act. Accessed 28 Aug 2025

Idoje G, Dagiuklas T, Iqbal M (2021) Survey for smart farming technologies: challenges and issues. Comput Electr Eng 92:107104. https://doi.org/10.1016/j.compeleceng.2021.107104

Javaid M, Haleem A, Singh RP, Suman R (2022) Enhancing smart farming through the applications of Agriculture 4.0 technologies. Int J Intell Netw 3:150–164. https://doi.org/10.1016/j.ijin.2022.09.004

Kim MJ, Mo C, Kim HT, Cho B-K, Hong S-J, Lee DH, Shin CS, Jang KJ, Kim YH, Baek I (2021) Research and technology trend analysis by big data-based smart livestock technology: a review. J Biosyst Eng 46(4):386–398. https://doi.org/10.1007/s42853-021-00115-9

Litvinova TN, Hakhverdyan DN, Popkova EG, Litvinova AI (2025) Management of the quality of agricultural products in the agrarian AI economy in support of food security. Int J Qual Res 19(1):211–222. Retrieved from http://ijqr.net/journal/v19-n1/14.pdf. Accessed 28 Aug 2025

Manucharyan M (2025) Climate change impacts on sustainable agriculture: evidence from Armenia. Unconv Resour 6:100159. https://doi.org/10.1016/j.uncres.2025.100159

Manukyan G (2023) New culture in agriculture: around 49 "smart" livestock buildings have been built and are operating in Armenia. Armenpress. Retrieved from https://armenpress.am/hy/article/1105079. Accessed 28 Aug 2025

Mohamed ES, Belal AA, Abd-Elmabod SK, El-Shirbeny MA, Gad A, Zahran MB (2021) Smart farming for improving agricultural management. Egypt J Remote Sens Space Sci 24(3–2):971–981. https://doi.org/10.1016/j.ejrs.2021.08.007

Moysiadis V, Sarigiannidis P, Vitsas V, Khelifi A (2021) Smart farming in Europe. Comput Sci Rev 39:100345. https://doi.org/10.1016/j.cosrev.2020.100345

Papadopoulos G, Papantonatou M-Z, Uyar H, Nychas K, Psiroukis V, Kasimati A, Nieuwenhuizen A, Van Evert FK, Fountas S (2025) Stakeholders' perspective on smart farming robotic solutions. Smart Agric Technol 11:100916. https://doi.org/10.1016/j.atech.2025.100916

Pivoto D, Waquil PD, Talamini E, Finocchio CPS, Dalla Corte VF, de Vargas Mores G (2018) Scientific development of smart farming technologies and their application in Brazil. Inf Process Agric 5(1):21–32. https://doi.org/10.1016/j.inpa.2017.12.002

Prashanth JS, Krishna GB, Prasad AVK, Rao PR (2025) Smart farming revolution: a cutting-edge review of deep learning and IoT innovations in agriculture. Oper Res Forum 6:35. https://doi.org/10.1007/s43069-025-00434-z

Sharma V, Tripathi AK, Mittal H (2022) Technological revolutions in smart farming: current trends, challenges & future directions. Comput Electron Agric 201:107217. https://doi.org/10.1016/j.compag.2022.107217

Statistical Committee Republic of Armenia (2024) Agriculture: statistical yearbook of Armenia. ARMSTAT, Yerevan, Armenia. Retrieved from https://armstat.am/file/doc/99552343.pdf. Accessed 28 Aug 2025

Terence S, Immaculate J, Raj A, Nadarajan J (2024) Systematic review on internet of things in smart livestock management systems. Sustainability 16(10):4073. https://doi.org/10.3390/su16104073

Virk AL, Noor MA, Fiaz S, Hussain S, Hussain HA, Rehman M, Ahsan M, Ma W (2020) Smart farming: an overview. In: Patnaik S, Sen S, Mahmoud M (eds) Smart village technology, pp 191–201. Springer, Cham, Switzerland. https://doi.org/10.1007/978-3-030-37794-6_10

Wolfert S, Ge L, Verdouw C, Bogaardt MJ (2017) Big data in smart farming—a review. Agric Syst 153:69–80. https://doi.org/10.1016/j.agsy.2017.01.023

The Labor Theory of Use Value in the Context of a Climate-Optimized Agricultural Economy in the Regional Space of the Russian State

Evgeniy S. Popov

Abstract The research aims to explore the theoretical prospects of a climate-optimized agricultural economy in the Russian Federation, within the framework of the economic laws of the labor theory of use value. The author employs the labor theory of use value, which is based on a series of economic laws, namely: the surplus of labor results over labor costs, the measure of labor becoming dependent on the needs of the population, and the efficiency of agricultural activities being reduced to labor savings and the reduction of working hours. The research demonstrates that Russia now faces interconnected climatic and economic challenges in agriculture. The positive resolution of these negative phenomena lies in transitioning to a climate-optimized agricultural economy based on the laws of the labor theory of value. Using a consumption-value analysis of agricultural land and cultivated plants, the article reveals indicators of labor savings and energy recovery for the population. It also identifies regions of the Russian state with the greatest contribution to utility potentials. Additionally, it considers the losses related to the labor costs of workers and the value indicators of fixed assets necessary to achieve the expected potential economic results in the country's agricultural sector.

Keywords Use value · Climate-optimized agriculture · Labor savings · Labor costs · Regions · Crops · Land resources · Person-hours

JEL Codes A19 · Q10 · Q19 · Q24 · Q29 · Q51 · Q54 · P00

1 Introduction

The dynamic changes in the climate of the territorial space of the Russian Federation at the regional level have become a significant endogenous factor influencing the efficiency and direction of national agricultural practices. For instance, in the

E. S. Popov (✉)
Don State Agrarian University, Persianovski, Russia
e-mail: povove@mail.ru

A. B. Karbekova (ed.), *Sustainable Development of Rural Territories Based on Climate-Smart Technologies*, Climate Change Management,
https://doi.org/10.1007/978-3-032-20106-5_15

first half of 2025, the Rostov Region experienced an atypical late temperature drop during spring and a subsequent drought in summer, which affected the overall grain crop yield. Meanwhile, agents in the agricultural sector, relying on optimal climatic indicators from previous seasons (e.g., the number of warm sunny days and precipitation), had long based their economic activities on minimizing the consumption of fixed capital elements, such as hydraulic engineering structures. As a result, these circumstances led to a negative economic and political situation, including the potential mass bankruptcy of farms, large agricultural enterprises, and a threat to the food security of the region.

In response, the Ministry of Agriculture of the Russian Federation and the Government of the Rostov Region took appropriate actions to stabilize the situation, including declaring a moratorium on bankruptcy proceedings, introducing "credit holidays," and providing preferential loans to agricultural producers. However, the negative trends that have already emerged in the agricultural sector are likely to intensify in the near future. While the adverse factors have manifested in the Rostov region, in other administrative units, they remain latent.

Thus, overcoming the interconnected climatic and economic problems in agriculture is an urgent issue in research, as it pertains to optimizing the agricultural sector based on the laws of objective social production.

2 Materials and Methods

In the scientific literature, overcoming the climatic challenges of the agricultural sector within the national economy is addressed within the framework of the concept of climate-optimized agriculture (Lipper et al. 2018). This concept essentially implies its practical implementation through the directed transformation of the agricultural system, with the focus on specific technical and planning measures, namely the introduction of systematic crop rotation, soil mulching, creation of protective forest zones, reduction of greenhouse gas emissions from stationary and mobile sources, and the construction of technological structures in crop farming and animal husbandry.

In the author's view, the use of agricultural technologies that increase the resilience of land resources in changing weather conditions is one-sided in the formation of climate-optimized agriculture. The second crucial aspect is the action of objective economic laws. If the law of value operates in agriculture as an economic law of commodity production, then the use of progressive agricultural technologies leads to a decline in the average rate of profit in the agricultural sector. In turn, this creates conditions for agents to minimize the use of fixed capital elements, which are capable of ensuring the sustainability of crop cultivation and redirecting production to the effective utilization of meteorological indicators specific to the regions.

Therefore, an important methodological step in the climate-optimized agricultural economy should be a shift towards a consumption-value paradigm. Founders of Soviet and Russian economic theory, such as Dolgov (1988) and Elmeev (2007), emphasized that the law of use value is reduced to a series of principles, namely, that the results of labor consumed exceed the labor costs of production, with labor costs becoming dependent on the measure of consumption by the population. The criterion for the efficiency of economic activity is the saving of labor in conditions of satisfying the population's needs.

Simultaneously, the comparison of labor savings in relation to energy equivalents in the labor theory of use value was addressed by (Bainev 2008). The economist highlighted that the utility energy properties of various natural resources could be translated into labor savings, namely by replacing and conserving the energy expended by the organism of the subject, depending on the severity of the work performed. Therefore, based on this theoretical premise, it can be argued that the mass of use values of crops, with natural indicators in kilocalories, serves the function of saving time to restore the physiological potential of the subject expended in performing a specific task.

3 Results

The territory of the Russian Federation encompasses a complex network of natural spatial conditions, divided into climatic zones that range from the Arctic to the subtropical. These endogenous factors determine the economic characteristics of the agricultural sector in different regions, particularly in crop production. For instance, areas with risky farming include the Krasnodar Territory, the Rostov Region, the Stavropol Territory, and regions of northern European Russia, Siberia, and the Far East. Thus, the country contains territories where agricultural production faces heightened risks of crop loss due to climate conditions and the low natural productivity of land resources.

Nevertheless, within the existing regional framework, productive consumption of land resources occurs in the context of the soil's fertility and the material and technical capacities of agricultural producers. According to the Ministry of Agriculture of the Russian Federation, as of the end of 2023, the total land area amounted to 1,712,519.2 thousand hectares. Of this, the forest fund (1,135,935.3 thousand hectares) is the largest category, followed by agricultural land (321,651.5 thousand hectares), land reserves (87,295.0 thousand hectares), protected areas (50,616.3 thousand hectares), water resources (28,082.6 thousand hectares), settlements (20,745.7 thousand hectares), and industrial zones and other uses (18,322.8 thousand hectares). In this context, we are particularly interested in agricultural land, which is actively used for growing crops, totaling 66,492.9 thousand hectares (Lipper et al. 2018).

According to the aforementioned methodological principles, it is necessary to convert the natural indicators of the agricultural sector into a consumption-value measurement. It is known that the land resource contains accumulated bioenergy in the fertile topsoil humus layer, and different types of work have a normative level of energy expenditure for the workforce, measured in terms of the intensity of labor functions, in person-hours (p/h). Consequently, it is possible to derive equivalent indicators depending on the type of work and energy expenditure, specifically: simple labor (Ia) has an indicator of 139 W × h, light labor (Ib) ranges from 140–174 W × h, moderate labor (IIa) is between 175 and 232 W × h, medium-heavy labor (IIb) is between 233 and 290 W × h, and heavy labor (III) is 291 W × h and above (Chief State Sanitary Inspector of the Russian Federation 2021).

From the above, it follows that in the context of energy equivalents, one can compare the use value of labor and soil complexes (GJ/hectare), which boils down to the ability to save human labor (person-hours). As a result, the regional space of the Russian Federation encompasses a variety of soil categories with corresponding utility potentials for labor savings, ranging from simple to heavy labor, depending on the intensity of the work. For example, the indicators for sod-podzolic soils range from 323,505 to 67,766 person-hours, typical chernozems, podzolized, and leached soils range from 930,138 to 1,947,267 person-hours, gray forest soils range from 471,065 to 986,187 person-hours, ordinary and southern chernozems range from 692,062 to 1,448,849 person-hours, dark chestnut and chestnut soils range from 510,310 to 1,068,345 person-hours, and light chestnut and brown semi-desert soils range from 192,440 to 402,878 person-hours (Kireycheva and Khokhlova 2019).

Based on this information, calculations can be made regarding the potential labor savings embedded in land resources across the federal districts of the Russian Federation. For instance, the leader in terms of the share of soil cover, characterized by the predominance of known bioenergy indicators ranging from simple to heavy labor, is the Volga Federal District, which includes a macroregion with ordinary and southern chernozems (33,183.2 thousand hectares), sod-podzolic soils (23,851.0 thousand hectares), and potential labor savings of 33.0 to 70.0 trillion person-hours. Next is the Central Federal District, which comprises sod-podzolic soils (31,210.0 thousand hectares), as well as chernozem, podzolized, leached, and typical soils (16,906.0 thousand hectares), with potential labor savings of 25 to 54 trillion person-hours. The Southern Federal District follows with ordinary and southern chernozems (17,018.0 thousand hectares), with potential savings of 11 to 24 trillion person-hours. The North Caucasus Federal District, with ordinary and southern chernozems (4321.4 thousand hectares) and light chestnut and brown semi-desert soils (4091.0 thousand hectares), has potential labor savings from 3.8 to 8.0 trillion person-hours. Notably, the information for the Northwestern Federal District, Ural Federal District, Siberian Federal District, and Far Eastern Federal District is not included due to the lack of theoretical indicators for the energy potential of the soil cover in these administrative units (Ministry of Natural Resources and Environment of the Russian Federation 2024).

It is important to emphasize that agricultural land has its own use value for subjects as a source of crop production. The food products grown on land resources participate in the simple reproduction of the biological existence of the population and fulfill the physiological needs, restoring the energy expended during working hours. The dynamics of the consumption of sown area, crop yields, caloric value of crops, and the recovery of energy from simple and complex labor are presented in Table 1.

The data presented demonstrate a significant contrast in the indicators of the utility efficiency of agricultural crops. There is a clear trend toward the consolidation of high-calorie crops (wheat, barley, corn, buckwheat, legumes, sunflower,

Table 1 Indicators of the mass of use values of agricultural crops in Russia as of January 1, 2024

Crop name	Sown area, thousand hectares	Total yield, million tons	Total caloric value of the crop, kcal	Energy compensation for simple labor to heavy labor (person × 8 h)
Wheat	29,753.9	94.6	2.74×10^{14}	$2.85 \times 10^{11} - 1.36 \times 10^{11}$
Rye	805.6	1.74	4.99×10^{12}	$5.19 \times 10^{9} - 2.48 \times 10^{9}$
Triticale	97.9	0.28	8.2×10^{11}	$8.54 \times 10^{8} - 4.08 \times 10^{8}$
Barley	7943.3	21.44	5.66×10^{13}	$5.89 \times 10^{10} - 2.81 \times 10^{10}$
Corn	2668.1	18.48	6.0×10^{13}	$6.25 \times 10^{10} - 2.98 \times 10^{10}$
Oats	1831.6	3.4	8.5×10^{12}	$8.85 \times 10^{9} - 4.23 \times 10^{9}$
Rice	190.3	1.1	3.11×10^{12}	$3.23 \times 10^{9} - 1.55 \times 10^{9}$
Buckwheat	1294.8	1.55	4.57×10^{12}	$4.76 \times 10^{9} - 2.27 \times 10^{9}$
Millet	294.6	0.47	1.46×10^{12}	$1.52 \times 10^{9} - 7.27 \times 10^{8}$
Legumes	2954.8	6.08	1.79×10^{13}	$1.86 \times 10^{10} - 8.9 \times 10^{9}$
Sugar Beet	1063.5	53.67	2.25×10^{13}	$2.34 \times 10^{10} - 1.12 \times 10^{10}$
Flax (long fiber)	35.7	0.02	1.06×10^{11}	$1.1 \times 10^{8} - 5.27 \times 10^{7}$
Sunflower	9868.8	18.25	1.09×10^{14}	$1.13 \times 10^{11} - 5.42 \times 10^{10}$
Soy	3667.8	7.04	6.37×10^{13}	$6.63 \times 10^{10} - 3.17 \times 10^{10}$
Rapeseed	2111.1	4.28	2.37×10^{12}	$2.46 \times 10^{9} - 1.18 \times 10^{9}$
Flax (curly)	1410.4	1.25	6.67×10^{12}	$6.94 \times 10^{9} - 3.32 \times 10^{9}$
Potatoes	313.7	8.62	6.89×10^{12}	$7.17 \times 10^{9} - 3.43 \times 10^{9}$
Vegetables	187.0	5.88	1.88×10^{12}	$1.95 \times 10^{9} - 9.36 \times 10^{8}$

Source Developed by the author based on (Ministry of Agriculture of the Russian Federation 2024b; Skurikhin and Volgareva 1987)

soy, rapeseed, curly flax) in relation to the sown area (61,673.0 thousand hectares). Furthermore, higher-value agricultural crops, in terms of their use value, have 3.6 times more bioenergetic value than low-calorie flora. However, such plant crops are necessary for consumption in accordance with healthy nutrition standards.

It should be noted that the calculations derived from Table 1 on the total caloric value of agricultural crops can be proportionally translated into corresponding coefficients for individual physiological energy recovery from expended physical effort during working hours. As of the beginning of 2024, the use value of land resources in agricultural land across the entire Russian territorial space provided, in absolute terms, the energy recovery of physiological needs ranging from 374 to 670 billion workers performing their labor functions for 8 h a day. Nevertheless, the structural specifics of the use of domestic agricultural land in various regions are reflected in the significant sown area of grain and legume crops (47,883.7 thousand hectares).

Importantly, plant production is carried out in all federal districts of Russia, and the distribution of the contribution to the overall fund for the recovery of the physiological potential of the consumer subject is presented in Table 2.

The data presented show that the largest share in the recovery of the population's physiological potential comes from the Central, Southern, and Volga federal districts. Simultaneously, contrasting trends are observed in achieving such results. Specifically, the Central and Southern macroregions, with 1.5 times less arable land compared to the Volga region, produce about twice the amount of crop production per unit area. When comparing the utility efficiency of land resource usage by regions, administrative entities that occupy leading positions in the production of crops essential for the reproduction of the population can be highlighted. These include Voronezh Region (4.94–14.8 billion person-days), Tambov Region (4.0–8.9 billion person-days), Kursk Region (4.35–9.72 billion person-days), and Krasnodar Territory (10.7–23.9 billion person-days).

Table 2 The structure of the distribution of Federal districts' contribution to the energy recovery of the population as of January 1, 2024

Federal district	Sown area, thousand hectares	Yield, centners per hectare	Energy potential, kcal	Contribution of the Federal district, %
Central	8231.6	47.7	1.13×10^{14}	26.42
Northwestern	319.1	38.0	3.5×10^{12}	0.82
Southern	9344.0	43.2	1.16×10^{14}	27.12
North Caucasus	3310.9	40.2	3.85×10^{13}	9.0
Volga	13,305.3	24.1	9.27×10^{13}	21.68
Ural	3541.3	17.2	1.76×10^{13}	4.12
Siberian	9303.3	15.6	4.19×10^{13}	9.8
Far Eastern	538.2	28.7	4.46×10^{12}	1.04

Source Developed by the author based on (Federal State Statistics Service of the Russian Federation (Rosstat) 2024)

Calculations have been made showing the absolute utility effect of agricultural land and crops in a consumption-value measure. However, achieving such a positive result is only possible within the context of the corresponding labor costs involved in transforming them into individual livelihoods. In other words, a significant factor in determining the economic efficiency of a climate-optimized agricultural economy is the consideration of total public labor costs, namely the labor costs of agricultural workers and the costs associated with the use of fixed assets. The 2024 figures for labor costs of agricultural sector workers, based on an average workforce size of 979,405 people, amount to 1.7 billion person-hours. The necessary costs for recovering the physiological potential of agricultural workers range from 2.0 to 4.3 billion kcal over the specified period (Federal State Statistics Service of the Russian Federation (Rosstat) 2024; Ministry of Agriculture of the Russian Federation 2024a).

Another major cost factor in agriculture is fixed assets. However, in the current situation, namely the lack of natural measurements of means of production in terms of past labor units, it is impossible to assess their weight coefficient. Nonetheless, it can be stated that fixed assets make up a significant portion of the labor savings in the agricultural sector, as statistical sources indicate that the value of fixed assets in the entire sector for 2023 amounted to 9,144,883.0 million rubles (Federal State Statistics Service of the Russian Federation (Rosstat) 2024).

4 Discussion

A climate-optimized agricultural economy, operating under the objective law of the labor theory of use value, provides qualitatively different assessments of the agricultural sector. The surplus of labor savings over labor costs, combined with the high energy recovery potential from crop production, which is based on the high use value of land resources, serves as a reliable criterion for economic efficiency, just as value indicators do.

Based on data from the territorial space of the Russian Federation, we have attempted to present theoretical calculations with heuristic results, as well as specific figures demonstrating the consumption-value potential of the regions of the country.

5 Conclusion

Currently, the issue of optimizing the agricultural sector in the context of a changing climate remains a relevant theoretical and practical challenge. In our view, the prospect of a positive resolution lies in the shift towards consumption-value principles in agricultural practices.

References

Bainev VF (ed) (2008) Scientific and technical progress and sustainable development: theory and practice of useful (consumer-value) assessment of the effectiveness of new technology. Belarusian State University, Minsk, Belarus

Chief State Sanitary Inspector of the Russian Federation (2021) Resolution No. 2 "on approval of sanitary rules and regulations SanPiN 1.2.3685-21 "hygienic standards and requirements for ensuring the safety and (or) harmlessness of human habitat factors," together with "SanPiN 1.2.3685-21. Sanitary rules and regulations...')" (28 Jan, 2021, as amended on 30 Dec 2022). Moscow, Russia. Retrieved from https://www.consultant.ru/document/cons_doc_LAW_375839/ba9db32f77fed99cb5ed5c8dfde8a1fb8da80a08/. Accessed 21 Sep 2025

Dolgov VG (1988) Management of scientific and technological progress: consumer-cost basis. Leningrad University Publishing House, Leningrad, USSR

Elmeev VY (2007) Social economy of labor: general principles of political economy. Saint-Petersburg State University Publishing House, St. Petersburg, Russia

Federal State Statistics Service of the Russian Federation (Rosstat) (2024) Regions of Russia. Socio-economic indicators, 2024: statistical collection. Rosstat, Moscow, Russia

Kireycheva LV, Khokhlova OB (2019) Estimation of energy resource of degraded soils of agricultural lands. Agrochemical Herald 3:21–27

Lipper L, McCarthy N, Zilberman D, Asfaw S, Branca G (2018) Climate-smart agriculture: building resilience to climate change. FAO and Springer, Cham, Switzerland. https://doi.org/10.1007/978-3-319-61194-5

Ministry of Agriculture of the Russian Federation (2024a) final report on the results of the Russian Ministry of Agriculture's activities for 2023. Retrieved from https://mcx.gov.ru/upload/iblock/60e/f3efndq2h1aju6jsas5j1kjavrgsj52s.pdf. Accessed 1 Oct 2025

Ministry of Agriculture of the Russian Federation (2024b) Report on the state and use of agricultural land in the Russian federation in 2023. Rosinformagrotekh, Moscow, Russia. Retrieved from https://cloud.mail.ru/public/k5yz/RJzLaBcqV. Accessed 1 Oct 2025

Ministry of Natural Resources and Environment of the Russian Federation (2024) Draft state report "on the state of the environment and environmental protection in the Russian federation in 2023". Ministry of Natural Resources and Environment of the Russian Federation; LLC "Intellectual Analytics"; "Directorate for Scientific and Technological Programs"; Foundation for Environmental Monitoring and International Technological Cooperation, Moscow, Russia. Retrieved from https://tinyurl.com/4v5xykmx. Accessed 1 Oct 2025

Skurikhin IM, Volgareva MN (eds) (1987) Chemical composition of food products: book 1: reference tables of the content of basic food substances and the energy value of food products, 2nd edn. Agropromizdat, Moscow, USSR

Current State and Prospects of Halal Business in Central Asian Countries in the Context of Ensuring Food Security Based on Climate-Smart Technologies

Murat Zh. Abdiev, Niyazali T. Aripov, Tulkin Kh. Farmanov, Ibrohim N. Ismanov, and Azizxon S. Xojayev

Abstract The research aims to theoretically and methodologically substantiate the state and prospects for the development of halal business, covering key sectors such as food, cosmetics, pharmaceuticals, and tourism in Central Asian countries, and to develop practical recommendations. The research examines factors influencing the development of halal businesses, identifies challenges in this area, and proposes prospects for halal business development in the region, including the need to improve standards and certification, as well as opportunities for further enhancing public–private cooperation. The findings of this research confirm the importance of the halal sector for the economic development of Central Asian countries and its potential on the international stage. The authors of this research aim to develop specific proposals for the development of halal businesses in Central Asian countries, based on innovative and investment-driven approaches, in accordance with international standards, within the context of enhancing integration relations.

M. Zh. Abdiev (✉)
Osh Technological University named After M.M. Adyshev, Osh, Kyrgyzstan
e-mail: mabdiev1977@mail.ru

M. Zh. Abdiev · I. N. Ismanov · A. S. Xojayev
Fergana State Technical University, Fergana, Uzbekistan
e-mail: i.ismanov@ferpi.uz

A. S. Xojayev
e-mail: azizxon.xojayev@fstu.uz

M. Zh. Abdiev
Andijan Institute of Agriculture and Agricultural Technologies, Andijan, Uzbekistan

N. T. Aripov
Osh State University, Osh, Kyrgyzstan

T. Kh. Farmanov
Tashkent State University of Economics, Tashkent, Uzbekistan

Tashkent State Agrarian University, Tashkent, Uzbekistan

A. B. Karbekova (ed.), *Sustainable Development of Rural Territories Based on Climate-Smart Technologies*, Climate Change Management,
https://doi.org/10.1007/978-3-032-20106-5_16

Keywords Halal business · Food · Certification · Investment · Support

JEL Codes D18 · D24 · E23 · F02 · F63 · Z12

1 Introduction

In the context of globalization, there are trends in the development of the international market for halal products and services, driven by the growing number of Muslims and increased spending on high-quality goods and services that comply with Islamic norms. By 2050, the total Muslim population is projected to increase from the current 25 to 30% of the world's population. Such a large market offers equally vast opportunities for halal product manufacturers. Muslims who live according to Islamic norms strive to use halal products in all areas of their lives, including food, beverages, medicine, and clothing.

Halal (Arabic: permitted)—"all that is permitted and permissible in Islam." The Koran, the holy book of Muslims, clearly states what is allowed and what is forbidden to a devout Muslim (Aripov et al. 2024).

These include food, clothing, cosmetics, and leisure products. Specially certified centers issue Halal certificates to hotels, shopping centers, cafes, restaurants, and other establishments, carefully checking them for quality and compliance with Islamic norms (International Halal Certification Center 2025).

Currently, the number of Muslims in the world reaches almost 1.8 billion people, which is almost 25% of the world's total population. Such a large market implies equally great opportunities for halal businesses. The growing number of consumers of this business, Muslims and non-Muslims, creates a global demand for basic goods and services that comply with Islamic principles and values (Uzenbaev et al. 2019).

Halal business is becoming an increasingly relevant topic in the global economic arena. In recent years, there has been a significant increase in interest in halal products and services, especially in countries with a predominantly Muslim population. Halal is a term that in Islam refers to what is allowed or permissible. Halal encompasses a wide range of industries, including food, financial services, pharmaceuticals, cosmetics, fashion, tourism, and others. Central Asian countries are also considered a promising market for the halal industry and are striving to improve it. In Central Asia, there are primarily specialized halal business sectors, including food production, tourism, and Islamic finance.

This research is dedicated to a comprehensive examination of the current state of halal business in Central Asia, encompassing Kazakhstan, Uzbekistan, Kyrgyzstan, Tajikistan, and Turkmenistan. Special attention will be given to key sectors, including food, cosmetics, pharmaceuticals, and tourism.

2 Methods

The research uses educational, methodological, and scientific publications of foreign and domestic scientists on halal business, as well as other sources, including reports of international organizations, market research, and statistical data from Central Asian countries, as well as publications in scientific journals and specialized publications. Research methods. The research uses economic-statistical, monographic, comparative-analytical, and computational-constructive research methods.

3 Results

In today's world, halal has ceased to be a purely religious issue, spreading to business and trade. It becomes a symbol of quality assurance, lifestyle choice, turning into a kind of postulate of how people should live, eat, and behave (Ismanov and Ermatov 2024).

Currently, there are more than 120 voluntary Halal certification bodies operating worldwide, which are represented by officially authorized bodies of individual countries that issue the Halal certificate.

For example, large pharmaceutical companies (e.g., the British Principle Healthcare and the Canadian Duchesnay) operate in the pharmaceutical market, selling vitamins and dietary supplements without animal gelatin and other products that comply with Islamic norms (Packaging Information Portal "Upakovano" 2011).

Despite being geographically distant from the Islamic world, several countries are considered global leaders in halal meat production. These include Australia, New Zealand, the Netherlands, Ireland, and others (Russian Muftis Council 2017).

The halal industry is becoming an increasingly important part of the global economy, covering a wide range of sectors and destinations. According to the latest data, the global halal market volume reaches approximately $2.3 trillion, and it is expected to grow to $4.5 trillion by 2028 (Research and Markets 2024).

This remarkable growth is driven by the rising number of Muslim consumers worldwide, as well as increasing interest in halal goods and services among non-Muslims, which continues to create new market niches. In Central Asia, the halal sector is expanding rapidly and plays an important role in economic development and job creation. The diversity of its segments, including halal food production, cosmetics, pharmaceuticals, and tourism, demonstrates the sector's strong potential. With a wide range of opportunities for local producers and international investors, the halal industry is becoming a key contributor to the region's sustainable economic development, fostering innovation and improving the quality of available services.

The halal industry is one of the least developed industries in Kazakhstan. On March 5, 2007, the President of Kazakhstan raised the issue of prospects for the development of the halal industry and instructed to increase the volume of exports of these products abroad, especially to Muslim countries. Since that day, the SAMK, together with the government, has started certifying Kazakhstani enterprises engaged in the production of halal products. Kazakhstan is showing strong growth in the halal industry, especially in the food and pharmaceutical sectors. In 2020, more than 600 halal products were certified, encompassing a diverse range of products, including meat, dairy, and confectionery. The products of these companies are exported to 43 countries worldwide (Ministry of National Economy of the Republic of Kazakhstan 2025).

This is a result of growing consumer interest in high-quality products that meet halal standards, as well as efforts by local producers to enhance the quality of their products and obtain international certifications.

In 2018, Kazakhstan was accredited by the Malaysian accreditation center "Jakim."

In 2021, the Government of Kazakhstan unveiled a strategic plan for developing the halal economy, which drew the attention of both local and international investors. This initiative includes stricter requirements for certification and the quality of halal products, as well as the creation of a convenient infrastructure for businesses, which contributes to the development of new industries and the entry of existing ones into international markets. Such attention to the halal industry creates favorable conditions for the country's economy to further grow and develop sustainably in this segment.

Uzbekistan also demonstrates an active development of the halal industry, especially in the production of halal food products, which was an important step in building the country's international reputation as a supplier of high-quality and safe products.

With the adoption and implementation of the Halal standard in the Republic of Uzbekistan, work is underway to create an appropriate direction in the national certification system. In the Republic of Uzbekistan, the application of the Halal standard is carried out on a voluntary basis.

Muzaffar Uzan, Head of the Personnel Certification Department of the Turkish Standards Institute (TSE), was invited as a lecturer. He personally participated in the development of the SMIIC 1 standard (Standards and Metrology for Islamic Country, applied in 37 countries), which served as the basis for the national standard of Uzbekistan. Representatives of the Uzstandard agency and the state-owned Republican Center for Testing and Certification regularly conduct training courses and seminars, and train auditors to monitor halal product certification. They also have branches in the regions (Uzbek Agency for Technical Regulation 2018).

In 2021, the Saxovat poultry farm received the Halal certificate of Uzbekistan. One of the leading enterprises of the food industry of Uzbekistan – poultry farm "Saxovat Broiler," founded in 2016, became the first poultry farm to receive the Uzbek Halal certificate in this direction (Uzbek Agency for Technical Regulation n.d.).

In June 2021, Saxovat received a certificate confirming compliance with Halol standards. The document was issued by the Center for Scientific Testing and Quality Control of Uzbekistan and the Department of Muslims of Uzbekistan. "Until then, the certificate of the Russian Mufti Council was valid for the company's product. Now we also have a "Halal" certificate from the Muslim countries of Uzbekistan," said Artur Dapkunas, the Executive Director of "Generousity Broiler."

This year, a South Korean grant of $2.8 million was allocated for the construction of the laboratory. Additionally, it is planned to establish nine laboratories in the country, thanks to a loan of approximately $11 million. After that, all laboratories will have the opportunity to organize tests in accordance with the criteria of the "Halal" standard (UZ24 Information Portal 2021).

Special attention will be paid to the composition of the product and the stages of cultivation, as well as issues such as livestock slaughtering, meat preservation, and waste prevention, which will also be in the spotlight.

Although not demonstrating the same scale of development of the halal industry as some of its neighbors, Kyrgyzstan is showing a growing interest in this market segment.

In 2020, several local certification bodies were established in the country, whose task was to support small and medium-sized enterprises in obtaining halal certificates, which greatly simplified the certification process for manufacturers (Halal Industry Development Center in Kyrgyzstan 2025).

The initiative focuses on the production of halal meat and dairy products, underscoring the country's commitment to improving product quality and ensuring compliance with Islamic standards. Success in this area can open up new opportunities for Kyrgyz producers, enhancing their competitiveness in both domestic and international markets.

In November 2024, a law regulating the certification of halal products, signed by the President of the Kyrgyz Republic, came into force. By the end of the same year, the Cabinet of Ministers approved the procedure for implementing a certification system to ensure compliance with halal standards. The certification process includes mandatory employee training conducted by the muftiate, as well as an audit to verify adherence to established standards. Companies that successfully pass the audit receive an official certificate. The halal industry in Kyrgyzstan is steadily gaining momentum. However, several factors continue to hinder its further development, including insufficient legislative support, an underdeveloped halal infrastructure, and a shortage of qualified personnel. To address these challenges, a set of tasks and measures has been outlined in the Concept for the Development of the Halal Industry in the Kyrgyz Republic for 2023–2027.

Nowadays, certification is obtained for a wide range of products, including food products, food additives, feed, soap and detergents, catering services, packaging, and even the tourism industry.

In recent years, Kyrgyzstan has remained dependent on imports of agricultural and food products. In this sector, efforts to modernize enterprises through innovative technologies, diversify production, and increase the competitiveness of domestic goods have not yet produced positive results. This situation is reflected in a decline in the country's export volumes.

In Tajikistan, the halal industry is in its initial stage of development. In 2024, 60 enterprises operating in the halal category were registered (Lestari et al. 2022).

This indicates the growing interest of businesses and consumers in products that meet Islamic standards. Considering the possibility of creating a legally sound and institutionally supported halal certification system opens new opportunities for local producers. Such a system would help improve product quality and strengthen consumer confidence in halal goods among the Muslim population.

Turkmenistan is also showing growing interest in the halal industry, particularly in halal tourism. In the future, these areas may become national priorities due to rising demand from both international visitors and the local population.

Food products are one of the key halal business sectors. The halal food sector in Central Asia is the largest and most dynamic, demonstrating significant scale and potential for further growth. According to the study "Halal Food Market in Central Asia," in 2024, the market volume of halal food products in Kazakhstan, Uzbekistan, and Kyrgyzstan amounted to about $1.6 billion (National Statistics Committee of the Republic of Uzbekistan 2021).

These figures highlight the rapid development of this industry, which is due to the growing number of manufacturers seeking to certify their products, and the increasing number of consumers looking for products that meet halal standards.

One of the most promising areas of the halal food industry is the processing and production of food products and groceries. However, to fully realize this potential, Central Asia must take several key steps. First, it is essential to conduct thorough research on national food markets to identify consumer demand and preferences. Second, governments should provide targeted support to regional producers of halal goods who supply this market segment. Third, establishing a regional halal certification authority would help strengthen consumer confidence by ensuring that products genuinely meet halal standards. Finally, creating a regional organization to connect Central Asian exporters with importers in target markets would facilitate trade expansion and improve market access.

Simultaneously, it is essential to engage Islamic financial institutions (e.g., the Islamic Development Bank) and development funds from partner countries in collaboration with government authorities to foster the growth of small and medium-sized enterprises in the region.

A major modernization of the packaging industry in the region should be carried out. In accordance with the requirements arising from environmental objectives, it is necessary to introduce environmentally friendly packaging materials and the concept of adequate and attractive, but minimal packaging.

Planners and managers of various transport corridors in the region should also support the idea of making Central Asia a major supplier of halal food. This will help to bring products produced in Central Asia to the world markets at competitive prices.

Another important sector is cosmetics and pharmaceuticals. The cosmetics and pharmaceutical sectors in Central Asia are also exhibiting growth trends, indicating a growing interest in halal products among consumers. It is estimated that the volume of halal cosmetic products in the region in 2021 reached $250 million, underscoring the active participation of local producers in creating and promoting products that meet halal standards. Brands offering natural and safe cosmetics are in high demand, which opens new opportunities for innovative solutions in this area (Research and Markets 2021).

The growing interest in halal pharmaceutical products is also noticeable among local and international customers. Consumers are increasingly conscious about choosing medications and medical products that match their religious and ethical beliefs. This creates a need for the development and certification of halal pharmaceutical products, which can significantly impact the health and well-being of the population. As is known, pharmaceutical companies operating under halal certification, in addition to replenishing the domestic market, strive to gain global market share with their higher-quality, safer products that are in high demand.

Tourism is one more important area. According to the World Tourism Organization, the number of Muslim tourists in the region is expected to grow by 20% over the next five years (Çıkı and Tanrıverdi 2025).

This is a prerequisite for improving the relevant infrastructure for the construction of hotels, restaurants, and tourist facilities according to the halal standard, which has recently begun to receive much attention.

Meanwhile, to realize the potential of halal tourism, it is also necessary to strengthen cooperation between government agencies, travel agencies, and local communities. The development and implementation of programs aimed at attracting Muslim tourists contributes to improving the quality of service and expanding the range of offers, which will positively affect the region's economy. Supporting local businesses and developing local attractions will increase the flow of tourists and create jobs, which will be an important contribution to the sustainable development of Central Asia as a center of halal tourism.

Despite the positive trends in the halal business in Central Asia, the industry faces several significant challenges. The main problems are the lack of clear Halal standards and certification, which makes it difficult for new companies to enter the market and reduces consumer confidence. Additionally, the lack of awareness of halal products and services among most consumers affects the pace of their consumption and distribution, requiring additional efforts from businesses for educational initiatives and marketing.

Despite these challenges, there are prospects for the development of halal businesses in Central Asia, driven by growing demand for halal goods and services, investment in infrastructure improvements, government support and partnerships with the private sector, the development of educational programs and certification initiatives, and other factors.

4 Conclusion

The research shows that the state of halal business in Central Asia is showing positive development. There is potential for further improvement in key sectors such as food, pharmaceuticals, and tourism, driven by growing demand. However, much work remains to be done, primarily through government regulation and support, the development and implementation of clear certification standards to ensure high-quality goods and services, and the establishment of effective collaboration between government agencies and the private sector to further advance the halal industry.

References

Aripov NT, Abdiev MZ, Baharuddin AS, Ubaidullaev MB (2024) Halal industry in Kyrgyzstan: current state and development prospects. Malays J Syariah Law 12(3):790–800. https://doi.org/10.33102/mjsl.vol12no3.874

Çıkı KD, Tanrıverdi H (2025) Understanding Muslim tourist behavior in non-Muslim countries through the stimulus–organism–response framework. J Islam Mark 16(11):3394–3412. https://doi.org/10.1108/JIMA-08-2024-0378

Halal Industry Development Center in Kyrgyzstan (2025) Research of the halal product market in Kyrgyzstan. Retrieved from https://www.halal.kg/. Accessed 2 Dec 2025

International Halal Certification Center (2025) Certification "halal". Retrieved from https://halal-certification.ru/. Accessed 2 Dec 2025

Ismanov IN, Ermatov IT (2024) Halal business in Uzbekistan: international experience and national development. Sci J Actuar Financ Account 4(07):8–14. Retrieved from https://finance.tsue.uz/index.php/afa/article/view/414/441. Accessed 2 Dec 2025

Lestari F, Mawardi M, Yola M, Muda Y, Abdul Hamid AB (2022) A model for assessment of halal good manufacturing practice in meat industry. Prod Manuf Res 10(1):666–695. https://doi.org/10.1080/21693277.2022.2117742

Ministry of National Economy of the Republic of Kazakhstan (2025) Report on the halal industry. Retrieved from https://forumspb.com/news/news/industrija-haljal-perspektivy-i-dostizhenija/. Accessed 2 Dec 2025

National Statistics Committee of the Republic of Uzbekistan (2021) Agency for export-import operations of Uzbekistan: statistical report. Retrieved from https://stat.uz/images/uploads/reliz2021/tashqi_savdo_ru_iyun.pdf. Accessed 2 Dec 2025

Packaging Information Portal "Upakovano" (2011) Halal steps around the world. Retrieved from http://www.upakovano.ru/analytics/4041. Accessed 2 Dec 2025

Research and Markets (2021) Global halal cosmetics market report 2021. Retrieved from https://www.prnewswire.com/news-releases/global-halal-cosmetics-market-report-2021-301279273.html. Accessed 2 Dec 2025

Research and Markets (2024) Global halal food market report 2021–24. Retrieved from https://www.researchandmarkets.com/reports/5437876/global-halal-food-market-report. Accessed 2 Dec 2025

Russian Muftis Council (2017) Features of production of "halal" on the Russian meat and poultry processing enterprises [report]. Retrieved from https://muslim.ru/articles/285/8751/. Accessed 2 Dec 2025

UZ24 Information Portal (2021) Uzbek company plans to export up to 30 tons of chicken products to Tajikistan for KFC chain. Retrieved from https://uz24.uz/ru/articles/saxovat-i-kfc. Accessed 2 Dec 2025

Uzbek Agency for Technical Regulation. (2018) Halal certification in Uzbekistan for unhindered access to Malaysian market. Retrieved from https://www.standart.uz/ru/news/view?id=1606. Accessed 2 Dec 2025

Uzbek Agency for Technical Regulation (n.d.) Information about Uzbek Institute of Standards of the Uzbek Agency for Technical Regulation (Institute of Standards). Retrieved from https://www.standart.uz/ru/map/view?id=15. Accessed 2 Dec 2025

Uzenbaev RA, Mardalieva L, Abdiev MJ, Umarov ST, Ergeshov KA (2019) Prospects for development of Kyrgyzstan's food market in the conditions of integration in the EAEU. In: Popkova E (ed) Ubiquitous computing and the internet of things: prerequisites for the development of ICT, pp 859–869. Springer, Cham, Switzerland. https://doi.org/10.1007/978-3-030-13397-9_88

Biological Productivity and Nutrient Cycling in Soils of Kyrgyzstan

Roza T. Orozakunova, Zulfiia I. Sakbaeva, Sovetbek A. Mamytkanov, Zhamila T. Pirmatova, and Uranbek A. Shergaziev

Abstract The research presents the results of a study of the biological productivity and nutrient cycling in two provinces of Kyrgyzstan that differ in their soil and climatic conditions, which consequently affect the biological productivity and nutrient turnover within the soil–plant system. Eastern Issyk-Kul (subboreal belt) is characterized by sufficient moisture and warmth, which favor robust root development and humus formation processes. In mountainous "paratypes" of soils, the total plant biomass accumulates more than in mountain-valley soils. The underground root fraction is dominated by live roots and humified residues, while annual litterfall increases with elevation. In the subtropical belt, in the Kok-Art River basin of the Jalal-Abad region, due to specific climatic conditions, biological productivity, especially the accumulation of underground biomass, is significantly greater than in the subboreal belt. Mountain soils are more susceptible to erosion processes because they are located on slopes of varying steepness and are influenced by numerous factors. One of the main factors protecting the soil from erosion is vegetation cover and the accumulation of root mass in the humus-accumulative horizon. Studies of the biological and ecological properties of soils have shown that soils in the Kok-Art River basin have varying potential fertility, determined by reserves of humus, nitrogen, and other elements.

Keywords Soil · Provinces · Plant biological productivity · Aboveground and underground plant biomass · Plant litter · Nutrient cycling · Soil fertility

JEL Classification Q1 · Q16 · Q54

R. T. Orozakunova (✉) · S. A. Mamytkanov · U. A. Shergaziev
Kyrgyz National Agrarian University Named After K.I. Skryabin, Bishkek, Kyrgyzstan
e-mail: orozakunovaroza@mail.ru

Z. I. Sakbaeva
Jalal-Abad State University Named After B. Osmonov, Manas, Kyrgyzstan

Z. T. Pirmatova
International University Named After K.Sh. Toktomamatov, Manas, Kyrgyzstan

A. B. Karbekova (ed.), *Sustainable Development of Rural Territories Based on Climate-Smart Technologies*, Climate Change Management,
https://doi.org/10.1007/978-3-032-20106-5_17

1 Introduction

In the mountainous regions of Kyrgyzstan, plant biomass development follows the laws of vertical zonation. This is reflected in the successive change of vegetation across altitudinal belts, from foothills to alpine highlands, ranging from lush subalpine, alpine steppes, and meadow-steppes to mountain forests, high-mountain wastelands, and tundra-like dwarf shrub formations. This diversity defines the types of soil formation.

Studies on the biological productivity of higher vegetation in Kyrgyz soils began in the mid-twentieth century. The works of Mamytov (1963, pp. 38–57), Blagobrazov (1966), Fedichkin (1971), and others characterize plant community productivity within vertical zonation systems. According to these researchers, the substantial variability in aboveground yield and root biomass makes it difficult to identify clear patterns of biomass formation in relation to specific genetic soil types or subtypes.

In subsequent stages (Karabaev 2000; Mamytov 1987; Mamytov et al. 1978, 1980; Voronov et al. 1981a, b), studies were conducted under semi-stationary conditions focusing on biomass composition, annual litterfall, and rates of mineralization and humification. S. I. Voronov also notes that the amount of biomass produced by different soil types varies significantly (Voronov 1986, pp. 105–113).

The founder of scientific soil science, Dokuchaev (1899), formulated key theories about the influence of living organisms on the Earth's crust and their role in soil formation. The roots and above-ground biomass of plants together produce organic litter, which serves as a source of energy-rich material for soil microbiological and biochemical processes, ultimately leading to the formation of humus substances.

2 Methodology

According to the general scheme of soil-geographical zoning, the soil and bioclimatic belts of Kyrgyzstan include three provinces: the mountainous Western Tien Shan subtropical belt (moderately warm), the Northern Tien Shan subboreal belt, and the Central Tien Shan boreal belt (Mamytov 1963; Roychenko 1960). The research objects are the mountain-valley and mountain soils of Eastern Issyk-Kul in the Central Tien Shan boreal belt and the mountain soils of the Kok-Art River basin in the Western Tien Shan subtropical belt. These provinces differ in key climatic parameters and conditions for plant biomass accumulation.

The Central Tien Shan province is a mountainous system distinguished by its high elevation and geographic isolation, which protects it from the penetration of moist and warm winds. As a result, this province is characterized by a harsh, dry, and sharply continental climate. Within this province, the Issyk-Kul soil subprovince stands out for its well-defined horizontal and vertical zonation and specific climatic features in the Eastern Issyk-Kul region. The climatic conditions of Eastern Issyk-Kul, due to the presence of Lake Issyk-Kul, are moderated and temperate, with significant moisture levels. Annual precipitation ranges from 400 to 500 (up to 600) mm. Even seasonal distribution of precipitation promotes the growth of lush grass cover, under which chestnut and chernozem soils develop (Jalilova et al. 2024).

In the mountainous Western Tien Shan province, the Fergana Valley is open to warm western air currents from Turan, giving the region characteristics of a dry subtropical and continental climate specific to the northern continental sector of the subtropical belt. A distinctive feature of the climate in the foothill zone is the contrast in moisture availability: a sharp transition from a moist autumn–winter-spring period to a hot and dry summer, where semi-desert ephemeral vegetation develops on sierozem soils. The vegetation cover is closely linked to hydrothermal conditions. Large-grain semi-savanna vegetation predominates in the low mountain zone. In the mid-mountain zone, where precipitation is higher, tall-grass shrub savannas are widespread, and typical mountain brown soils develop. The forest belt consists of juniper stands and unique walnut–fruit forests, under which characteristic black-brown soils form. Field and laboratory work are conducted using methodologies officially approved in the Kyrgyz Republic (Grishina and Samoilova 1971).

3 Results and Discussion

The soil and climatic conditions of Eastern Issyk-Kul, representing the Northern Tien Shan subboreal belt, are characterized by sufficient moisture and warmth, which favor the robust development of plant root systems and humus formation processes (Orozakunova and Baibagyshov 2019). The total biomass of higher vegetation in mountain-valley soils averages between 201.56 and 243.23 centners per hectare. In mountain soil "paratypes," it ranges from 206.88 to 316.40 centners per hectare, exceeding by up to 23%. The dominant vegetation consists of sainfoin fields cultivated by farmers growing fodder crops for livestock development in the region. The above-ground biomass in mountain-valley soils averages between 105.0 and 107.0 centners per hectare, with some farmers achieving yields of up to 150.0–160.0 centners per hectare depending on their agricultural skills (Table 1).

The primary mass of roots is concentrated in the 0–25 cm soil layer. Under otherwise equal conditions, the distribution of roots within the soil profile is strongly influenced by soil aridity. In sainfoin crops, the largest share of biomass is concentrated in the underground root system. Root biomass ranges from 96.56 to 137.23 centners per hectare in mountain-valley soils and from 186.69 to 263.0 centners per hectare in mountain "paratype" soils. This demonstrates the high productivity of underground biomass in the latter, where moisture availability plays a crucial role. In mountain meadow-steppe subalpine soils, underground biomass is slightly lower due to climatic factors and amounts to 179.9 centners per hectare.

The main portion of underground root biomass in mountain-valley soils consists of living and partially humified root residues (Table 1 and Fig. 1), indicating favorable nutrient conditions and optimal physical properties of the region's soils.

Table 1 Composition of the biological mass of higher vegetation (sainfoin) on mountain-valley soils of eastern Issyk-Kul

Depth, cm	Total phytomass (c/ha)	Including (c/ha)		Root mass from total (c/ha)			Annual litterfall (c/ha)	
		Aboveground	Underground	Live	Dead and partially decomposed	Humified residues	Total	Including root
Mountain-valley light chestnut soils (irrigated)								
B-01 0–25	239.0	76.0	163.0	61.37	41.61	60.01	96.46	20.46
B-02 0–25	232.48	140.0	92.48	18.40	35.52	39.2	146.13	6.13
B-07 0–25	188.96	80.0	118.96	53.28	32.48	33.2	97.76	17.76
B-08 0–25	312.48	128.0	184.48	41.28	40.16	103.04	141.76	13.76
Average	243.23 100	106.0 43.6	137.23 56.4	43.5 17.9	37.4 15.4	56.3 23.1	120.53	14.52
Mountain-valley dark chestnut soils (irrigated)								
B-04 0–25	228.40	138.0	90.40	37.12	20.0	33.93	150.37	12.37
B-09 0–25	286.04	124.0	162.24	31.04	48.96	82.24	134.35	10.35
B-11 0–25	152.80	60.0	92.80	17.6	35.2	40.0	65.87	5.87
B-12 0–25	184.56	106.0	78.56	32.64	29.12	16.8	116.88	10.88
Average	212.95	107.0	106.0	37.12	33.32	43.24	116.87	9.87
Mountain-valley low-humus chernozem soils (irrigated)								

(continued)

Table 1 (continued)

Depth, cm	Total phytomass (c/ha)	Including (c/ha)		Root mass from total (c/ha)			Annual litterfall (c/ha)	
		Aboveground	Underground	Live	Dead and partially decomposed	Humified residues	Total	Including root
B-05 0–25	172.80	88.0	84.80	22.72	29.76	32.32	95.57	7.57
B-10 0–25	230.32	122.0	108.32	30.4	28.0	49.92	132.13	10.13
Average	201.56	105.0	96.56	26.56	28.88	41.12	113.85	8.85
Mountain-valley medium-humus chernozem soils (irrigated)								
B-06–20	209.16 100	94.0 44.9	115.16 55.1	41.44 19.8	32.16 15.4	41.56 19.9	107.81	13.81

Source Developed by the authors

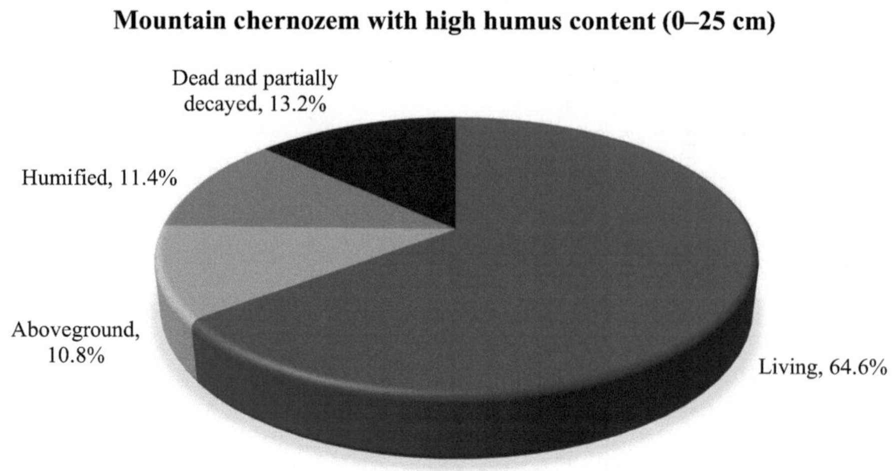

Fig. 1 Plant biomass of mountainous humus chernozem. *Source* Developed by the authors

Despite the quantitative reduction of fractions in mountain-valley light and dark chestnut soils, sainfoin crops contribute to the gradual restoration of the soil, creating conditions for the accumulation of humified root residues, from 43.24 to 56.36 c/ha. In the composition of the biomass of mountain soils, the underground mass increases from mountain light chestnut soils to mountain chernozems from 186.69 to 263.0 c/ha, and from the total reserves of roots, the largest part is made up of living and humified residues, from 102.56 to 158.09 c/ha and from 49.56 to 62.63 c/ha, respectively. In the meadow-steppe subalpine soil, the composition of the underground biomass consists mainly of living and dead, semi-decomposed fractions.

Annual litter consists of the sum of aboveground and underground root litter. There is a pattern of increasing annual litter from mountain light chestnut soils to chernozems, then decreasing in their meadow-steppe subalpine soils. Thus, the total annual litter increases when moving up the slope.

Relatively small masses of dead semi-decomposed residues in the soils of the Eastern Issyk-Kul region indicate optimal oxidation processes in these soils (FAO 2022).

As a result of the conducted research, it can be noted that in the soils of the Eastern Issyk-Kul region, the accumulation of underground plant biomass indicates optimal conditions for the formation of humified residues, which contribute to the restoration of soil fertility in this region.

The Kok-Art River basin, where studies of biological productivity were also carried out, belongs to the mountainous Western Tien Shan subtropical belt. The energy of the sunbeam concentrated on the Earth's surface of the Kok-Art River basin has a significant impact on the process of soil formation through chemical transformations of organic matter in the corresponding environmental conditions (Sakbaeva et al. 2025).

Soil represents one of the primary habitats for living matter, the predominant component of which is plant roots. Root systems play a crucial role in soil formation and in the development of soil fertility. They are central to the accumulation of organic matter, the biological redistribution of nutrients, the formation of soil structure, and the regulation of microbial activity.

Slope aspect significantly influences the hydrothermal regime, the accumulation of litterfall and forest floor layers, as well as the composition and nature of soil organic matter. Despite significant differences in slope exposure, mountain soils are characterized by specific features in terms of biological productivity.

In the southern part of Kyrgyzstan, the majority of precipitation falls during the winter-spring period, while summers are characterized by a lack of moisture. The results of our research (Sakbaeva 2013; Sakbaeva et al. 2013) show that mountain soils in the Kok-Art River basin accumulate relatively high reserves of phytomass. Dark sierozem soils on foothill pastures produce 32.92 tons/ha, sierozem soils formed under pistachio open woodlands—22.80 tons/ha, meadow-steppe subalpine soils—32.59 tons/ha, mountain brown soils—42.19 tons/ha and mountain forest black-brown soils of walnut-fruit forests—62.37 tons/ha of total phytomass.

The data obtained indicate that the enrichment of soils with organic matter occurs primarily through the root systems of plants, as the root biomass significantly exceeds the above-ground biomass.

According to the data, in mountain soils, most roots are concentrated in the upper 0–25 cm layer, where they account for 71.2–92.5% of the total root mass. At a depth of 25–50 cm, their amount decreases sharply, which is a significant characteristic feature of root distribution under mountainous conditions (Table 2).

Table 2 Number of roots in the main soil types of the Kok-Art River basin

Soil types	Depth (cm)	Roots (c/ha)	Root mass from total (c/ha)	Total (c/ha)
Mountain forest black-brown soils of walnut-fruit forests	0–25	403.2	92.5	553.3
	25–50	150.1	40.5	
Mountain brown soils	0–25	276.0	83.8	389.5
	25–50	113.5	31.2	
Pasture lands of the foothills, dark sierozem soils	0–25	212.0	79.1	301.0
	25–50	89.0	28.6	
Pistachio woodlands, typical sierozems	0–25	120.0	71.2	134.0
	25–50	24.0	11.9	
Summer pastures on mountain meadow-steppe soils	0–25	252.0	79.8	305.9
	25–50	53.9	20.2	

Source Developed by the authors

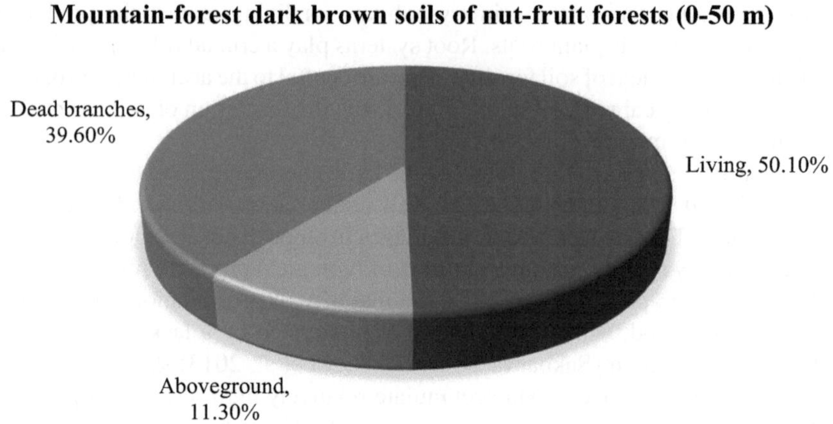

Fig. 2 Total biomass of mountain forest black-brown soils of walnut-fruit forests. *Source* Developed by the authors

The main portion of the biomass of higher vegetation is represented by roots, which are located in the 0–50 cm soil layer and make up 90%–95% of the total biomass reserve. In this layer, root stocks amount to 55.3 t/ha in mountain-forest dark brown soils of walnut-fruit forests, 38.9 t/ha in mountain brown soils, 30.1 t/ha in dark sierozems; 14.4 t/ha in pistachio open woodlands with typical sierozems, and 30.6 t/ha in mountain meadow-steppe subalpine soils.

The biological productivity of mountain-forest dark brown soils in walnut-fruit forests (623.7 c/ha) exceeds that of mountain-valley soils (329.2 c/ha), and even more so that of plain sierozems. The share of living roots in mountain-forest dark brown soils of walnut-fruit forests makes up 50.10% of their total amount (Fig. 2).

In the mountain-forest dark brown soils of nut-fruit forests, due to large reserves of biomass, especially root biomass, the greatest amount of concentrated chemical elements is retained and involved in the small biological cycle of substances, thereby creating conditions for their participation in the large geological cycle.

4 Conclusions

In the plant biomass of Eastern Priiisyk-Kul, within the mountain-valley and mountainous "paratypes" of soils, the main share is accounted for by underground roots. In mountain soils, their proportion exceeds 52.0%, which indicates the high productivity of these soils.

In the composition of underground root biomass, living and humified residues dominate, which indicates the best nutritional regimes and optimal physical soil conditions for the distribution of these soils in the region.

There is a consistent pattern of increasing annual litterfall from mountain light chestnut soils to chernozems of Issyk-Kul in Northern Kyrgyzstan, and from typical sierozems to mountain-forest dark brown soils in the walnut-fruit forests of the Kok-Art River basin in southern Kyrgyzstan, followed by a decrease in litterfall on the meadow-steppe subalpine soils, annual litterfall increases as one moves up the slope until the meadow-steppe subalpine soils are reached.

Soils of the Kok-Art River basin of Jalal-Abad region of Kyrgyzstan have unequal potential fertility, which is determined by the reserves of humus, nitrogen, and other elements. Simultaneously, biological productivity is reflected in the increasing root biomass of grass vegetation, ranging from 30.1 tons per hectare in mountain dark sierozems to 55.3 tons per hectare in mountain-forest dark brown soils of walnut-fruit forests.

References

Blagobrazov VA (1966) Biomass of some meadow and steppe plant communities of the Tien Shan. In: Blagobrazov VA, Mikhailov IS (eds) Materials on the biogeography of the Issyk-Kul Basin. ILIMPP, Frunze, Kirghiz SSR, pp 15–31

Dokuchaev VV (1899) To the study of natural zones. St. Petersburg, Russian Empire

FAO (2022) Global status of black soils. FAO, Rome, Italy. https://doi.org/10.4060/cc3124en

Fedichkin GY (1971) The role of root biomass in preventing soil erosion in the mountains of Kyrgyzstan. Biol Sci 3:181–200

Grishina LA, Samoilova EM (1971) Biomass accounting and chemical analysis of plants. Moscow University Press, Moscow

Jalilova G, Orozakunova R, Baibagyshev E, Karabaev N, Shergaziev U (2024) Farmers' adaptation to climate change in Southern Issyk-Kul. Ekonomika APK [agricult Econ] 31(4):23–32. https://doi.org/10.32317/ekon.apk/4.2024.23

Karabaev NA (2000) Agrochemical and ecological foundations of fertility and productivity of mountain soils of Kyrgyzstan (Abstract of the Dissertation of Candidate of Agricultural Sciences). Kyrgyz Agricultural Academy, Bishkek, Kyrgyzstan

Mamytov AM (1963) Soil resources and land cadastre issues of the Kyrgyz Republic, 3rd ed. Bishkek, Kirghiz SSR

Mamytov AM (1987) Soils of the Central Tien Shan. Publishing House of the Academy of Sciences of the Kirghiz SSR, Frunze, Kirghiz SSR

Mamytov AM, Voronov SI, Karabaev NA (1978) Biological productivity of higher vegetation on major types of mountain soils. Abstracts of the 4th republican conference of soil scientists (part I). Nauka, Alma-Ata, Kazakh SSR, pp 143–150

Mamytov AM, Voronov SI, Ashirakhmanov ShA, Karabaev NA (1980) Biological productivity of higher vegetation on the main soil types of Kyrgyzstan. In Akimaliev JA (ed) Peculiarities of soil formation and properties of mountain soils of the Tien Shan. ILIM, Frunze, Kirghiz SSR, pp 203–215.

Orozakunova RT, Baibagyshov EM (2019) Chernozem soils of Kyrgyzstan. In: Moşoi I (ed) Proceedings of the international scientific conference "Eastern European chernozems-140 years after V. Dokuchaev". Nicolae Dimo Institute of Pedology, Agrochemistry and Soil Protection "Nicolae Dimo", Chisinau, Republic of Moldova, pp 219–223

Roychenko GI (1960) Soils of Southern Kyrgyzstan. Academy of Sciences of the Kirghiz SSR, Frunze, Kirghiz SSR

Sakbaeva ZI (2013) Biological productivity and nutrient cycling in soils of the Kok-Art River basin of the Jalal-Abad Region. Soil Sci Agrichem 4:82–87

Sakbaeva ZI, Nurdinov ShSh, Kalaeva GS, Pirmatova ZhT, Abdykalykov ZhA (2025) Ecological condition of walnut forests during the development of green economy. Int J Minor Fruits Med Arom Plants 11(1):297–300. https://doi.org/10.53552/ijmfmap.11.1.2025.297-300

Sakbaeva Z, Schroetter S, Karabaev NA, Avazov A, Rogasik J, Schnug E (2013) Soils of nut-fruit forests in southern Kyrgyzstan—important ecosystems worthy of protection. Landbauforschung—Appl Agric Forest Res 63(1):93–102. https://doi.org/10.3220/LBF_2013_93-102

Voronov SI (1986) Microbiological and biochemical features of mountain soils of Kyrgyzstan. ILIM, Frunze, Kirghiz SSR

Voronov SI, Mamytova BA, Umetov AU (1981a) Relationship between biological productivity of higher vegetation and the quality and composition of humus. Proc Kyrgyz Res Inst Soil Sci 13:232–244

Voronov SI, Pak NM, Mamytova BA, Zhanuzakova ZI (1981) Biological productivity and composition of organic matter in the main soils of Kyrgyzstan. In: Tangaeva NG, Voronov SI et al (eds) Current issues in soil science in Kyrgyzstan. ILIM, Frunze, Kirghiz SSR, pp 115–123

Economic Advantages of Using Solar Power Plants in the Conditions of Kyrgyzstan

Ainura M. Zhoroeva, Zhypara Asilbek Kyzy, Ularkan G. Ganybaeva, Kanybek N. Kenzhekulov, and Zhyldyzgul A. Bokoeva

Abstract The research examines the economic feasibility and key benefits of deploying solar power plants within the Kyrgyz Republic. The research explores the country's solar resource potential, evaluates the current stage of sectoral development, and identifies the main economic and institutional factors influencing the expansion of solar energy technologies. The research provides an overview of the natural potential of solar energy, considering regional climatic conditions, the intensity of solar radiation, and the geographical distribution of promising sites for solar installations. The authors analyzed the current state of the renewable energy sector, with particular attention paid to the regulatory framework, existing pilot projects, and investment trends. The research methodology integrates analytical, statistical, and comparative approaches, allowing for the assessment of the cost-effectiveness and payback periods of solar power plants under different economic scenarios. The research identifies key constraints hindering the accelerated development of the sector: high initial capital costs, insufficient development of grid infrastructure, limited access to financing, and the lack of unified long-term state policies. Simultaneously, the results demonstrate that with targeted government support, improved regulatory mechanisms, and the attraction of international investors, solar energy can significantly strengthen the national energy balance. It can reduce dependence on hydropower, ensure greater energy security, lower greenhouse gas emissions, and stimulate the creation of new high-tech jobs.

A. M. Zhoroeva (✉) · K. N. Kenzhekulov
Jalal-Abad State University Named After B. Osmonov, Manas, Kyrgyzstan
e-mail: zoroevaajnura28@gmail.com

Z. A. Kyzy
Osh State University, Osh, Kyrgyzstan

U. G. Ganybaeva
International University Named After K.Sh. Toktomamatov, Manas, Kyrgyzstan

Z. A. Bokoeva
Kyrgyz State Technical University Named After I. Razzakov, Bishkek, Kyrgyzstan

A. B. Karbekova (ed.), *Sustainable Development of Rural Territories Based on Climate-Smart Technologies*, Climate Change Management,
https://doi.org/10.1007/978-3-032-20106-5_18

Keywords Kyrgyz Republic · Solar energy · Renewable energy sources · Economic efficiency · Investments · Sustainable development

JEL Classification O33 · Q42 · Q48

1 Introduction

Over the past decade, the global energy landscape has undergone a steady transformation, largely driven by the growing adoption of renewable energy technologies. These changes are associated with the need to reduce reliance on conventional fuel sources and address the environmental challenges created by traditional energy production. Among the available renewable options, solar energy has emerged as a particularly attractive direction for future development, especially for countries that benefit from abundant sunlight and favorable climatic conditions for photovoltaic generation.

Kyrgyzstan possesses considerable natural conditions for the development of solar energy because many regions in the country receive a high number of sunny days throughout the year. This creates a favorable environment for the deployment of solar power plants and small photovoltaic systems. Despite this potential, the actual contribution of solar energy to the national electricity mix remains very limited. The slow pace of adoption is largely explained by technological constraints, insufficient investment activity, and the absence of long-term financial instruments that would support the growth of the sector.

Kyrgyzstan, with its significant solar energy potential, is gradually developing this sector. According to the International Renewable Energy Agency (IRENA), the average annual number of sunny days in the country is more than 250, which creates favorable conditions for the widespread introduction of solar power plants (SPPs). However, despite the favorable climatic conditions, the share of solar energy in the total energy balance of the country remains insignificant due to several economic, technological, and institutional factors (Ministry of Energy of the Kyrgyz Republic 2023).

The primary objective of this research is to evaluate the economic advantages associated with the introduction of solar energy systems in the Kyrgyz Republic. The analysis focuses on a range of factors that influence the evolution of the sector, including technological, financial, and institutional conditions that determine the pace of renewable energy adoption (National Statistical Committee of the Kyrgyz Republic 2022). The relevance of studying the economic aspects of solar energy in Kyrgyzstan is reinforced by the ongoing challenges in the national energy system, including rising electricity demand, seasonal fluctuations in hydropower generation, and increasing pressure on aging infrastructure. These issues highlight the necessity of identifying alternative and more resilient energy sources capable of ensuring long-term stability. Solar power, given its availability and scalability, represents one of the most viable pathways for addressing these structural imbalances (World Bank 2021).

2 Materials and Methods

The methodological framework of this research is built upon a comprehensive assessment of the economic potential of solar power plants under the conditions of the Kyrgyz Republic. The research relies on a wide set of empirical materials, including statistical information provided by the Ministry of Energy of the Kyrgyz Republic, data from the National Statistical Committee, and analytical reports of international organizations (e.g., the World Bank and the International Renewable Energy Agency (IRENA)). These sources made it possible to obtain reliable indicators characterizing the current state and dynamics of the renewable energy sector. A combination of analytical, statistical, and comparative methods was applied to ensure the validity of the results.

The comparative method was employed to contrast Kyrgyzstan's experience with that of countries where solar energy has already achieved significant progress, enabling the identification of patterns and practices that could be effectively adapted to local conditions. This integrated methodological approach allowed for a balanced and objective assessment of the economic prospects for solar energy deployment in Kyrgyzstan and provided a solid basis for formulating practical recommendations.

Overall, the combination of quantitative and qualitative methods ensured the reliability of the results and enabled the formation of a comprehensive methodological base for analyzing the prospects of solar energy development in the Kyrgyz Republic.

3 Results

Currently, the electricity deficit in Kyrgyzstan has reached 6 billion kWh. Electricity consumption increases annually by 4–6%. During the Soviet era, scientists substantiated that the use of hydropower is efficient for electricity generation. To provide Central Asia with electricity, several hydroelectric power plants (HPPs) were built on the Naryn River. Unfortunately, the capacity of the units at the HPPs, designed for 9 MW, was reduced to 3 MW, resulting in 80% of Kyrgyzstan's energy system being in disrepair. The use of fuel and energy resources leads to global warming, melting of polar ice, and nuclear energy leads to the destruction of the ecological balance (International Renewable Energy Agency (IRENA) 2020; World Bank 2021).

Looking ahead, the continued modernization of solar panel manufacturing, coupled with favorable regulatory policies, can enhance Kyrgyzstan's role in the renewable energy market (Asian Development Bank (ADB) 2020). By taking advantage of these opportunities, Kyrgyzstan can achieve energy independence and become a leading player in the regional renewable energy sector.

Currently, thanks to advanced technologies, the production of solar panels is actively developing, which can be found in the Data Availability section.

In June 2022, the Kyrgyz Republic adopted a new regulatory framework aimed at promoting the development and use of renewable energy technologies. Law No.

49, which addresses the integration of renewable energy sources into the national energy system, outlines several state priorities (Government of the Kyrgyz Republic 2022; Masdar 2023). In early 2023, the Government of the Kyrgyz Republic began implementing a social initiative designed to expand access to renewable energy for vulnerable households (Bishkek Solar Company 2023). Altogether, more than one thousand homes across the country are expected to benefit from this support measure, allowing residents to reduce energy expenses and improve the reliability of electricity supply (United Nations Economic Commission for Europe (UNECE) 2021).

A 300 MW solar power plant is being built in the Issyk-Kul Region on a 500-hectare area in the village of Toru-Aigyr at a cost of $300 million. The project is financed by the Kyrgyz-Russian Development Fund; construction is being carried out by Bishkek Solar, the only limited liability company specializing in solar power generation. According to calculations, the power plant will produce 600 million MW·h of electricity per year (Toktogul Hydropower Plant Directorate 2022).

The President of the Kyrgyz Republic has been focusing on the energy sector for the past two years, planning to use solar energy to turn Kyrgyzstan into an electricity exporter. During a visit to China in 2023, he emphasized that energy is a priority and a promising area for Kyrgyzstan and encouraged Chinese investors to invest in this industry. As a result, agreements on cooperation in power generation were signed.

In the village of Toru-Aigyr in the Issyk-Kul Region, the Kyrgyz-Kazakh enterprise Kүn Bulagy, together with the National Electric Holding Company Taza Energiya, plans to build a solar power plant worth $35 million. It is expected that the plant will generate 90 million kW h of electricity per year; in the long term, its capacity may reach 300 MW h.

Kyrgyzstan has a huge potential for solar energy development due to high solar activity and vast free territories for the location of solar power plants. Thus, the development of solar energy in Kyrgyzstan contributes to the energy security of the country and opens new prospects for sustainable economic growth and environmental protection (International Energy Agency (IEA) 2023).

These values illustrate the substantial difference in scale between residential and industrial-level installations, while also emphasizing the long-term economic advantages derived from reduced operating expenses and stable electricity production. The average annual electricity bill for a household is about 50,000 soms. When switching to solar energy, the investment is recouped within ten years, after which electricity becomes practically free. The service life of solar panels is 25–30 years, which saves more than 1 million soms for the entire operation period (Kyrgyz-Russian Development Fund 2022). Table 1 presents a comparative analysis of solar energy, hydropower, and thermal power plants based on key technical, economic, and environmental parameters.

The comparison presented in Table 1 highlights the advantages and limitations of each energy source within the context of Kyrgyzstan's energy sector.

Service life also varies by technology. Hydropower plants have the longest operational lifespan, often exceeding 50 years. Solar systems typically operate efficiently for 25–30 years, while thermal plants have a service life of 30–40 years.

Table 1 Comparative analysis of solar, hydropower, and thermal energy systems

Parameter	Solar energy	Hydropower	Thermal power plants
Initial costs	High	High	Average
Operating expenses	Low	Average	High (fuel costs)
Environmental friendliness	High (no emissions)	High	Low (high CO_2 emissions)
Service life	25–30 years	50 + years	30–40 years
Payback	10 years	15–20 years	7–10 years old
Energy dependence	Fuel independence	Dependence on water resources	Imports of coal and gas

Source Developed by the authors

Payback periods reflect investment returns: solar power typically offers a 10-year payback, hydropower requires 15–20 years, and thermal plants provide a faster return of 7–10 years due to lower upfront costs (Global Solar Atlas n.d.).

In 2024, the Chinese company SunEnergyCo announced the construction of a 350 MW solar photovoltaic plant in the Batken Region with an investment of $230 million.

Table 2 presents the key economic indicators that characterize the efficiency of using solar power plants under the conditions of the Kyrgyz Republic. It summarizes the main parameters affecting the investment attractiveness and financial feasibility of solar energy deployment both at the household level and in large industrial-scale projects.

One of the most important indicators is the payback period. For households, it ranges from 8 to 11 years, whereas for large-scale projects it is between 7 and 10 years, confirming the economic viability of solar energy deployment. By reducing electricity expenses, a household can save 45,000–55,000 KGS annually; the total economic benefit over the full service life of solar panels reaches 900,000–1,200,000 KGS.

A chart representing the main economic dynamics related to the development of solar power plants in the Kyrgyz Republic for 2018–2023 can be found in the Data Availability section. It displays two quantitative indicators: the volume of electricity produced (GW), shown as vertical columns on the left axis, and the overall investment in solar installations ($ million), represented by a line graph on the right axis. This visualization makes it possible to trace how changes in generation volumes correspond to shifts in capital expenditures during the observed years.

In 2018, the contribution of solar energy amounted to approximately 5 GW. Over the following years, generation levels steadily increased, reaching nearly 28 GW by 2023. This more than fourfold rise highlights the continuous expansion of solar power capacity and the growing role of renewable technologies in the national energy mix.

Table 2 Economic indicators of solar power plant efficiency in Kyrgyzstan

Indicator	Value/range	Description
Initial investment for household SPP	450,000–550,000 KGS	Cost of installing a 3 kW home solar system
Initial investment for 1 MW SPP	$1.1–$1.3 million	Average investment into a utility-scale solar power plant
Payback period (households)	8–11 years	Depends on the consumption level and tariff savings
Payback period (large solar plants)	7–10 years	Depends on solar irradiation and financing conditions
Operational expenses	Very low	No fuel costs; minimal maintenance costs
Lifetime of solar panels	25–30 years	Average period of efficient operation
Annual savings for a household	45,000–55,000 KGS	Reduction in electricity expenditure after installation
Total savings over system lifetime	900,000–1,200,000 KGS	Savings over 25–30 years of use
Contribution to reducing energy imports	High	Replaces imported fossil fuels and reduces pressure on hydropower
Environmental impact	No emissions	Zero CO_2 emissions during electricity generation

Source Compiled by the authors based on data from the Ministry of Energy, IRENA, and the World Bank (International Renewable Energy Agency (IRENA) 2020; Ministry of Energy of the Kyrgyz Republic 2023; World Bank 2021)

By increasing the share of electricity generated from solar power plants, the country can reduce its reliance on imported hydrocarbons and ease the burden on hydropower facilities, especially in periods of water scarcity.

4 Discussion

The development of solar energy in Kyrgyzstan is a promising area that promotes economic growth, energy security, and a reduced carbon footprint. However, the successful implementation of this sector requires addressing several issues.

From an economic point of view, the main challenge remains the high cost of initial investment. Despite the decline in prices for solar panels, the construction of power plants requires significant investment. Attracting international investment and government subsidies can accelerate the development of the sector.

In the technological aspect, it is necessary to improve infrastructure and develop energy storage systems. The lack of local production of solar panels increases dependence on imports, which leads to additional costs. The introduction of advanced technologies, such as perovskite solar cells and battery systems, can improve the efficiency of the industry.

Government policy plays a key role in the development of the industry. The adoption of incentive tariffs, tax incentives, and legal mechanisms to attract investors will create conditions for the active growth of solar energy. It is important to consider the experience of other countries that have already achieved success in this area.

Thus, solar energy has high potential. However, its development requires a comprehensive approach—economic support, technological innovations, and competent regulation.

5 Conclusions

The development of solar energy in Kyrgyzstan is a promising area that contributes to increasing the country's energy independence, improving the environmental situation, and ensuring sustainable economic growth. The analysis showed that the use of solar power plants has several significant advantages, including reducing the cost of electricity production in the long term, reducing carbon dioxide emissions, and increasing the availability of electricity in remote areas. Despite the high potential of solar energy, its development faces some challenges, such as the high cost of initial investment, the need to modernize infrastructure, and insufficient legislative regulation.

The research identified key economic factors influencing the profitability of solar power plants in Kyrgyzstan and possible ways to optimize costs. The analysis shows that the integration of solar energy into the country's energy system requires a comprehensive approach, including government support, financial mechanisms to attract investment, and the introduction of modern technologies. Thus, solar energy has significant development potential in Kyrgyzstan. Effective government policy and active attraction of investment will increase the share of solar generation in the country's overall energy balance, create new economic opportunities, reduce dependence on traditional energy sources, and improve the environmental situation.

Data Availability An example of a solar panel installation in an open area is available at https://figshare.com/ with the identifier https://doi.org/10.6084/m9.figshare.30857711.

The growth in electricity production and the cost of solar power plants in Kyrgyzstan for the period 2018–2023 is presented at https://figshare.com/ with the identifier https://doi.org/10.6084/m9.figshare.30911870.v1.

References

Government of the Kyrgyz Republic (2022) Law "On Renewable Energy Sources" (30 June 2022 No. 49). Bishkek, Kyrgyzstan. Retrieved from https://cbd.minjust.gov.kg/112570. Accessed 11 Dec 2025

Asian Development Bank (ADB) (2020) Solar energy opportunities in Central Asia. Asian Development Bank, Manila, Philippines

Bishkek Solar Company (2023) Project overview: 1 GW solar plant investment proposal. Bishkek, Kyrgyzstan

Global Solar Atlas (n.d.) Solar resource data for Kyrgyzstan. World Bank Group. Retrieved from https://globalsolaratlas.info/download/kyrgyz-republic. Accessed 11 Dec 2025

International Energy Agency (IEA) (2023) Solar PV global market outlook 2023. IEA, Paris, France

International Renewable Energy Agency (IRENA) (2020) Renewable energy prospects: Central Asia. IRENA, Abu Dhabi, UAE

Kyrgyz-Russian Development Fund (2022) Annual investment briefing. Bishkek, Kyrgyzstan

Masdar (2023) Masdar and the Kyrgyz Republic sign agreement on renewable energy cooperation. Official press release. Retrieved from https://masdar.ae/en/news/newsroom/masdar-and-edf-sign-major-agreement-to-develop. Accessed 11 Dec 2025

Ministry of Energy of the Kyrgyz Republic (2023) Energy development report 2023. Bishkek, Kyrgyzstan

National Statistical Committee of the Kyrgyz Republic (2022) Energy production and consumption bulletin. Bishkek, Kyrgyzstan

Toktogul Hydropower Plant Directorate (2022) Technical report on reservoir-based solar installations. Bishkek, Kyrgyzstan

United Nations Development Programme (UNDP) (2021) Sustainable energy solutions for Kyrgyzstan. UNDP, Bishkek, Kyrgyzstan

United Nations Economic Commission for Europe (UNECE) (2021) Energy security in Central Asia. UNECE, Geneva, Switzerland

World Bank (2021) Kyrgyz Republic energy sector diagnostic. World Bank Group, Washington, DC

Sustainable Development of the Tourism Business in Rural Areas of Kyrgyzstan: Trends and Priorities in the Use of Climate-“Smart” Technologies

Kobogon A. Atyshov, Antonina V. Babosiuk, Altynbek K. Ergeshov, and Anar K. Atyshova

Abstract This research highlights that Kyrgyzstan possesses considerable potential for tourism development, owing to its mountainous landscapes, recreational resources, and historical-cultural heritage. The country’s tourism sector is structured in accordance with altitudinal zoning: resorts and recreational facilities are concentrated in the mid-mountain region. Ecotourism along the route of the Great Silk Road is also situated within this belt, while high-altitude zones accommodate ski tourism, hunting, and resource-based tourism. The national strategy envisions development precisely along these three directions. As a solution, the authors propose introducing a cluster-based approach and a two-stage development model. This approach should focus on existing, low-cost yet profitable areas such as mountaineering, jailoo tourism, and hiking. An econometric review of the empirical experience of green rural tourism in Kyrgyzstan was conducted for the ten-year period from 2014 to 2023 using trend and correlation analysis methods. The research substantiates a neutral development trend in green rural tourism in Kyrgyzstan, with its level remaining relatively stable throughout the considered decade. Quantitative characteristics of the relationships between the sustainability of rural development, supported by climate-smart technologies, and the implementation of various types of tourism in Kyrgyzstan are presented. The theoretical contribution of the identified relationships lies in their formation of a systemic understanding of how tourist flows in Kyrgyzstan influence the sustainable development of rural areas, both organizationally and geographically. The practical contribution is determined by the justification of key priorities: the development of unorganized tourism and tourism from CIS countries, because these types of tourism exhibit the strongest correlation with the sustainable development of rural territories in Kyrgyzstan.

K. A. Atyshov (✉) · A. V. Babosiuk · A. K. Ergeshov · A. K. Atyshova
Research University “Kyrgyz Economic University named after M. Ryskulbekov”, Bishkek, Kyrgyzstan
e-mail: atyshov.keu@mail.ru

A. B. Karbekova (ed.), *Sustainable Development of Rural Territories Based on Climate-Smart Technologies*, Climate Change Management,
https://doi.org/10.1007/978-3-032-20106-5_19

Keywords Cluster · Trend · Main direction · Personnel · Infrastructure · Innovation · "Green" rural tourism · Sustainable rural territories · Climate-"smart" technologies · Kyrgyzstan

JEL Classification C38 · L83 · O18 · O32

1 Introduction

The tourism sector has become one of the largest industries in the global economy. According to the United Nations World Tourism Organization, approximately 1.4 billion international trips were recorded in 2024. This represents an increase of 11%, or 140 million, compared to 2023, indicating that the global industry has recovered to 99% of its pre-pandemic level (prior to COVID-19) (Tadviser 2025).

Within this system, the tourism business stands out for its substantial contribution to employment, income growth, and the resolution of socio-economic and environmental challenges. It plays a key role by generating economic benefits and creating jobs within the tourism sector and across many related sectors. This industry is characterized by its high level of dynamism, creates opportunities for small and medium-sized enterprises, and encourages entrepreneurial innovation.

In business, especially in tourism, success depends not only on the ability to adapt to change but also on the capacity to shape future conditions by anticipating how production systems in the sector will evolve. Achieving this requires a more rational approach to both economic structures and management practices.

Tourism is inherently connected with innovation from the creation of services to the implementation of technologies and management practices. The need to meet diverse demands in this field underscores the importance of entrepreneurial activity. Given Kyrgyzstan's untapped tourism potential, capital investment in this sector is critically important.

The tourism business in the republic is an evolving system where new market forms and solutions to problems are continuously sought. Its development contributes to economic openness and societal harmonization. Tourism activity differs fundamentally from many other forms of business. Its managerial principles extend beyond classical approaches, requiring creativity and sector-specific solutions for this unique field.

2 Methodology

This research is grounded in the theoretical framework of the concept of green rural tourism, whose scientific and economic foundations are presented in the works of Ding and Jiang (2025) and Ristanović et al. (2025), and consist of the following propositions:

- Rural tourism is gaining popularity worldwide due to its environmental sustainability (Ekua and Lee 2025; Hosen et al. 2023)
- The development of rural tourism is facilitated by the introduction of climate-smart technologies in rural areas (Li and Wang 2025; Liu et al. 2025)
- There are two main classifications of modern tourism (i.e., the structuring of tourist flows): (1) an organizational classification distinguishing between unorganized (Akylbekova et al. 2024) and organized (Tolonov et al. 2023) tourism; (2) a geographical classification, which in the Kyrgyz Republic includes tourism from distant foreign countries (Atyshov et al. 2025), tourism from CIS countries (Akbar et al. 2024), and domestic tourism (Li et al. 2024).

The literature review shows that, despite the robustness of its theoretical foundations, the empirical experience of green rural tourism in Kyrgyzstan remains understudied. Particularly, a gap has been identified concerning the development trends of green rural tourism in Kyrgyzstan and the relationship between the sustainability of rural development supported by climate-smart technologies and the implementation of different types of tourism in the country. The present research aims to fill this gap.

The methodology of this research is based on a systematic analysis of economic statistics. Trends and assess key economic indicators of the tourism business are identified. Using trend analysis, the research identifies the ten-year trajectory of changes in the degree of sustainability of rural development in Kyrgyzstan, supported by climate-smart technologies, over the period 2014–2023. The statistical indicator of this sustainability is the "Sustainable land use" index (measured on a scale from 1 to 100), taken from the materials of the Global Green Growth Institute (Global Green Growth Institute (GGGI) 2025).

To quantify (in percentage terms) the relationship between tourist flows and the sustainable development of rural areas in Kyrgyzstan supported by climate-smart technologies during 2014–2023, the research evaluates the correlation between these flows (the total flow, its organizational structure, and its geographical structure) and the "Sustainable land use" index. Positive correlation coefficients indicate a positive relationship.

An economic assessment of the republic's natural and recreational potential, along with the current state of its tourism sector, was conducted. To achieve a detailed sectoral segmentation and clearly define three contrasting altitudinal zones for specialized tourism, systems-based research methods were employed. The analysis drew on publicly available scientific and economic literature, as well as official statistical data.

3 Results

The active development of the tourism business has stimulated the emergence of a significant number of academic studies, which have given rise to several specialized areas. They include the theoretical and methodological foundations of tourism

(Ordzhonikidze 2011; Vukolov 2018), the analysis of the competitive environment (Zhidkov et al. 2012), quality management (Baumgarten 2010), innovation implementation (Malakhova and Ushakov 2010; Yakimovich 2011), and project management in the field of tourism (Birzhakov et al. 2014).

Simultaneously, aspects directly related to the development of business activity within the tourism complex remain insufficiently explored. This determines the need for intensifying and expanding academic inquiry in this field. This is particularly important, as a persistent misconception among practicing professionals reduces their primary function to routine document processing.

In reality, a successful entrepreneur in the tourism industry must possess deep professional competencies in tourism and related academic disciplines. In this context, a key role is assigned to such areas as marketing research and logistics.

As research shows, the primary centers of tourism development are emerging in the regions and major cities. Their advantage lies in the concentration of consumers and a developed urban environment, including well-developed districts and retail or recreational zones, which generates demand for tourism services. Consequently, the most effective strategy for tourism businesses is to operate in densely populated or resort areas with favorable natural conditions. This enables the successful development of core tourism segments and, in turn, enhances the financial performance of the sector.

A scientific approach to the utilization of recreational resources and the assessment of their impact on tourism contributes to the proper development of the sector while avoiding environmental issues (Atyshov 1992). The economic assessment of regional potential enables the efficient placement of tourism clusters, the construction of infrastructure, and the management of environmental load in high-demand tourist areas (Government of Kyrgyzstan 2013).

Since Kyrgyzstan gained independence, significant changes have occurred in this sector, driven by economic transformations that have led to the formation of a complex functioning under market conditions. Tourism, including inbound and domestic tourism, is developing dynamically and serves as an important source of government revenues. Thus, over the five-year period, the unorganized segment developed at a significantly higher rate.

Despite this, the organized sector plays a key role in the tourism industry. In 2023, it accounted for 63% (1805.2 thousand tourists) of the total tourist flow. Its appeal to travelers has been steadily increasing (Table 1).

According to statistical data, the development of the service sector within the tourism complex has had a significant positive impact on the economy. In 2023, the export of tourism services generated $994.4 million; the number of visitors to the country reached 2.86 million people. Particularly notable growth was recorded in the services of travel companies, the volume of which amounted to 4008.9 million soms.

The trends in the sustainable development of rural areas in Kyrgyzstan supported by climate-smart technologies over the ten-year period from 2014 to 2023 are

Table 1 The number of tourists in Kyrgyzstan growth, thousand persons

Year	Total tourists served	Unorganized sector	Organized sector	of which		
				Long-Haul abroad	CIS countries	Domestic tourism
2014	1132.2	460.6	671.6	43.1	62.3	566.3
2015	1245.0	547.0	698.0	43.0	89.3	565.6
2016	1265.1	558.7	706.4	54.0	90.5	562.2
2017	1273.2	645.6	627.6	66.0	66.2	496.0
2018	1375.1	647.3	727.8	65.8	93	568.5
2019	1380.4	630.5	750.0	86.5	98.6	564.8
2020	1778.90	809.6	969.3	73	94	643
2021	463.9	209.0	254.9	70.349	12.349	236.876
2022	1148.2	1 133.9	837.0	84.4	139.3	924.5
2023	2864.2	1 059.0	1805.2	161.3	257.0	1414.2

Source Developed by the authors based on National Statistical Committee of the Kyrgyz Republic (2024)

presented in Fig. 1. The statistical indicator of this development is the "Sustainable land use" index (measured on a scale from 1 to 100), calculated by the Global Green Growth Institute (Global Green Growth Institute (GGGI) 2025).

According to Fig. 1, a ten-year trend of a slight decline in the sustainability of rural development in Kyrgyzstan supported by climate-smart technologies has been identified: a decrease of 0.16%, from 63.09 points in 2014 to 62.99 points in 2023. The relationship between the activities of the tourism business, broken down by its

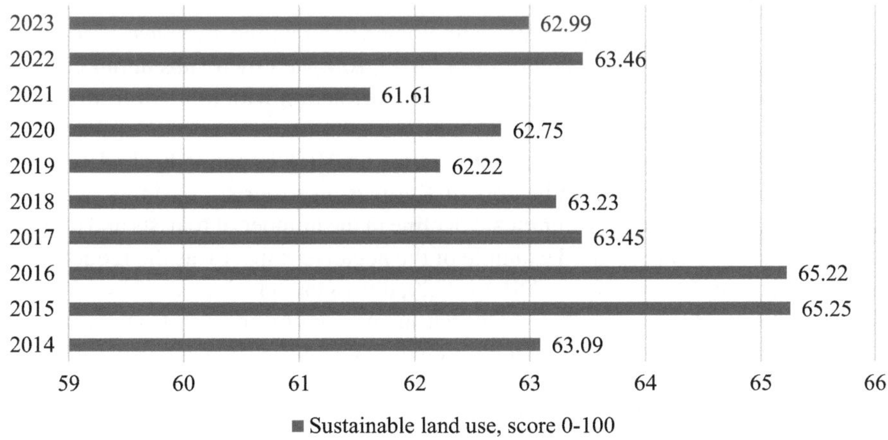

Fig. 1 Sustainable development of rural areas in Kyrgyzstan supported by climate-smart technologies, 2014–2023, points 0–100. *Source* Developed by the authors based on Global Green Growth Institute (GGGI) (2025)

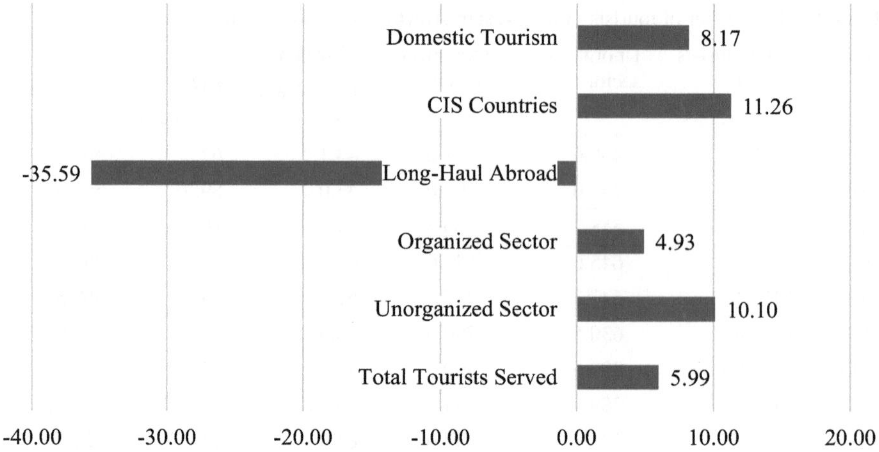

Fig. 2 Correlation between tourist flows and sustainable development of rural areas in Kyrgyzstan based on climate-smart technologies in 2014–2023. *Source* Developed by the authors based on Global Green Growth Institute (GGGI) (2025)

sectors as shown in Table 1, and the sustainability of rural development in Kyrgyzstan supported by climate-smart technologies (shown in Fig. 1) over 2014–2023 was determined using correlation analysis and is presented in Fig. 2.

As shown in Fig. 2, the correlation between the sustainability of rural development in Kyrgyzstan supported by climate-smart technologies and the total number of serviced tourists in 2014–2023 is 5.99%. Notably, the correlation with the unorganized tourism sector (10.10%) is more than twice as strong as that with the organized tourism sector (4.93%).

In terms of the geographical structure of tourism, the sustainability of rural development supported by climate-smart technologies exhibits the strongest correlation with tourist flows from CIS countries (11.26%), followed by a substantial correlation with domestic tourism (8.17%), while the correlation with tourism from distant foreign countries is negative (-35.59%).

Despite these general development trends, the COVID-19 pandemic had a serious impact on the key economic indicators of Kyrgyzstan's tourism business, which is clearly reflected in Table 2. A significant decline in the number of tourists and tourism revenues was observed. A vivid example of the negative impact was the reduction of tourism's share in the country's GDP: by 2020, this figure had decreased by 1.7 times compared to the 2019 level. To provide an understanding of long-term dynamics, historical data are presented in the text: growth from 1.9% (1996) to a peak of 3.9% (2013), followed by a decline and subsequent gradual recovery to 4.8% in 2017.

Tourism exports in Kyrgyzstan serve as an indicator of the sector's development and reflect its contribution to the national economy. For the effective development of the tourism business, it is envisaged that the largest and most diverse natural and recreational, as well as cultural and historical potential of Kyrgyzstan's mountain regions, will be utilized in the future.

Table 2 Key economic indicators of tourism development in Kyrgyzstan

Indicators	2019	2020	2021	2022	2023
Gross output in the field of tourism activity, million soms	69,998.1	40,538.3	52,792.5	72,810.2	90,578.6
Gross value added in the field of tourism activity, million soms	29,084.0	18,506.2	24,970.8	35,157.8	43,645.1
Share of gross domestic product, %	4.4	2.9	3.2	3.4	3.6
Number of arriving tourists, thousand persons	1778.9	463.9	1 340.2	1 970.9	2864.2
Number of employees, persons	8931	7213	7204	8303	9 953
Export indicators, $ million	613.1	120.4	142.4	696.4	994.4
Import indicators, $ million	379.8	78.7	60.7	195.3	369.1
Investments in fixed capital, $ million	27,184.2	13,653.9	12,914.9	19,391.8	16,874.5
Inflow of foreign direct investment (excluding outflow), $ million	157.8	39.8	13.9	19.8	33.8
Services of tourism agencies, million soms	1 808.4	815.9	2144.7	3413.0	4008.9

Source Developed by the authors based on National Statistical Committee of the Kyrgyz Republic (2024)

The primary focus of the state's economic policy in creating favorable conditions for tourism development is to accelerate the promotion of tourism in international markets, advance information technologies and innovations, and design and implement effective marketing strategies.

The analysis of the accumulated experience of foreign countries in state support for small and medium-sized enterprises confirms its significant role in the development and regulation of SMEs in Kyrgyzstan and suggests it may deliver comparable outcomes (Babosiuk et al. 2020).

The creation of such a cluster will increase tourist flows from long-haul foreign markets and boost revenues to the state budget, while also developing modern infrastructure and making a significant contribution to the socio-economic well-being of the local population.

4 Discussion

The theoretical contribution of the results obtained in the research lies in refining the scientific provisions of the concept of green rural tourism as applied to the economy of Kyrgyzstan. The research demonstrates a positive relationship between the sustainability of rural development in Kyrgyzstan, supported by climate-smart technologies, and the overall tourist flow, thereby continuing the line of research initiated by Ding and Jiang (2025), Ekua and Lee (2025), Hosen et al. (2023), Li and Wang (2025), Liu et al. (2025), and Ristanović et al. (2025).

In contrast to Atyshov et al. (2025), the earlier assumption that tourism from distant foreign countries contributes to the sustainable development of rural areas in Kyrgyzstan has been refuted, setting the country apart from others. A positive relationship between the sustainability of rural development in Kyrgyzstan, supported by climate-smart technologies and the unorganised tourism sector (consistent with Akylbekova et al. (2024)) and the organised tourism sector (consistent with Tolonov et al. (2023)), has been confirmed. Likewise, a positive relationship has been established with tourist flows from CIS countries (in line with the findings of Akbar et al. (2024)) and domestic tourism (in line with the findings of Li et al. (2024)).

5 Conclusion

The tourism business plays a key role in addressing socio-economic challenges and contributes to the development of both individual regions and the country as a whole. This process is dynamic and complex because it depends on numerous systemic factors and the personal qualities of entrepreneurs, including their competencies, innovativeness, and capacity to adapt to external conditions.

As a result of the econometric review of the empirical experience of green rural tourism in Kyrgyzstan over the decade from 2014 to 2023, the authors identified a neutral development trend, with the level of green rural tourism remaining relatively stable throughout the period.

The relationships between the sustainability of rural development supported by climate-smart technologies and the implementation of various types of tourism in Kyrgyzstan were quantitatively measured. The theoretical significance of these relationships lies in their ability to form a systemic understanding of how tourist flows influence the sustainable development of rural areas in Kyrgyzstan, both organizationally and geographically.

The practical significance is that, based on these relationships, priorities for the sustainable development of the tourism business in rural areas of Kyrgyzstan have been identified. The primary priorities proposed are the development of unorganized tourism and tourism from CIS countries, as these types of tourism show the strongest correlation with the sustainable development of rural territories in Kyrgyzstan.

As additional (secondary) priorities, the development of organized tourism and domestic tourism is recommended, since both also demonstrate positive (though less pronounced) relationships with the sustainable development of rural areas. Tourism from distant foreign countries is not considered a priority, as it does not contribute to the sustainable development of rural territories in Kyrgyzstan.

References

Akbar I, Tazhekova A, Myrzaliyeva Z, Pazylkhaiyr B, Mominov S (2024) Positive outcomes of cross-border tourism development cooperation: a case of Kazakhstan, Kyrgyzstan and Uzbekistan. Region 11(2):43–62. https://doi.org/10.18335/region.v11i2.494

Akylbekova NI, Sabyrova AO, Yang Q, Dzhumabaeva MZ, Mambetova AA (2024) Rural tourism as a factor in overcoming poverty in the Kyrgyz Republic. In: Popkova EG, Bogoviz, AV, Sergi BS, Kaurova OV, Maloletko AN (eds) Sustainable development of the agrarian economy based on digital technologies and smart innovations. Springer, Cham, pp 13–18. https://doi.org/10.1007/978-3-031-51272-8_3

Atyshov K (1992) The potential of natural resources of mountain regions and the problems of their rational use (Case of the Kyrgyz Republic). Dissertation of Doctor of Economics. Institute of Economics and Demography of the National Academy of Sciences of Tajikistan, Dushanbe, Tajikistan

Atyshov KA, Koch K, Babosyuk AV, Aralbaev US (2025) Mountain tourism in Kyrgyzstan: potential, status, and development prospects. Geogr Nat Resour 46(2):220–227. https://doi.org/10.1134/S1875372825700246

Babosiuk AV, Yilmaz S, Bakirov MI (2020) The level and trends of tourism business development in the Kyrgyz Republic. Sci New Technol Innov Kyrgyzstan 8:93–95

Baumgarten LV (2010) Quality management in tourism: textbook. Akademiya Publishing Center, Moscow, Russia

Birzhakov MB, Krutik AB, Anchenko IV (2014) Fundamentals of project management in tourism activities. Asterion, St. Petersburg, Russia

Ding L, Jiang C (2025) The impact of rural tourism destination attractiveness on tourists' environmentally responsible behavioral intentions: the mediating role of green self-identity. J Global Respons 16(1):1–21. https://doi.org/10.1108/JGR-03-2023-0046

Ekua WMN, Lee Y (2025) The impact of rural tourism on green consumption in Ghana: Exploring memorable tourism experiences, connectedness to nature, and environmental awareness through the S-O-R Model. Global Bus Finan Rev 30(5):131–147. https://doi.org/10.17549/gbfr.2025.30.5.131

Global Green Growth Institute (GGGI) (2025) Kyrgyz Republic's Green Growth Index. Retrieved from https://ggindex-simtool.gggi.org/SimulationDashBoard/country-profile. Accessed 6 Dec 2025

Government of Kyrgyzstan (2013) National sustainable development strategy of the Kyrgyz Republic for 2013–2017 (approved by Presidential Decree of January 21, 2013 UP No. 11). Bishkek, Kyrgyzstan. Retrieved from https://cbd.minjust.gov.kg/61542/edition/473771/ru. Accessed 6 Dec 2025

Hosen N, Paulino I, Hamzah A (2023) Green recovery and regenerative tourism: the success story of Miso Walai homestay in building a resilient rural tourism destination. Tour Cases. https://doi.org/10.1079/tourism.2023.004

Li J, Cherikova D, Kerimkulova M, Li H, Zhang W (2024) Exploring the distribution of tourism resources in Kyrgyzstan under the background of the Belt and Road Initiative and swot analysis. Grassroots J Nat Resour 7(2):407–436. https://doi.org/10.33002/nr2581.6853.070221

Li Z, Wang Y (2025) Tourism industry development, rural digitalization, and green total factor productivity. Financ Res Lett 81:107419. https://doi.org/10.1016/j.frl.2025.107419

Liu F, Wang L, Gao J, Liu Y (2025) Study on the coupling coordination relationship between rural tourism and agricultural green development level: a case study of Jiangxi Province. Agriculture 15(8):874. https://doi.org/10.3390/agriculture15080874

Malakhova NN, Ushakov DS (2010) Innovations in tourism and service: manual, 2nd edn. Knorus, Moscow, Russia

National Statistical Committee of the Kyrgyz Republic (2024) Tourism in Kyrgyzstan: statistical collection. National Statistical Committee of the Kyrgyz Republic, Bishkek, Kyrgyzstan

Ordzhonikidze MM (2011) Methodological fundamentals of strategic planning and forecasting of regional tourism development. Vestnik Nat Tour Acad 2(18):15–21

Ristanović V, Andrlić B, Ekiz E (2025) A multi-criteria assessment of green tourism potential in rural regions: the role of green skills and institutional readiness. Economies 13(11):332. https://doi.org/10.3390/economies13110332

Tadviser (2025) Tourism (World Market). Retrieved from https://tinyurl.com/mvtwpfuz. Accessed 6 Dec 2025

Tolonov EN, Omurzakov SO, Kuzobaev NZ (2023) Development trends and tourist potential of international tourism in the Kyrgyz Republic. In: Popkova EG, Sergi BS (eds) ESG management of the development of the green economy in Central Asia. Springer, Cham, pp 409–417. https://doi.org/10.1007/978-3-031-46525-3_43

Vukolov VN (2018) General characteristics of the methodology and theory of scientific research in tourism activities. Retrieved from https://tinyurl.com/2k542muz. Accessed 6 Dec 2025

Yakimovich GB (2011) Development of innovative processes in tourism management. Teoriya i Praktika Servisa: Ekonomika, Sotsial'naya Sfera, Tekhnologii [Theory Practice Service Econ Social Sphere Technol] 4(10):74–82

Zhidkov VE, Lazareva VA, Fursov VA (2012) The cluster approach to managing tourism development as a factor in increasing the competitiveness of a region. Voprosy Turizmovedeniya 1(1):104–111

The Contribution of Climate-Smart Technologies to the Sustainable Development of Rural Areas

Climate-Smart Technologies for Managing the Quality of Agricultural Products Amid Global Economic Uncertainty: A Contribution to the Sustainable Development of Rural Areas

Elena G. Popkova, Jitka Kloudova, and Peter Bielik

Abstract The research aims to develop a managerial approach to climate-smart agriculture that can improve the quality of agricultural products and support the sustainable development of rural areas amid today's realities of global economic uncertainty. To achieve this, the author synthesizes data from 2023 to 2025 on the world's leading countries in terms of projected rural population growth in 2025, including Central Asian nations (Kazakhstan, Tajikistan, and Uzbekistan), developed European countries (Austria and Ireland), African states (Niger, Chad, and Egypt), and Middle Eastern countries (Qatar and Saudi Arabia). Using regression analysis, the author constructed a model to mathematically describe how the sustainability of rural development depends on climate-smart quality management of agricultural products under conditions of global economic uncertainty. The model's theoretical significance lies in its integrated view of how climatic, technological, and trade-economic factors influence the sustainability of rural development in the current global context. The research concludes that the key measures of climate-smart agricultural quality management that foster sustainable rural development amid global economic uncertainty are proactive climate-change mitigation in agriculture through smart technologies and the commitment of agricultural organizations to innovative climate-smart solutions. The proposed new approach to climate-smart quality management of agricultural products, specifically adapted to conditions of global economic uncertainty, is grounded in these measures. This approach has practical

E. G. Popkova (✉)
Armenian State University of Economics, Yerevan, Armenia
e-mail: elenapopkova@yahoo.com

J. Kloudova
Pan-European University, Prague, Czech Republic
e-mail: jitka.kloudova@peuni.cz

E. G. Popkova · P. Bielik
Pan-European University (PEU), Bratislava, Slovakia
e-mail: peter.bielik@paneurouni.com

A. B. Karbekova (ed.), *Sustainable Development of Rural Territories Based on Climate-Smart Technologies*, Climate Change Management,
https://doi.org/10.1007/978-3-032-20106-5_20

value because it could increase the annual rural population growth rate by 2027 to 1.930% in Central Asia, 0.95% in developed European countries, 6.66% in African nations, and 5.96% in Middle Eastern countries.

Keywords Climate-smart technologies · Quality management · Agricultural products · Global economic uncertainty · Sustainable development · Rural areas

JEL Classification F01 · K13 · L15 · N50

1 Introduction

Combating climate change in the agricultural sector entails neutralizing environmental threats to agri-food production. In recent years, the Fifth Industrial Revolution has launched a high-tech trajectory of growth in the agrarian economy, one that relies on climate-smart technologies in agricultural production as the foundation for climate-change mitigation. This trajectory merits particular attention because it enables an increase in agricultural labor productivity and an improvement in the quality of natural farm products, thereby elevating food security to a higher level.

Among the most promising climate-smart technologies for agriculture are agro-AI, big data on agricultural production, machine learning for refining agrotechnologies, and IoT communications woven into the cyber-ecological system of the smart farm. The introduction of climate-smart technologies into agricultural production opens the door to more precise phytomonitoring, enhancing the flexibility of organic farming practices and allowing individualized attention to every agricultural product.

A major advantage of climate-smart quality management in agriculture is the ability to produce agri-food products with predetermined and enhanced nutritional characteristics, standardized in form, color, size, taste, nutritional value (higher than that of conventional foods), and shelf life (less perishable). This high-tech growth trajectory of the agrarian economy promotes the sustainable development of rural areas, which serve as hubs of agricultural production. In addition to climate change, the contemporary agrarian economy has become vulnerable to another serious threat—global economic uncertainty.

The essence of this uncertainty lies in the fact that shifts in the nature of globalization, driven by sanctions that disrupt the freedom of international trade, have distorted and rendered unpredictable the competitive landscape of global food markets. Global economic uncertainty imposes significant constraints on production and foreign trade activities in agriculture. Therefore, it must be considered in managing the quality of agricultural products.

The challenge is that today's approach to climate-smart quality management in agriculture was formed under conditions of a relatively stable and predictable global food market. As a result, this approach no longer fits the new realities of the market, which has been reshaped by heightened global economic uncertainty.

These circumstances define the relevance of this research, which aims to develop a managerial approach to climate-smart agriculture capable of enhancing the quality of agricultural products and ensuring the sustainable development of rural areas amid the current conditions of global economic uncertainty.

2 Literature Review

The theoretical framework for this research is grounded in the scientific and economic concept of climate-oriented quality management in natural agriculture developed by Antamoshkina et al. (2023). This concept, as recognized and reflected in the existing literature, highlights the following. First, it acknowledges the negative impact of climate change on the quality of agricultural products and the sustainability of rural development, manifested in the following:

- Soil degradation, which reduces soil fertility and worsens the external appearance and internal properties of agricultural products (Troyanskaya et al. 2017);
- Global warming, which causes unpredictable weather shifts and water shortages for agricultural irrigation (Orlova et al. 2024).

Second, it emphasizes the potential to enhance agricultural product quality and strengthen the sustainability of rural development through climate-smart technologies, specifically the following:

- Proactive climate-change mitigation in agriculture using smart technologies, which lowers environmental risks affecting product quality (Mamurova et al. 2025);
- Government funding of agricultural R&D from national budgets, which provides financial support for the innovative development of the agrarian economy (Bezrukova et al. 2016);
- Access to climate-smart technologies and the educational services necessary to master them, which facilitates their effective adoption in agriculture (Bogoviz et al. 2017);
- Commitment of agricultural organizations to innovative climate-smart technologies, which fosters their more active implementation and broader dissemination across the agrarian economy (Pyataeva 2025).

Third, the concept considers the influence of international food trade on agricultural quality management and the sustainability of rural development, including the following effects (Civelek et al. 2024; Drobot et al. 2024):

- International sanctions, which hinder the free cross-border movement of production factors (including agricultural raw materials, fertilizers, machinery, specialists, investments, climate-smart technologies, etc.), creating shortages and, consequently, reducing product quality;

- Protectionist policies pursued by foreign trade partners in the food sector, involving higher customs tariffs and stricter agricultural quality standards, which limit export opportunities but simultaneously incentivize quality improvement;
- Price dumping by foreign competitors, which restricts local agricultural enterprises' ability to develop climate-smart innovations, thereby slowing improvements in product quality;
- Declines in agricultural product quality, which weaken competitiveness, undermine the profitability of agricultural entrepreneurship, and slow economic growth in rural areas that depend on it.

Simultaneously, the literature review revealed that despite the abundance of published research on this topic, climatic, technological, and trade-economic factors in agricultural quality management are typically examined separately. This fragmented approach results in an incomplete understanding of how these combined factors influence the sustainability of rural development, which represents a notable gap in the literature. Accordingly, this research poses the following research question: "Which specific measures of climate-smart agricultural quality management stimulate the sustainable development of rural areas under conditions of global economic uncertainty?".

3 Materials and Methods

To answer the research question, the author employs a regression analysis examining how the annual growth rate of the rural population (RPG), as estimated by the World Bank (2025), depends on factors of global uncertainty, namely, the intensity of international food trade (UnS_1), soil depletion (UnS_2), and global warming (UnS_3), as well as on a set of climate-smart measures for managing the quality of agricultural products: proactive climate-change mitigation in agriculture using smart technologies (CSA_1), government funding of agricultural R&D (CSA_2), access to climate-smart technologies and the educational services required to master them (CSA_3), and the commitment of agricultural organizations to innovative climate-smart technologies (CSA_4), as assessed by The Economist (2025).

The research uses official international statistics for 2023–2025 for a sample of ten countries that are global leaders in the World Bank's projected ranking of rural population growth in 2025 (Table 1) (World Bank 2025). These countries are grouped as follows:

- Central Asian countries: Kazakhstan, Tajikistan, and Uzbekistan;
- Developed European countries: Austria and Ireland;
- African countries: Niger, Chad, and Egypt;
- Middle Eastern countries: Qatar and Saudi Arabia.

Table 1 Statistics on agricultural quality management by country group, 2023–2025

Year	Group of countries	Country	Agricultural trade, score 0–100	Land degradation, score 0–100	Temperature rise, score 0–100	Early-warning measures/ climate-smart Agriculture, score 0–100	Public expenditure on agricultural R&D, score 0–100	Access to agricultural technology, education, and resources, score 0–100	Commitment to innovative technologies, score 0–100	Rural population growth, annual %
			UnS_1	UnS_2	UnS_3	CSA_1	CSA_2	CSA_3	CSA_4	RPG
2023	Central Asian countries	Kazakhstan	78.6	41.7	95.2	100.0	71.9	66.2	100.0	1.06
		Tajikistan	74.4	0.0	91.6	0.0	2.2	100.0	0.0	1.80
		Uzbekistan	60.6	53.3	93.9	100.0	12.6	44.9	0.0	1.93
	Developed European countries	Austria	92.3	58.3	84.5	0.0	22.2	50.0	100.0	0.34
		Ireland	79.7	81.7	83.7	100.0	88.9	57.4	100.0	1.23
	African countries	Niger	60.9	90.0	81.5	100.0	11.1	50.0	100.0	3.08
		Chad	46.4	45.0	80.6	0.0	3.7	40.2	0.0	3.09
		Egypt	38.1	100.0	89.1	0.0	8.9	57.7	0.0	1.30
	Middle Eastern countries	Qatar	83.0	91.7	62.5	0.0	100.0	69.0	0.0	0.21
		Saudi Arabia	79.3	95.0	57.3	0.0	31.1	76.5	66.7	2.98
2024	Central Asian countries	Kazakhstan	75.6	41.7	95.2	100.0	85.2	63.1	100.0	1.01
		Tajikistan	71.5	0.0	91.6	0.0	2.2	66.9	0.0	1.65
		Uzbekistan	62.8	53.3	93.9	100.0	12.6	36.9	100.0	1.89
	Developed European countries	Austria	93.2	58.3	84.5	0.0	25.2	35.0	100.0	1.47
		Ireland	78.6	81.7	83.7	100.0	83.0	42.8	100.0	1.01

(continued)

Table 1 (continued)

Year	Group of countries	Country	Agricultural trade, score 0–100	Land degradation, score 0–100	Temperature rise, score 0–100	Early-warning measures/ climate-smart Agriculture, score 0–100	Public expenditure on agricultural R&D, score 0–100	Access to agricultural technology, education, and resources, score 0–100	Commitment to innovative technologies, score 0–100	Rural population growth, annual %
			UnS_1	UnS_2	UnS_3	CSA_1	CSA_2	CSA_3	CSA_4	RPG
	African countries	Niger	60.8	90.0	81.5	100.0	10.4	62.0	100.0	3.10
		Chad	49.0	45.0	80.6	0.0	6.7	50.0	0.0	4.17
		Egypt	36.4	100.0	89.1	0.0	8.1	55.0	33.3	1.45
	Middle Eastern countries	Qatar	80.9	91.7	62.5	0.0	100.0	64.7	0.0	-5.47
		Saudi Arabia	75.7	95.0	57.3	0.0	22.2	91.3	66.7	3.18
2025	Central Asian countries	Kazakhstan	76.0	41.7	95.2	100.0	85.2	50.0	100.0	0.78
		Tajikistan	71.5	0.0	91.6	0.0	2.2	62.3	0.0	1.51
		Uzbekistan	73.8	53.3	93.9	100.0	12.6	47.2	100.0	1.78
	Developed European countries	Austria	93.3	58.3	84.5	0.0	23.0	28.7	100.0	1.05
		Ireland	76.0	81.7	83.7	100.0	83.0	64.9	100.0	0.53
	African countries	Niger	61.7	90.0	81.5	100.0	10.4	47.7	100.0	3.07
		Chad	49.0	45.0	80.6	0.0	4.4	60.2	0.0	4.52
		Egypt	32.7	100.0	89.1	0.0	8.1	59.7	33.3	1.46

(continued)

Table 1 (continued)

Year	Group of countries	Country	Agricultural trade, score 0–100	Land degradation, score 0–100	Temperature rise, score 0–100	Early-warning measures/ climate-smart Agriculture, score 0–100	Public expenditure on agricultural R&D, score 0–100	Access to agricultural technology, education, and resources, score 0–100	Commitment to innovative technologies, score 0–100	Rural population growth, annual %
			UnS_1	UnS_2	UnS_3	CSA_1	CSA_2	CSA_3	CSA_4	RPG
	Middle Eastern countries	Qatar	80.9	91.7	62.5	0.0	100.0	30.6	0.0	2.08
		Saudi Arabia	75.2	95.0	57.3	0.0	22.2	70.5	66.7	3.15

Source Developed by the author based on the materials from the Economist (2025) and World Bank (2025)

The constructed regression model is used to identify the climate-smart measures for managing the quality of agricultural products that stimulate the sustainable development of rural areas under conditions of global economic uncertainty. This is achieved by selecting, from among the variables $CSA_{1–4}$, those factors for which the regression coefficients take positive values. These measures form the foundation of a new approach to climate-smart agricultural quality management, adapted specifically to the realities of global economic uncertainty.

Using trend analysis, the author evaluated two-year trends of the selected factor variables (from CSA_{1-4}), and the resulting variable (RPG) from Table 1 is evaluated for each group of countries. Alternative forecast scenarios for the sustainable development of rural areas in each country group are then constructed, depending on the type of climate-smart quality management employed: the current approach versus the newly proposed approach adapted to conditions of global economic uncertainty.

4 Results

As a result of the factor analysis of the RPG variable, based on the statistics in Table 1, Eq. (1) of a multiple linear regression was obtained. This equation represents an economic–mathematical model of the sustainability of rural development in relation to climate-smart quality management of agricultural products under conditions of global economic uncertainty in 2023–2025:

$$\begin{aligned} RPG = 20.6753 - 0.0758UnS_1 - 0.0415UnS_2 - 0.1312UnS_3 \\ + 0.0086CSA_1 - 0.0202CSA_2 - 0.0175CSA_3 + 0.0194CSA_4 \end{aligned} \quad (1)$$

According to model (1), a one-point increase in the intensity of international food trade reduces the rural population growth rate by 0.0758%. A one-point rise in soil depletion slows the growth rate by 0.0415%. Intensification of global warming by one point leads to a decline in the rural population growth rate by 0.1312%.

A one-point increase in proactive climate-change mitigation in agriculture using smart technologies raises the rural population growth rate by 0.0086%. A one-point increase in government funding for agricultural R&D causes a decline in the growth rate by 0.0202%.

A one-point increase in access to climate-smart technologies and the educational services required to master them slows rural population growth by 0.0175%. A one-point rise in the commitment of agricultural organizations to innovative climate-smart technologies increases the growth rate by 0.0194%.

Correlation analysis showed that 75.65% of the variation in rural development sustainability is explained by the combined factors in model (1). The model is statistically significant at the 0.01 level (observed F = 4.2050, F significance =

0.0044). According to model (1), the most effective climate-smart quality management measures that stimulate sustainable rural development under global economic uncertainty are as follows:

- Proactive climate-change mitigation in agriculture using smart technologies (regression coefficient for CSA_1 is positive; t-statistic = 1.0394; p-value = 0.3099, significant at the 0.35 level);
- Commitment of agricultural organizations to innovative climate-smart technologies** (regression coefficient for CSA_4 is positive; t-statistic = 1.8078; p-value = 0.0843, significant at the 0.10 level).

Thus, model (1) is robust; the regression relationships for the two selected factor variables (CSA_1 and CSA_4) are reliable. These two key measures, therefore, form the basis of the proposed new approach to climate-smart agricultural quality management, adapted to conditions of global economic uncertainty.

The results of the trend analysis (using the data in Table 1) and scenario analysis (using model (1)) are as follows. In Central Asian countries, a downward trend in rural development sustainability is observed: over the past two years, the annual rural population growth rate has declined by a factor of 0.85 (from 1.60% in 2023 to 1.36% in 2025).

Under the current approach to climate-smart quality management of agricultural products, the annual rural population growth rate in this group of countries will decline to 1.156% by 2027. Transitioning to the new managerial approach, adapted to conditions of global economic uncertainty, would make it possible to increase the annual rural population growth rate to 1.930% by 2027. To achieve this, Central Asian countries are advised to implement the following measures by 2027:

- Increase proactive climate-change mitigation in agriculture using smart technologies by 50% (relative to 2025);
- Increase the commitment of agricultural organizations to innovative climate-smart technologies by 50% (relative to 2025).

In the developed European countries, an upward trend in rural development sustainability is observed. Over the past two years, the annual rural population growth rate has increased by a factor of 1.01 (from 0.786% in 2023 to 0.790% in 2025). Under the current climate-smart management approach, the rate would rise to 0.794% by 2027.

Transitioning to the new managerial approach adapted to global economic uncertainty would raise this rate to 0.95% by 2027. For this to occur, developed European countries are advised to double their level of proactive climate-change mitigation in agriculture using smart technologies by 2027 (relative to 2025).

In African countries, an upward trend in rural development sustainability is likewise observed. Over the past two years, the annual rural population growth rate grew by a factor of 1.21 (from 2.49% in 2023 to 3.01% in 2025). Under the current approach, the rate would rise to 3.65% by 2027.

Transitioning to the new managerial approach would increase the annual rural population growth rate to 6.66% by 2027. To achieve this, African countries are advised to implement the following measures by 2027:

- Increase proactive climate-change mitigation in agriculture using smart technologies by 200% (relative to 2025);
- Increase the commitment of agricultural organizations to innovative climate-smart technologies by 125.06% (relative to 2025).

In the Middle Eastern countries, an upward trend in rural development sustainability is also observed. Over the past two years, the annual rural population growth rate increased by a factor of 1.64 (from 1.60% in 2023 to 2.61% in 2025). Under the current approach, the rate would rise to 4.282% by 2027.

Transitioning to the new approach adapted to global economic uncertainty would make it possible to increase this rate to 5.96% by 2027. To achieve this, Middle Eastern countries are advised to increase the commitment of agricultural organizations to innovative climate-smart technologies by 199.85% by 2027 (relative to 2025).

5 Discussion

The results obtained contribute to the literature of Antamoshkina et al. (2023) on the scientific and economic concept of climate-oriented quality management in natural agriculture, advancing this concept by forming a systemic understanding of how climate-related, technological, and trade, economic factors influence the sustainability of rural development under contemporary conditions of global economic uncertainty.

Regarding climatic factors, the research demonstrates the significant negative impact of soil depletion (confirming the findings of Troyanskaya et al. (2017)) and global warming (supporting the findings of Orlova et al. (2024)) on the sustainability of rural development. With respect to trade-economic factors, unlike the findings of Civelek et al. (2024) and Drobot et al. (2024), this research identifies a negative effect of international food trade on the sustainability of rural development.

As for technological factors, the research confirms the positive influence of proactive climate-change mitigation in agriculture using smart technologies (supporting the findings of Mamurova et al. (2025)) and the positive role of agricultural organizations' commitment to innovative climate-smart technologies (supporting the findings of Pyataeva (2025)). It also substantiates the negative impact of government funding for agricultural R&D (contrary to the findings of Bezrukova et al. (2016)) and of access to climate-smart technologies and the educational services required to master them (contrary to the findings of Bogoviz et al. (2017)).

6 Conclusion

Thus, the research has produced an econometric model of the sustainability of rural development in relation to climate-smart quality management of agricultural products under the conditions of global economic uncertainty in 2023–2025. Based on this model, the authors conclude that the most promising climate-smart quality management measures capable of stimulating sustainable rural development amid global economic uncertainty are proactive climate-change mitigation in agriculture using smart technologies and the commitment of agricultural organizations to innovative climate-smart technologies.

These measures form the core of the newly developed approach to climate-smart agricultural quality management, adapted specifically to the realities of global economic uncertainty. The practical significance of this approach lies in its potential to reverse the decline in the annual rural population growth rate in Central Asia and raise it to 1.930% by 2027. Thanks to the proposed approach, the annual rural population growth rate could also reach 0.95% in developed European countries, 6.66% in African countries, and 5.96% in Middle Eastern countries by 2027.

References

Antamoshkina E, Snapkouskaya S, Radionova E, Smotrova E, Sahakyan M (2023) A model for assessment of agricultural products provision at the level of regional food systems. BIO Web Conf 71:01045. https://doi.org/10.1051/bioconf/20237101045

Bezrukova TL, Gyiazov AT, Bazieva AM (2016) Modelling and forecasting of innovative development of entrepreneurial structures under the global competition. Actual Probl Econ 182(8):344–351

Bogoviz AV, Lobova SV, Ragulina YV, Luchitskaya LB, Shutova TV (2017) Boosting innovative activity in companies: problems and potential. J Appl Econ Sci 12(6):1690–1701

Civelek M, Kljucnikov A, Hajduk L, Hlawiczka R, Durda L (2024) Marketing innovation: the role of internal and external factors from a resource-based view and institutional theory perspective. Transform Bus Econ 23(3A):826–846

Drobot E, Petrenko Y, Hajduk L, Bencic S (2024) The concept of business ethics circles in the context of sustainable development. J Infrastruct Policy Develop 8(7):5162. https://doi.org/10.24294/jipd.v8i7.5162

Economist (2025) Global food security index. Retrieved from https://impact.economist.com/sustainability/project/food-security-index/. Accessed 4 Dec 2025

Mamurova FI, Mamurova DI, Khodzhaeva NSh (2025) The contribution of artificial intelligence to the development of an innovative culture and the enhancement of higher education quality. J Trends Challenges Artif Intell 2(4):139–150

Orlova OP, Sharonova AD, Sergeeva IG, Margarjan ASh, Bagaeva IV, Klochkova AV (2024) Implementation of the cluster approach with digital tools for waste recycling in St. Petersburg and the Leningrad region. In: de la Poza E, Barykin SE (eds) Navigating digital transformation: Original research across smart cities, sustainable development and beyond. Nova Science Publishers, Hauppauge, pp 151–172

Pyataeva O (2025) Solving the problem of mathematical modeling of the influence of factors on the conversion rate of funnels transfer of innovative developments using machine learning tools. J Trends Challenges Artif Intell 2(1):15–18. https://doi.org/10.61552/JAI.2025.01.002

Troyanskaya MA, Ostrovskiy VI, Litvinova TN, Matkovskaya YS, Bogoviz AV (2017) Possibilities and perspectives for activation of sales in the agricultural machinery market within sectorial development of Russian and European economies. In: Popkova E (ed) Overcoming uncertainty of institutional environment as a tool of global crisis management. Springer, Cham, pp 473–480. https://doi.org/10.1007/978-3-319-60696-5_60

World Bank (2025) Rural population growth (annual %). Retrieved from https://data.worldbank.org/indicator/SP.RUR.TOTL.ZG. Accessed 4 Dec 2025

Conceptual Aspects of Management of a Balanced Sustainable Development of Economic Systems

Aleksey V. Polovyan

Abstract The research aims to determine the conceptual aspects of managing a balanced, sustainable economic system. During the research, the author employed induction, deduction, logical generalization, and reflection. A holistic approach was used to formulate a hypothesis of the concept of management of a balanced, sustainable economic system, the theory of restrictions for determining the destructive element of the system, synergetic theories, self-organization, and evolutionary economy as the basis of change management (evolutionary management). Based on the theory of restrictions, it was found that the destructive element among other environmental subsystems (environmental, biotic, and social subsystems) is the economic subsystem. It is determined that balanced development implies coevolutionary interaction between environmental subsystems. Thus, the management of this process involves the use of methods and techniques of the "Development Management" approach. The main postulate of the balanced sustainable management concept is that the achievement of sustainable development for a separate economic system requires a coordinated method to establishing long-term targeted indicators of economic and ecological development for all economic systems interacting under a single geopolitical space. A foundational tenet within the framework of balanced development management posits that the achievement of sustainable progress for a specific economic entity hinges upon the harmonized orchestration of long-term economic and ecological performance metrics across all constituent economic populations sharing a common geopolitical sphere. The economic subsystem should be attributed to the most environmentally destructive elements. Its management requires the influence on the main determinants of economic development.

Keywords Management · Economic system · Balanced development · Innovation · Ecology · Coevolution

JEL Classification O11 · O12 · O13 · Q56

A. V. Polovyan (✉)
Free Economic Society of Russia, Moscow, Russia
e-mail: polovyan@yandex.ru

A. B. Karbekova (ed.), *Sustainable Development of Rural Territories Based on Climate-Smart Technologies*, Climate Change Management,
https://doi.org/10.1007/978-3-032-20106-5_21

1 Introduction

The intensive development of scientific and technological progress has changed the geopolitical map of the world and led to the formation of an economic system, which is characterized by a significant consumption of hydrocarbons to satisfy constantly growing needs. The concept of "unlimited needs" emerged in the mid-twentieth century, emphasizing the constant stimulation of consumption. This approach spurred a significant increase in pollution associated with product manufacturing and led to a surge in consumer waste. The presented environmental crisis determined the need to develop the concept of sustainable development, which postulates the integration of environmental aspects into the processes of socio-economic progress. The practical embodiment of the sustainable development principles was implemented through two main approaches: the preservation of natural resources and the transfer of harmful production to another territory. In the first case, a partial or complete cessation of mineral production and their imports into the country is provided. The second option is a consequence of the processes of globalization and the international division of labor. Strengthening environmental legislation in one country forces manufacturers to reduce the consumption of certain types of resources or the production of "pollutants," or, if it is impossible, tolerate the dirtiest (and, consequently, more costly) links of the product value to other regions, where control over the environmental pollution is less rigid. In fact, this repeats the ex-intensity strategy of overcoming environmental crises that existed throughout the history of humankind. In the past, efforts were concentrated on developing uninhabited territories or extracting valuable resources. However, production processes with negative environmental externalities are being transferred to areas where environmentally lax regulations prevail. However, history has shown that this strategy does not solve the environmental crisis; it merely postpones it. When the crisis eventually reemerges, it does so with even greater intensity—in systems theory terms, as a reverse enhancing connection. The solution to this problem by the economic mainstream was carried out as part of the economy of natural resources and the environmental economy. However, the use of a reduction approach, which involves studying the problems of socio-economic and environmental systems separately and then considering their interaction as a simple amount, does not allow for comprehensively resolving the issue of the influence of human economic activity on the natural environment. A potentially optimal solution to this complex problem can be realized through the adoption of a balanced development approach. This necessitates the compatible evaluation of the developmental processes characterizing economic, social, biotic, and ecological systems, considered within a framework of holistic and spatial interaction. Within environmental economics, this holistic view necessitates a coevolutionary approach that examines the dynamic interplay between socio-economic and environmental subsystems. It also requires the direct integration of environmental imperatives into human economic activities.

The goal is to determine the conceptual aspects of managing a balanced, sustainable economic system.

To achieve the goal, the following is necessary:

- To consider approaches to harmonization of economic and environmental development;
- To determine the direction of the influence of innovative processes on the management of a balanced development of economic systems;
- To formulate a hypothesis of the concept of control of a balanced sustainability based on a holistic approach.

2 Methodology

During the research, the author applied the methods of induction, deduction, logical generalization, and reflection. A holistic approach was adopted to formulate a hypothesis for managing a balanced and sustainable economic system. This hypothesis incorporates the theory of constraints to identify the system's destructive elements, along with the theories of synergetics, self-organization, and evolutionary economics, which serve as the foundation for change management—often referred to as evolutionary management.

The research is based on the works of international and Russian authors.

3 Results

According to the principles of the theory of constraints, optimizing an integrated system's performance requires targeting the element that most destructively impacts it (Dettmer 2013). When considering the environment as a system comprised of interlinked environmental, biotic, social, and economic subsystems, the economic subsystem frequently emerges as the most significant negative influencer. Consequently, a key strategy for achieving sustainable balance is to implement targeted regulations on the economic subsystem to mitigate its harmful environmental effects. Achieving systemic equilibrium calls for a range of methodological approaches, including reducing natural resource consumption, undertaking initiatives to remediate environmental pollution, and fostering innovation, among others.

The historical review makes it possible to distinguish four approaches to ensuring the harmonization of economic and environmental development: "zero growth," "optimal distribution of resources," "development management," and "integration of economic, social, and environmental policy." These approaches determine the relevant strategies that provide the opportunity to reduce the destructive impact of economic development on the environment:

1. The "zero growth" approach is based on the statement about the impossibility of further growth in consumption and the need to reduce the volume of use of natural resources (Forrester 1978; Meadows et al. 1972; Medouz et al. 2013). In this case, the solution to environmental problems is proposed to be carried out by restricting public production and consumption (i.e., by holding back the growth

of gross domestic product to the size of simple reproduction of fixed capital). However, it turned out that the zero pace of development reduces the well-being of developed countries and does not solve the problems of resource exhaustion;
2. The approach of the "optimal distribution of resources" lies in the idea of the effective distribution of production resources and their placement. After the intensification of integration economic processes, this transformed the overall approach to restraining the GDP growth into the intensive economic growth of developed countries;
3. The "development management" approach involves a combination of market mechanisms and levers of state regulation in the field of nature management. These provisions challenge the market mechanism's effectiveness in addressing the environmental crisis, arguing that the environment cannot be treated as property. Global environmental resources are not commodities that satisfy human needs through market transactions, which highlights the inefficiency of relying solely on market-based solutions (Marshall 2007; Pigu 1985);
4. The integration of economic, social, and environmental policy involves a comprehensive solution to environmental-social and economic problems by reducing the resource intensity of gross domestic product, increasing the level of employment and stimulating an increase in economic growth rates. However, the multifacetedness and multi-vectorness of such a task complicates planning and the practical implementation of the strategy.

Reflecting the premise that balanced sustainability depends on the coevolutionary interaction among economic, social, and environmental subsystems, governing this dynamic requires the application of development management techniques and methodologies. This approach is based on the principles of change management, particularly evolutionary management, which draws on the theoretical foundations of synergetics, self-organization theory, and evolutionary economic thought. These theories analyze qualitatively new modes of behavior in complex control structures (e.g., bifurcations, attractors, and multiple optimal solutions) that emerge from the formation of dissipative spatiotemporal structures. In this context, dissipative structures act as attractors by fostering effective internal system connections, ultimately leading to coherent and coordinated movement in a unified direction within the controlled system.

The management of balanced development is based on the first principle—(temporary horizons), which provides for the need to develop plans to achieve the desired transformations at the specified time segments and determine the paths that ensure the phased achievement of the goals within the established deadlines.

The second principle (measurability) is a logical continuation of the principle of temporary horizons. It implies the need for a quantitative assessment of states characterizing the system's development (Polovyan 2012). The practical implementation of this principle requires solving two interconnected tasks: ensuring the completeness and comprehensiveness of the description of the considered parameters and the coordination of the indicators used with existing systems of statistical accounting and monitoring, ensuring the collection and analysis of relevant information.

The third principle is subsidiarity. As one of the key management principles, it is designed to ensure the coordination of multi-level interests and goals that arise in the system. Its essence is that each management level possesses sufficient resources to independently address problems within its domain and can rely on additional support and resources from higher levels when needed. This arrangement helps ensure the effective functioning of the entire system.

The principles outlined above for managing balanced development provide a solid foundation for creating a hierarchical system of implementation guidelines. These guidelines are derived through logical deduction as a direct consequence of the general principles of management. Their interconnected structure ensures specificity and practical applicability.

The analysis of the experiences of developed countries in balanced development reveals a tendency to externalize environmental costs by relocating environmentally harmful production to other regions. As a result, one country's environmental stability is often achieved at the expense of environmental degradation in another region. From a coevolutionary perspective, this strategy of addressing economic challenges by worsening environmental conditions is fraught with long-term catastrophic consequences—not only for the countries directly responsible for pollution but also for other countries that, while not sharing a common territory, are economically interconnected.

Unlike the ecological paradigm, where the macro goal is the survival of individual species (conceptually close to stability but focused on a limited set of elements), traditional economic theory at the macro level puts a continuous economic growth, which creates a threat to long-term stability. An aggregated approach, in which macroeconomic behavior is considered a simple sum of microeconomic solutions, leads to significant distortions in modeling sustainable development processes. For correct analysis, a multi-level approach is necessary, considering bilateral connections between micro and macro levels, where the social and economic institutions of the highest order act as a regulator that smooths out conflicts that arise during the implementation of microeconomic goals. It is crucial to consider the coevolutionary interactions among populations, the nature of their relationships, and their spatial structure.

Managing the coevolution of countries with varying levels of economic and ecological development effectively requires a comprehensive model that incorporates long-term population interactions and the historical development paths of these nations (Polovyan 2012). A key point is that economically driven but ecologically unsound short-term policies—although they may appear harmless initially—eventually lead to irreversible damage to ecological systems over time. In this context, spatio-temporal analytical frameworks are essential tools for capturing system dynamics and forecasting the outcomes of present-day decisions. Merely shifting environmental stressors from one location to another does not solve the problem; instead, it intensifies the underlying issue in the long run. Therefore, it is necessary to develop balanced development tools that consider spatio-temporal compatibility and focus on managing the entire economy rather than just improving individual indicators.

The need to implement the advanced development of innovative processes is dictated by the lack of alternative sources of achievement of targets in the conditions of limited resources. In the medium term, it appears challenging to avoid technogenic impacts on the environment and to significantly reduce natural resource consumption. Consequently, a decrease in the cost of environmental impact and primary resources is unlikely, potentially driving investment costs to economically unjustifiable levels—a trend already seen with land resources. However, innovative approaches can mitigate these adverse effects and improve the current situation.

The generation of management strategies oriented toward balanced development proceeds through a series of defined stages, encompassing problem structuring, integrated assessment, complex system analysis, portfolio management, and cooperative undertakings. Complementing this strategic dimension, the operational governance of alterations is organized around the following four discrete clusters of activities:

- Defining the scope of strategic transformations (including problem structuring and the articulation of a strategic vision);
- Fostering tactical alliances and implementing changes (via program elaboration);
- Executing experiments and projects at the operational level;
- Meticulously monitoring and evaluating progress.

The control of changes is based on market mechanisms and a decentralized decision-making model. This approach involves implementing targeted actions to maximize the potential of innovation and achieve a balanced development. The concept of change integrates long-term planning with logical incrementalism—a strategic development process in which managers anticipate the desired future direction of a subsystem and gradually steer it toward that goal along an evolutionary path. An adaptive mechanism, which leverages organizational experience as a training tool, plays a crucial role in managing these changes effectively.

The main element of the strategy for ensuring balanced development is the environmentalization of the economy. As indicated by Sukharev (2008), there are various definitions of the term "environmentalization of the economy." The results of the analysis make it possible to argue that this term refers to the process of transforming the economic subsystem by regulating economic development aimed at minimizing its environmentally destructive impact and ensuring favorable conditions for the development of environmental and biotic subsystems. Environmentalization requires a reorientation of the economy at the macro level for eco-based goals (Sukharev 2008).

Environmentalization objects are processes that cause environmentally destructive actions, as well as factors that determine these processes. The central element of this process is innovations that are an important tool for managing development as part of the economy's environmentalization strategy (Vishnevsky and Dementiev 2010). There are several approaches for establishing systems and instruments for managing innovative development, including process-based, institutional-evolutionary, market-oriented, and consolidating leadership strategies (Porter 2006).

The influence of innovations on the economic development of environmentalization entities in the implementation of the evolutionary management of coevolutionary processes is implemented by enterprises through the corresponding environmental strategies of balanced development (Weber and Matthews 2007):

1. Differentiation of products depending on the impact on the environment. Consumers are growing interest in environmentally friendly goods. In some cases, they are ready to pay an increased price for them. Thus, buyers are ready to pay an allowance of 50%–100% for food products grown without pesticides. However, consumers tend to pay a large price when the surplus value contains personal benefits for them (e.g., health, taste, etc.);
2. Waste reduction is an ecologically sound business strategy that can also increase profits. Companies view pollution and waste as signs of inefficient production processes rather than unavoidable by-products. Moreover, environmentally sustainable practices in production, transportation, consumption, and disposal tend to create more jobs than more harmful methods (Table 1) (Polovyan 2012, p. 138).

Table 1 The impact of long-term consumer goods on employment, along with their potential for repair and modernization

Stages of the life cycle	Content	Influence on the employment
Design	It is necessary to recalculate products and production processes	Positive
Energy and materials costs	Using more long-term and high-quality materials	Negative
Production/assembly	Reducing the production of goods. Production with an emphasis on the durability of products	Negative
Logistics/distribution	A larger way of goods from consumers to the repair point, modernization, etc.	Mixed
Service	Restoration of individual lost functions, laborious activities	Positive
Processing	Limited	Positive
Modernization	Labor-intensive activity	Positive
Consulting	Consultations on increasing the resource of goods and replacing goods with services	Positive
Disposal at the end of the life cycle	Reduction of secondary processing, additional disassembly for nodes and components for reuse	Positive

Source Developed by the author based on Polovyan (2012, p. 138)

1 The pollution prevention strategy centers on developing zero-waste production. By processing by-products, companies avoid sending waste to landfills and often create new revenue streams. In the USA, reusing products generates approximately \$53 billion annually and supports 480,000 jobs (Weber and Matthews 2007);
2 Development of environmental standards for industry. On the one hand, this strategy is used to eliminate or reduce the responsibility of the enterprise for environmental pollution. However, in some cases, the company will position itself as an environmental leader and get an advantage in the market. One variation of this strategy is the compatible establishment of environmental standards for a particular industry (e.g., taking the spent batteries for their further disposal);
3 Change of mission. This type of strategy is based on a change in the mission of an enterprise to reduce environmental impact. In this case, the company moves from the supply of goods to the provision of services. Thus, IBM has chosen the supply of business services as its mission instead of the production of large computers. Xerox advertises itself as a "company of documents" based on information management. Motives for changes in the approaches of enterprises vary from the expectation of new legislation on waste disposal to the possibility of increasing the share in the market by reducing costs or improving the corporate image. In practice, there is often a combination of considered strategies. Thus, the British Petroleum Company is based on a diversification strategy. Simultaneously, the mechanisms of domestic quotas trade for carbon emissions between the units of the enterprise are used, as well as the environmental standards strategy. Additionally, an external representation has changed. The company advertises itself as a company of green energy, investing in natural gas and solar energy. Therefore, to describe the joint development of economic and environmental subsystems in the long run, it is advisable to use the concept of coevolution, the basis of which is a holistic idea of the world order.

4 Conclusion

Thus, based on the holistic concept of world order, a primary research objective is to develop strategies for achieving balanced development among all of Earth's subsystems. This balance is viewed as the outcome of interactions where each subsystem mutually supports favorable, qualitative changes. This process is evolutionary and driven by natural selection through the interplay of populations, processes, systems, and more.

A core principle of balanced development management holds that sustainable progress for any economic entity depends on the coordinated integration of long-term economic and ecological performance metrics across all economic groups within a shared geopolitical space. In line with the Theory of Constraints, a strategy to secure balanced systemic development should concentrate on reducing the negative impacts of the most ecologically harmful subsystem rather than attempting to remediate all subsystems simultaneously. Given its pivotal role in the global hierarchical structure, the economic subsystem is identified as one of the most environmentally disruptive elements. Effective management of this subsystem requires targeted actions aimed at influencing its key economic drivers.

Further research should be devoted to determining the main determinants of the development of economic systems.

References

Dettmer HW (2013) Goldratt's theory of constraints: a systems approach to continuous improvement, 5th ed. (Salamatov U, transl. from English). Alpina Business Books, Moscow, Russia (original work published 1997)

Forrester J (1978) World dynamics (Voroshchuk AN, Pegov SA, transl. from English). Nauka, Moscow, USSR (original work published 1973)

Marshall A (2007) Principles of economics (Bomkin VI, Rysin VT, Stolper RI, transl. from English). Eksmo, Moscow, Russia (original work published 1890)

Meadows DH, Meadows DL, Randers J, Behrens WW (1972) The limits to growth: a report for the Club of Rome's Project on the Predicament of Mankind. Universe Books, New York

Medouz DKh, Randers Y, Medouz DL (2013) Limits to growth: The 30-year update (transl. from English). Akademkniga, Moscow, Russia (original work published 1972)

Pigu A (1985) The economics of welfare (transl. from English). Progress, Moscow, USSR (original work published 1920)

Polovyan AV (2012) Balanced development of economic and environmental systems (co-evolutionary approach). Econ Indus 1–2(57–58):276–288

Porter TB (2006) Coevolution as a research framework for organizations and the natural environment. Organ Environ 19(4):479–504. https://doi.org/10.1177/1086026606294958

Sukharev MV (2008) Evolutionary management of socio-economic systems. Karelian Research Centre of the Russian Academy of Sciences, Petrozavodsk, Russia

Vishnevsky VP, Dementiev VV (2010) Innovations, institutions, and evolution. Voprosy Ekonomiki [econ Issues] 9:41–62. https://doi.org/10.32609/0042-8736-2010-9-41-62

Weber CL, Matthews HS (2007) Quantifying the global and distributional aspects of American household carbon footprint. Ecol Econ 66(2–3):379–391. https://doi.org/10.1016/j.ecolecon.2007.09.021

Formation of a Mechanism for Green Investment in the Economy

Alexey V. Erpelev, Viktor A. Ilyin, Yuliya S. Stashina, and Irina S. Kholina

Abstract The relevance of this research is determined by the growing importance of the green investment market in achieving the UN Sustainable Development Goals (SDGs). With proper government regulation, these investments can become a valuable tool for their implementation. This concept is grounded in an integrated approach to ESG (environmental, social, and corporate governance), which considers the interrelated and complementary aspects of environment, society, and governance. Recently, issues surrounding environmental and economic development, along with the financing and support of the green investment market, have gained increasing importance for domestic and international scholars, researchers, and practitioners. Nevertheless, despite the rising interest, numerous unresolved challenges remain regarding the establishment of an effective sustainable development model and the assurance of green security within the Russian economy and its regions. This research examines the primary challenges and opportunities for advancing green financing in Russia. The authors offer practical recommendations to improve the investment climate for environmental projects, which is crucial for promoting sustainable economic growth. Despite growing interest in green investments, this segment of the Russian financial market remains underdeveloped and understudied, creating obstacles to its further progress. The practical significance of this research lies in the potential to utilize the obtained results to evaluate the current investment landscape in the green finance market and develop strategies and measures to attract potential investors and stimulate their activity.

Keywords Financing · Ecology · Economic development · Green instruments · Financial market

A. V. Erpelev (✉) · V. A. Ilyin · Y. S. Stashina · I. S. Kholina
Russian State Social University, Moscow, Russia
e-mail: ErpelevAV@rgsu.net

Y. S. Stashina
e-mail: DrygoJUS@rgsu.net

I. S. Kholina
e-mail: EvreinovaIS@rgsu.net

A. B. Karbekova (ed.), *Sustainable Development of Rural Territories Based on Climate-Smart Technologies*, Climate Change Management,
https://doi.org/10.1007/978-3-032-20106-5_22

JEL Classification E22 · E63 · G00 · G31

1 Introduction

In today's world, environmental issues have become an integral part of implementing international and national sustainable development goals, as well as shaping a green economy. This emerging model prioritizes environmentally oriented strategies, opening new opportunities for the future development of the global financial system and responsible investment. There is a noticeable shift in the investment paradigm toward so-called variable or impact investing, which considers economic returns and social and environmental outcomes. According to expert forecasts, by 2030, global investment in green projects could reach approximately $93 billion.

Simultaneously, each country is developing its mechanisms for financing innovative projects and investments in the green sector to drive fundamental changes in approaches to sustainable production and the smart consumption of natural resources, as well as to advance national development priorities. The significance of these issues is evident, given the serious challenges facing society and the renewed socio-political interest in green investment within Russia. Among the various financial instruments in the green sector (e.g., loans, credits, issuance rights, green procurement, and others), green bonds stand out in particular. The total volume of the global green bond market is approximately $390 billion. According to the Climate Bonds Initiative, the volume of green bonds reached $167.3 billion by the end of 2021 and is projected to grow to $250 billion by 2023. The analysis of expert perspectives on this key green financing instrument provides insight into the development prospects of the green bond market in Russia and helps define areas for dialogue among stakeholders.

Of particular importance is Russia's international commitment to the environmental goals enshrined in the Paris Agreement on climate change, the transformation of the global economic system under the influence of investments in environmentally sustainable technologies and industries, and the growing importance of environmentally friendly and low-carbon business models. Simultaneously, there is a risk that Russia will fall behind global trends, which contradicts the justification for some countries to create their green development financing schemes.

Internal incentives for the development of green financial resources are formed within the framework of integrating environmental and economic policy instruments, which is associated with environmental reforms and the institutionalization of Russian economic structures, including the activities of federal bodies and companies.

Among the key factors contributing to the development of green bonds in Russia and shaping their prospects, session participants emphasized the directive issued by the President of the Russian Federation following the extended meeting of the State Council Presidium. The directive called for the prompt implementation of a green bond mechanism, which is expected to enhance the appeal of direct investments from both domestic and foreign sources and reduce borrowing costs.

2 Methodology

During the research, the authors employed several methods to identify the core principles underlying the formation of the green investment mechanism within the economy. These methods included the following:

- Study and analysis of literature. Electronic libraries and scientific articles have all the necessary information to achieve the goals set in this research;
- The descriptive method was used to reveal the essence of the formation of the green investment mechanism in the economy;
- Induction was used to make logical conclusions based on the data obtained;
- Synthesis was used to create a single work material from several pieces of information provided on various scientific resources.

The research employs analysis and synthesis of economic literature, along with various empirical methods, including abstraction, generalization, comparison, and observation. The concepts of socially sustainable development and green investment emerged in the last century; the ESG framework has deep historical roots. In recent years, environmental, social, and governance factors have gained increasing significance in the business world due to the global challenges the world is facing. There are also specialized green funds that invest in companies committed to environmental sustainability. A key aspect of this research is the analysis of the current state of Russian research in this field, with particular emphasis placed on the role of green investors and the integration of public policy tools to support the development of the green investment market.

3 Results

The issue of green financing encompasses all aspects of human activity. Most international organizations, in response to current challenges and the pursuit of sustainable development, incorporate the environmental dimension into their strategies and programs (Kazantseva 2020). The transition to a green economy (i.e., a model aimed at improving the quality of life and social justice while significantly reducing environmental risks and degradation of nature) requires careful theoretical analysis. Green investments are understood as investments in technologies and projects that promote resource conservation and environmental friendliness, supporting the development of a green economy through the implementation of relevant initiatives. UNEP defines a green economy as one that simultaneously improves the standard of living, ensures social justice, and significantly reduces environmental threats (Agarkov and Matviishin 2019).

Existing statistical systems at national and international levels apply varying approaches to classifying investments. Under the umbrella of green investments, they include projects in environmentally friendly industries and initiatives focused on environmental protection and the efficient use of resources. This diversity in classification makes it difficult to accurately assess the total volume of green investments. Currently, the green economy accounts for approximately 2.7–3% of global GDP, with revenues from such investments reaching around $1.5 billion and employment in the sector totaling about ten million people. Higher figures are observed in countries such as the UK, the USA, Germany, and Japan, where most funds are directed toward the development of eco-friendly technologies and infrastructure. The growth of this sector is supported by two key pillars: institutional and investment frameworks. The highest concentration of green investments is found in G20 countries. The primary objective of policy in this area is to ensure sustainable and balanced economic growth (Agarkov and Matviishin 2019).

In addition to material resources, a high level of education, and a functioning healthcare system, essential components of sustainable development include clean air, access to clean water, and a stable energy supply, areas in which many countries face significant challenges. Special attention is given to the temporal adaptability of green financing instruments, which have demonstrated resilience during times of crisis and pandemics. Notably, there has been a steady increase in the value of such instruments between 2021 and 2023.

Environmental issues (i.e., pollution, loss of natural habitats, reduction of biodiversity, and resource scarcity) make environmental issues increasingly important for the global economy. Many countries are now paying significant attention to the green economy, which involves using clean energy and caring for the environment. The development of advanced technologies and support for environmental initiatives create opportunities for growth in this area (Borkova et al. 2019; Borkova 2013).

Sustainable development has key international benchmarks. The main one is the "2030 Agenda for Sustainable Development," adopted by the UN in 2015, which includes 17 Sustainable Development Goals (SDGs), including environmental protection, clean energy, and responsible use of resources. In parallel, the Paris Agreement aims to combat climate change by reducing greenhouse gas emissions (Bagiev et al. 2021).

Green finance plays a key role in implementing international and national sustainable development strategies, easing the transition to an environmentally sustainable economy, and improving the quality of life of people worldwide. The Paris Agreement requires each country to develop and implement its own national climate change strategy, which aims to reduce emissions and adapt to their impacts. These strategies must be regularly reviewed and updated every five years to improve their effectiveness and ambition. Additionally, the agreement creates a mechanism for financial and technological support for countries that need help in mitigating climate change and reducing emissions.

To protect the environment, countries are implementing various measures, including transitioning to renewable energy sources and enhancing energy efficiency. Nature conservation, improving water supplies, developing innovative solutions, and utilizing land more sustainably are also important. Developed countries have committed to distributing $100 billion a year until 2025 to support environmental projects in developing countries, which will help achieve the goals of the agreement.

The International Energy Agency estimates that keeping the global temperature rise to 2 °C by 2100 will require investments of around $53 trillion by 2035 (Borkova 2013).

In Russia, sustainable development and the transition to a green economy are important tasks. However, despite a significant increase in companies' spending on environmental protection (more than 4.5 times from 2000 to 2018), the share of investments in environmental projects (green investments) in the total volume of capital investments stays extremely low, only 0.9%. Moreover, this share has decreased almost by half compared to the figures of the early 2000s, which shows insufficient support for green growth (Myagkova et al. 2024).

Environmental investments in Russia are distributed unevenly across regions. The analysis revealed that regions can be categorized into five groups based on the volume of green financing. The leaders are Moscow, St. Petersburg, the Tyumen Region, the Yamalo-Nenets Autonomous Area, and the Krasnoyarsk Region. The regions actively working in this area also include the Komi Republic, Volgograd Region, Bashkortostan, Tatarstan, and Perm and Sverdlovsk Territories. Most regions are just beginning to develop green financing; a significant portion is still in the early stages, with minimal investments in environmental protection. The gap between the leaders and the laggards is enormous: the former attracts about a quarter of all environmental funds, while the latter attracts only 1%.

The analysis of data for 2000–2023 showed that increased investment in green technologies is associated with higher rates of economic growth in Russia. This confirms the importance of green investment for sustainable economic development and the transition to a green economy. The degree of influence of numerous factors on economic growth is assessed using a statistical method called multiple regression (Yamshchikov et al. 2024).

Investments in ecology and resource conservation have a positive effect on the Russian economy. According to the study, every million rubles distributed for these purposes increases the country's GDP by 0.1 billion rubles. Similarly, if the current expenses of organizations on environmental protection increase by one million rubles, this contributes to a 0.3 billion ruble increase in the country's GDP. These results show a significant, although not yet maximum, impact of green investments on the country's economic development.

The underdevelopment of green investments in Russia is attributed to several factors, the most significant of which is the insufficient development of the institutional environment and national financial markets. The absence of effective mechanisms for regulating and supporting green projects hinders the flow of investment into environmentally oriented sectors of the economy. Another important aspect is the assessment of the transparency of non-financial disclosures by issuing companies, along with the analysis of independent sources that may highlight discrepancies or violations. This approach enables a more objective evaluation of companies' adherence to the principles of socially sustainable development.

The experience of foreign countries demonstrates that a successful transition to sustainable development necessitates an active role for the state and business. At the international level, new norms are emerging that assign government bodies the primary role in reforming national economies. To promote environmentally sustainable development and reduce negative environmental impacts, governments are formulating comprehensive strategies, programs, and action plans. However, it is the private sector (not public funding) that plays the leading role in financing the transition to a green economy.

Instead of directly funding environmental initiatives, the state should focus on creating an effective regulatory system and providing information. This system should encourage the adoption of environmentally friendly technologies and the development of green production practices, while limiting the activities of enterprises that have a negative impact on the environment. The goal of such measures is to redirect capital flows to sustainable sectors of the economy that ensure environmentally responsible production and consumption.

To overcome the resistance of traditional, unstable sectors of the economy, the state must develop and implement a comprehensive policy to reform the national financial system. This includes providing businesses with access to market resources, tools, and technologies for financing green investments. Such an approach will foster conditions for sustainable development and enhance the country's environmental security (Bagiev et al. 2021).

Active participation of the state is necessary to successfully transition private investment to environmentally sustainable industries. Simultaneously, state support instruments are developed considering global trends and specific conditions, problems, and development features of each country (Maloletko 2024).

The analysis of the effectiveness of measures taken by foreign governments to attract private capital to green projects shows that the key factor is government funding and expenditures aimed at supporting such initiatives. Additionally, active environmental policies contribute to the growth of private investment in green sectors. Nevertheless, in some cases, excessive strictness of environmental standards can function as a deterrent to business (Bagiev et al. 2021).

The EU has proved leadership in sustainable finance with its Action Plan on Sustainable Development Financing adopted in 2018. The plan aims to direct investments toward environmentally and socially responsible projects, encourage financial institutions to consider sustainability risks, and enhance the transparency of financial transactions.

Despite considerable progress in developing a green economy, the EU countries and Russia face familiar challenges that slow down the transition to sustainable growth and hinder the attraction of private investment. In this regard, a corresponding initiative was developed (Vertakova and Plotnikov 2020).

To kick-start the implementation of the sustainable finance plan, a common language and classification system in this area have been created. This includes defining the terms "sustainable" and "green," as well as setting up common rules for "green" financial products, including their labelling. A special EU commission will determine which activities can be considered sustainable; these decisions will be enshrined in EU law, ensuring that standards apply to all sectors of the economy, starting with investments (Vertakova and Plotnikov 2020).

Thus, the state plays a crucial role in promoting green investments through direct financing, establishing a regulatory framework, developing strategies, and creating conditions that facilitate the attraction of private capital into environmentally sustainable projects. The experience of the European Union and other countries demonstrates that only a comprehensive approach (i.e., combining political will, legislative initiatives, and active government involvement) can ensure a successful transition to a green economy and sustainable development. Investments in green projects and related financial instruments represent a long-term strategy aimed at addressing global challenges in the environmental, economic, and social sectors.

Green investments serve as a platform for raising capital from private and public investors, channeling it into innovative initiatives that improve environmental, social, and governance practices. The primary goal of such investments is to ensure sustainable development, rather than to make a quick profit.

Green finance is a rapidly growing field that combines financial considerations with environmental concerns. Socially responsible investments support projects that generate profits and improve the environment and society. It is an effective way to stimulate the development of environmentally friendly technologies, reduce harmful impacts on nature, and build a sustainable economy.

The exchange of scientific and technical knowledge between countries is critical for effective environmental protection and the transition to a green economy. This exchange stimulates the creation of a market for innovations—valuable knowledge and information, which may include equipment and documentation, as well as experience and advanced methods. Given the disparities in scientific and technological development and the varying economic levels of different countries, along with the insufficient public awareness of environmental issues (particularly in developing nations), it is essential to intensify efforts in this area. Addressing global environmental challenges is only possible through the concerted efforts of the international community. Therefore, the study and advancement of technology transfer mechanisms within countries and across borders are crucial components of green projects.

Technology transfer is a valuable tool for disseminating new developments. It can be done in various ways, for example, through licenses or the exchange of experience. Although the global market for environmentally friendly technologies is growing rapidly, Russia remains significantly behind leading countries in their development and implementation. This is due to an insufficient number of patents, limited funding for green research and development, and lagging in key areas of environmental protection (e.g., air and water purification and waste recycling). Russian companies also invest less in innovation than their foreign competitors.

Investing always involves certain risks, including potential loss of capital. It is impossible to cut all possible risks. Making investment decisions, especially in socially responsible projects (e.g., green investments), requires careful analysis and an understanding of long-term prospects that go beyond pure financial profit (Kazantseva 2020).

4 Conclusion

Thus, the extractive sector accounts for about 25% of all investments, of which green investments make up approximately 1.23%. About 44–45% of investments are concentrated in the manufacturing industry. The share of green investments in this sector reaches 2.16%. Metallurgy, as well as the production and distribution of electricity, gas, and water, account for approximately 26% of investments, with green investments comprising around 2.24%.

In 2010, investments in fixed capital for the protection and rational use of water resources amounted to 445.719 billion rubles. By 2020, this figure had decreased to 46.674 billion rubles. In 2023, it dropped further to 12.146 billion rubles. Notably, there were no investments recorded for air protection during this period.

Russia's transition to a green economy and the adoption of sustainable development principles are becoming critically important for the country's long-term stability. The growing dependence on raw material exports, exacerbated by falling hydrocarbon prices and the global shift toward alternative energy, poses significant economic risks. To mitigate these risks and avoid future crises, it is essential to actively develop environmentally friendly industries and diversify the national economy (Pashkus et al. 2011).

The depletion of domestic natural resources, the rising cost of their extraction, and the ongoing search for new deposits threaten to reduce the profitability of the mining industry. The Russian economy's vulnerability to fluctuations in global oil prices and geopolitical tensions further heightens these risks, potentially hindering long-term economic growth. To mitigate these challenges and enhance the country's economic security, it is essential to stimulate the development of alternative industries and actively pursue import substitution (Kaurova et al. 2024).

The acute environmental situation in Russia requires a transition to a green economy. A sizable part of the urban population (about 17%) lives in areas with severe air pollution, which negatively affects human health and overall well-being. The problem is exacerbated by the fact that nearly a fifth of wastewater is discharged untreated; most of it (70%) is inadequately treated. Effective water treatment systems cover only 60% of the total water volume. Additionally, soil pollution and the prevalence of a consumer approach in forestry and agriculture lead to deterioration of public health and increased social tension. Sociological studies indicate that most Russians (over half) are concerned about the deterioration of the environmental situation; a third (35%) are willing to express their position through participation in environmental campaigns, which reflects growing public anxiety.

Government green bonds are a promising way to stimulate socially responsible investing in Russia. An example is the Moscow government's 70-billion-ruble bond issue in 2021. Due to the government guarantee, these instruments are considered more dependable than corporate securities. When choosing green bonds, it is important to evaluate the environmental and social responsibility of the issuing company for conscious investing.

Theoretical analysis confirms the high importance of ESG standards in global practice. In Russia, there is a renewed and growing interest in these principles. During epidemics, green stocks and bonds, including examples of companies like Aptor, prove steady growth, which is associated with investment principles and investor behavior in the conditions of uncertainty and a general decline in stock markets. The growing popularity of sustainable investment criteria contributes to a significant increase in the volume of green investments. According to VEB RF forecasts, the volume of green bonds in Russia may more than double in the coming years, exceeding 500 billion rubles, which shows the prospects of this area. It is also important to understand that green investments are not a source of high profits; they are primarily a tool for solving social problems aimed at improving the quality of life for the population.

References

Agarkov SA, Matviishin DA (2019) The effect of the hydrocarbons regional resources development on the Pechora sea ecological state. Bull St Petersburg State Univ Econ 2(116):58–67

Bagiev GL, Cherenkov VI, Cherenkova NI (2021) Marketing for implementing the concept of sustainable development: essence and terminological paradigm. Bull St Petersburg State Univ Econ 4(112):139–152

Borkova EA (2013) Dynamic model of investment climate evaluation when managing investment processes. Bull INZHEKON Ser: Econ 5(64):19–24

Borkova EA, Timchenko MN, Markova AA (2019) Investments in green technologies as a tool of the Russian economic growth. Bus Educ Law Bull Volgogr Bus Inst 3(48):87–91. https://doi.org/10.25683/VOLBI.2019.48.341

Kaurova OV, Balalova EI, Maloletko AN, Tkach AV, Kovshova MV (2024) Role of crop production for sustainable agricultural development. In: Popkova EG, Kaurova OV, Maloletko AN (eds) Sustainable cooperation for the creation of green supply chains based on environmental technologies and responsible innovations. Springer, Cham, Switzerland, pp 117–124. https://doi.org/10.1007/978-3-031-70962-3_14

Kazantseva AN (2020) Issues of development and implementation of state policies in the sphere of production and applying environmentally pure products. Theory Pract Serv: Econ, Soc Sphere, Technol 2(36):16–20

Maloletko AN (2024) The role of agricultural and consumer cooperation in ensuring food security. OpenScience 6(3):66–77

Myagkova MV, Zakharkina RA, Abelova LA, Kataykina NN, Markova AA (2024) Trends of the development of the green economy in Russia. In: Popkova EG, Kaurova OV, Maloletko AN (eds) Sustainable cooperation for the creation of green supply chains based on environmental technologies and responsible innovations. Springer, Cham, Switzerland, pp 455–463. https://doi.org/10.1007/978-3-031-70962-3_51

Pashkus NA, Alcksandrova AI, Aleksunin VA, Archegov AB, Akhobadze DT, Balueva MV, Freidkina EM (2011) Competitiveness and breakthrough positioning in the post-crisis period. Herzen State Pedagogical University of Russia, St. Petersburg, Russia

Vertakova Y, Plotnikov V (2020) Problems of sustainable development worldwide and public policies for green economy. Econ Ann XXI 166(7–8):4–10. https://doi.org/10.21003/ea.V166-01

Yamshchikov AS, Karabitsina EF, Guts DS (2024) State support for non-profit organizations of the social sphere in the ESG concept. OpenScience 6(1):86–100

The Role of Climate-Smart Technologies in Agricultural Development and Enhancing the Resilience of Rural Areas

Nikolay I. Litvinov, Aktam U. Burkhanov, Zalina B. Tedeyeva, and Nuraiym Timur kyzy

Abstract The research examines the role of climate-smart technologies in the development of agriculture and strengthening the resilience of rural areas. Moreover, the research offers recommendations for enhancing the effectiveness of these technologies. To this end, the authors conducted trend and regression analyses using statistical data from five developed and five developing countries for 2021 and 2025. As a result, the authors constructed econometric models to describe the sustainable development of rural areas under the influence of factors related to the use of climate-smart technologies in agriculture, separately for developed and developing countries. According to these models, the authors conclude that the sustainability of rural development is determined by different factors of implementing and using climate-smart technologies in developed (the level of development of electronic communities) and developing countries (the robustness of financial support and the innovative activity of agricultural enterprises in adopting and using climate-smart technologies). Based on the constructed models, the prospects for sustainable rural development driven by the growth of climate-smart agriculture through 2030 are identified for developed and developing countries, considering their specific characteristics. Applied solutions are proposed to optimize the role of climate-smart technologies in agricultural development in both groups of countries.

N. I. Litvinov
Volgograd State Agricultural University, Volgograd, Russia
e-mail: kolialitvinov97@yandex.ru

A. U. Burkhanov (✉)
Alfraganus University, Tashkent, Uzbekistan
e-mail: burxanov.a@afu.uz; a.burkhanov@nuu.uz

National University of Uzbekistan named after Mirzo Ulugbek, Tashkent, Uzbekistan

Z. B. Tedeyeva
Vladikavkaz Branch of the Financial University under the Government of the Russian Federation, Vladikavkaz, Russia

N. Timur kyzy
Osh State University, Osh, Kyrgyzstan

A. B. Karbekova (ed.), *Sustainable Development of Rural Territories Based on Climate-Smart Technologies*, Climate Change Management,
https://doi.org/10.1007/978-3-032-20106-5_23

Keywords Climate-smart technologies · Agriculture · Territorial resilience · Rural areas · Developed countries · Developing countries

JEL Classification O13 · O18

1 Introduction

The global fight against climate change aims to create a more favorable environment for living and conducting economic activity. These efforts are primarily focused on reducing the severity of environmental pollution in major cities (i.e., industrial agglomerations) by cutting waste generation and lowering the consumption of natural resources in industrial production to slow their depletion.

Against this backdrop, significantly less attention is given to the development of rural areas, because they are less populated and the economic activities carried out there contribute less to the economic growth of today's countries. Simultaneously, rural areas are, for the most part, green agglomerations with the most favorable environmental conditions and are therefore highly desirable places to live.

It is worth noting that prolonged, systemic, large-scale urbanization (a clearly pronounced global trend in domestic migration) has by now created a serious imbalance within national economies. Deep-rooted disparities caused by overcrowded cities and the depopulation of rural areas necessitate a reconsideration of national migration policies, which should aim to correct these disproportions.

To achieve a more even distribution of population between urban and rural territories, ecological advantages alone are insufficient; the economic development potential of rural areas must also be more fully realized. Since these areas are typically characterized by economies specializing in agricultural production, climate change mitigation in rural regions largely involves adapting traditional agriculture to shifting climatic conditions. The emergence of the digital economy, shaped by the Fourth and later the Fifth Industrial Revolution, has opened broad new opportunities in this regard.

Digitalization initially affects major urban centers and is accompanied by the creation of smart cities and smart industries, yet technological innovations are now increasingly spreading to rural territories as well. Nevertheless, the impact of technological transformation on rural areas remains insufficiently studied. As a result, the development of these territories, one of the potential applications of digital innovation, often lacks a solid scientific and methodological foundation and is therefore unfolding in a largely spontaneous manner.

The above underscores the importance of a deeper exploration of how digitalization can contribute to the climate adaptation of agricultural economies in support of sustainable rural development. Motivated by the desire to address this issue, the research presented in this article aims to identify the role of climate-smart technologies in agricultural development and in enhancing the resilience of rural areas, as well as to develop recommendations for improving the effectiveness of these technologies.

2 Literature Review

The fundamental basis of this research is the scientific and theoretical concept of climate adaptation in agriculture, interpreted in the works of authors such as Rogachev et al. (2018) as a source of economic development for rural areas.

According to this concept, the following three key areas of sustainable rural development, closely interconnected, are identified in the available literature:

- The share of the rural population in the total population (Vechkinzova et al. 2019);
- The share of agriculture in GDP structure (Huseynov et al. 2024);
- The share of employment in agriculture in total employment (Tovmasayn and Gevorgyan 2024).

A review of the available scholarly literature on the topic also made it possible to identify the following key factors influencing the implementation and use of climate-smart technologies in agriculture:

- The prevalence and level of digital skills in society, which determine the availability of digital agri-specialists for agricultural enterprises adopting and using climate-smart technologies (Szpilko et al. 2023);
- The robustness of financial support for the introduction and application of climate-smart technologies in agriculture (Salnazaryan et al. 2022);
- The level of development of electronic communities, which shapes demand for climate-smart agricultural products (Hayrapetyan and Isayan 2022);
- The accessibility of climate-smart agrotechnologies for agricultural enterprises (Burkhanov et al. 2024);
- The innovative activity of agricultural enterprises in adopting and utilizing climate-smart technologies (Bogoviz et al. 2019).

However, despite the abundance of accumulated scientific sources, the impact of these identified factors on the sustainability of rural development remains insufficiently explored. This uncertainty represents a significant gap in the literature. It is also important to note the widely recognized differences in the economic development of developed and developing countries, which makes it advisable to study the sustainable development of rural areas in these groups separately.

The resulting research question posed is: "Which factors of implementing and using climate-smart technologies in agriculture contribute to the sustainable development of rural areas, and do these factors differ between developed and developing countries?" Drawing on previous works, including Krutilin et al. (2022), which provide evidence of divergent approaches to the digitalization of agricultural economies in these country groups, this research advances the hypothesis that the factors determining sustainable rural development through climate-smart agricultural technologies differ between developed and developing countries.

3 Materials and Methods

In the first stage of the research, conducted to test the proposed hypothesis, the role of climate-smart technologies in agricultural development and strengthening the resilience of rural areas is examined using a dataset covering the years 2021 and 2025 in developing (Argentina, China, Kazakhstan, Malaysia, and Saudi Arabia) and developed countries (Austria, Belgium, Bulgaria, Canada, and Croatia).

Using regression analysis, the research evaluates the dependence of rural sustainability, measured through the share of the rural population in total population (RTA_1, according to World Bank (2025c)), the share of agriculture in GDP (RTA_2, according to World Bank (2025a)), and the share of agricultural employment in total employment (RTA_3, according to World Bank (2025b)), on factors associated with the implementation and use of climate-smart technologies in agriculture (based on IMD statistics (2025)):

- The prevalence and level of digital skills in society (CST_1);
- The share of R&D funding in GDP (CST_2);
- The level of development of electronic communities (CST_3);
- The accessibility of smart technologies (CST_4);
- The innovative activity of enterprises (CST_5).

The empirical foundation of this research is presented in Table 1.

Based on the results of the regression analysis, those factors ($CST_{1–5}$), that exert a positive influence, i.e., those with positive regression coefficients, on the outcome variables ($RTA_{1–3}$) are selected. The proposed hypothesis is considered confirmed if the set of selected factors differs between developed and developing countries.

The second stage of this research identifies the prospects for increasing the sustainability of rural areas through a fuller realization of the potential of the selected climate-smart technologies in agricultural development for developed and developing countries through 2030. To achieve this, the authors used trend analysis to determine the growth rate of the variables from Table 1 over the previous five-year period (2025 compared to 2021).

Forecast values of the selected factor variables for the subsequent five-year period are then calculated by projecting their established growth rates to 2030. These projected values are substituted into the regression equations separately for each country group. The resulting values are compared with forecast values obtained under the assumption of continuing past trends.

4 Results

In the first stage of the research, to verify the validity of the proposed hypothesis and clarify the role of climate-smart technologies in agricultural development and strengthening rural sustainability, the authors conducted a regression analysis using

Table 1 Climate-smart technologies, agricultural development, and rural sustainability in 2021 and 2025

Year	Country	Digital/ Technological skills, score 0–10	Total expenditure on R&D, % GDP	E-Participation, Shares of 1	Use of big data and analytics, score 0–10	Digital transformation in companies, score 0–10	Rural population, % of total population	Agriculture, forestry, and fishing, value added, % of GDP	Employment in agriculture, % of total employment
		CST_1	CST_2	CST_3	CST_4	CST_5	RTA_1	RTA_2	RTA_3
Developing countries (i)									
2021	Argentina	5.28	0.52	0.65	4.51	4.96	7.89	6.36	0.61
	China	7.74	2.38	0.86	5.97	6.67	38.57	7.54	24.72
	Kazakhstan	6.11	0.13	0.81	6.56	6.45	42.33	5.39	13.70
	Malaysia	7.19	0.97	0.68	5.48	6.43	22.84	8.17	10.22
	Saudi Arabia	7.65	0.44	0.69	5.38	6.70	15.71	2.96	3.47
2025	Argentina	6.15	0.61	0.63	5.14	5.45	7.42	5.98	0.61
	China	7.77	2.58	0.93	6.54	6.90	34.46	6.78	22.33
	Kazakhstan	6.57	0.14	0.85	6.49	6.91	41.61	3.94	12.13
	Malaysia	6.70	1.00	0.70	4.99	6.16	20.80	8.16	9.84
	Saudi Arabia	8.25	0.56	0.96	6.40	7.26	14.83	2.54	2.83
Developed countries (j)									
2021	Austria	6.37	3.26	0.77	5.38	5.85	41.25	1.07	3.66
	Belgium	7.00	3.41	0.45	4.93	5.59	1.92	0.71	0..92
	Bulgaria	6.70	0.77	0.74	4.04	4.58	24.31	3.43	6.62
	Canada	8.00	1.87	0.83	6.21	6.62	18.44	1.84	1.54
	Croatia	6.14	1.24	0.74	3.75	4.46	42.45	3.13	6.19
2025	Austria	6.78	3.29	0.78	4.87	5.65	40.18	1.23	3.12

(continued)

Table 1 (continued)

Year	Country	Digital/ Technological skills, score 0–10	Total expenditure on R&D, % GDP	E-Participation, Shares of 1	Use of big data and analytics, score 0–10	Digital transformation in companies, score 0–10	Rural population, % of total population	Agriculture, forestry, and fishing, value added, % of GDP	Employment in agriculture, % of total employment
		CST_1	CST_2	CST_3	CST_4	CST_5	RTA_1	RTA_2	RTA_3
	Belgium	7.33	3.48	0.51	5.54	6.13	1.78	0.80	1.09
	Bulgaria	6.58	0.79	0.67	3.63	4.43	22.96	2.07	5.68
	Canada	8.12	1.81	0.92	6.84	7.16	18.02	1.60	1.27
	Croatia	6.33	1.39	0.92	4.25	4.81	41.06	3.41	4.68

Source Developed by the authors based on the materials from IMD (2025) and World Bank (2025a, 2025b, 2025c)

the statistical data from Table 1 for each group of countries separately. The model of sustainable rural development under the influence of factors related to the use of climate-smart technologies in agriculture for the group of developing countries (i) took the following form:

$$\begin{aligned}
RTA_{i1} &= -45.00 - 23.51CST_{i1} + 12.23CST_{i2} - 44.15CST_{i3} + 4.98CST_{i4} \\
&\quad + 35.54CST_{i5}; \\
RTA_{i2} &= 13.95 - 2.35CST_{i1} + 2.69CST_{i2} - 10.46CST_{i3} - 0.70CST_{i4} \\
&\quad + 2.78CST_{i5}; \\
RTA_{i3} &= -22.83 - 10.84CST_{i1} + 11.29CST_{i2} - 24.16CST_{i3} + 1.57CST_{i4} \\
&\quad + 16.80CST_{i5}.
\end{aligned} \tag{1}$$

According to Model (1), in developing countries a one-point increase in the prevalence and level of digital skills in society leads to a reduction in the share of the rural population in the total population by 23.51%, a reduction in the share of agriculture in GDP by 2.35%, and a reduction in the share of agricultural employment in total employment by 10.84%.

An increase of 1% in the share of R&D funding in GDP contributes to an increase in the share of the rural population in the total population by 12.23%, the share of agriculture in GDP by 2.69%, and the share of agricultural employment in total employment by 11.29%.

A rise of 1 unit in the level of development of electronic communities results in a decrease in the share of the rural population in the total population by 44.15%, a decrease in the share of agriculture in GDP by 10.46%, and a decrease in the share of agricultural employment in total employment by 24.16%.

A one-point increase in the availability of smart technologies leads to a 4.98% increase in the share of the rural population in total population, a 0.70% decrease in the share of agriculture in GDP, and a 1.57% increase in the share of agricultural employment in total employment.

A one-point increase in the innovative activity of enterprises is associated with a 33.54% increase in the share of the rural population in total population, a 2.78% increase in the share of agriculture in GDP, and a 16.80% increase in the share of agricultural employment in total employment.

According to the results of the correlation analysis, the combined factors of climate-smart technology adoption and use in agriculture explain 96.44% of the variation in the share of the rural population in total population (the equation for RTA_{i1} is significant at the 0.05 level, F-value = 0.0199), 87.39% of the variation in the share of agriculture in GDP (the equation for RTA_{i2} is significant at the 0.20 level, F-value = 0.1891), and 97.31% of the variation in the share of agricultural employment in total employment (the equation for RTA_{i3} is significant at the 0.05 level, F-value = 0.0116).

The Fisher test, conducted and successfully passed, demonstrated that Model (1) is reliable and suitable for further use in this study as the mathematical basis for forecasting in developing countries. Consequently, in developing countries, the variables CST_{i2} and CST_{i5} exert a positive influence on all three resulting variables (RTA_{i1-3}) because their regression coefficients in Model (1) are positive.

The model of sustainable rural development under the influence of climate-smart agricultural technologies for the group of developed countries (j) was as follows:

$$\begin{aligned} RTA_{j1} &= 59.37 - 23.35CST_{j1} - 1.19CST_{j2} + 77.16CST_{j3} - 30.31CST_{j4} \\ &\quad + 40.46CST_{j5}; \\ RTA_{j2} &= 3.81 + 1.17CST_{j1} + 0.20CST_{j2} + 3.92CST_{j3} + 3.47CST_{j4} \\ &\quad - 5.51CST_{j5}; \\ RTA_{j3} &= 19.06 - 1.84CST_{j1} - 1.06CST_{j2} + 1.12CST_{j3} - 0.12CST_{j4} \\ &\quad - 0.15CST_{j5}. \end{aligned} \tag{2}$$

Model (2) indicates that in developed countries, a one-point increase in the prevalence and level of digital skills in society leads to a 23.34% decrease in the share of the rural population in total population, a 1.17% increase in the share of agriculture in GDP, and a 1.84% decrease in the share of agricultural employment in total employment.

An increase of 1% in the share of R&D funding in GDP contributes to a 1.19% rise in the share of the rural population in total population, a 0.20% decline in the share of agriculture in GDP, and a 1.06% increase in the share of agricultural employment in total employment.

A one-point increase in the level of development of electronic communities is accompanied by a 77.16% rise in the share of the rural population in total population, a 3.92% increase in the share of agriculture in GDP, and a 1.12% increase in the share of agricultural employment in total employment.

A one-point increase in the availability of smart technologies leads to a 30.31% reduction in the share of the rural population in total population, a 3.47% increase in the share of agriculture in GDP, and a 0.12% decrease in the share of agricultural employment in total employment.

A one-point increase in the innovative activity of enterprises is associated with a 40.46% increase in the share of the rural population in total population, a 5.51% decline in the share of agriculture in GDP, and a 0.15% reduction in the share of agricultural employment in total employment.

According to the results of the correlation analysis, the combined factors of climate-smart technology adoption and use in agriculture explain 97.84% of the variation in the share of the rural population in total population in developed countries (the equation for RTA_{j1} is significant at the 0.01 level, $F = 0.0077$), 96.73% of the variation in the share of agriculture in GDP (the equation for RTA_{j2} is significant at the 0.05 level, $F = 0.0169$), and 96.22% of the variation in the share of agricultural employment in total employment (the equation for RTA_{j3} is significant at the 0.05 level, $F = 0.0224$).

The Fisher test, conducted and successfully passed, demonstrated that Model (2) is reliable and suitable for further use in this study as the mathematical basis for forecasting in developed countries. Consequently, in developed countries, only one

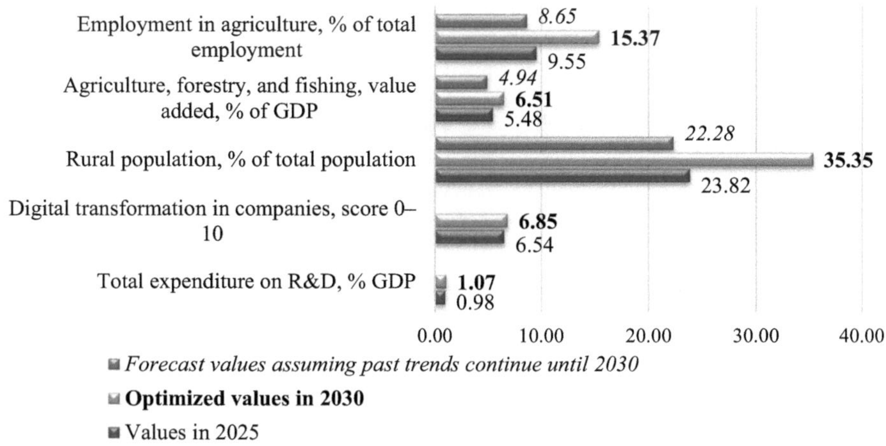

Fig. 1 Prospects for sustainable development of rural areas through the growth of climate-smart agriculture in developing countries by 2030. *Source* Developed by the authors

factor variable (CST_{j3}) exerts a positive influence on all three resulting variables (RTA_{j1-3}) because its regression coefficients in Model (2) are positive.

In the second stage of the research, to identify the prospects for enhancing rural sustainability through a fuller realization of the potential of the selected climate-smart technologies in agricultural development, a trend analysis of the data from Table 1 was conducted. Alternative forecasts to 2030, one based on the continuation of past trends and the other on the optimization of the selected factor variables, for developing countries, constructed using Model (1), are presented and compared with each other and with the baseline data for 2025 in Fig. 1.

According to Fig. 1, to optimize the role of climate-smart technologies in agricultural development in developing countries, it is recommended to increase the share of R&D funding in GDP by 9.43% (from 0.98% in 2025 to 1.07% by 2030, assuming the previous five-year trend continues) and raise the innovative activity of enterprises by 94.78% (from 6.54 points in 2025 to 6.85 points by 2030, also following the past trend). Thanks to this optimization, more substantial progress is projected across all areas of sustainable rural development in developing countries, specifically the following:

- An increase in the share of the rural population in the total population by 48.37%, reaching 35.35% (instead of declining to 22.28% if past trends continue);
- A rise in the share of agriculture in GDP by 18.79%, reaching 6.51% (instead of decreasing to 4.94% under the continuation of past trends);
- An increase in the share of agricultural employment in total employment by 60.92%, reaching 15.37% (instead of dropping to 8.65% if past trends persist).

Alternative forecasts to 2030 (one based on the continuation of past trends and the other on the optimization of the selected factor variables) for developed countries,

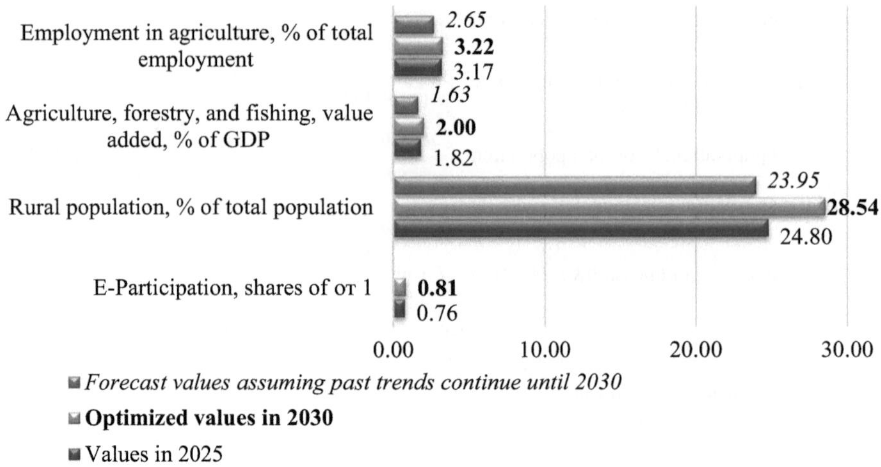

Fig. 2 Prospects for sustainable development of rural areas through the growth of climate-smart agriculture in developed countries by 2030. *Source* Developed by the authors

constructed using Model (2), are presented and compared with each other and with the baseline 2025 data in Fig. 2.

Based on Fig. 2, to optimize the role of climate-smart technologies in agricultural development in developed countries, it is recommended to increase the level of development of electronic communities by 6.74% (from 0.76 points in 2025 to 0.81 points by 2030, assuming the continuation of the previous five-year trend). Owing to this optimization, more substantial progress is projected across all areas of sustainable rural development in developed countries, specifically the following:

- An increase in the share of the rural population in the total population by 15.09%, reaching 28.54% (instead of declining to 23.95% if past trends continue);
- A rise in the share of agriculture in GDP by 10.02%, reaching 2.00% (instead of decreasing to 1.63% under unchanged trends);
- An increase in the share of agricultural employment in total employment by 1.59%, reaching 3.22% (instead of dropping to 2.65% if past patterns persist).

5 Discussion

The findings of this research extend the series of studies by Huseynov et al. (2024), Rogachev et al. (2018), Tovmasayn and Gevorgyan (2024), and Vechkinzova et al. (2019), contributing to the development of the concept of climate adaptation in agriculture by clarifying the influence of factors related to the implementation and use of climate-smart technologies on the sustainability of rural development. In support of Krutilin et al. (2022), the hypothesis that these factors differ between developed and developing countries has been confirmed.

In developing countries, the sustainable development of rural areas is determined by the robustness of financial support for the introduction and application of climate-smart technologies in agriculture (supporting the findings of Salnazaryan et al. (2022)) and by the innovative activity of agricultural enterprises in adopting and using climate-smart technologies (supporting the findings of Bogoviz et al. (2019)). In developed countries, it is determined by the level of development of electronic communities (supporting the findings of Hayrapetyan and Isayan (2022)).

Simultaneously, the research demonstrated that the prevalence and level of digital skills in society (contrary to the findings of Szpilko et al. (2023)), as well as the availability of climate-smart agricultural technologies for agricultural enterprises (contrary to the findings of Burkhanov et al. (2024)), do not support the sustainable development of rural areas in either developed or developing countries.

6 Conclusion

The principal theoretical result of the research is the development of models of sustainable rural development under the influence of factors associated with the use of climate-smart technologies in agriculture for developed and developing countries.

Drawing on these models, the author concludes that the sustainability of rural development is shaped by different factors of climate-smart technology adoption and use: in developed countries, by the level of development of electronic communities, and in developing countries, by the robustness of financial support and the innovative activity of agricultural enterprises in adopting and employing climate-smart technologies.

The research outlines the prospects for sustainable rural development driven by the expansion of climate-smart agriculture through 2030 in developed and developing countries, considering their specific characteristics. The practical value of the work lies in the applied recommendations offered for optimizing the role of climate-smart technologies in agricultural development in both groups of countries.

References

Bogoviz AV, Sandu IS, Demishkevich GM, Ryzhenkova NE (2019) Economic aspects of formation of organizational and economic mechanism of the innovational infrastructure of the EAEU countries' agro-industrial complex. In: Popkova E, Ostrovskaya V (eds) Perspectives on the use of new Information and Communication Technology (ICT) in the modern economy. Springer, Cham, Switzerland, pp 108–117. https://doi.org/10.1007/978-3-319-90835-9_13

Burkhanov AU, Popkova EG, Galoyan DR, Mkrtchyan TM, Sergi BS (2024) Deep learning, irrigation enhancement, and agricultural economics for ensuring food security in emerging economies. Glob Transit 6:164–172. https://doi.org/10.1016/j.glt.2024.06.002

Hayrapetyan A, Isayan L (2022) Social capital and its changes in Armenia: Challenges and expectations. J Lib Int Aff 8(1):81–97. https://doi.org/10.47305/JLIA2281081h

Huseynov A, Guliyev R, Suleymanov A, Ostrovskaia O (2024) Assessment of regional innovation activity. J Trends ChallS Artif Intell 1(3):97–112. https://doi.org/10.61552/JAI.2024.03.004

IMD (2025) World competitiveness online. Retrieved from https://worldcompetitiveness.imd.org. Accessed 30 Nov 2025

Krutilin AA, Bazieva AM, Dugina TA, Giyazov AT (2022) Sustainable agriculture for food security: Conceptual framework and benefits of digitalization. In: Popkova EG, Sergi BS (eds) Sustainable agriculture. Springer, Singapore, pp 35–42. https://doi.org/10.1007/978-981-16-8731-0_4

Rogachev AF, Ostrovskaya VN, Natsubidze AS, Litvinova TN, Yakovleva EA (2018) Tools for sustainability management of socio-ecological systems in the globalizing world. In: Popkova E (ed) The impact of information on modern humans. Springer, Cham, Switzerland, pp 241–247. https://doi.org/10.1007/978-3-319-75383-6_31

Salnazaryan G, Hayrapetyan A, Avetisyan M, Avetisyan L (2022) The efficiency of ensuring the financial security of Armenia: New approaches. J Lib Int Aff 8(3):22–33. https://doi.org/10.47305/JLIA2283022s

Szpilko D, Budna K, Drmeyan H, Remiszewska A (2023) Sustainable and smart mobility—research directions. A systematic literature review. Econ Environ 86(3):31–61. https://doi.org/10.34659/eis.2023.86.3.584

Tovmasayn G, Gevorgyan M (2024) Evaluating the role of sustainable tourism in regional development: evidence from Armenia. Reg Sect Econ Stud 24(1):23–40

Vechkinzova E, Petrenko Y, Benčič S, Ulybyshev D, Zhailauov Y (2019) Evaluation of regional innovation systems performance using Data Envelopment Analysis (DEA). Entrep Sustain Issues 7(1):498–509. https://doi.org/10.9770/jesi.2019.7.1(35)

World Bank (2025a) Agriculture, forestry, and fishing, value added (% of GDP). Retrieved from https://data.worldbank.org/indicator/NV.AGR.TOTL.ZS. Accessed 30 Nov 2025

World Bank (2025b) Employment in agriculture (% of total employment) (modeled ILO estimate). Retrieved from https://data.worldbank.org/indicator/SL.AGR.EMPL.ZS. Accessed 30 Nov 2025

World Bank (2025c). Rural population (% of total population). Retrieved from https://data.worldbank.org/indicator/SP.RUR.TOTL.ZS. Accessed 30 Nov 2025

Regional-Industrial Approach to Innovative Development of Agriculture

Ainagul T. Mamyralieva, Zhandaraly Sadyraliev, Baktygul A. Ashiralieva, and Chynygul S. Orozova

Abstract The development of agriculture in the Kyrgyz Republic requires a shift toward an innovation-driven model that ensures production sustainability and food security amid regional disparities, climate challenges, and limited resources. In the context of globalization and the digital transformation of the economy, the integration of innovations aimed at improving the efficiency of land, water, and labor resource use becomes especially important. Differences in technological capacity, infrastructural support, and investment opportunities underscore the need for a region- and sector-specific approach to managing innovation processes. The research aims to substantiate the organizational, economic, and methodological foundations of innovative development in the agrarian sector, considering the country's natural-climatic and socio-economic characteristics. The methodological basis includes systemic, institutional, and reproductive approaches, as well as comparative, structural-dynamic, and factor analysis. Based on the assessment of innovation potential and constraints, the authors identified the key determinants of development, which are technological readiness, infrastructure accessibility, human capital capacity, and the quality of the institutional environment. The outcome of the research is a hierarchical model of priorities for the innovative development of agriculture, focused on building a competitive production system capable of ensuring sustainable economic growth, food security, and the social well-being of rural areas. The findings may be used in the development of national and regional strategies for modernizing agriculture.

Keywords Innovative development · Agrarian sector · Regional-sectoral approach · Sustainable agriculture · Institutional environment · Technological readiness · Food security

JEL Classification O30

A. T. Mamyralieva (✉) · Z. Sadyraliev · B. A. Ashiralieva · C. S. Orozova
Jalal-Abad State University named after B. Osmonov, Manas, Kyrgyzstan
e-mail: aynagul.mamyralieva@mail.ru

A. B. Karbekova (ed.), *Sustainable Development of Rural Territories Based on Climate-Smart Technologies*, Climate Change Management,
https://doi.org/10.1007/978-3-032-20106-5_24

1 Introduction

Agriculture plays a key role in the socio-economic development of the Kyrgyz Republic, ensuring food security, employment, and a significant share of export revenues. For nearly 65% of the country's population living in rural areas, land and agricultural activities remain the primary source of income, making the agrarian sector a strategic component of the national economy and a foundation for sustainable regional development.

The current state of the sector is marked by several systemic challenges: a shortage of irrigated land and water resources, high transportation costs, low levels of processing, fragmented farm structures, and limited access to modern technologies. Climate change also exerts a significant influence, increasing the risks of drought, soil degradation, and unstable yields. These factors reduce the competitiveness of domestic agricultural production and hinder its innovative development.

In the face of global challenges, further growth of the agrarian sector is impossible without innovation, modernization of production processes, and improved management systems. A transition is required toward a new model of agriculture grounded in digitalization, technological renewal, producer cooperation, and effective institutional support mechanisms. One promising pathway for this transition is a regional-sectoral approach to innovation-driven development, which involves adapting technologies and organizational-economic solutions to the natural-climatic, socio-economic, and production-specific characteristics of individual regions and branches of the agrarian sector. The application of this approach promotes the rational use of resources, the formation of sustainable farming models, and the enhancement of regional agricultural competitiveness.

The relevance of this research is determined by the need to identify effective tools for managing innovation processes in agriculture that ensure its sustainability and long-term growth. The research aims to substantiate a regional-sectoral approach to the innovative development of the agrarian sector of the Kyrgyz Republic and determine priority areas for its practical implementation. To achieve this aim, the research includes an analysis of regional characteristics and operating conditions of agricultural production, an assessment of the factors limiting innovation among agricultural producers, and the development of recommendations to improve the efficiency of innovation processes and strengthen organizational-economic mechanisms for agricultural development.

The scientific novelty of this research lies in the systematization of the organizational and economic conditions for innovation-driven development in the agrarian sector, considering regional and sectoral specifics. This allows for the substantiation of practical directions for shaping a sustainable and technologically oriented agricultural policy in the Kyrgyz Republic.

2 Materials and Methods

Understanding the specifics of innovation-driven development is impossible without analyzing the methodological foundations that shape the theoretical and practical orientation of the research. It is a methodology that enables the systematization of information and the formulation of scientifically grounded conclusions about the factors and mechanisms influencing innovation processes in the agrarian sector.

The methodological basis of this research consists of systemic and regional-sectoral approaches, which view the innovative development of agriculture as a multidimensional process dependent on natural-climatic, economic, and institutional factors. The research employs analysis, synthesis, comparative and structural–functional analysis, and economic and statistical methods for processing official statistical data and identifying patterns in sectoral development. The empirical base includes data from the National Statistical Committee of the Kyrgyz Republic, the Ministry of Agriculture, FAO, and the World Bank. The use of these methods made it possible to identify key trends and barriers to innovative development, as well as to formulate directions for its stimulation.

Additionally, the research draws on the results of expert surveys and analytical publications by Kyrgyz and international researchers dedicated to issues of innovation and sustainable agrarian development.

Overall, the methodological framework of the research is aimed at developing a conceptual model of a regional-sectoral approach to the innovative development of Kyrgyzstan's agrarian sector, focused on enhancing its sustainability, efficiency, and competitiveness.

3 Results and Discussion

The results of this research show that the innovative development of agriculture in the Kyrgyz Republic is a key factor in ensuring its sustainability, competitiveness, and adaptability to recent challenges. In the context of intensifying climate change, limited natural resources, and rising market demands, innovation serves as a systemic mechanism for renewing production capacity and increasing the efficiency of the agrarian sector.

Innovation processes in agriculture encompass technological modernization and new forms of management organization, cooperative development, and improvements in logistics and marketing. Their implementation contributes to higher labor productivity, reduced costs, improved product quality, and the rational use of natural resources. Simultaneously, the effectiveness of innovations directly depends on regional characteristics: farm structure, soil and climatic conditions, infrastructure levels, and access to knowledge (Jailov and Zhumaliev 2023).

The contemporary model of innovation-driven development in the agrarian sector is based on ecosystem and spiral approaches, which emphasize interaction among the state, business, science, civil society, and environmental institutions. Such integration creates a balanced innovation environment in which each participant plays a specific role: the state provides regulatory and financial frameworks, science develops technologies and trains specialists, business implements innovative solutions, and civic organizations help disseminate and adapt them to local conditions.

For Kyrgyzstan, the key principles of innovation-led development include regional adaptation, clustering, scalability, inclusiveness, and environmental sustainability (Cabinet of Ministers of the Kyrgyz Republic 2021). This means that innovations must be tailored to natural-resource conditions, ensure the participation of small and medium-sized producers, stimulate the development of cooperatives, and minimize negative environmental impacts.

Innovation in the agrarian sector involves the introduction of new technologies, the use of modern and more productive equipment, and improvements in production processes and methods (Musaeva 2021; Parpieva 2020). Ultimately, these innovations contribute to increased labor productivity, measured as growth in the value of gross output produced over a certain period or per unit of input (Afanasyeva 2021; Shpak and Rusakovich 2020; Shumskaia 2020).

The analysis of the dynamics of key indicators confirms positive trends in the country's agricultural sector. Currently, the National Statistical Committee of the Kyrgyz Republic does not have the capacity to provide regular, precise quantitative indicators of labor productivity growth attributable specifically to innovation. Nevertheless, indirect indicators used in statistical practice make it possible to assess overall trends (Table 1).

According to data from the National Statistical Committee of the Kyrgyz Republic (Table 1), between 2018 and 2022, the gross agricultural output per 100 hectares of

Table 1 Economic efficiency of agricultural enterprises in the Kyrgyz Republic in 2018–2022

Indicator	2018	2019	2020	2021	2022
Gross agricultural output per 100 hectares of agricultural land, 1000 soms	1929.0	2079.8	2349.9	3056.8	3375.7
Labor productivity, 1000 soms per person	267.3	291.5	371.5	444.5	628.2
Capital-labor ratio, 1000 soms per person	1183.3	1144.1	1359.3	998.7	1162.8
Capital productivity, soms	0.2	0.3	0.3	0.4	0.5
Capital intensity, soms	4.4	3.9	3.7	2.3	1.9
Production of major agricultural products per capita, kg					
Meat (in slaughter weight)	37.6	37.7	37.8	37.9	39.4
Raw milk	270.3	271.5	273.4	274.3	275.2
Grain (after processing)	321.1	322.1	329.1	235.9	316.2
Potatoes	245.9	229.1	217.6	208.2	202.4
Vegetables	228.5	230.0	228.3	216.2	220.6

Source Developed by the authors based on the materials from the National Statistical Committee of the Kyrgyz Republic (National Statistical Committee of the Kyrgyz Republic 2023)

agricultural land increased from 1929.0 to 3375.7 1000 soms; labor productivity rose from 267.3 to 628.2 1000 soms per worker. During this period, capital productivity grew from 0.2 to 0.5 soms, indicating more efficient use of fixed assets. Despite some uneven dynamics across specific product categories, the overall trend demonstrates a 174% increase in labor productivity over five years. These results confirm the presence of an innovation component expressed in the modernization of equipment, improvements in cultivation technologies, and increased efficiency of production processes.

Technological renewal aimed at improving the quality of agricultural products also plays a crucial role. Further progress in this area will require intensified scientific research. The state should devote greater attention to developing sectoral institutions, specialized laboratories, and research centers, equipping them with modern infrastructure. This will provide a foundation for generating innovative solutions aimed at increasing yields and improving product quality, as well as enhancing the competitiveness of agricultural output both domestically and internationally.

A significant direction of innovation in Kyrgyzstan's agrarian sector lies in the breeding achievements of domestic scientists. According to the Kyrgyz Research Institute of Agriculture under the Ministry of Water Resources, Agriculture, and Processing Industry, several new crop varieties and hybrids have been developed in recent years. These include five wheat varieties, three barley varieties, two maize hybrids, three sugar beet hybrids, two cotton varieties, two potato varieties, five tomato varieties, and three pepper varieties (CIS Internet portal "e-CIS.info." 2024).

Substantial progress has also been made in horticulture. The implementation of a joint project by the FAO and the Ministry of Agriculture of the Kyrgyz Republic in 2023 resulted in the establishment of infrastructure for two fruit-and-berry nurseries with a total area of seven hectares, the planting of more than 26,000 seedlings, and the introduction of drip irrigation systems. These initiatives strengthen the technological foundation of the sector and enhance the export potential of domestic production (Food and Agriculture Organization of the United Nations (FAO) 2024).

The use of advanced technologies and machinery significantly reduces production costs. For example, according to the Ministry of Agriculture of the Kyrgyz Republic, the introduction of drip irrigation systems in the southern region (Batken Region) enabled farmers to reduce water consumption by an average of 30–35% while simultaneously increasing vegetable yields by 15–20%. This lowers production costs and enhances the competitiveness of agricultural products in the market.

In the context of global warming and increasingly frequent droughts, new crop varieties and hybrids resistant to heat and water scarcity have become particularly important. Their effectiveness confirms the value of a sectoral–regional approach tailored to the needs of southern regions. For instance, new drought-resistant wheat and barley varieties developed by breeders at the Kyrgyz Research Institute of Agriculture allow farmers to maintain stable yields even under the challenging climatic conditions of Chuy and Jalal-Abad Regions. Furthermore, the adoption of digital technologies, such as satellite monitoring of crops and the use of drones to assess field conditions, enables timely responses to environmental changes and improves production management efficiency.

Innovative methods and technologies also facilitate the entry of domestic products into new markets. Certification of environmentally friendly products, the development of organic agriculture, and the implementation of "Halal" and "Global G.A.P." standards create opportunities for Kyrgyz producers to supply goods not only to the domestic market but also to countries of the Eurasian Economic Union, the Middle East, and the European Union. According to the Association of Organic Producers of the Kyrgyz Republic, exports of organic beans have nearly doubled in recent years, demonstrating strong demand for innovation-oriented products.

However, comparative analysis shows that the level of innovation activity in the Kyrgyz Republic remains relatively low. According to the National Statistical Committee, in 2020, only 5.6% of surveyed enterprises engaged in innovation, and the volume of exported innovative products amounted to just 2.4 million soms (Alybaev 2022). The country's position in the Global Innovation Index (WIPO) between 2019 and 2024 ranged from 90th to 106th, significantly lower than Kazakhstan (77th–83rd) and the Russian Federation (45th–59th) (Sultanova 2020). These figures indicate the need to strengthen institutional and financial mechanisms for stimulating innovation, as well as to develop infrastructure for technology transfer and commercialization of scientific developments.

Overall, the research results show that the sustainable development of agriculture in the Kyrgyz Republic requires a comprehensive approach combining support for scientific research, the development of cooperation, digitalization, the formation of regional clusters, and improvements in state policy. The implementation of these measures will strengthen food security, enhance export potential, and ensure the long-term competitiveness of the country's agrarian sector.

4 Conclusion

The analysis conducted shows that the successful realization of the innovative potential of agriculture requires a comprehensive approach that includes both strategic and institutional solutions.

In the authors' view, financial support remains a critical factor in innovation-driven development. The implementation of innovative projects is impossible without stable access to investment, credit resources, and government subsidies. The creation of a multilevel financing system that combines public and private sources establishes the conditions necessary to increase innovation activity and scale up successful practices.

Human capital and educational capacity also play a decisive role. The training of qualified specialists equipped with modern knowledge and competencies, as well as ensuring farmers and entrepreneurs have access to up-to-date information and advisory services, contributes to the effective implementation of innovations.

Thus, the innovative development of agriculture in Kyrgyzstan is possible only through systemic interaction among the state, the private sector, and the scientific community. The state must act as an initiator in establishing regulatory and financial support mechanisms; the business sector must invest in modernization and the adoption of innovations; and scientific institutions must provide technical expertise and workforce training.

References

Afanasyeva Yu S (2021) On the issue of state regulation of innovation in the agro-industrial complex. In: Bespalova SV (ed) VI Donetsk readings 2021: education, science, innovation, culture, and contemporary challenges, Donetsk National University, Donetsk, Ukraine, pp 242–244. Retrieved from https://dspace.lgpu.org/bitstream/123456789/5743/1/2298-22_s.pdf. Accessed 6 Sept 2025

Alybaev AS (2022) The current state of innovation in the economy of the Kyrgyz Republic and the quaternary sector of the economy. Sci Rev Econ Sci 1: 27–33. Retrieved from https://s.science-economy.ru/pdf/2022/1/1095.pdf Accessed 6 Sept 2025

Cabinet of Ministers of the Kyrgyz Republic (2021) Concept of Agricultural Development of the Kyrgyz Republic for 2021–2025. Retrieved from https://www.gov.kg/ru/npa/s/3145. Accessed 6 Sept 2025

CIS Internet portal "e-CIS.info." (2024) New varieties of agricultural crops have been developed in Kyrgyzstan. Retrieved from https://e-cis.info/news/569/123945/. Accessed 6 Sept 2025

Food and Agriculture Organization of the United Nations (FAO) (2024) FAO trains farmers in Kyrgyzstan in modern horticultural techniques. Retrieved from https://www.fao.org/europe/news/detail/fao-trains-farmers-in-kyrgyzstan-in-modern-horticultural-techniques/ru. Accessed 7 Sept 2025

Jailov DS, Zhumaliev EK (2023) Development and regulation of beet sugar production in the Kyrgyz Republic. Econ Bus: Theory Pract 4-1(98):122–126. https://doi.org/10.24412/2411-0450-2023-4-1-122-126

Musaeva BM (2021) Features of the agricultural sector of the Russian economy. Entrepreneur's Business Herald 3(1):48–50. https://doi.org/10.24412/2687-0991-2021-11021

National Statistical Committee of the Kyrgyz Republic (2023) Agriculture of the Kyrgyz Republic 2018–2022, Bishkek, Kyrgyzstan. Retrieved from https://stat.gov.kg/ru/publications/sbornik-selskoe-hozyajstvo-kyrgyzskoj-respubliki/. Accessed 6 Sept 2025

Parpieva NR (2020) State regulation of innovative development in the agricultural sector of the Kyrgyz Republic. Bulletin of the Kyrgyz National Agrarian University 3(54):140–145

Shpak A, Rusakovich A (2020) Foreign experience in supporting investment and innovation activities in the agricultural sector of the economy. Sci Innov 10(212):57–62. https://doi.org/10.29235/1818-9857-2020-10-57-62

Shumskaia EI (2020) Economic policy measures for stimulating innovative development. Econ Revival Russ 4(66): 163–176. Retrieved from https://cyberleninka.ru/article/n/mery-ekonomicheskoy-politiki-dlya-stimulirovaniya-innovatsionnogo-razvitiya. Accessed 6 Sept 2025

Sultanova G T (2020) Problems and prospects for the development of agricultural production in the Republic of Kazakhstan in modern conditions (dissertation of doctor of philosophy), Nur-Sultan, Kazakhstan: Financial Academy JSC Retrieved from https://repository.apa.kz/handle/123456789/162. Accessed 6 Sept 2025

Tourism Clusters in Rural Areas: A Climate-Smart Approach to their Formation and Sustainable Development

Kobogon A. Atyshov, Antonina V. Babosiuk, Urmat S. Aralbaev, and Yrysgul A. Mamanova

Abstract The main objective of this research is to study the cluster approach as a tool for the development of tourism in the region. The formation of tourism clusters is one of the important directions in the development of the industry. Almost all tourism development programs at the state and regional levels discuss the need to create and develop such clusters. Clusters contribute to the effective development of tourism in the regions; the full support of government initiatives at various levels has a significant impact on this process. The cluster approach is one of the most effective methods of stimulating the economic growth of regions. This research examines the essence of a tourist cluster, its structure, and its role in strengthening the economic sustainability of the region. Existing conceptual projects of tourism clusters in the Kyrgyz Republic were also studied; recommendations were offered for improving the cluster approach in the formation of a tourism product. The authors developed an innovative approach to stimulate the sustainable development of rural areas in Kyrgyzstan: the creation of tourism clusters and the organization of cluster management using climate-smart technologies. The advantage of the proposed approach lies in increasing the effectiveness of managing climate risks in the development of rural territories. The approach has been tested using the example of the rural areas of Ak-Suu, Jeti-Oguz, and Tyup within the Karakol International Cluster of the Issyk-Kul region of Kyrgyzstan.

Keywords Tourism economy · Rural areas · Climate-smart technologies · Tourism clusters · Tourism in Kyrgyzstan · Combating climate threats

---

K. A. Atyshov (✉) · A. V. Babosiuk
Research University "Kyrgyz Economic University named after M. Ryskulbekov", Bishkek, Kyrgyzstan
e-mail: atyshov.keu@mail.ru

U. S. Aralbaev
Academy of Public Administration under the President of the Kyrgyz Republic in the Field of Mountain Tourism Development, Bishkek, Kyrgyzstan

Y. A. Mamanova
Kyrgyz National Agrarian University named after K.I. Skryabin, Bishkek, Kyrgyzstan

A. B. Karbekova (ed.), *Sustainable Development of Rural Territories Based on Climate-Smart Technologies*, Climate Change Management,
https://doi.org/10.1007/978-3-032-20106-5_25

JEL Classification Q01 · Q54 · Z32

1 Introduction

The global importance of tourism continues to grow steadily, having a significant impact on the regional and national economy every year. Particularly, the growth of international tourism businesses plays a critical role in economic dynamics, serving multiple purposes: generating foreign exchange and employment opportunities, facilitating the emergence of new tourism-related industries that are displaced, strengthening the country's balance of payments and GDP, thereby increasing the overall welfare of its population (Ferreiram and Estevao, 2009).

In 2023, international tourism recovered to 88% of pre-pandemic levels, reaching approximately 1.286 billion arrivals (1.3 billion people). This is 34% more than in 2022, but 12% less than in 2019, before the outbreak of the pandemic (Travel, 2025). The level of tourism development in the countries of the world varies significantly due to different socio-economic conditions. It is expected that these promising trends will have a beneficial impact on the development of tourism in Kyrgyzstan.

This method has already proved its effectiveness in increasing productivity, efficiency, and innovativeness of various economic sectors and territories, including rural territories. In the context of globalization of economic relations, transnational production, and digitalization of the economy, cluster initiatives are becoming increasingly significant. Kyrgyzstan still ranks among the lowest globally in terms of utilizing its tourism potential, attracting tourists, and generating income from tourism, despite its regions possessing unique natural and cultural attractions, such as high mountains, historical, and cultural resources.

2 Methodology

International experience consistently emphasizes that regional clusters are crucial tools for enhancing competitiveness in countries with market economies. Michael Porter's notable contribution highlights the clustering of tourism at the regional level as an important factor for competitiveness. In today's interconnected world, intensive economic growth plays a crucial role in enhancing the competitiveness of various economic sectors and shaping the regional characteristics of each country's economy (Kruzhalin et al. 2014).

To achieve this goal, it is necessary to utilize progressive approaches that promote regional development, thereby increasing economic efficiency through the enhancement of innovative technology (Porter 2000). It can be viewed as the most effective tool for supporting the economic development mechanism of business at the state level in the international competitive environment. They contribute to the formation of regional clusters in states and directly participate in the analysis and implementation of each competitive initiative.

Studies show that such experience is developed in the UK, Germany (biotechnology clusters), Belgium (aerospace), Norway (marine industry development), Finland (forestry), and Canada (regional cluster development) (Shvets 2014). The cluster approach has been used for 30 years in industry development programs worldwide, and for over 20 years in European countries.

A regional tourism cluster is an organization that creates, develops, and implements tourism products for a specific region, utilizing resources from the tourism industry and related economic sectors. This process involves interaction between business structures, government authorities, public institutions, and non-governmental organizations.

These efforts are aimed at enhancing the competitiveness of products produced within the clusters, increasing revenues, and improving the overall economic potential of the region. Moreover, market participants in clusters aim to establish the necessary regulatory framework and infrastructural institutions that support the effective functioning of interconnected business processes and their further development (Gulyaev 2003).

In the operation of a regional tourism cluster, two elements have been established. First and foremost, within the cluster, certain methods used by companies should be interconnected and compatible. In this case, the connections are typically direct (e.g., purchase and sales lists) and indirect (e.g., complementary goods and services, similar specialized expenses, technologies and institutions, and other relationships). Most of this communication is related to social media or networks. In turn, they contribute to generating corresponding profits for the firms associated with them. Therefore, a cluster is considered a form of network. Experience shows that this is often done within regional boundaries. Firms and institutions provide some form of commonality, and the influence of one on another increases (President of the Kyrgyz Republic 2000).

Companies within a regional tourism cluster are closely located geographically and significantly differ in their strong connections with each other in various aspects. The location of companies working together in a specific area positively impacts their formation and intensifies competition, thereby increasing their revenues.

After studying the potential of emerging regional tourism clusters and their scientific justification, specific measures will be taken at the state level to develop business structures.

International experience shows that strengthening relationships between the government and the tourism business is a crucial aspect that requires a careful and persistent approach. Joint and mutually beneficial solutions to key issues contribute to the harmonious development of business and the economy.

A tourism cluster is effectively formed in regions with high levels of technical and technological advancements or where new discoveries are anticipated. International experience confirms the importance of the cluster approach for developing and regulating national innovation programs, especially in developed economies that utilize market mechanisms. The government's cluster programs are aimed at increasing competitiveness in tourism and other sectors of the economy (Government of the Kyrgyz Republic 2019).

It is crucial to clearly understand the strategy of the innovation system planned to be implemented. Experts emphasize that regional innovation structures are often the result of new systems for the development and management of tourism, highlighting the need to focus on economic sectors capable of leveraging innovations (Chontoev 2013).

Tourism is crucial for the economy of vast regions around the world, supporting sectors such as transportation, communication, trade, infrastructure, and agriculture. Thus, it contributes to economic growth, job creation, and entrepreneurial activity.

In today's era of globalization, tourism has a profound impact on the economic development of various countries and regions. Moreover, tourism thrives as a powerful industry and represents a widespread cultural phenomenon, as evidenced by the increasing influx of tourists from around the world each year.

Particular attention should be given to the development of tourism economics through the promotion of rural tourism as a driver of sustainable rural development (Geng et al. 2025; Li et al. 2025; Samper-Mendívil et al. 2025). The important role of natural and climatic factors in the development of rural tourism has been highlighted in the works of Muda et al. (2025) and Tian and Jiang (2025). A promising means of advancing this direction is the introduction of climate-smart technologies into the management of tourism service provision in rural areas (Yang and Ning 2025; Zhang and Zhang 2025).

The literature review conducted for this research revealed that, although tourism clusters have been extensively examined (Bašić et al. 2025; Nie and Cronin 2025), their application to the sustainable development of rural areas remains insufficiently addressed, as does the use of climate-smart technologies for managing such clusters. This research fills that gap in the literature.

3 Results

When examining the tourism-based mechanism for ensuring the sustainable development of rural areas, it is important to consider the geographical, particularly climatic, features of rural territories in Kyrgyzstan. Over 90% of the country's territory in Kyrgyzstan lies at elevations above 1000 m, featuring some of the world's highest peaks, such as Pobedy Peak (7439 m) and Lenin Peak (7134 m), which draw climbers from around the globe.

Clearly, the establishment of ski resorts in these areas, along with the development of professional and amateur mountaineering and specialized hunting for tourists, could significantly contribute to the growth of a substantial independent sector within the national economy. These needs often require a tailored approach and are addressed by both tourism resources and enterprises within the sector.

Therefore, the development of mountain tourism plays a crucial role in global integration processes and holds significant economic importance (National Statistical Committee of the Kyrgyz Republic 2023). For over 20 years, the tourism sector in Kyrgyzstan has been regarded as a promising field with the potential to impact the country's overall development positively. Despite this, its development over the years has left much to be desired. Everyone agrees that tourism holds significant potential, but its full realization has yet to be achieved.

Currently, the country hosts a diverse range of large and small travel companies. The residents of Issyk-Kul Region mainly depend on the influx of tourists during the seasonal period. To stimulate the influx of foreign tourists, a visa-free regime was introduced for 44 developed countries in the summer of 2012.

Kyrgyzstan's tourism sector is described as chaotic and fragmented, with its development occurring slowly. There is a lack of a unified strategy and vision for sector development, as well as consistency in policy, which indicates weaknesses in government management. The quality of service and human resources involved in the organization and provision of travel services require significant improvement. Underdeveloped infrastructure and a lack of qualified personnel are obstacles to attracting foreign tourists to the country. Furthermore, the level of investment is not high enough to attract foreign capital for the development of the tourism sector.

Tourism has been viewed as a potential driver of economic growth in Kyrgyzstan for more than two decades. However, during this time, its development has slowed down significantly. Although the significant potential of tourism is widely recognized, it remains largely untapped.

There are many medium and small tourism enterprises operating in the country. In the Issyk-Kul region, the livelihood of many residents depends on the tourist season. Recently, a 60-day visa-free regime was introduced for citizens of 52 countries.

The tourism sector of Kyrgyzstan is characterized as unorganized and slowly developing. Government policy demonstrates shortcomings related to the lack of a holistic vision and a unified strategy for the development of the sector (Table 1).

In Kyrgyzstan, there is currently no real opportunity for tourism development funded by government resources. Service standards and the quality of human resources involved in the tourism sector also pose challenges. Inadequate infrastructure and a shortage of qualified personnel hinder the attraction of foreign tourists.

Table 1 Tourism in Kyrgyzstan

Indicators	2018	2019	2020	2021	2022
Gross output in the tourism sector activities, million soms	69153.0	69998.1	40538.3	52792.5	74373.2
Share of tourism activities in GDP, %	5.0	4.5	2.9	3.2	3.6
Number of people who had a rest-total, thousand people	1380.4	1778.9	463.9	1340.2	1970.9
In the organized sector	749.9	809.6	254.9	664.7	1133.9
In the unorganized sector	630.5	969.3	209.0	675.5	837.0
Receipt of foreign direct investment (excluding outflow), $ million	59.7	157.8	39.8	43.9	49.8
Average salary number of tourism workers, people	8485	8931	7212	7204	8303
Export of tourism services, $ million	453.4	613.1	120.4	142.4	696.4

Source Developed by the authors based on (National Statistical Committee of the Kyrgyz Republic 2023).

Currently, the private sector plays a crucial role in developing the tourism industry. However, it faces significant challenges from government structures. These issues include unfair competitive conditions and insufficient support from the government. Foreign tourists perceive the tourism infrastructure as underdeveloped, service offerings as limited, service quality as low, and prices as high.

Currently, many studies focus on developing models related to the formation of regional tourism clusters and their characteristics (Atyshov et al. 2021). To ensure competitiveness, interaction among the four determinants is essential:

- Human capital and natural resources, conditions for scientific research, factors supporting infrastructure capabilities aligned with demand;
- The creation of demand conditions for stable growth in domestic market volumes;
- Efficient placement of interconnected service networks that are closely connected to each other;
- The definition of goals, main directions, progressive innovative methods of organization, strategies and management structures, and intra-industry competition.

As noted earlier, each component of the cluster represents a set of tools that influence elements at various levels (Fig. 1).

After studying and analyzing the existing scientific works [Government of the Kyrgyz Republic (2019); President of the Kyrgyz Republic (2000); Chontoev (2013); Gulyaev (2003); Kruzhalin et al. (2014); Porter (2000); Shvets (2014)] on models of tourism cluster formation, the authors developed a regional tourism cluster model (Fig. 2).

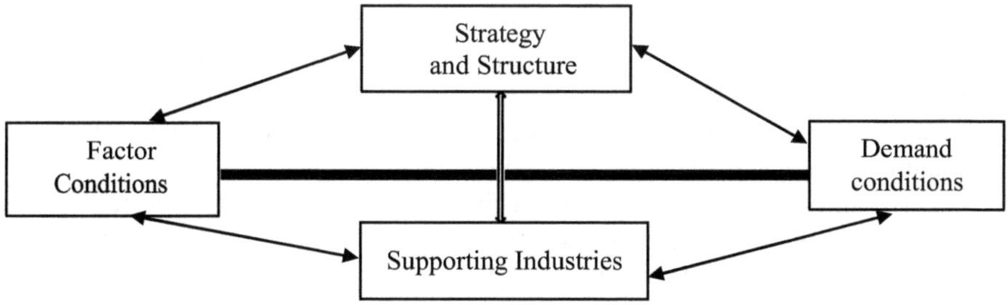

Fig. 1 Porter's diamond model of competitiveness. *Source* Developed by the authors based on (Porter 2000)

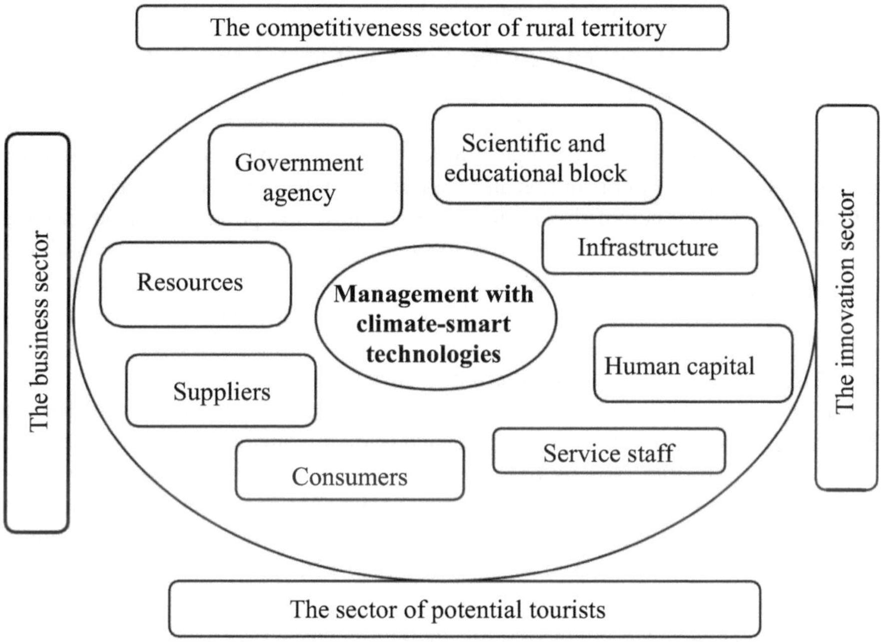

Fig. 2 The model of managing a tourism cluster in rural areas using climate-smart technologies. *Source* Compiled by the authors

The division of Kyrgyzstan into ten tourism clusters was researched. Let us explore the regional tourism clusters of Kyrgyzstan (Tazabek 2025):

- Cholpon-Ata (Issyk-Kul; Ton)—specialization by type: mountain, ecological, wellness, business and professional, ski, beach and recreational, educational, guest;
- Karakol (Ak-Suu; Jeti-Oguz; Tyup)—specialization by type: mountain, ecological, wellness, extreme, adventure, ski, beach and recreational, educational, guest;

- Song-Kul (Naryn, Kochkor; Ak-Talaa Jumgal; Toguz-Toro)—specialization by type: mountain, ecological, wellness, extreme, adventure, ski, educational, caving, guest;
- Arslanbob (Nooken; Suzak; Toktogul; Bazaar-Korgon)—specialization by type: mountain, ecological, Wellness, ski, educational, guest;
- Aksy (Chatkal; Aksy; Ala-Buka)—specialization by type: mountain, ecological, ski, educational, adventure, sightseeing;
- Bishkek (Panfilov, Jaiyl, Moskovsky; Sokuluk Alamudun)—specialization by type: mountain, environmental, business and professional, ski, guest, educational;
- Tokmok (Chui; Kemin; Ysyk-Ata)—specialization by type: mountain, ecological, wellness, ski, educational, adventure, guest;
- Talas (Bakai-Ata; Kara-Buura; Manas, Talas)—specialization by type: mountain, ecological, ski, educational, guest, excursion.

Special attention is given to the competitiveness of the cluster and the analysis of its strengths and weaknesses. Conducting a comprehensive SWOT analysis allows for an assessment of the current state of cluster development, effectively aligning strengths and weaknesses with opportunities and risks. These results help identify the region's potential opportunities and the cluster's advantages over competitors.

Given the advantages of this region, one can infer the potential for developing tourism clusters, provided that the government adopts sound economic policies, creates a favorable image, and utilizes regional advantages effectively. This will significantly boost the development of the tourism industry in the Issyk-Kul region, enable it to achieve leading positions in the global market, diversify the services offered to consumers, and attract international events.

An important task is to stimulate tourism development in the Issyk-Kul Region through the application of a cluster approach, which contributes to achieving high socio-economic results. This is based on the close interaction of various areas of tourism. The Issyk-Kul region not only has a favorable geographical location and rich tourism resources but also receives government support in the development of this sector. It is important to note the availability of sufficient labor potential, which is a key factor for the formation of clusters.

A SWOT analysis was conducted for the region, based on a systematic approach to evaluating its tourism potential, considering both internal and external factors. The main aspects of natural and recreational potential were analyzed, priorities for identity were defined, and negative aspects and trends that slow down progressive development and emerging problems were also considered. The results of the Issyk-Kul region tourism cluster study are presented in Table 2. Based on this analysis, it is possible to conclude the feasibility and potential.

Table 2 SWOT analysis of tourist clusters in the Issyk-Kul region

Strengths	Weaknesses
1. The presence of a mountain lake with natural conditions, as well as recreational resources, which positively impact people's health 2. Warm summer days and snowy winters create favorable conditions for tourism throughout the year 3. The presence of cultural and historical heritage 4. The presence of resort and wellness complexes 5. The presence of sports complexes for skiing and water sports 6. The abundance of high-quality mineral water sources and therapeutic muds 7. Convenient geographical location considering economic factors 8. The route of the northern silk road passing through this region 9. The hospitality of the local population, rich historical heritage, cultural features, and unique national artifacts 10. Increasing the number of tourists, especially from distant countries	1. Slow development of infrastructure capabilities 2. Pronounced fluctuations in seasonal activity 3. Uneven environmental impact 4. Low level of living comfort among the local population 5. Lack of specialists with a foreign language, especially during periods of high demand 6. Migration outflow of the most motivated and competent residents 7. Lack of high-quality international centers for large events that meet all standards 8. Inadequate functioning of some elements of the regulatory and legal framework 9. The negative impact of climate change on tourism in rural areas

(continued)

Table 2 (continued)

Strengths	Weaknesses
Opportunities	*Threats*
1. Development of various forms of tourism, including climate services, balneology, mud therapy, agroecological tourism, cultural and educational recreation, hunting, fishing, ethnography, and water and mountain-extreme tourism 2. Adoption of the tourism development plan in the Kyrgyz Republic for the period from 2019 to 2023 3. Improvement of the population's living standards, leading to increased purchasing power 4. Creating new tours to reduce seasonality 5. Further cooperation of clusters with other regions of the country and the world 6. Entering new markets 7. Ensuring year-round operation of most resorts 8. Improving infrastructure within clusters 9. Improvement of conditions and provision of additional services 10. Improvement of sanatorium and resort facilities in accordance with international standards 11. Accommodation of large ski complexes 12. Increase in the number of tourists in winter 13. Construction of small tourism complexes in high-altitude areas 14. Large-scale infrastructure development projects implemented in the region 15. Continuous increase in the use of information technology, allowing tourists to receive digital information from service providers 16. Financial support for profitable tourism initiatives from donors 17. Tourism, mountaineering, and skiing routes, both existing and under development, possess uniqueness Kyrgyz people have rich cultural traditions as descendants of Manas, one of the oldest peoples 18. Introduction of climate-smart technologies into the management of tourism clusters in rural areas	1. Increasing global health threats and tense geopolitical conflicts with potential serious consequences for tourism 2. Climate changes, accompanied by global warming and deteriorating environmental conditions 3. Natural disasters and technological emergencies 4. Shortage of highly qualified specialists 5. Deterioration of infrastructure quality 6. Insufficient efforts to preserve cultural, historical, and natural heritage, which leads to its destruction 7. Degradation of the cultural level of the population 8. Lack of a legislative framework for the development of the tourism industry 9. Low demand for tourist services from the local population due to limited solvency 10. Utilization of unique tourism resources at an intensity exceeding the allowable level for their preservation 11. The intensification of climate threats to tourism in rural areas

Source Developed by the authors.

As a result of the SWOT analysis, it became evident that the Issyk-Kul region has significant potential for creating a profitable tourism cluster offering high-quality services. However, to mitigate associated risks, it is essential to consistently attract clients, inform them about the region's unique features and the superior quality of its services, and build trust through high service standards. Furthermore, increasing tourist inflows will require substantial infrastructure development and active marketing efforts to strengthen the region's overall attractiveness.

The development of a tourism cluster in a particular region is vital for the expansion of the tourism sector. It is heavily influenced by the recreational opportunities and advantages available. Unique natural conditions play a key role in shaping local tourism by influencing preferences for various types of recreational activities that correspond to the region's potential. The introduction of climate-smart technologies into the management of rural tourism clusters will ensure more sustainable development of these areas in Kyrgyzstan by reducing their vulnerability to climate-related risks.

4 Discussion

The research shows that among the emerging regional tourism clusters in Kyrgyzstan, the Karakol cluster holds a prominent and privileged position on the global stage. Despite its modest size, this region has ideal conditions for seamless integration into the global economy by leveraging its unique mountain landscapes and rich, eco-friendly recreational offerings. In the future, the development of the cluster will primarily focus on resort and mountain-adventure tourism, as well as tourism related to the historical Silk Road, which will expand its international reach.

The region is rich in high-quality therapeutic muds and mineral waters, as well as pristine sandy beaches around the lake. Travel and mountaineering opportunities make this region attractive for tourism development. The resort and recreational complex includes renowned sanatoriums, which have long successfully utilized natural resources for therapeutic purposes. In the future, there is potential for further intensive development of resort tourism.

The cluster has a significant diversity and high quality of natural and recreational potential, which makes it one of the regions of world tourism. In the future, the primary focus of the tourism industry will be the development of resorts around the lake, specializing in beaches, mineral springs, and therapeutic muds.

Currently, all conditions are in place, and there is significant potential to make the resort treatments in this region a leading destination in the Asian region, meeting international standards. Research shows that the cluster's territories have a variety and quality of natural and recreational potential, beneficial to human health, and are among the world's tourist centers.

The development of tourism in the Great Silk Road cluster and the expansion of international relations will address many issues in the tourism sector. Currently, only a small portion of this potential has been developed—about 15–20%. Depending on the preferences of tourists, choosing specific types of adventures (Government of the Republic of Kazakhstan 2023).

The glacier stretches for 60.5 km, with a total glaciated area exceeding 657 km^2. The thickness of the ice cover in the lower parts of the Engilchek Glacier reaches up to 150–200 m, making this ice giant a significant reservoir of fresh water. These locations are challenging for ascents, requiring high physical fitness and skill from athletes.

The modernization of existing facilities and the construction of new, modern structures at climbing bases, the creation of necessary infrastructure for the convenience of foreign climbers and tourists, as well as the implementation of required environmental measures, cannot be accomplished solely by the climbing bases themselves. This requires the active participation of government agencies and foreign investors.

As previously mentioned, a large part of the Karakol cluster area is occupied by high mountainous regions with rich natural and recreational resources. To fully utilize this potential, significant investments will be required to develop various types of mountain tourism and to create infrastructure that meets international standards. Financing of such large-scale projects from the republic's budget is currently impossible. In this context, it seems reasonable to start constructing tourist facilities that can meet the needs of foreign tourists while minimizing capital expenditures.

Considering the allowable recreational ecological load, amounts to over 3000 foreign visitors per year, which could generate more than $7.0 million in profit. The Government of the Kyrgyz Republic adopted a special resolution on this major project seven years ago (TASS Russian News Agency 2024). It is estimated that these complexes will sell an average of 1.0 million tickets per year during the season:

- Location 1: Ak-Suu, Titles: Jyrgalan (2023–2025), Boz-Uchuk (2026–2028), Chelpek (2024–2026), Kerege-Tash (2027–2029), tickets to year: 200.0 thousand people;
- Location 2: Jeti-Oguz, Title: Jeti-Oguz (2028–2030), tickets to year: 200.0 thousand people;
- Total quantityof tickets for the year: 1000.0 thousand people.

For the implementation of the pilot ski cluster project, a total of $3.3 billion is required, with $1.3 billion allocated for infrastructure and the remaining $2.0 billion for the development of five ski resorts (TASS Russian News Agency 2024). In the future, the ski complexes will fully comply with contemporary standards. Each of them will include slopes with ski lifts, an ice rink, hotels, cafes, restaurants, and medical service points.

In the Karakol cluster, mountain adventure tourism has a longer season compared to resort and recreational tourism. Although the number of tourists here is smaller, they come in groups and spend significantly more. This type of tourism shows rapid growth compared to traditional resorts, including activities such as trekking, rafting, skiing, and mountain biking.

This research contributes to the literature of Geng et al. (2025), Li et al. (2025), and Samper-Mendívil et al. (2025), by revealing the organizational and managerial aspects of applying tourism clusters to the sustainable development of rural areas.

The results obtained are consistent with prior studies by Bašić et al. (2025), Nie and Cronin (2025), Yang and Ning (2025), and Zhang and Zhang (2025), confirming the potential of climate-smart technologies as effective tools for managing tourism clusters.

Building on Muda et al. (2025) and Tian and Jiang (2025), the research demonstrates the important role of natural and climatic factors in the development of rural tourism in Kyrgyzstan. The core of the scientific novelty lies in substantiating the appropriateness of using climate-smart technologies in the management of tourism clusters in rural areas of Kyrgyzstan to strengthen the sustainability of their development.

5 Conclusion

Research shows that the successes of the Karakol International Cluster will have a significant impact on the overall development of the Kyrgyz Republic. This success will directly affect the economic and social indicators of the country. Investments in the tourism sector, which lead to job growth, can be achieved by leveraging the synergistic effects of the comprehensive development of tourism infrastructure. The cluster will become a center of attraction for investments and stimulate the development of small and medium-sized enterprises.

As an original approach to stimulating the sustainable development of rural areas in Kyrgyzstan, the creation of tourism clusters is proposed, along with the organization of cluster management using climate-smart technologies. The advantage of this approach lies in enhancing the effectiveness of managing climate risks affecting rural development. This has been demonstrated using the example of the rural areas of Ak-Suu, Jeti-Oguz, and Tyup within the Karakol International Cluster of the Issyk-Kul region of Kyrgyzstan.

References

Atyshov KA, Koch H, Kadyraliev MT, Toktorov AA (2021) Regional tourism clusters of the Kyrgyz Republic: prerequisites for creation and development. Eurasian Sci Assoc 10–3(80):177–180

Bašić M, Barišić P, Franc S (2025) The effect of tourism on income inequality reduction in welfare states: evidence from the European country clusters. Tourism 73(2):264–277. https://doi.org/10.37741/t.73.2.6

Chontoev DT (2013) The potential of mountain tourism in Kyrgyzstan. Nauka, Bishkek, Kyrgyzstan

Ferreiram JM, Estevao C (2009) Regional competitiveness of tourism cluster: a conceptual model proposal. Retrieved from https://mpra.ub.uni-muenchen.de/14853/. Accessed 6 Dec 2025

Geng Y, Zhao S, Zhang X, Li J, Yan Y, Gao J (2025) Bibliometric analysis of sustainable rural tourism. Humanit Soc Sci Commun 12(1):788. https://doi.org/10.1057/s41599-025-05213-z

Government of the Kyrgyz Republic (2019) Resolution "On Approval of the Program of the Government of the Kyrgyz Republic for the Development of the Tourism Sector for 2019–2023" (31

Jan, 2019 no. 36), Bishkek, Kyrgyzstan. Retrieved from https://online.zakon.kz/Document/?doc_id=34384891. Accessed 6 Dec 2025

Government of the Republic of Kazakhstan (2023) Resolution "on approval of the concept for the development of the tourism industry of the Republic of Kazakhstan for 2023–2029" (28 Mar 2023 no. 262), Astana, Kazakhstan. Retrieved from https://tinyurl.com/yjy5ya2v. Accessed 6 Dec 2025

Gulyaev VG (2003) Tourism: economics and social development. Finance and Statistics, Moscow, Russia

Kruzhalin VI, Mironenko NS, Zigern-Korn NV, Shabalina NV (2014) Geography of tourism: textbook. Federal Agency for Tourism, Moscow, Russia

Li K, Zou Y, Wang H, Chen S (2025) Exploring the relationship between tourist behavior and the spatial characteristics of rural tourism. Sci Rep 15(1):9229. https://doi.org/10.1038/s41598-025-93517-0

Muda I, Vohra SK, Gangwar VP, Haralayya B, Singh P, Pandey AK et al (2025) Evaluation of economic, environmental, and social impacts of COVID on rural tourism. Environ Dev Sustain 27(12):29565–29579. https://doi.org/10.1007/s10668-024-04884-z

National Statistical Committee of the Kyrgyz Republic (2023) Statistical yearbook of the Kyrgyz Republic 2018–2022. National Statistical Committee of the Kyrgyz Republic, Bishkek, Kyrgyzstan

Nie Z, Cronin S (2025) Co-opetition as a pathway to sustainability: how bed and breakfast clusters achieve competitive advantage in high-density tourism destinations. Sustainability 17(21):9562. https://doi.org/10.3390/su17219562

Porter M (2000) On competition: textbook (Transl. From English). Publishing House "Williams", Moscow, Russia

President of the Kyrgyz Republic (2000) Decree "on the concept of development of the tourism industry in the Kyrgyz Republic until 2010" (4 Sep 2000, no. 255), Bishkek, Kyrgyzstan. Retrieved from https://online.zakon.kz/Document/?doc_id=30253996. Accessed 6 Dec 2025

Samper-Mendívil M, Aramendia-Muneta ME, Alarcón-López R (2025) Assessing sustainability in rural tourism: insights from accommodation managers and residents in Navarre. J Rural Stud 120:103866. https://doi.org/10.1016/j.jrurstud.2025.103866

Shvets YV (2014) State management of a tourist cluster. Mod Sci Bull 1:77

TASS Russian News Agency (2024) The French company intends to build the largest ski resort in the region in Kyrgyzstan. Retrieved from https://tass.ru/ekonomika/20682133. Accessed 6 Dec 2025

Tazabek (2025) The state-owned company ala-too resort will be transformed into a joint-stock company to accelerate the construction of an all-season mountain cluster. Retrieved from https://www.tazabek.kg/news:2348285?from=tazabek&place=archive. Accessed 6 Dec 2025

Tian X, Jiang Y (2025) Exploring behavioral determinants of residents' ecological conservation in rural tourism development. Sci Rep 15(1):1826. https://doi.org/10.1038/s41598-025-85571-5

WP Travel (2025) World tourism ranking by country 2024. Retrieved from https://wptravel.io/world-tourism-ranking-by-country/. Accessed 6 Dec 2025

Yang L, Ning W (2025) Mechanisms and effects of the sustainable integration of digital-driven rural cultural tourism from the perspective of symbiosis. Sustain Futur 10:100867. https://doi.org/10.1016/j.sftr.2025.100867

Zhang W, Zhang L (2025) Research on the coupling and coordination of harmonious rural construction and integration of agriculture and tourism. Sci Rep 15(1):33804. https://doi.org/10.1038/s41598-025-01354-y

Climate-Smart Agriculture: Legal Foundations, Benefits for the Sustainable Development of Rural Areas, and Contributions to Food Security

Agnessa O. Inshakova and Denis E. Matytsin

Abstract The research aims to develop a scientific and methodological approach to the legal and regulatory framework governing the mechanism of climate-smart agriculture in Russia, enabling a more complete realization of its potential to ensure the sustainable development of rural areas and strengthen food security. To this end, the authors conducted a systemic analysis of official statistics related to this mechanism in Russia for the period 2013–2025. The approach developed as a result of the research includes, first, a quantitative assessment of the dependence of rural sustainability on the development of climate-smart agriculture, as well as the susceptibility of this development to the influence of e-commerce and e-government factors in Russia. Second, the authors proposed a public administration instrument for managing the climate-smart agriculture mechanism in Russia, namely the promotion of e-government services among agricultural organizations. Third, the authors presented a comprehensive vision of future opportunities for the sustainable development of rural areas and Russia's food security based on climate-smart agriculture through 2030 is presented. The research concludes that a regulatory-neutral scenario will lead to restrained development of climate-smart agriculture, whereas its potential can be more fully realized under a regulatory-enhanced scenario. The theoretical contribution of this research lies in clarifying the current legal and regulatory platform underpinning the implementation of the climate-smart agriculture mechanism in Russia. The empirical contribution consists of practical recommendations for strengthening this platform in support of more sustainable rural development and more comprehensive food security.

Keywords Climate-smart agriculture · Legal foundations · Sustainable development · Rural areas · Food security · Russian agriculture · Government regulation

JEL Classification K23 · K32 · O13 · O18 · Q54

A. O. Inshakova (✉) · D. E. Matytsin
Kuban State University, Krasnodar, Russia
e-mail: ainshakova@list.ru

A. B. Karbekova (ed.), *Sustainable Development of Rural Territories Based on Climate-Smart Technologies*, Climate Change Management,
https://doi.org/10.1007/978-3-032-20106-5_26

1 Introduction

Climate change has the greatest impact on rural areas, where economic activity is largely determined by the favorability of the natural environment. In recent years, prolonged urbanization driven by intensive industrialization of urban territories has been compounded by environmental degradation caused by climate change. This has further slowed the pace of economic development in rural areas and deepened their lag behind urban territories.

Overcoming territorial development imbalances is one of the key objectives of state policy in regulating the spatial economy. In contemporary Russia, equalizing the scale of economic activity between urban and rural areas is a strategic priority enshrined by the Government of the Russian Federation (2015). Accordingly, the implementation of the climate agenda requires particular attention to strengthening the climate resilience of rural areas to stimulate their economic development.

The technological foundation for this has been formed by green digitalization, understood as an economic practice in which smart technologies are applied to addressing climate change. The spread of this practice in rural areas is accompanied by a transition to climate-smart agriculture, which can significantly accelerate the economic progress of these territories. However, like any innovative economic practice, climate-smart agriculture requires a carefully designed system of legal and regulatory measures.

Due to the reduced investment attractiveness of rural areas, the establishment of climate-smart farming represents a "market failure": its development is hindered and yields minimal benefits in the absence of sufficient and strengthened government intervention. The challenge lies in the fact that the effectiveness of regulatory measures is largely determined by the institutional environment, which is specific to each country. Moreover, the limited study of empirical experience in public administration of climate-smart agriculture development in Russia creates uncertainty regarding the outcomes of applying available measures, resulting in an underdeveloped set of governance instruments.

The above highlights the relevance of an in-depth scientific examination of practical experience in the legal and regulatory governance of climate-smart agriculture in Russia. To contribute to addressing this problem, the research aims to develop a scientific and methodological approach to the legal and regulatory regulation of climate-smart agriculture in Russia, facilitating a more complete realization of its potential to ensure the sustainable development of rural areas and strengthen food security.

To achieve this goal, the research addresses the following research tasks. The first task is to formulate a set of public administration instruments that enable the application of the climate-smart agriculture mechanism to the sustainable development of rural areas and the achievement of food security in Russia. The second task is to identify the prospects for enhancing the sustainability of rural development and strengthening Russia's food security based on the proposed public administration toolkit for managing the climate-smart agriculture mechanism.

2 Literature Review

This research is conducted within the theoretical framework of the scientific concept of climate-focused digitalization of agriculture (Inshakova et al., 2022a, b; Matytsin et al., 2022), whose main postulates are as follows:

- Climate-smart agriculture implemented in rural areas activates a high-tech mechanism for the development of these territories, facilitating the transition of the spatial economy toward a more balanced state (Gyiazov et al. 2023);
- This mechanism is set in motion through an appropriate combination of organizational conditions (the climate orientation of agricultural digitalization) and financial conditions (adequate funding for climate-smart technologies in rural areas) (Bogoviz and Mezhov 2015);
- By triggering a new phase of economic development in rural areas, the climate-smart agriculture mechanism increases the volume of natural food production, the distribution of which makes it possible to meet domestic demand for food products while also expanding exports, thereby strengthening food security (Vanchukhina et al. 2022).

The literature review conducted on the subject reveals a relatively high degree of scholarly attention to the climate-smart agriculture mechanism from environmental perspectives (as a source of "green" growth in economic systems and a domain of agricultural economics), technological perspectives (as an automated farming practice), and economic perspectives (as a practice requiring organizational and financial support).

Simultaneously, the mechanism remains insufficiently studied from a legal perspective. The lack of clarity regarding the legal and regulatory conditions for the diffusion of climate-smart agriculture in Russia constitutes a gap in the existing literature. In view of this gap, the research poses the research question of what contribution the climate-smart agriculture mechanism makes to the implementation of the government's course toward sustainable rural development and enhanced food security, as well as which legal and regulatory measures of state regulation determine the functioning of this mechanism in Russia.

Based on a content analysis of existing publications (Margaryan and Terzyan 2023; Tadevosyan 2023) related to the topic of this study, which, drawing on international experience, note that e-commerce in rural areas (Litvinova et al. 2017) and e-government, as a digital platform for the provision of public services to rural households (Vasilikhina 2025) and agricultural organizations (Sngryan 2022) create a favorable legal and regulatory environment for the development of climate-smart agriculture, the research advances the hypothesis that the climate-smart agriculture mechanism in Russia is influenced by e-commerce and e-government. The testing of this hypothesis constitutes the focus of this research.

3 Materials and Methods

The research subject is the legal and regulatory framework of the climate-smart agriculture mechanism in Russia. To ensure the most comprehensive coverage of the selected subject area, the research adopts a broad time horizon, encompassing the period from 2013 to 2025, which represents the entire duration of the mechanism's operation in Russia. An appropriate methodology is selected for each research task.

The first task is the development of a public administration toolkit that enables the application of the climate-smart agriculture mechanism to the sustainable development of rural areas and the achievement of food security in Russia.

The methodology for addressing this task is based on the use of regression analysis. First, a regression relationship is identified between the sustainability of rural development, measured by the outcome of SDG 11 implementation (Rs) as assessed by UNDP (2025), and the expenditures of agricultural organizations on climate-smart technologies (Df) as estimated by HSE University (2025), as well as the overall climate resilience of the economy, measured by the outcome of SDG 13 implementation (Cl) according to UNDP (2025).

Second, the dependence of Df and Cl on three factors is examined using HSE University statistics (HSE University 2025): the share of e-commerce service users among rural residents (Dc) as an e-commerce factor, and the share of e-government service users among rural residents (Lh) and among agricultural organizations (Lo) as e-government factors. Statistics for the analyzed indicators are presented in Table 1.

Factors are identified (from among Dc, Lh, and Lo) that exert a favorable simultaneous influence on both outcome variables (Df and Cl), that is, factors for which the regression coefficients are positive. These factors constitute the proposed public administration toolkit for managing the climate-smart agriculture mechanism.

The second task is to identify the prospects for enhancing the sustainability of rural development and strengthening Russia's food security based on the proposed public administration toolkit for managing the climate-smart agriculture mechanism. To address this task, the method of trend analysis is employed, through which five-year trends (the ratio of values in 2025 to those in 2020) of the indicators presented in Table 1 are determined.

In addition, a forecasting method is applied to construct the following alternative scenarios for the development of climate-smart agriculture in Russia in relation to state management of its mechanism through 2030:

- A regulatory-neutral scenario, which assumes the preservation over the next five years of the previous five-year trends in the development of climate-smart agriculture, changes in the sustainability of rural areas, and food security;
- A regulatory-enhanced scenario, associated with the fullest realization of the potential for sustainable rural development and strengthened food security through the maximum utilization of the public administration toolkit for climate-smart agriculture.

Table 1 Climate-smart agriculture, the sustainability of rural development, and Russia's food security in 2013–2025

Year	Share of e-commerce service users among rural residents, %	Share of e-government service users among rural residents, %	Share of e-government service users among agricultural organizations, %	Expenditures of agricultural organizations on climate-smart technologies, billion RUB	Climate resilience of the economy based on SDG 13 implementation outcomes, score 1–100	Sustainability of rural development based on SDG 11 implementation outcomes, score 1–100
	Dc	Lh	Lo	Df	Cl	Rs
2013	7.50	27.80	40.75	5.33	69.74	92.63
2014	8.30	31.77	40.75	5.33	70.37	92.45
2015	10.60	35.74	40.75	5.33	71.53	92.97
2016	12.90	46.30	45.16	5.33	71.73	93.07
2017	18.40	58.03	45.16	5.33	71.01	92.91
2018	23.70	67.50	47.01	5.98	70.55	93.35
2019	23.80	70.03	50.19	7.51	70.60	93.47
2020	25.80	73.19	49.16	6.93	71.63	93.57
2021	31.50	76.80	51.01	9.03	70.45	93.60
2022	40.40	80.00	54.40	9.80	69.07	93.97
2023	48.50	83.10	58.40	10.10	69.08	93.94
2024	48.50	83.10	58.40	10.10	69.08	93.94
2025	48.50	83.10	58.40	10.10	69.08	93.94

Source Developed by the authors based on the materials from the HSE University (2025) and UNDP (2025)

Scientific and practical recommendations are proposed for implementing the regulatory-enhanced scenario in Russia, aimed at improving the effectiveness of public administration of the climate-smart agriculture mechanism based on the developed toolkit.

4 Results

The Public Administration Toolkit for Climate-Smart Agriculture in Russia

To address the first task of this research, the authors conducted a regression analysis of the statistical data from Table 1. As a result, the authors identified the dependence of rural sustainability on the development of climate-smart agriculture in Russia, expressed by the following regression equation:

$$\mathrm{Rs} = 77.5345 + 0.2969\mathrm{Df} + 0.1940\mathrm{Cl},\ \mathrm{e} = 0.1820,\ \mathrm{R}^2 = 0.9485,\ \mathrm{p} = 0.01 \quad (1)$$

Equation (1) shows that an increase in expenditures by agricultural organizations in Russia on climate-smart technologies by 1 billion rubles leads to an increase in the sustainability of rural development, measured by the SDG 11 implementation outcome, by 0.2969 points. Simultaneously, a one-point increase in the overall climate resilience of the economy in Russia, measured by the SDG 13 implementation outcome, contributes to an increase in the sustainability of rural development, measured by the SDG 11 implementation outcome, by 0.1940 points.

Overall, 94.85% of the sustainability of rural areas is determined by the development of climate-smart agriculture in Russia. The standard error is low (0.1820); Eq. (1) is statistically significant at the 0.01 level (F-significance = 0.00001; the Fisher F-test is passed), confirming its reliability. Second, the dependence of the development of climate-smart agriculture in Russia on e-commerce and e-government factors was identified, taking the form of the following regression equations:

$$\mathrm{Df} = -0.3226 + 0.1001\mathrm{Dc} - 0.0181\mathrm{Lh} + 0.1255Lo,\ \mathrm{e} = 0.6593,\\ R^2 = 0.9631,\ p = 001 \tag{2}$$

$$\mathrm{Cl} = 69.047600.1259\mathrm{Dc} + 0.0570\mathrm{Lh} + 0.0212\mathrm{Lo},\ \mathrm{e} = 0.6067,\\ \mathrm{R}^2 = 0.8552,\ \mathrm{p} = 0.01 \tag{3}$$

According to Eqs. (2) and (3), an increase in the share of e-commerce service users among rural residents by 1% leads to an increase in expenditures by agricultural organizations in Russia on climate-smart technologies by 0.1001 billion rubles, but a decrease in the overall climate resilience of the economy, measured by the SDG 13 implementation outcome, by 0.1259 points.

An increase in the share of e-government service users among rural residents by 1% results in a rise in the overall climate resilience of the economy, measured by the SDG 13 implementation outcome, by 0.0570 points, but a reduction in expenditures by agricultural organizations on climate-smart technologies in Russia by 0.0181 billion rubles.

An increase in the share of e-government service users among agricultural organizations by 1% leads to an increase in expenditures by agricultural organizations in Russia on climate-smart technologies by 0.1255 billion rubles and an increase in the overall climate resilience of the economy, measured by the SDG 13 implementation outcome, by 0.0212 points.

Simultaneously, 96.31% of expenditures by Russian agricultural organizations on climate-smart technologies are determined by the development of e-commerce and e-government, while 85.52% of the overall climate resilience of the Russian economy, measured by the SDG 13 implementation outcome, is explained by these factors. The standard errors are low, amounting to 0.6593 in Eq. (2) and 0.6067 in Eq. (3). Both equations are statistically significant at the 0.01 level (the Fisher F-test is passed): F-significance equals 0.00002 in Eq. (2) and 0.0062 in Eq. (3), confirming their reliability.

Thus, the only factor that has a favorable simultaneous effect on both outcome variables (Df and Cl) (i.e., for which the regression coefficients are positive) is Lo. Therefore, the promotion of e-government services among agricultural organizations is proposed as the key public administration instrument for managing the climate-smart agriculture mechanism in Russia.

Prospects for the Sustainable Development of Rural Areas and Food Security in Russia Based on Climate-Smart Agriculture

To address the second research task, the authors conducted a trend analysis of the empirical data from Table 1. It showed that the share of e-government service users among Russian agricultural organizations increased by a factor of 1.88, from 49.16% in 2020 to 58.40% in 2025. The regulatory-neutral scenario developed on this basis, assuming the preservation in Russia over the next five years of the previous five-year trends in the development of climate-smart agriculture, changes in rural sustainability, and food security, is presented in Fig. 1.

As shown in Fig. 1, if the previous five-year trend is maintained over the next five years, the share of e-government service users among Russian agricultural organizations will increase (relative to 2025) to 69.37%. This will naturally lead to a 15.67% increase in expenditures by agricultural organizations on climate-smart technologies in Russia (to 15.67 billion rubles), as well as a 0.28% rise in the overall climate resilience of the economy, measured by the SDG 13 implementation outcome (to 69.27 points).

Accordingly, an increase in the sustainability of rural development, measured by the SDG 11 implementation outcome, of 0.41% (to 94.32 points) by 2030 is projected. An alternative regulatory-enhanced scenario, associated with the fullest realization of the potential for sustainable rural development and strengthened food security

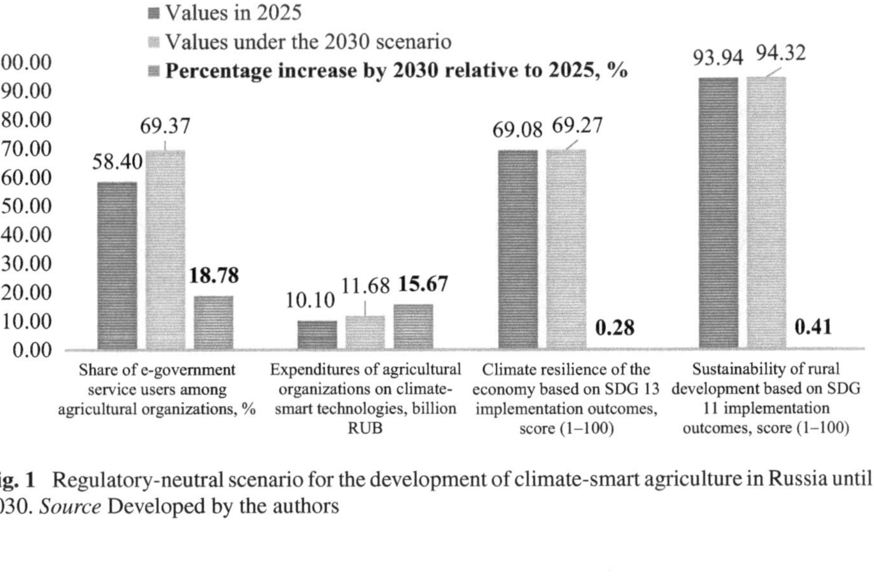

Fig. 1 Regulatory-neutral scenario for the development of climate-smart agriculture in Russia until 2030. *Source* Developed by the authors

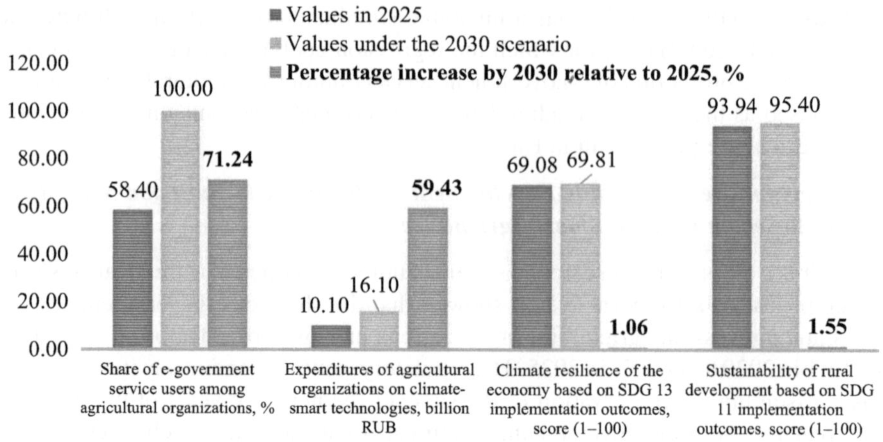

Fig. 2 The regulatory-enhanced scenario for the development of climate-smart agriculture in Russia through 2030. *Source* Developed by the authors

in Russia through the maximum utilization of the public administration toolkit for climate-smart agriculture, is presented in Fig. 2.

Figure 2 shows that increasing the share of e-government service users among Russian agricultural organizations to 100% (+71.24% relative to 2025) would result in a 59.43% increase in expenditures by agricultural organizations on climate-smart technologies in Russia (to 16.10 billion rubles), as well as a 1.06% rise in the overall climate resilience of the economy, measured by the SDG 13 implementation outcome (to 69.81 points). As a result, an increase in the sustainability of rural development, measured by the SDG 11 implementation outcome, of 1.55% (to 95.40 points) is projected.

5 Discussion

The authors' conclusions support the scholarly discussion advanced by Bogoviz and Mezhov (2015), Gyiazov et al. (2023), Inshakova et al., (2022a, b), Matytsin et al., (2022) and Vanchukhina et al. (2022), and further develop the concepts of climate-focused digitalization of agriculture by clarifying the legal and regulatory conditions for the diffusion of climate-smart agriculture in Russia (Table 2).

The results summarized in Table 2 confirm that the climate-smart agriculture mechanism in Russia is indeed generally influenced by e-commerce and e-government, thereby supporting the proposed hypothesis [in line with Margaryan and Terzyan (2023) and Tadevosyan (2023)]. Simultaneously, the analysis reveals the ambiguous nature of this influence and the greater significance of e-government factors.

Table 2 The Russian mechanism for climate-smart agriculture compared with international experience from past literature

Factors of the regulatory environment		Literature that examines factors	Impact on the conditions for the development of climate-smart agriculture in Russia	
			Financial	Organizational
E-commerce factor	The prevalence of e-commerce in rural areas	Litvinova et al. (2017)	Improvement of conditions	Deterioration of conditions
E-government factors	Mass consumption of e-government services among rural residents	Vasilikhina (2025)	Deterioration of conditions	Improvement of conditions
	Coverage of agricultural organizations by e-government services	Sngryan (2022)	Improvement of conditions	Improvement of conditions

Source Developed by the authors

Only an increase in the coverage of agricultural organizations by e-government services simultaneously improves both the financial conditions (financing of climate-smart technologies in rural areas) and the organizational conditions (the climate orientation of agricultural digitalization) in Russia for activating the climate-smart agriculture mechanism [consistent with Sngryan (2022)].

By contrast, the effects of the remaining factors are contradictory: the diffusion of e-commerce in rural areas improves only the financial conditions but not the organizational ones [contrary to Litvinova et al. (2017)], while a broader uptake of e-government services among rural residents improves only the organizational conditions but not the financial ones [contrary to Vasilikhina (2025)].

6 Conclusion

Thus, the stated objective has been achieved: a scientific and methodological approach to the legal and regulatory governance of climate-smart agriculture in Russia has been developed, enabling the fullest possible realization of its potential to ensure the sustainable development of rural areas and strengthen food security. This approach includes the following:

- A mathematical description of the relationship between the sustainability of rural areas and the development of climate-smart agriculture, as well as the susceptibility of this development to the influence of e-commerce and e-government factors in Russia;

- Substantiation of the most promising public administration instrument for managing the climate-smart agriculture mechanism in Russia, namely, the promotion of e-government services among agricultural organizations;
- An articulated outlook for the sustainable development of rural areas and food security in Russia based on climate-smart agriculture, demonstrating that a regulatory-neutral scenario would lead to restrained development of climate-smart agriculture, whereas its potential can be more fully realized under a regulatory-enhanced scenario.

The theoretical significance of the article's results lies in identifying the current legal and regulatory platform on which the climate-smart agriculture mechanism is implemented in Russia, while the practical significance consists in offering applied recommendations to strengthen this platform in support of more sustainable rural development and more comprehensive food security. The implementation of the authors' recommendations will support the realization of Russia's strategy for the sustainable development of rural areas through 2030 (Government of the Russian Federation 2015).

Acknowledgements The research was financially supported by the Russian Science Foundation, project No. 25-28-00269 "Smart Agriculture and LegalTech in the Agro-Industrial Complex: Agro 4.0 Legal Model in Sustainable Development".

References

Bogoviz A, Mezhov S (2015) Models and tools for research of innovation processes. Mod Appl Sci 9(3):159–172. https://doi.org/10.5539/mas.v9n3p159

Government of the Russian Federation (2015) Strategy for the Sustainable Development of Rural Areas in the Russian Federation for the Period up to 2030 (approved by Order on 2 Feb 2015, No. 151-r), Moscow, Russia. Retrieved from http://static.government.ru/media/files/Fw1kbNXVJxQ.pdf. Accessed 11 Dec 2025

Gyiazov AT, Gamulinskaya NV, Markeev SA, Popkova ZV (2023) Case experience of conducting climate-responsible entrepreneurship in the digital economy markets in Russia. In: Popkova EG (ed) Smart green innovations in Industry 4.0. Springer, Cham, Switzerland, pp 97–109. https://doi.org/10.1007/978-3-031-45830-9_11

HSE University (2025) Digital economy indicators in the Russian Federation, 2025: Data Book, Moscow, Russia, HSE University. Retrieved from https://issek.hse.ru/mirror/pubs/share/1026726402.pdf. Accessed 11 Dec 2025

Inshakova AO, Anisimov AP, Matytsin DE (2022a) Agricultural technologies as a factor in the development of organic farming: regulation of foreign trade turnover in Russia and the EAEU. Geo Econ Futur Sustain Agric Altern Energy Volume II 2:395-407. https://doi.org/10.1007/978-3-030-92303-7_43

Inshakova AO, Matytsin DE, Kazachenok SY (2022b) Modern problems of digitalization of environmental protection in the russian federation. Smart Innov Syst Technol 288:105–110. https://doi.org/10.1007/978-981-16-9808-8_12

Litvinova TN, Kulikova ES, Kuznetsov VP, Taranov PM (2017) Marketing as a determinant of the agricultural machinery market development. In: Popkova E (ed) Overcoming uncertainty of

institutional environment as a tool of global crisis management. Springer, Cham, Switzerland, pp 465–471. https://doi.org/10.1007/978-3-319-60696-5_59
Margaryan A, Terzyan H (2023) Interaction of inflation and corruption in developing markets in the context of technological changes. Econ Ann-XXI *203*(5–6):16–24. https://doi.org/10.21003/ea.V203-02
Matytsin DE, Anisimov AP, Inshakova AO (2022) Legal regulation of the turnover of biotechnologies in Russia, in countries of the BRICS and of the EAEU: certification and labeling of GMO products. Geo Econ Futur Sustain Agric Altern Energy Volume II 2:409–420. https://doi.org/10.1007/978-3-030-92303-7_44
Sngryan S (2022) Assessing the effects of trade liberalization with third countries: The case of the Eurasian Economic Union. J Lib Int Aff *8*(2):35–46. https://doi.org/10.47305/JLIA2282035s
Tadevosyan R (2023) Innovations and international competitiveness: Country-level evidence. Econ Sociol *16*(3):248–260. https://doi.org/10.14254/2071-789X.2023/16-3/13
UNDP (2025) Sustainable Development Goals Report 2025, UN Publications, New York, NY. Retrieved from https://desapublications.un.org/publications/sustainable-development-goals-report-2025. Accessed 11 Dec 2025
Vanchukhina LI, Andronova IV, Sysoev OE, Smetanina AI (2022) The mechanism of innovative development of agribusiness based on AI, Big Data, and IoT for transitioning to expanded reproduction in AgroTech. In: Popkova EG, Sozinova AA (eds) Agrotech: AI Big Data Iot. Springer, Singapore, pp 241–247. https://doi.org/10.1007/978-981-19-3555-8_25
Vasilikhina I (2025) Artificial intelligence: Problems of ethics and law. J Trends ChallS Artif Intell *2–1*(05):33–36. https://doi.org/10.61552/JAI.2025.01.005

Sustainable Development of Rural Areas in Central Asia Through Climate-Smart Agricultural Technologies

Gulmirakhon T. Akhmedova, Nasipakhan Kozhomberdieva, Aktam U. Burkhanov, and Rano A. Mannapova

Abstract The research explores a promising pathway for the sustainable development of rural areas in Central Asia based on the use of climate-smart technologies in agriculture. To this end, regression and trend analyses were conducted using statistics from five Central Asian countries (Kazakhstan, Kyrgyzstan, Tajikistan, Turkmenistan, and Uzbekistan) for the years 2022–2025. As a result, an economic–mathematical econometric model was constructed to determine the extent to which the sustainable development of rural areas in Central Asia depends on the spread of climate-smart technologies in agriculture. The theoretical significance of the model lies in its ability to quantify the impact of GDP carbon intensity on the growth of value-added in agricultural production, assess the effectiveness of regulatory efforts to combat climate change in the agrarian economy of Central Asia, and support the development of its rural territories. Using the model, the research identifies the most promising instrument of state regulation for advancing the dissemination of climate-smart technologies in agriculture: regulatory support and government subsidies for the adoption of climate-smart technologies by agricultural producers. The model also substantiates the cyclical nature of the development of the climate-oriented agrarian economy in Central Asia, showing that reductions in the carbon footprint, occurring as the rural population grows, initiate successive cycles of sustainable development. Each new cycle accelerates the annual growth rate of value added generated in agricultural production, thereby increasing the pace at which rural areas in Central Asia become more populated. A road map for rural development in Central Asia, contingent upon strengthening the climate resilience of the agrarian economy, has been

G. T. Akhmedova · R. A. Mannapova
Tashkent State University of Economics, Tashkent, Uzbekistan
e-mail: r.mannapova@tsue.uz

N. Kozhomberdieva
Osh Technological University named after M.M. Adyshev, Osh, Kyrgyzstan

A. U. Burkhanov (✉)
Alfraganus University, Tashkent, Uzbekistan
e-mail: burxanov.a@afu.uz; a.burkhanov@nuu.uz

National University of Uzbekistan named after Mirzo Ulugbek, Tashkent, Uzbekistan

A. B. Karbekova (ed.), *Sustainable Development of Rural Territories Based on Climate-Smart Technologies*, Climate Change Management,
https://doi.org/10.1007/978-3-032-20106-5_27

developed. It includes medium-term (to 2029) and long-term (to 2033) forecasts, as well as recommendations for improving state regulation aimed at combating climate change in agriculture.

Keywords Sustainable development · Rural areas · Rural population growth · Central Asia · Climate-smart technologies · Agriculture

JEL codes N55 · O18 · Q16

1 Introduction

Climate change is occurring worldwide and poses a serious common threat to all socio-economic systems, disrupting and constraining their economic activities and reducing their efficiency. Simultaneously, the specific damage caused by environmental degradation varies considerably among groups of countries distinguished by economic and geographic criteria. Therefore, although the fight against climate change is a global task, it must be approached differently across these groups of countries and must be targeted, precisely oriented toward the types of damage characteristic of each group.

A focused approach to environmental problem areas enables many groups of countries to achieve visible progress in combating climate change. For example, in Western European countries united within the EU, the main climate-related damage is concentrated in the energy sector and is associated with an unpredictable and growing demand for fuel and energy resources, which are scarce for this group of countries. The progress achieved in recent years in reducing the energy intensity of economic processes and in adopting advanced practices for generating and using clean energy has allowed the EU to make significant strides in combating climate change.

Another illustrative example is the successful experience of the EAEU. In this group of Eurasian countries, climate-related damage primarily manifests in the industrial economy through the accumulation of large volumes of industrial waste, posing environmental hazards and degrading the ecological conditions of major industrial centers. Thanks to the coordinated and systematic efforts of the EAEU countries in implementing green production technologies, substantial progress has been made in Eurasia in addressing climate change, as reflected in reduced waste generation, environmentally sound waste disposal, the development of circular economic practices, and improved environmental conditions.

Central Asian countries deserve particular attention in the study of climate economics. This group of five neighboring states is undergoing dynamic development in social, economic, and technological dimensions. In the Central Asian economic and geographic space, climate change is most strongly manifested in the agrarian economy, hindering agricultural development, reducing the attractiveness of rural areas specializing in agriculture, and making these territories increasingly less populated compared to previous periods and relative to urban areas.

Accordingly, combating climate change in Central Asia must involve the adoption of climate-smart technologies in agriculture to enhance the environmental resilience of agricultural production, expand agricultural output in rural areas, and promote more active settlement of these territories. The challenge lies in the lack of clarity about what exactly drives the spread of climate-smart technologies in the agriculture of Central Asia, which limits planning, forecasting, and regulatory capacities in the region's climate policy. Recognizing the high relevance of this issue, the research aims to identify a promising pathway for the sustainable development of rural areas in Central Asia based on climate-smart technologies in agriculture.

2 Literature Review

The theoretical foundation of this research is the concept of a climate-resilient agrarian economy, in which the fight against climate change is carried out through the spread of climate-smart technologies in agriculture (Sozinova et al. 2022). Previous scholarly publications identify the following as key manifestations of the development of a climate-resilient agrarian economy:

- Acceleration of the annual growth rate of value added generated in agricultural production (Melkamu et al. 2024);
- Increase in the annual population growth rate in rural areas (Atoyan et al. 2022).

Existing studies also highlight the significant influence of GDP carbon intensity on the growth of the agrarian economy. As the carbon footprint of an economy increases, conditions for agricultural activity deteriorate, thereby slowing the pace of agrarian development (Mayilyan and Torosyan 2023). According to the available literature, the main instruments of state regulation that support the spread of climate-smart technologies in agriculture include the following:

- Regulatory frameworks and government subsidies for the adoption of climate-smart technologies in agricultural production (Butorin and Bogoviz 2019; Margaryan and Margaryan 2022);
- Incentives for the innovative development of the agrarian economy and for import substitution of climate-smart technologies for agriculture (Vanchukhina et al. 2022);
- Informational and infrastructural support for mastering climate-smart technologies in agricultural practices (Litvinova 2020a, 2020b).

A review of the scientific literature on the subject (Lisina et al. 2022) shows that several published sources partially address the practical experience of climate-change mitigation in the agrarian economies of Central Asian countries. However, this experience remains insufficiently examined, and its scientific interpretation is incomplete. This reveals a gap in the literature, namely, the uncertainty surrounding the causal links between regulatory climate-action measures in the agrarian economy of Central Asia and the development of its rural areas.

The need to fill this gap arises from the methodological fragility it creates in the state regulation of a climate-resilient agrarian economy in Central Asia. In light of this, the research question is: "How significant is the influence of GDP carbon intensity on the growth of value added in agricultural production, and which instruments of state regulation supporting the spread of climate-smart technologies in agriculture can weaken this influence and ensure more sustainable development of rural areas in Central Asia?"

3 Materials and Methods

The research object is the achievement of climate resilience in the agrarian economy of Central Asia. The research sample includes all five Central Asian countries (Kazakhstan, Kyrgyzstan, Tajikistan, Turkmenistan, and Uzbekistan). The study period covers 2022–2025, with statistical data for these years combined into a continuous dataset (20 observations).

The research begins with the application of regression analysis to determine the following relationships:

- The annual growth rate of value added generated in agricultural production (Avag, based on World Bank data (World Bank 2025a)) as a function of GDP carbon intensity (CIgdp, based on World Bank data (World Bank 2025b), and the application of regulatory instruments (based on Oxford Insights data (Oxford Insights 2025)): regulatory and subsidy support for the adoption of climate-smart technologies in agricultural production (G_1), incentives for the innovative development of the agrarian economy and for import substitution of climate-smart agricultural technologies (G_2), and informational and infrastructural support for mastering climate-smart technologies in agricultural practices (G_3);
- the annual population growth rate in rural areas (RPrg, based on World Bank data (World Bank 2025c)), as a function of the regulatory instruments (G_{1-3});
- GDP carbon intensity (CIgdp) as a function of the annual rural population growth rate (RPrg).

The data used for the regression analysis are presented in Table 1.

Next, using trend analysis, the authors assessed the four-year growth of the indicators presented in Table 1. Based on the identified regression relationships and established trends, the authors compiled forecasts for the development of rural areas in Central Asia under the condition of strengthening the climate resilience of the agrarian economy for the medium term (to 2029) and the long term (to 2033). Taken together, these forecasts form a road map for this development.

Table 1 Statistics on the climate resilience of the agricultural economy in Central Asia, 2022–2025

Year	Central Asian country	Agriculture, forestry, and fishing, value added (annual % growth)	Rural population growth (annual %)	Carbon intensity of GDP (kg CO_2e per 2021 PPP $ of GDP)	Government (score 0–100)	Techno-logy sector (score 0–100)	Data and infrastructure (score 0–100)
		Avag	RPrg	CIgdp	G_1	G_2	G_3
2022	Kazakhstan	− 2.20	0.98	0.36	48.80	32.38	64.10
	Kyrgyzstan	− 4.54	1.38	0.25	35.16	23.62	54.04
	Tajikistan	7.90	1.90	0.26	35.85	26.29	53.31
	Turkmenistan	0.10	1.00	0.61	22.43	29.43	43.99
	Uzbekistan	3.95	1.92	0.42	37.95	31.32	51.13
2023	Kazakhstan	9.10	1.06	0.38	48.40	29.22	59.72
	Kyrgyzstan	7.31	1.29	0.27	33.73	22.49	45.50
	Tajikistan	− 0.30	1.80	0.24	41.81	19.79	45.90
	Turkmenistan	0.10	0.87	0.56	22.43	29.43	43.99
	Uzbekistan	3.61	1.93	0.42	49.90	24.45	59.17
2024	Kazakhstan	− 7.40	1.01	0.36	48.56	30.97	66.13
	Kyrgyzstan	0.65	1.22	0.24	33.53	22.86	45.90
	Tajikistan	− 4.50	1.65	0.21	53.18	20.41	42.77
	Turkmenistan	19.30	0.74	0.54	20.05	30.51	42.94
	Uzbekistan	4.06	1.89	0.41	49.07	24.80	57.49
2025	Kazakhstan	13.70	0.78	0.34	54.75	33.54	65.93
	Kyrgyzstan	6.30	1.15	0.22	34.68	24.49	50.49
	Tajikistan	12.20	1.51	0.20	51.05	19.79	39.31

(continued)

Table 1 (continued)

Year	Central Asian country	Agriculture, forestry, and fishing, value added (annual % growth)	Rural population growth (annual %)	Carbon intensity of GDP (kg CO_2e per 2021 PPP $ of GDP)	Government (score 0–100)	Techno-logy sector (score 0–100)	Data and infrastructure (score 0–100)
		Avag	RPrg	CIgdp	G_1	G_2	G_3
	Turkmenistan	13.96	0.61	0.50	17.03	32.92	47.96
	Uzbekistan	3.11	1.78	0.39	64.71	33.50	62.14

Source Developed by the authors based on Oxford Insights (2025) and World Bank (2025a, b, c)

Table 2 The specifics of the climatic agricultural economy in Central Asia compared with global experience from previous literature

Instruments of state regulation of the dissemination of climate-smart technologies in agriculture	Reflection in previous literature	Consequences of applying tools in Central Asia for (regression):	
		Annual growth rate of value added generated in agricultural production	Annual growth rate of the population living in rural areas
Regulatory and subsidy support for the adoption of climate-smart technologies in agricultural production	Butorin and Bogoviz (2019), Margaryan and Margaryan (2022)	Acceleration (0.0114)	Increase (0.0156)
Incentives for the innovative development of the agrarian economy and for import substitution of climate-smart technologies for agriculture	Vanchukhina et al. (2022)	Acceleration (0.9431)	Decrease (− 0.0417)
Informational and infrastructural support for mastering climate-smart technologies in agricultural practices	Litvinova (2020a, b)	Slowdown (− 0.4383)	Decrease (− 0.0003)

Source Developed by the authors

4 Results

The regression analysis of the statistics presented in Table 1 made it possible to construct the following Model (1), which reflects the determinants of the sustainable development of rural areas in Central Asia driven by the spread of climate-smart technologies in agriculture:

$$\begin{cases} \text{Avag} = 5.5060 - 12.1982\text{CIgdp} + 0.0114\text{G}_1 + 0.9431\text{G}_2 - 0.4383\text{G}_3, \\ \text{RPrg} = 1.8446 + 0.0156\text{G}_1 - 0.0417\text{G}_2 - 0.0003\text{G}_3, \\ \text{CIgdp} = 0.5067 - 0.1117\text{RPrg}. \end{cases} \quad (1)$$

Model (1) shows that a 1 kg CO_2e increase in GDP carbon intensity per 2021 PPP dollar of GDP leads to a 12.1982% slowdown in the annual growth rate of value added generated in agricultural production in Central Asia. In turn, a 1% acceleration in the annual population growth rate in rural areas results in a reduction of GDP carbon intensity by 0.1117 kg CO_2e per 2021 PPP dollar of GDP.

Strengthening regulatory and subsidy support for the adoption of climate-smart technologies in agricultural production by one point ensures an acceleration of 0.0114% in the annual growth rate of agricultural value added in Central Asia, while simultaneously increasing the annual population growth rate in rural areas by 0.0156%.

Intensifying incentives for the innovative development of the agrarian economy and for the import substitution of climate-smart agricultural technologies by one point leads to a 0.9431% acceleration in the annual growth rate of agricultural value added. However, it also simultaneously results in a 0.0417% slowdown in the annual rural population growth rate.

Strengthening informational and infrastructural support for mastering climate-smart technologies in agricultural practices by one point causes a 0.4383% slowdown in the annual growth rate of agricultural value added in Central Asia and a 0.0003% slowdown in the annual growth rate of the rural population.

The trend analysis carried out using the data from Table 1, combined with Model (1), enabled the development of a "road map" for the sustainable development of rural areas in Central Asia under conditions of strengthened climate resilience in the agrarian economy. This road map includes medium-term projections (to 2029, Fig. 1) and long-term projections (to 2033, Fig. 2).

As shown in the forecast in Fig. 1, the climate-oriented agrarian economy of Central Asia exhibits the following four-year trends (2022–2025):

- A decline in the carbon footprint by 12.25% (a factor of 0.87);

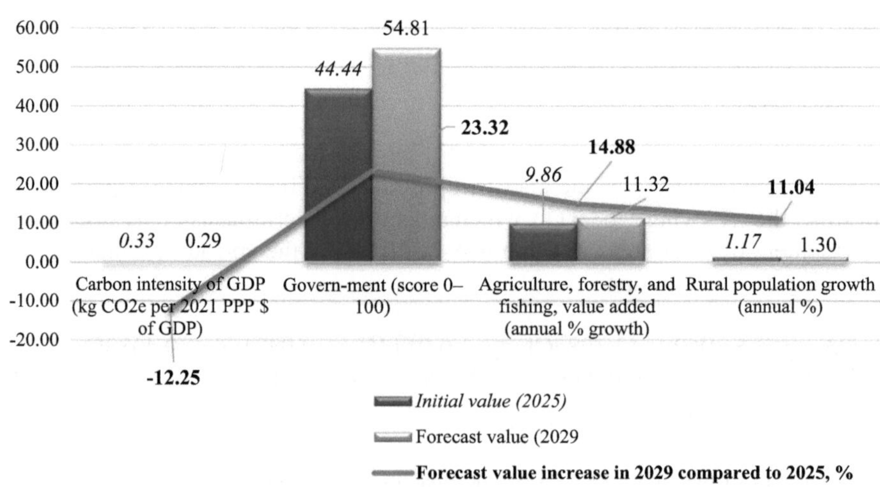

Fig. 1 Forecast for the development of rural areas in Central Asia through strengthening the climate resilience of the agricultural economy until 2029. *Source* Developed by the authors

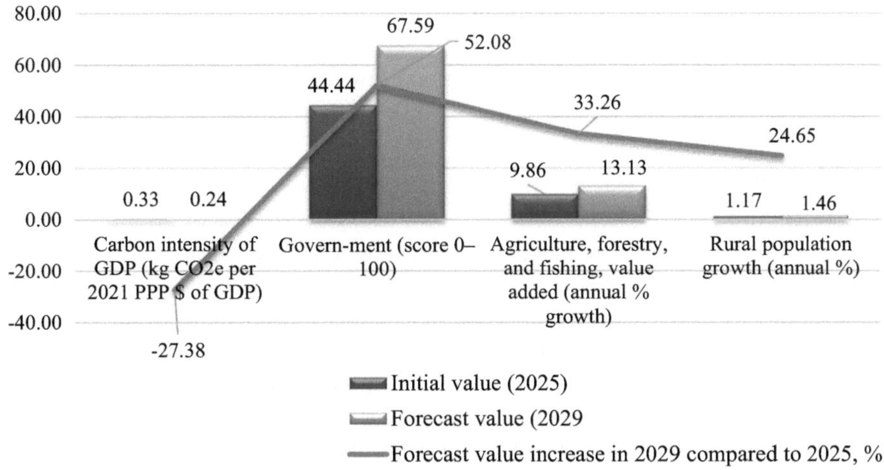

Fig. 2 Forecast for the development of rural areas in Central Asia through strengthening the climate resilience of the agricultural economy until 2033. *Source* Developed by the authors

- Strengthening of the regulatory and subsidy support for the adoption of climate-smart technologies in agricultural production by 23.32% (a factor of 1.23).

Continuation of these trends in Central Asia into the next four-year period (2026–2029) implies the following:

- A decrease in GDP carbon intensity from 0.33 kg CO_2e per 2021 PPP $ of GDP in 2025 to 0.29 kg CO_2e per 2021 PPP $ of GDP by 2029 (0.33 × 0.87);
- A strengthening of the regulatory and subsidy framework supporting the adoption of climate-smart technologies in agricultural production from 44.44 points in 2025 to 54.81 points by 2029 (44.44 × 1.23).

As a result, by the medium-term horizon of 2029, the climate-oriented agrarian economy of Central Asia is projected to accelerate the annual growth rate of value added generated in agricultural production from 9.86% in 2025 to 11.32%. An increase in the annual rural population growth rate from 1.17% in 2025 to 1.30% is also anticipated. This will lead to an additional reduction of GDP carbon intensity by 4.53% (a factor of 0.95), further reinforcing the downward trend in the carbon footprint of Central Asia during the subsequent four-year period (2030–2033).

As shown in Fig. 2, the reinforced trend of reducing the carbon footprint in Central Asia will lead to a decline in GDP carbon intensity to 0.29 kg CO_2e per 2021 PPP $ of GDP by 2033 (0.29 × 0.95 × 0.87). Simultaneously, the strengthening of the regulatory and subsidy support for the adoption of climate-smart technologies in agricultural production will continue, reaching 67.59 points by 2033 (54.81 × 1.23). Owing to this, the long-term outlook for the climate-oriented agrarian economy of Central Asia to 2033 includes the following:

- An acceleration of the annual growth rate of value added generated in agricultural production to 13.13%;
- An increase in the annual rural population growth rate to 1.46%.

5 Discussion

The contribution of the results achieved in this research to the literature (Atoyan et al. 2022; Lisina et al. 2022; Melkamu et al. 2024; Sozinova 2022) lies in advancing the scientific principles of the concept of a climate-resilient agrarian economy by revealing previously unknown causal relationships between regulatory climate-action measures in the agrarian economy of Central Asia and the development of its rural areas (Table 2).

As reflected in Table 2, consistent with the findings of Butorin and Bogoviz (2019), as well as Margaryan and Margaryan (2022), regulatory and subsidy support for the adoption of climate-smart technologies in agricultural production is the key instrument of state regulation driving the spread of climate-smart technologies in agriculture in Central Asia. The application of this instrument systematically accelerates both the annual growth rate of value added generated in agricultural production and the annual growth rate of the rural population.

In contrast to Vanchukhina et al. (2022), the research demonstrates the limited effectiveness of another regulatory instrument, the stimulation of innovative development in the agrarian economy and the import substitution of climate-smart agricultural technologies. While this tool accelerates the annual growth rate of agricultural value added, it does not accelerate the growth rate of the rural population.

In contrast to Litvinova (2020a, b), informational and infrastructural support for mastering climate-smart technologies in agricultural practice is not an effective regulatory instrument for promoting the spread of climate-smart technologies in agriculture in Central Asia because it contributes neither to agricultural growth nor to the settlement of rural areas.

In contrast to Mayilyan and Torosyan (2023), the research established that the climate-oriented agrarian economy in Central Asia develops not linearly but in a cyclical manner. This is evidenced by the newly identified reverse relationship between GDP carbon intensity and the sustainability of rural development, in addition to the previously known direct relationship between carbon intensity and agrarian growth (where an increase in the carbon footprint slows this growth).

This newly identified reverse relationship manifests in the fact that an increase in the rural population leads to a reduction in the carbon footprint, which initiates the next cycle of sustainable development of the climate-oriented agrarian economy. With each new cycle, the annual growth rate of value added in agricultural production accelerates further, thereby increasing the rate at which rural areas in Central Asia become more populated.

6 Conclusion

The primary results of the research can be summarized as follows:

1. An econometric model was developed to determine the sustainability of rural development in Central Asia under the influence of the spread of climate-smart technologies in agriculture. The theoretical contribution of this model lies in its ability to quantify the impact of GDP carbon intensity on the growth of value added in agricultural production, as well as the effectiveness of regulatory climate-action measures within the agrarian economy of Central Asia and their influence on the development of rural areas;
2. The research concludes that the most promising instrument of state regulation for promoting the spread of climate-smart technologies in agriculture is the regulatory and subsidy support for the adoption of climate-smart technologies in agricultural production;
3. The cyclical nature of the climate-oriented agrarian economy in Central Asia is substantiated: as the rural population grows, the carbon footprint decreases, triggering a new cycle of sustainable development. With each successive cycle, the annual growth rate of value added generated in agricultural production accelerates, thereby increasing the pace of rural settlement across Central Asia;
4. A road map for the development of rural areas in Central Asia, conditional upon the strengthening of climate resilience in the agrarian economy, has been developed. It includes medium-term (to 2029) and long-term (to 2033) forecasts and recommendations for improving state regulation in the fight against climate change in agriculture. Implementing these recommendations will help weaken the impact of GDP carbon intensity on agricultural value-added growth. Following the proposed road map will stimulate the sustainable development of rural areas through accelerated rural settlement in Central Asia.

References

Atoyan V, Ohanyan S, Malkjyan A, Movsisyan N (2022) On identification, identity, and security issues in modern Armenia. WISDOM 2(1):5–14. https://doi.org/10.24234/wisdom.v2i1.783

Butorin SN, Bogoviz AV (2019) The innovational and production approach to management of economic subjects of the agrarian sector. In: Popkova E, Ostrovskaya V (eds) Perspectives on the use of new Information and Communication Technology (ICT) in the modern economy. Springer, Cham, Switzerland, pp 758–773. https://doi.org/10.1007/978-3-319-90835-9_88

Lisina LM, Giyazov AT, Oudah ALMMY, Dubova YI (2022) Agriculture in developing countries: cultural differences, vectors of sustainable development, digitalization, and international experience. In: Popkova EG, Sergi BS (eds) Sustainable agriculture. Springer, Singapore, pp. 65–74. https://doi.org/10.1007/978-981-16-8731-0_7

Litvinova TN (2020a) Agricultural lease as a perspective mechanism of development of infrastructure of entrepreneurship in the agricultural machinery market. In: Popkova E, Sergi B (eds) The 21st century from the positions of modern science: Intellectual, digital and innovative aspects. Springer, Cham, Switzerland, pp 624–630. https://doi.org/10.1007/978-3-030-32015-7_70

Litvinova TN (2020b) Managing the development of digital infrastructural provision of entrepreneurial activities in the agricultural machinery market. In: Popkova E, Sergi B (eds) Digital economy: complexity and variety vs. rationality. Springer, Cham, Switzerland, pp 424–431. https://doi.org/10.1007/978-3-030-29586-8_49

Margaryan A, Margaryan A (2022) Convergent technologies in the system of modernisation of economy: conceptual-methodological comparisons. WISDOM 2(1):106–114. https://doi.org/10.24234/wisdom.v2i1.791

Mayilyan FN, Torosyan AA (2023) The impact of minimum wage on income inequality in developing countries. In: Popkova EG, Sergi BS (eds) ESG management of the development of the green economy in Central Asia. Springer, Cham, Switzerland, pp 355–364. https://doi.org/10.1007/978-3-031-46525-3_38

Melkamu M, Fisseha T, Dugassa E (2024) Maximization profit of agricultural crop production through decision making application. J Trends ChallS Artif Intell 1(2):51–60. https://doi.org/10.61552/JAI.2024.02.002

Oxford Insights (2025) *Government AI readiness index*. Retrieved from https://oxfordinsights.com/ai-readiness/framework-2025/. Accessed 1 Dec 2025

Sozinova AA, Sofiina EV, Petrenko YS, Bencic S (2022) International features of using smart technology in agriculture: overview of innovative trends. Smart Innov, Syst Technol 264:167–173. https://doi.org/10.1007/978-981-16-7633-8_18

Vanchukhina LI, Andronova IV, Sysoev OE, Smetanina AI (2022) The mechanism of innovative development of agribusiness based on AI, Big Data, and IoT for transitioning to expanded reproduction in AgroTech. In: Popkova EG, Sozinova AA (eds) AgroTech: AI Big Data IoT. Springer, Singapore, pp. 241–247. https://doi.org/10.1007/978-981-19-3555-8_25

World Bank (2025a) Agriculture, forestry, and fishing, value added (annual % growth). Retrieved from https://data.worldbank.org/indicator/NV.AGR.TOTL.KD.ZG. Accessed 1 Dec 2025

World Bank (2025b) Carbon intensity of GDP (kg CO_2e per 2021 PPP $ of GDP). Retrieved from https://data.worldbank.org/indicator/EN.GHG.CO2.RT.GDP.PP.KD. Accessed 1 Dec 2025

World Bank (2025c) Rural population growth (annual %). Retrieved from https://data.worldbank.org/indicator/SP.RUR.TOTL.ZG. Accessed 1 Dec 2025

The Potential for Developing Organic Agriculture in Kyrgyzstan Within the Framework of Sustainable Rural Development

Murat Zh. Abdiev, Tulkin Kh. Farmanov, Svetlana B. Makeeva, Kubanych K. Toktorov, and Bayana T. Mametova

Abstract The research aims to provide a theoretical and methodological justification for the potential development of organic agriculture in the Kyrgyz Republic (KR) and develop practical recommendations. Based on the analysis and assessment of the current state of organic farming in the Kyrgyz Republic, the research identifies the key challenges in this area, the level of development, and the underlying factors. It also proposes improvements to the regulatory and legal framework, adherence to ecological principles in national agriculture, and the production of environmentally friendly products through the application of innovations and scientific and technological advancements in this sector, such as drip irrigation and biohumus. Other proposed measures include preserving and enhancing soil fertility, abandoning synthetic substances, ensuring proper crop rotation, promoting efficient and targeted use of natural resources, and safeguarding public health and that of future generations.

M. Zh. Abdiev (✉) · K. K. Toktorov · B. T. Mametova
Osh Technological University named after M.M. Adyshev, Osh, Kyrgyzstan
e-mail: mabdiev1977@mail.ru

K. K. Toktorov
e-mail: toktorovk70@oshsu.kg

B. T. Mametova
e-mail: bayana310597@list.ru

M. Zh. Abdiev · K. K. Toktorov
Fergana State Technical University, Fergana, Uzbekistan

M. Zh. Abdiev
Andijan Institute of Agriculture and Agrotechnology, Andijan, Uzbekistan

T. Kh. Farmanov
Tashkent State University of Economics, Tashkent, Uzbekistan

S. B. Makeeva
Kyrgyz National University named after Jusup Balasagyn, Bishkek, Kyrgyzstan

K. K. Toktorov
University of Economics and Pedagogy (UNIEP), Andijan, Uzbekistan

A. B. Karbekova (ed.), *Sustainable Development of Rural Territories Based on Climate-Smart Technologies*, Climate Change Management,
https://doi.org/10.1007/978-3-032-20106-5_28

The object and subject of this research are the organizational and economic mechanisms related to the development of organic agriculture under conditions of sustainable rural development. The research proposes specific mechanisms for improving organic farming in the context of innovation- and investment-driven growth. Based on the conducted research, the theoretical and methodological foundations, as well as the current state and development prospects of organic agriculture, are substantiated; practical recommendations are presented. The implementation of these theoretical and practical recommendations will help create favorable conditions for the further advancement of organic agriculture in the Kyrgyz Republic, enable effective solutions to major challenges in this field, and strengthen state support and regulation. The research results serve as a scientific basis for the continued development of organic farming. The research methods employed include system analysis, economic and statistical analysis, monographic and comparative methods, and others.

Keywords Agriculture · Organic farming · Food security · Innovation · Organic products · Market · Production · Implementation · Efficiency · Regulation · Strategy · Support · Mechanism

JEL codes D13 · L23 · O13 · Q10

1 Introduction

Agriculture oriented toward an organic development path holds competitive advantages and significant potential in strengthening the country's and its regions' socio-economic standing, as well as ensuring food and environmental security amid the transition to a green economy and the widespread adoption of information technologies and innovations. From a socio-economic and environmental perspective, agriculture is an industry closely linked to the processing sector, where the preservation of the country's ecology and the safety of food production must be maintained. The research results show that agriculture occupies a competitive position due to the production of organic products and serves as a driving force for the dynamic development of the country's regions in the context of economic integration. This sector of the economy also possesses strong potential for further improvement of organic farming, particularly through the preservation of the traditional and sustainable foundations of the agricultural sector.

2 Methodology

From a theoretical and methodological perspective, this research draws upon the research findings of Kyrgyz and international scholars who have studied organic agriculture. The empirical, legal, and informational foundations of this research include legislative acts and resolutions of the Government of the Kyrgyz Republic, the Ministry of Agriculture of the Kyrgyz Republic, and the National Statistical Committee of the Kyrgyz Republic, as well as scientific articles, monographs, textbooks, teaching materials, and online sources.

The research methods employed include comparative-analytical, monographic, economic-statistical, and computational-constructive methods.

3 Results

In the context of globalization and growing integration, one of the key challenges facing the Kyrgyz Republic (KR) is ensuring the country's food security.

The supply of food and the achievement of its ecological safety have always been among humanity's most vital concerns (Aripov et al. 2024; Momunov 2017).

During the reforms in the Kyrgyz Republic, relevant legal and regulatory documents were developed, addressing the country's food security issues in greater detail. With the adoption of the Law "On food security," the main directions for ensuring food security in the Kyrgyz Republic were outlined, establishing it as an integral and crucial component of national security. According to this law, "food security is a state of the Kyrgyz Republic's economy in which the country's food independence is ensured, and physical and economic access to food for the population is guaranteed in accordance with established minimum consumption norms" (Government of the Kyrgyz Republic 1999, 2019; Kyrgyz Republic 2008; President of the Kyrgyz Republic 2018).

The national food market of the Kyrgyz Republic remains highly dependent on imports, largely due to the decline in production of socially significant food products, the disruption of linkages between stages of the agro-industrial cycle, and the growing share of small farms (peasant and family holdings) in the agricultural structure. These factors have contributed to low production efficiency, reduced competitiveness of the sector, diminished export potential, and the lack of a coherent system of state support and regulation for the agro-industrial complex.

Small-scale agricultural production remains unattractive to investors. It is generally considered more effective and feasible to invest in large-scale production, which ultimately offers greater advantages than small-scale operations. Therefore, the country is currently taking steps to restore close integration among producers, processors, and distributors through the creation of agro-industrial cooperative clusters.

Within the framework of a green economy, agriculture in the Kyrgyz Republic should remain a state priority, given that as of 2024, approximately 65% of the population lived in rural areas, and agriculture accounted for about 9–10% of the GDP (compared to 12.2% in 2020). The GDP of Kyrgyzstan in 2024 amounted to 1.2 trillion soms. That same year, gross agricultural output increased by 6.2%; the volume of processed agricultural products reached 79.2 billion soms. Thus, agriculture continues to serve as a key pillar for the country's strategic development and a cornerstone for strengthening its long-term sustainability. Land restoration is a strategic objective, forming the foundation for the future growth of the agricultural sector, including its environmentally oriented development (National Statistical Committee of the Kyrgyz Republic 2021).

Approximately 70% of arable land in Kyrgyzstan is irrigated. The country's potential irrigated land resources amount to 2044.7 thousand hectares, of which 636.3 thousand hectares lie within irrigated zones and 1408.4 thousand hectares in foothill regions. About 94% of the national territory is mountainous, with 41% classified as harsh highlands located above 3000 m.a.s.l. Much of the mountainous region is covered by ice and snow, with glaciers occupying roughly 4% of the country's total area. Only 6.8% of the land is used for cultivating crops, while 44% serves as pasture for livestock.

It is worth noting that Kyrgyz researchers have concluded that the existing intensive model of agricultural development, along with the underdeveloped processing of agricultural raw materials, provides a foundation for the further advancement of organic farming. This is largely because the country has managed to preserve much of its natural ecology and untouched environmental resources.

The current National Development Program of the Kyrgyz Republic up to 2030 serves as the primary document defining the country's future development, considering its present state and historically established achievements. Within this program, one of the key priorities in agricultural production is the design and implementation of measures aimed at advancing organic agricultural practices for certain types of products. This includes large-scale production of environmentally friendly goods in accordance with international standards, the manufacturing of halal products, diversification of product ranges, mandatory certification, and incentives for export promotion (President of the Kyrgyz Republic 2025).

Currently, the principal organization promoting and supporting the development of organic agriculture worldwide is the International Federation of Organic Agriculture Movements (IFOAM). The Federation has established regulations that define this field as a production system supporting ecological balance and contributing to the health of nations. The main goal of organic agriculture is to preserve environmental balance while improving the state of the natural environment, thereby enhancing the overall quality of life in society (Kharitonov and Kharitonov 2014; Turdiev 2013, 2019).

In the agricultural sector, organic farming is rooted in the historical traditions of the Kyrgyz people, the nation's natural features, diverse landscapes, territorial variations, and its unique flora and fauna. It also integrates scientific achievements, practical experience, innovations, investments, and the country's agro-resource potential in order to produce environmentally safe agricultural products.

In the Kyrgyz Republic, the industrial sector significantly lags behind developed countries in the level of innovation use. However, in terms of its potential to develop a "green economy," the country could occupy a leading position within the system of international economic relations. Over the past three years, global prices for agricultural products, including organic goods, have increased due to the processes of globalization, inflation, and other factors. Considering its internal reserves and external potential, the Kyrgyz Republic has the opportunity to strengthen and further develop its agricultural sector in this direction.

According to Professor M. Zh. Abdiev, new market segments have emerged in the food sector, such as eco-supermarkets and bio-hypermarkets, designed to sell environmentally friendly products from the agricultural and processing industries that comply with organic farming standards. In recent years, these products have seen growing domestic demand and have also been exported abroad, as they are free from GMOs and beneficial to public health. The state's role in this context is to provide comprehensive support through various mechanisms, including tax and organizational incentives, as well as penalties for the production of GMO-containing goods. Such measures will facilitate the gradual transition toward food safety and environmental responsibility in both agriculture and the processing industry (Abdiev et al. 2024). Given that the combined profits of transnational corporations engaged in the development, cultivation, and sale of genetically modified plants currently amount to \$3–\$5 billion, and are projected to reach \$50–\$100 billion by 2030, the measures to promote the green market and suppress the "harmful" one must be genuinely substantial; otherwise, the substitution effect will fail to take hold (Balatsky and Ekimova 2011, p. 244).

For example, in neighboring China, fruits grown in Kyrgyzstan are sold at twice the price, as they are regarded as environmentally pure and naturally produced goods.

In the long term, the authors believe that by supporting the organic products market, the state can ensure the stability and continued development of this sector.

Organic agriculture is based on the following core principles:

- Production of environmentally clean products;
- Rational and targeted use of natural resources;
- Preservation and improvement of soil fertility;
- Mandatory crop rotation; concern for public health and that of future generations;
- Socially and economically efficient production.

Taking these principles into account, organic agriculture should be considered a priority direction in the strategic development of the agrarian sector. The greening of agricultural production represents an essential step toward the country's prosperity, particularly through the production of environmentally safe goods.

Compared to other sectors of the economy, the agrarian sector occupies a leading competitive position due to its ecological orientation. Nowadays, consumers around the world are paying increasing attention to "green economy" brands and halal certification. Shopping centers, stores, supermarkets, and hypermarkets now offer a wide variety of organic products. For example, a young entrepreneur from the Kadamjay district of the Batken Region has been engaged in the production of "Ak Turpak" rice for three years. Through participation in various development projects, she has purchased new equipment and launched a new brand, selling her products both domestically and abroad, including exports to the USA and other countries. Her organically produced rice enjoys high demand. In European countries, prices for such products exceed those of conventional goods by 10–50%, depending on the type of product, while in Russia, organic products can cost two to three times more (Kharitonov 2021, p. 90).

The endemism rate of plant species in the Kyrgyz Republic is close to 70%, meaning that a large portion of the country's flora has preserved its original natural characteristics. In contrast, the endemism rate in the USA is approximately 30%, indicating that much of its flora has lost its natural properties due to degradation and mutation (Adamaliev 2011, p. 7).

In Kyrgyzstan, a top priority should be the certification of the country's territories in accordance with international standards for organic agriculture and the green economy. As international certification practices show, land areas undergo certification to qualify as organic, and the resulting products are recognized as organically clean. Unfortunately, due to the absence of such standards, household farms cannot label their products as organic. Furthermore, these types of products are not yet regulated by national legal frameworks, which underscores the need for the development of relevant regulations and oversight mechanisms.

Globally, organic agricultural lands cover about 35 million hectares, while in the Kyrgyz Republic, only 2.5 thousand hectares are currently certified. This indicates that the country has both the prerequisites and the experience necessary to expand and further develop organic farming.

Agricultural producers in the Kyrgyz Republic are striving to increase their income, expand their market share, and, in the future, develop the production of environmentally friendly products. By 2025, the number of farmers engaged in organic agriculture in Kyrgyzstan exceeded 1200. Some farmers have even begun adopting innovative technologies. For instance, in the Talas Region, farmers cultivating beans achieved a total planting area of 60,000 hectares in 2025, with a gross yield of 89.3 thousand tons and a productivity rate of 18.4 centners per hectare. Kyrgyz beans have gained strong popularity in European markets, confidently establishing their niche abroad. In 2024, a new enterprise, "Kyrgyz Fasol," was launched under the JSC "Kyrgyz Industry," specializing in bean processing and exports.

Currently, organic agriculture in the Kyrgyz Republic is still in its formative stage. State authorities, agricultural producers, and local self-government bodies are making efforts to promote its development for the benefit of national and regional economic growth, food security, and public health. However, it is still too early to claim that comprehensive progress has been achieved in this field, as public awareness of the benefits of organic farming remains limited. Many farmers continue to rely on mineral fertilizers and agrochemicals, and have yet to transition away from them fully. There is also a general lack of understanding about the value of the "green economy." Therefore, expanding the certified production and sale of environmentally clean products remains a key objective of the modern agrarian sector.

Research data indicate that enterprises producing organic goods under "bio" or "eco" labels are experiencing high demand from consumers. Unfortunately, many of these producers lack official certification and do not fully meet established quality standards.

From a social standpoint, the agricultural sector has a significant impact on the environment and public health, influenced by numerous factors. One of the main challenges is the reduced effectiveness of efforts to intensify this sector of the economy.

The research results show that many government programs and policies remain focused on supporting and regulating traditional agriculture, while organic farming continues to receive secondary attention. The key incentives that could stimulate its growth include tax benefits, subsidies, and access to concessional loans.

It is now time for Kyrgyzstan to prioritize the expansion of ecologically oriented agricultural production.

It is important to emphasize that a combination of pasture degradation, soil salinization, and erosion has disrupted the ecological functions of the soil cover. As a result, land degradation has worsened the overall environmental situation across the country, significantly limiting the use of the agro-resource potential and, consequently, hindering socio-economic development.

The adoption of advanced innovative agrotechnical methods has become imperative. Ultimately, these should help preserve soil fertility. In this regard, the experience of foreign countries with centuries of expertise in sustainable and organic farming practices could serve as a valuable example for Kyrgyzstan.

4 Conclusion

Thus, based on the research results, the authors conclude that globally and in Kyrgyzstan, the demand for organic products is steadily increasing. In this regard, the state should actively support the improvement and development of organic agriculture as one of the strategic directions of the national economy. An environmentally oriented agrarian sector should become a defining feature of the country's "national agriculture."

References

Abdiev MZ, Toktorov KK, Abdrakhmanov OE, Momunov UN, Ergeshov KA (2024) Ways to improve the innovative development of Kyrgyz agriculture. In: Popkova EG, Bogoviz AV, Sergi BS, Kaurova OV, Maloletko AN (eds) Sustainable development of the agrarian economy based on digital technologies and smart innovations. Springer, Cham, Switzerland, pp 51–54. https://doi.org/10.1007/978-3-031-51272-8_9

Adamaliev U (2011) Organic farming—the future of Kyrgyzstan. Echo of Osh 5:2–3

Aripov NT, Abdiev MZ, Baharuddin AS, Ubaidullaev MB (2024) Halal industry in Kyrgyzstan: current state and development prospects. Malays J Syariah Law 12(3):790–800. https://doi.org/10.33102/mjsl.vol12no3.874

Balatsky E, Ekimova N (2011) The qualitative component of Russia's food security. Soc Econ 11–12:224–245

Government of the Kyrgyz Republic (1999) National food security policy of the Kyrgyz Republic (adopted by Resolution No. 585 of October 26, 1999). Bishkek, Kyrgyzstan. Retrieved from https://cbd.minjust.gov.kg/33866/edition/1144190/ru. Accessed 23 Sept 2025

Government of the Kyrgyz Republic (2019) Food security and nutrition program in the Kyrgyz Republic for 2019–2023 (adopted by Resolution No. 320 of June 27, 2019). Retrieved from https://cbd.minjust.gov.kg/14562/edition/958753/ru. Accessed 23 Sept 2025

Kharitonov SA (2021) Organic farming and production of environmentally friendly products in Russia. AIC: Econ Manag 8:88–93

Kharitonov NS, Kharitonov CA (2014) The development of organic farming in Baden-Württemberg. AIC: Econ Manag 1:80–87

Kyrgyz Republic (2008) Law On food security (August 4, 2008 No. 183). Bishkek, Kyrgyzstan. Retrieved from https://cbd.minjust.gov.kg/202397/edition/1244333/ru. Accessed 23 Sept 2025

Momunov UN (2017) Mechanisms for ensuring food security in the context of deepening integration (based on materials from the Kyrgyz Republic) (Dissertation of Candidate of Economic Sciences). Osh State University, Osh, Kyrgyzstan

National Statistical Committee of the Kyrgyz Republic (2021) Agriculture of the Kyrgyz Republic. National Statistical Committee of the Kyrgyz Republic, Bishkek, Kyrgyzstan. Retrieved from https://stat.gov.kg/en/publications/sbornik-selskoe-hozyajstvo-kyrgyzskoj-respubliki/. Accessed 23 Sept 2025

President of the Kyrgyz Republic (2018) National development strategy of the Kyrgyz Republic for 2018–2040 (adopted by Presidential Decree No. 221 of October 31, 2018). Bishkek, Kyrgyzstan. Retrieved from https://www.gov.kg/ru/programs/8. Accessed 23 Sept 2025

President of the Kyrgyz Republic (2025) National development program of the Kyrgyz Republic until 2030 (adopted by Presidential Decree No. 178 of June 5, 2025). Bishkek, Kyrgyzstan. Retrieved from https://www.gov.kg/files/news/froala/dd08061311362a4faa610d681b49a92ae4f73a55.pdf. Accessed 23 Sept 2025

Turdiev TI (2013) Threats to food security in Kyrgyzstan: environmental and economic aspects. Natl Interes Priorities Secur 1:43–52

Turdiev TI (2019) The concept of strategizing socio-economic and environmental security (based on materials from the Kyrgyz Republic) (Dissertation of Doctor of Economic Sciences). Lomonosov Moscow State University, Moscow, Russia

International Experience in the Sustainable Development of Rural Areas Based on Climate-Smart Technologies in Agriculture

Regional Socio-Economic Policy in the Field of Housing and Its Impact on Population Migration

Saltanat E. Imankulova, Bakytbek K. Kozhogulov, Nurgul S. Aldzhembaeva, Elmira D. Derbisheva, and Akylai Zh. Kozhogulova

Abstract The research considers the most topical issues for improving the quality of life of the population using methods such as providing housing conditions to improve the socio-economic development of the regions in the Kyrgyz Republic. In the modern world, the improvement of living conditions and overall quality of life remains a pressing issue. This area has demonstrated significant positive outcomes, largely due to the meticulous approach taken by state governance in addressing the political, economic, social, and environmental dimensions of national development. In the regional context, this careful consideration has led to tangible improvements that directly influence the daily lives of citizens. One of the fundamental aspects contributing to social well-being is access to private housing. Homeownership provides individuals and families with a stable living environment and serves as a crucial factor in fostering social, political, and economic independence. A person who owns a home is more likely to engage actively in social processes, contribute to the economy, and establish a strong foundation for family development. Moreover, the availability of adequate housing creates essential conditions for a secure and stable life, enabling individuals to focus on personal growth and professional aspirations.

Keywords Housing · Housing conditions · Government programs · Housing policy · Lending · Favorable conditions · Mortgage · Regions · Quality of life of the population

JEL Classification J1 · O18 · R11 · R21 · R23 · R28

S. E. Imankulova · N. S. Aldzhembaeva (✉) · E. D. Derbisheva · A. Zh. Kozhogulova
Kyrgyz State Technical University Named After I. Razzakov, Bishkek, Kyrgyzstan
e-mail: ms.chik8080@mail.ru

B. K. Kozhogulov
Design and Research Center "Ken-Too", Bishkek, Kyrgyzstan

A. B. Karbekova (ed.), *Sustainable Development of Rural Territories Based on Climate-Smart Technologies*, Climate Change Management,
https://doi.org/10.1007/978-3-032-20106-5_29

1 Introduction

Given the strategic focus on regional development, the state has been implementing policies to enhance living standards, particularly in less-developed and problem-prone areas. Prioritizing regional growth has become a central objective of national policy, ensuring that all parts of the country experience balanced progress. The efforts to address disparities between regions have led to improved infrastructure, increased investment, and targeted programs designed to uplift marginalized communities (Imankulova 2024).

The social sector of the economy is the basis for achieving stability and development of the country, especially in the regions of the country. The priority task is to create favorable living conditions and improve the human environment. The problem of quality housing also determines the state of health. Providing the population with comfortable housing has become particularly important and has significantly changed the situation in the housing sector. This is one of the main reasons for internal migration in the country (Ergeshbaev and Chamashev 2018; Ergeshbaeva and Maksudova 2019). The main reason for changing place of residence is the high cost of housing,most of the country's population changed their place of residence for this reason. This reason is especially relevant in the Chui Region, where 62.5% of migrants surveyed changed their housing because of this. If we take the statistics of the survey, every fourth interviewee changed their place of residence due to living conditions, especially in the Chui Region; in Bishkek and Jalal-Abad, one out of three interviewed migrants moved because of this. This reason is irrelevant in the Osh and Batken Regions, where only 3% of surveyed migrants changed housing due to these reasons (Saitov 2009). Overall, almost half of internal migrants are homeowners or live with close relatives who are property owner. Simultaneously, they are renters,some live with relatives. There is also a part of the population that lives in housing provided by the employer free of charge. To a greater extent, young people live in rented housing; respondents over 50 years of age often live in their own housing (Valentey and Kvasha 1989).

A larger number of internal migrants who have their own housing live in the Naryn (77.8%) and Batken (73.8%) Regions. In the cities of Osh (53.4%), Bishkek (45.7%), and Jalal-Abad (42%), internal migrants more often rent housing. Internal migrants who own housing mainly live in rural areas (77.1%). In the capital and small towns, they more often rent housing (Saitov 2009). Therefore, the country faces a housing problem,housing conditions are also important (Eurasian Economic Commission 2023).

2 Methodology

This research employs a comprehensive research methodology incorporating statistical analysis, comparative assessments, and correlation-regression models. By utilizing these analytical tools, the research establishes a solid foundation for examining the intricate relationship between internal migration patterns and housing accessibility. The findings provide critical insights into how state policies influence population distribution and housing provision, offering valuable data for future socio-economic planning and development strategies. The authors processed data for the past 15 years (from 2008 to 2022) and compared it using the "Data Analysis" add-in of the MS Excel program.

To determine the projected indicators of housing commissioning until 2030, the authors also used the MS Excel program. The forecast assumes three scenarios: a normal forecast, an optimistic forecast, and a pessimistic forecast, which provides valuable information for the development of strategic planning.

To prepare statistical data for the research, the authors used grouping, classification, and systematization methods.

The methodological framework of this research integrates a diverse set of analytical approaches, ensuring a multidimensional examination of the correlation between internal migration patterns and housing provision, particularly in the context of newly commissioned residential properties. This research aims to generate valuable insights that contribute to the discourse on sustainable national development while serving as a foundational reference for shaping state housing policy at the national and regional levels.

3 Results

A significant milestone in the formulation of housing policy was marked by adopting the "My Home 2021–2026" program, as per the resolution of the Cabinet of Ministers of the Kyrgyz Republic on July 13, 2021 (Rhythm of Eurasia 2023). Recognizing the constraints imposed by reduced budgetary allocations for housing construction and distribution, the program underscored the pivotal role of alternative financing mechanisms.

Ensuring adequate living conditions for citizens residing in dilapidated housing or lacking personal property remains a critical issue requiring immediate intervention. The authors developed various proposals to enhance and integrate solutions within state housing programs, aiming to address one of the most urgent socio-economic concerns (Government of the Kyrgyz Republic 2017). A fundamental aspect of this initiative is the creation of a structured support system for individuals and families in need of improved housing conditions. This will solve the problem of internal migration, which is one of the most important problems of the country. Let us consider

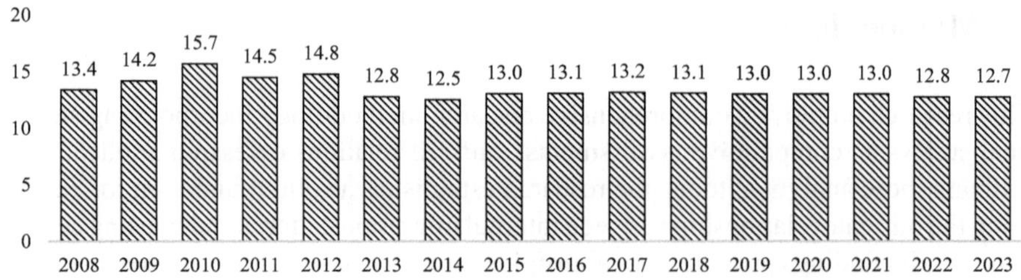

Fig. 1 Dynamics of housing provision per person in the Kyrgyz Republic in 2008–2023, in square meters. *Source* Compiled by the authors based on data from the national statistics committee of the Kyrgyz Republic for 2008–2023 (National Statistical Committee of the Kyrgyz Republic (n.d.))

the dynamics of housing provision in the Kyrgyz Republic over the past 15 years (Fig. 1).

According to Fig. 1, the average living space per person has remained relatively stable over the past 15 years. In 2008, there were 13.4 square meters per person, followed by a slight increase in 2009; peaking at 15.7 square meters during the period. However, between 2019 and 2021, this figure declined to 13 square meters per person; it further dropped to 12.8 and 12.7 square meters in 2022 and 2023, respectively. This trend is attributed to the population growth rate outpacing the rate of new housing construction.

If we look at the regions, we can conclude that there has recently been an increase in the provision of housing in Bishkek. In all regions, except for the Talas Region, there has been a decrease in the indicator. In the Talas Region, the housing supply increased by 0.4 square meters over the review period. In Osh, housing provision grew from 12.2 square meters per person in 2022 to 12.3 square meters in 2023 (Table 1).

Table 1 Dynamics of housing provision per person by region and cities, from 2019 to 2023, in square meters

Region	2019	2020	2021	2022	2023
Batken region	12.2	12.4	12.2	12.0	11.9
Jalalabad region	14.1	14.0	13.9	13.8	13.6
Ysyk-Kol region	15.5	15.7	15.7	14.9	14.8
Naryn region	12.7	12.8	12.8	12.3	12.2
Osh region	10.6	10.5	10.4	10.2	10.2
Talas region	14.0	14.1	14.1	14.4	14.4
Chui region	14.2	14.2	14.2	13.8	13.8
Bishkek city	13.0	13.1	13.2	13.1	13.5
Osh city	12.9	12.8	12.7	12.2	12.3

Source Compiled by the authors based on data from the National Statistics Committee of the Kyrgyz Republic for 2019–2023 (National Statistical Committee of the Kyrgyz Republic (n.d.))

Providing housing is a primary goal for nearly every family in the Kyrgyz Republic. However, due to the current wage levels, the working population struggles to maintain a decent standard of living and save enough to purchase a home. As a result, many are forced to work abroad. The main target audience for housing programs (i.e., entrepreneurs, farmers, and migrants) are those who earn money but cannot document their income. With over 2 million citizens working overseas, their inability to officially confirm their earnings poses a significant barrier, even though many of these migrants ultimately aim to buy their own homes.

Nowadays, numerous attractive programs are available to help individuals purchase their own homes. For migrants, applying for a state mortgage has been simplified through the People's Housing Company, a subsidiary of the State Mortgage Company (SMC). Under the "My Home" program, SMC facilitates housing purchases using a model based on German mortgage lending practices—a system that has been successfully implemented in European countries and Kazakhstan for more than 20 years (Cabinet of Ministers of the Kyrgyz Republic 2021). The main advantage is that clients do not need to prove a stable income. To obtain a mortgage, migrants need to open a housing deposit account with the state-owned NLC company and collect 50% of the cost of housing. The minimum accumulation period is three years. An individual accumulation schedule is drawn up, taking into account the financial capabilities of the migrant. After a three-year accumulation period, once the client has saved a sufficient amount, they become eligible for a 15-year mortgage with interest rates ranging from 6 to 8%.

Currently, there are programs to provide and expand housing financing, which will create many opportunities to improve the living conditions of the population. SMC offers the following areas of housing financing (State Mortgage Company (n.d.)):

- Shared construction. For all categories of citizens, with a down payment of at least 50% of the cost of the purchased housing for a period of up to 15 years at 8% per annum;
- Social mortgage. For socially vulnerable segments of the population, without a down payment, for a period of up to 25 years at 4% per annum;
- Preferential mortgage. For citizens working in government agencies, state and municipal educational institutions, and healthcare organizations without a down payment for a period of up to 25 years at 4% per annum;
- Affordable mortgage. For citizens working in the private sector, there is an initial payment of 20% for up to 25 years at 8% per annum.

Among the proposed solutions, particular emphasis is placed on facilitating mortgage-based homeownership and implementing state-backed rental housing with a subsequent purchase mechanism. Under this framework, ready-built housing, constructed as part of government initiatives, would be made available to eligible citizens. Furthermore, mortgage loans under the KFW program continue to operate across various regions, providing financial accessibility to a broader demographic.

Facilities are under construction in Bishkek, Naryn, Jalal-Abad, Karakol, and Cholpon-Ata. This year, it is planned to issue 2184 apartments to citizens in line for a mortgage. Currently, 31,630 applications have been registered and approved. They are awaiting a turn on the GEC website; a total of 9420 families have been provided with housing; funding has been issued for 15.0 billion soms.

3.1 The Relationship Between Housing Provision and Population Migration

One of the key factors driving population migration in the country is the shortage of available housing. We analyzed 15 years of statistical data to explore the relationship between housing construction and migration patterns (see Table 2). Many individuals migrate abroad primarily to earn money, with the ultimate goal of buying or building their own home.

Correlation and regression analysis revealed the following equation for the dependence of housing commissioning and external migration of the population $y = -51.831x + 75{,}377.9$. The correlation coefficient between housing commissioning and external migration of the population was 0.77, which means that there is a moderate inverse relationship between these indicators. Thus, the more housing is commissioned, the less external migration, namely the number of people leaving

Table 2 Indicators of housing commissioning and population migration for 2008–2022

Years	Housing commissioning, 1000 square meters	Number of people leaving the Kyrgyz Republic, people	Migration increase in the Chui region, people	Migration increase in Bishkek, people
2008	828.7	41,287	4677	4834
2009	875.8	33,380	3593	4860
2010	735.0	54,531	2809	5424
2011	865.2	45,740	3220	5785
2012	850.5	13,019	3298	4648
2013	937.0	11,552	2642	4739
2014	1082.3	11,685	3395	4573
2015	1225.0	7788	2731	4513
2016	1242.1	7125	2698	5646
2017	1471.7	5899	2771	5991
2018	1311.6	7077	2320	4196
2019	1380.1	7560	1925	3965
2020	1063.3	5822	1956	1172
2021	1313.2	8998	358	4752
2022	1462.6	6527	411	8416

Source Compiled by the authors based on data from the National Statistics Committee of the Kyrgyz Republic for 2008–2022 (Eurasian Economic Commission 2018; National Statistical Committee of the Kyrgyz Republic (n.d.); National Statistical Committee of the Kyrgyz Republic 2023)

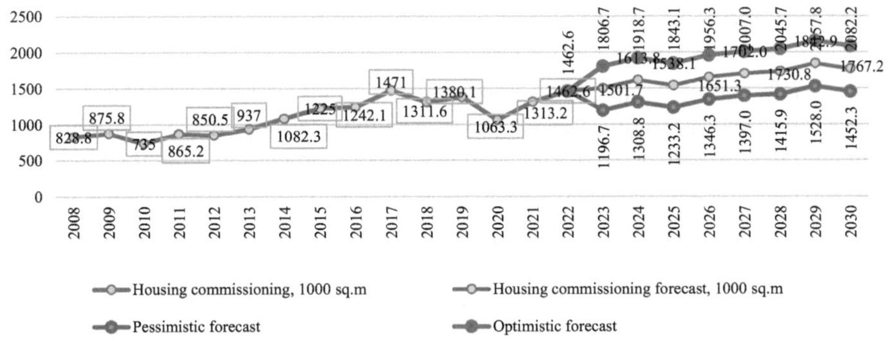

Fig. 2 Forecast of housing commissioning from 2022 to 2030. *Source* Compiled by the authors based on data from the national statistics committee of the Kyrgyz Republic for 2008–2022 (National Statistical Committee of the Kyrgyz Republic (n.d.))

the country. The analysis was conducted with a significance level of 95%. To determine the significance of the coefficients of the regression equation, it is necessary to assess their typicality using Student's t-test. With an accepted error probability of 5% and a degree of freedom of 13, the critical value $t_{crit} = 2.1604$. The calculated t-value, $t_{calc} = |-4.3779|$, exceeds the critical value in absolute terms, confirming that the housing commissioning parameter (x) is significant in the model. To evaluate the overall significance of the model, the authors employed the Fisher F-criterion. With a significance level of 95% and degrees of freedom (1, 13), the critical value is 4.6672. Since the calculated Fisher F-statistic is 19.17 (well above the critical value), the model, given by $y = -51.831x + 75{,}377.9$, is statistically significant and can be reliably used for analysis and forecasting.

For a reliable and accurate study of housing security, the forecast was made from 2022 to 2030, which will be shown in the prospective period for the choice of priorities and justification of strategic and tactical decisions (Fig. 2).

As can be seen from Fig. 2, with an average statistical forecast, the supply of housing for the population during the reviewed period in the country will increase to 1767.2 thousand square meters, which indicates the housing supply is almost double. From an optimistic perspective, the total housing supply could potentially increase to 2082.2, significantly benefiting those in need of stable housing solutions. Conversely, in a pessimistic scenario, the likelihood of newly commissioned housing remains low, indicating minimal improvements in the overall standard of living for affected populations.

4 Discussion

An in-depth analysis of the findings reveals a complex interplay between housing accessibility and migration trends. Despite ongoing efforts, the overall situation remains challenging, necessitating a reassessment of current policies (Valentey 1967):

- Declining housing provision per capita: Over the past two years, the housing provision rate has shown a downward trend. Long-term analysis indicates a decline from 13.4 square meters per person to 12.7 square meters, reflecting persistent inadequacies in residential infrastructure;
- Unmet targets in state housing programs: Despite a nominal increase in the national housing stock, the government's target of 18 square meters per capita remains unachieved. The primary reason for this shortfall lies in the disproportionate growth rates of the population versus housing expansion, exacerbated by insufficient state investment in construction projects;
- Migration as a driving force behind housing demand: The research underscores regional disparities in migration patterns. Over the past 15 years, Bishkek and the Chui Region have exhibited positive net migration, in stark contrast to other parts of the country. This influx of residents has fueled a sharp rise in housing demand and prices, making homeownership less attainable for young families. To mitigate this issue, large-scale government-led housing initiatives should be expanded in urban centers and across regional areas to promote balanced demographic distribution;
- Policy recommendations and future directions;
- The findings emphasize the need for a comprehensive and adaptive housing strategy at the national and regional levels. Effective policy measures should focus on the following:
- Enhancing living conditions through targeted infrastructure projects;
- Expanding access to affordable housing via subsidies and financial mechanisms;
- Increasing budget allocations for large-scale housing developments.

By implementing a strategic, well-funded, and regionally inclusive housing policy, the state can address migration-driven housing pressures and broader socio-economic challenges, ensuring sustainable urban and rural development in the years to come.

5 Conclusion

The findings of this research underscore the intricate relationship between population migration, housing development, and overall accessibility to housing resources. This research focuses on solving the problem of housing shortages and internal and external migration of the population, improving living conditions in all regions of the country.

The forecast shows a positive result of housing security until 2030, which will improve living standards and reduce migration in the country.

The recommendations provided by the authors are as follows:

- Increase government funding for housing construction: To increase the availability of housing for the population, the state must build houses to eliminate the growth of housing prices when buying housing on a mortgage through banking institutions;
- Support for regional development: A holistic approach is necessary to address these interconnected factors and establish long-term solutions that contribute to sustainable urban and regional development;

The housing and social problems of the population in the country are solved through existing programs, which makes it possible to implement them for better living conditions of the population. One of the core challenges driving internal and external migration is the lack of adequate living conditions and employment opportunities in rural areas and smaller towns. To curb large-scale population shifts toward major urban centers and foreign labor markets, comprehensive regional development strategies must be prioritized. These strategies should focus on enhancing infrastructure in villages and small towns, encouraging job creation through targeted economic policies, and ensuring accessible and affordable housing for local populations. By implementing these measures, rural populations can be incentivized to remain in their home regions, reducing the strain on metropolitan areas while fostering balanced national development.

Recognizing the urgent need for a more structured and far-reaching housing policy, the new State Housing Program, "Menin Uyum," was officially adopted on August 6, 2024. This program is designed to implement a multi-faceted approach to housing provision, incorporating policies that will stimulate large-scale residential construction to meet growing demand, reduce overall housing costs by increasing market supply, and enhance housing finance mechanisms, making mortgages and loans more accessible (Sarbaeva 2016).

Through integrated policy implementation, the program aims to bridge existing gaps in housing accessibility, ensuring that a greater number of citizens can secure permanent and affordable homes. A well-structured state housing policy, coupled with regional economic support measures, can significantly reduce migration pressures while improving the overall quality of life for citizens.

The program contains a new vision for the further development of state housing policy, considering the results and experience of implementing state housing programs.

References

Cabinet of Ministers of the Kyrgyz Republic (2021) Housing program of the Kyrgyz Republic my home for 2021–2026 (Annex 1 to the resolution no. 75 of July 13, 2021). Bishkek, Kyrgyzstan. Retrieved from https://cbd.minjust.gov.kg/158357/edition/5595/ru. Accessed 1 Oct 2024

Ergeshbaev UZ, Chamashev MK (2018) External labor migration of Kyrgyzstan after joining the EAEU. Questions Econ Sci 1(89):102–104

Ergeshbaeva RD, Maksudova IS (2019) The influence of external labor migration on the development of entrepreneurship in Kyrgyzstan. Bull RGSU Branch Osh Kyrgyz Repub 2(20):104–108

Eurasian Economic Commission (2018) Eurasian economic union in figures: brief statistics yearbook. Eurasian Economic Commission, Moscow, Russia. Retrieved from https://eec.eaeunion.org/upload/iblock/996/Brief_Statistics_Yearbook_2018.pdf. Accessed 1 Oct 2024

Eurasian Economic Commission (2023) Demography in numbers 2018–2022: statistics of the Eurasian economic union. Eurasian Economic Commission, Moscow, Russia. Retrieved from https://eec.eaeunion.org/upload/iblock/929/DEMOGR_2018_2022.pdf. Accessed 1 Oct 2024

Government of the Kyrgyz Republic (2017) Concept of regional policy of the Kyrgyz Republic for the period 2018–2022 (approved by decree no. 194 on March 31, 2017). Bishkek, Kyrgyzstan. Retrieved from https://www.gov.kg/wp-content/uploads/2018/12/con_reg_rus.pdf. Accessed 1 Oct 2024

Imankulova S (2024) The impact of regional socio-economic policy on housing accessibility for different population groups. Bull Sci Pract 10(10):1–7. https://doi.org/10.33619/2414-2948/107/33

National Statistical Committee of the Kyrgyz Republic (n.d.) Kyrgyzstan in figures, 2008–2023: statistical collection. Bishkek, Kyrgyzstan. Retrieved from https://stat.gov.kg/ru/publications/sbornik-kyrgyzstan-v-cifrah/. Accessed 1 Oct 2024

National Statistical Committee of the Kyrgyz Republic (2023) Demographic yearbook of the Kyrgyz Republic, 2018–2022: annual publication. National Statistical Committee of the Kyrgyz Republic, Bishkek, Kyrgyzstan. Retrieved from https://www.stat.gov.kg/en/publications/demograficheskij-ezhegodnik-kyrgyzskoj-respubliki/. Accessed 1 Oct 2024

Rhythm of Eurasia (2023) The housing program my home 2021–2026 is beginning to be implemented in reality in Kyrgyzstan. Retrieved from https://www.ritmeurasia.ru/news--2023-12-11--v-kyrgyzstane-nachinaet-realno-voploschatsja-zhilischnaja-programma-moj-70353. Accessed 1 Oct 2024

Saitov ZK (2009) To the problem of the quality of housing conditions in the Kyrgyz Republic. Med Kyrgyzstan 2:41–45

Sarbaeva IY (2016) Government programs as a tool for the implementation of housing policy. Azimuth Sci Res: Pedagogy Psychol 5(3):349–352

State Mortgage Company (n.d.) Official website. Retrieved from https://gik.kg/ru/. Accessed 1 Oct 2024

Valentey DI (1967) Theory and policy of population. Higher School, Moscow, USSR

Valentey DI, Kvasha AY (1989) Fundamentals of demography: Textbook. Mysl, Moscow, USSR

Foreign Direct Investment in Zambia's Mining Sector: The Influence of Political Stability and Contestation

Oscar Kaonga, Lubinda Haabazoka, and Bwalya Chilolo

Abstract Foreign Direct Investment (FDI) serves as a vital catalyst for economic growth, particularly in resource-rich developing economies such as Zambia, where the mining sector plays a pivotal role in national development. This research investigates the impact of political stability and contestation on FDI inflows within Zambia's mining industry, addressing a critical gap in understanding how governance dynamics influence investment decisions. Utilizing a mixed-methods approach with a convergent parallel design, the research integrates quantitative and qualitative data to provide a comprehensive analysis. An Autoregressive Distributed Lag (ARDL) model is applied to assess long-term and short-term relationships between political variables and FDI trends. Historical data, policy shifts, and investor perceptions are examined to contextualize the findings. The results reveal that sustained political stability enhances investor confidence, encouraging long-term capital commitments in the mining sector. Conversely, periods of heightened political contestation during elections, leadership transitions, or major policy reforms generate uncertainty that can deter investment. The research highlights the balance required in governance frameworks, where democratic processes must coexist with measures that safeguard investor confidence. To address these challenges, the authors propose policy interventions, including stronger legal protections for foreign investors, greater transparency in regulatory practices, and risk mitigation strategies to insulate the mining sector from political volatility. These insights contribute to the broader academic and policy discourse on governance, political risk, and economic development in resource-dependent economies, offering valuable lessons for policymakers, investors, and development institutions seeking to optimize FDI attraction while balancing political competition and governance reforms.

O. Kaonga (✉) · L. Haabazoka · B. Chilolo
Graduate School of Business, University of Zambia, Lusaka, Zambia
e-mail: oscarkaonga@gmail.com

L. Haabazoka
e-mail: lubinda.haabazoka@unza.zm

B. Chilolo
e-mail: bwalyachilolo@yahoo.com

A. B. Karbekova (ed.), *Sustainable Development of Rural Territories Based on Climate-Smart Technologies*, Climate Change Management,
https://doi.org/10.1007/978-3-032-20106-5_30

Keywords Foreign direct investment · Political stability · Political contestation · Mining sector · Zambia · Investor confidence · Governance

JEL Classification E61 · H5 · H11 · H83 · P26 · P33

1 Introduction

In resource-abundant developing economies, particularly African countries, Foreign Direct Investment (FDI) represents a vital catalyst for economic progress. Zambia, endowed with vast mineral resources, has long attracted significant FDI, particularly in its mining sector, which remains the backbone of the country's economy. The mining sector contributes significantly to Zambia's Gross Domestic Product (GDP), government revenue, and employment (World Bank 2019). However, the relationship between FDI inflows and political dynamics, specifically political stability and contestation, remains complex and requires in-depth analysis. Political stability fosters an environment conducive to investment. Political contestation, often characterized by electoral disputes, policy inconsistencies, and governance uncertainties, may deter investors (Asiedu 2006; Jensen 2008). Zambia has experienced fluctuating levels of political stability and contestation since gaining independence in 1964. The transition from a one-party state to a multiparty democracy in 1991 marked a significant shift in the political landscape, influencing investment patterns. Although Zambia has remained relatively stable compared to other African nations, periodic electoral tensions, policy reversals, and governance challenges have raised investor concerns. Consequently, understanding how these political dynamics influence FDI in Zambia's mining sector is essential for policymakers and stakeholders in the investment and governance domains.

Despite Zambia's rich mineral endowment and the government's efforts to attract FDI, inconsistencies in political stability and rising political contestation have introduced uncertainties in the investment climate. Investment flows in the mining sector have fluctuated, influenced by evolving government policies, changes in political leadership, and disputes over taxation and ownership frameworks (Fraser and Lungu 2007). Some research indicates that political stability boosts investor trust. Other studies argue that excessive political competition can trigger policy instability, risks of nationalization, and regulatory unpredictability, discouraging foreign investment (Kolstad and Wiig 2011). Empirical research on the relationship between political dynamics and FDI in Zambia's mining sector remains limited. Although existing literature examines the general determinants of FDI in Africa, there is insufficient focus on the specific impact of political stability and contestation in Zambia's mining industry. This research aims to fill this gap by exploring how political stability and contestation influence FDI inflows into Zambia's mining sector and the implications for sustainable economic development.

2 Literature Review

FDI is vital for the economic growth of resource-dependent countries such as Zambia. However, political stability and competition significantly impact FDI flows, especially in the mining sector. The authors analysed current research on how political stability and contestation affect FDI in Zambia's mining industry and evaluated theoretical frameworks, empirical findings, and major discussions while highlighting research gaps that require deeper investigation.

2.1 Overview of Political Stability and Contestation in Zambia

Zambia's political landscape has been characterized by a mixture of stability and contestation since gaining independence from Britain in 1964. As a democratic state, Zambia has undergone significant political transformations, shifting from a one-party system to a multiparty democracy in 1991. However, political stability in the country has been periodically challenged by electoral disputes, governance concerns, and party competition (Burnell 2005).

Political stability in Zambia has largely been maintained through peaceful transitions of power, strong state institutions, and adherence to constitutional frameworks. The country's democratic consolidation has been shaped by periodic elections, with key political actors largely respecting electoral outcomes. For example, the peaceful transfer of power from the Movement for Multiparty Democracy (MMD) to the Patriotic Front (PF) in 2011 and later from PF to the United Party for National Development (UPND) in 2021 demonstrates Zambia's commitment to democratic governance (Cheeseman 2022). Additionally, Zambia has avoided major armed conflicts, a factor that distinguishes it from some of its regional counterparts in Sub-Saharan Africa (Bratton and Walle 1997). Despite these achievements, Zambia's political stability has faced several challenges. Issues such as electoral violence, suppression of opposition parties, and judicial interference have periodically threatened democratic processes. For instance, the 2016 elections were marred by allegations of irregularities and political intimidation, raising concerns about democratic backsliding (Goldring and Wahman 2016). Furthermore, concerns over executive overreach and restrictions on press freedom have also contributed to a fragile democratic environment (House 2023; Siachiwena 2021).

Meanwhile, political contestation in Zambia has been a defining feature of its multiparty democracy. The competition between ruling and opposition parties has often been intense, particularly during elections. The rise of the UPND as a formidable opposition party under President Hakainde Hichilema, leading to his victory in the 2021 elections, underscores the dynamic nature of Zambia's political contestation (Larmer 2023). The opposition has frequently accused ruling parties of electoral malpractices, such as vote rigging, media bias, and the use of state security apparatus to intimidate opponents (Rakner 2017).

Zambia has also experienced shifting alliances within the political landscape, with parties merging or fragmenting based on electoral prospects. This phenomenon reflects the evolving nature of political contestation and the need for strategic coalition-building among political actors (Resnick 2011). However, the dominance of the executive branch and the limited independence of key democratic institutions, such as the judiciary and the electoral commission, have sometimes undermined the credibility of political competition (Simutanyi 1996). Zambia's political stability has been maintained through constitutionalism and relatively peaceful transitions of power. However, it remains vulnerable to electoral disputes, governance issues, and political polarisation. The country's multiparty democracy has fostered intense political contestation, with opposition parties playing a critical role in challenging the ruling establishment. Addressing issues such as electoral integrity, media freedom, and institutional independence will be key to sustaining Zambia's democratic trajectory in the future.

2.2 Foreign Direct Investment (FDI) in Zambia

According to the International Monetary Fund (IMF), Foreign Direct Investments (FDIs) are defined as cross-border investments where the investor holds over 10% of the shares in a foreign enterprise (UNCTAD 2021). These investments typically involve multinational enterprises (MNEs) directing capital into foreign-controlled entities, including subsidiaries or affiliated companies. This demonstrates that the investor has substantial control over the organisation's administration in the destination nation (Bodea and Ye 2020). MNEs participate in FDI projects for a variety of strategic factors. Four primary elements influence foreign direct investment (FDI): markets, assets, natural resources, and efficiency-seeking (Buckley et al. 2007). Ownership enables a company to take full advantage of its competitive advantages abroad. Exploiting geographical advantages globally is part of location advantages (e.g., supply of human capital or natural resources).

FDI can be categorised in several ways. It can be FDI in flow or FDI outflow, depending on whether the investment is inside or outside the country. It can also be categorised as greenfield investment or brownfield investment depending on the way in which the investment is made in the host nation. In this case, greenfield FDI occurs when a parent company that controls a subsidiary company invests in that subsidiary company in a different country. Brownfield investment occurs when an investor from outside a country merges with a local company or acquires a significant share of a local company (Kaulu and Haabazoka 2023).

Only the most productive businesses can make international investments. Thus, high productivity is essential. Accessibility of company benefits that are crucial to the company's production and challenging to transfer to outside parties, as well as a relatively strong market position in the home nation, are other criteria to consider (Bodea and Ye 2020; Buthe and Milner2008; Pinto and Zhu2016).

The most commonly employed proxies in FDI studies are FDI flows and FDI stocks. Data on greenfield investments is preferred as an FDI indicator because it shows investment in real capital (current assets) linked by positive externalities for the country hosting foreign investors (e.g., job creation and knowledge transfer are some of the externalities). Due to the ease of access to data on these proxies in developing countries, FDI stocks and flows are commonly employed indicators. FDI stocks are less volatile than FDI flows since capital prices have been rising steadily. According to Bellak et al., a better indicator is FDI flow because it represents decisions made by multinational corporations (MNEs) or even an exit from a certain region (Bellak et al. 2009). This research employs FDI inflow in the mining sector of Zambia as the dependent variable.

Investment managers employ a two-step strategy in choosing the location choice of FDI (Buckley et al. 2007). The ultimate selection of investment location is scarcely linked with economic theories because the characteristics of the host nation are significant in FDI location selection. The examples of these characteristics are political stability, quality human capital, and infrastructure. Various studies have indicated that various variables can influence FDI location decisions. Most research has used economic and non-economic variables to determine FDI drivers (Bodea and Ye 2020; Buthe and Milner 2008; Pinto and Zhu 2016). Economic factors of FDI are what motivate multinational corporations to invest in the recipient nation.

FDI has been pivotal in shaping Zambia's economic landscape, particularly within its mining sector. Historically, Zambia's economy has been heavily reliant on mining, with the sector contributing approximately 13.2% to the nation's GDP in 2019 (Zambia Chamber of Mines, Association of Zambian Mineral Exploration Companies 2024). The liberalization of the economy in the 1990s and subsequent privatization of state-owned enterprises created a conducive environment for FDI, leading to significant capital inflows aimed at revitalizing the mining industry.

In recent years, Zambia has experienced fluctuations in FDI inflows. According to the United Nations Conference on Trade and Development (UNCTAD), FDI flows into Zambia rose to $116 million in 2022, a notable recovery from a negative inflow of $352 million in the previous year (United Nations Conference on Trade and Development (UNCTAD) 2023). Despite this improvement, the 2022 FDI levels remained below the 2018–2020 average of $504.3 million, indicating ongoing challenges in attracting consistent foreign investment. The mining sector continues to be a major recipient of FDI in Zambia, primarily due to the country's rich deposits of copper and other minerals. In 2024, Zambia's copper output increased by 12%, reaching approximately 820,670 metric tons, up from 732,580 tons in the previous year (Zambia Chamber of Mines, Association of Zambian Mineral Exploration Companies 2024). This surge in production was largely attributed to the recovery of key mines, including Barrick Gold's Lumwana mine and the resumption of operations at Vedanta Resources' Konkola Copper Mines. Additionally, acquiring a 51% stake in Mopani Copper Mines by International Resources Holding, a firm based in the United Arab Emirates, underscores the growing diversification of investment sources in Zambia's mining industry (Zambia Chamber of Mines, Association of Zambian Mineral Exploration Companies 2024).

The Zambian government has set an ambitious target to elevate annual copper production to 3 million tons within the next decade (70). To achieve this, significant investments are being funneled into the sector. Vedanta Resources has announced plans to invest approximately $1.3 billion to rejuvenate operations at Konkola Copper Mines. Similarly, Canadian mining companies Barrick Gold and First Quantum Minerals are augmenting their production capacities. Moreover, KoBold Metals, a US-backed firm, has unveiled intentions to develop a $2 billion copper mine in Zambia, reflecting heightened investor confidence in the country's mining prospects (Zambia Chamber of Mines, Association of Zambian Mineral Exploration Companies 2024).

To summarize, FDI continues playing a pivotal role in Zambia's economic growth, particularly in the mining industry. Although recent data shows renewed foreign investor interest, ongoing improvements to the business environment remain crucial for attracting and maintaining FDI inflows. This will support the long-term expansion and diversification of Zambia's mining sector.

2.3 Political Contestation and Its Impact on FDI

Political contestation can have dual effects on FDI. On the one hand, high contestation may increase political risk and policy uncertainty, discouraging investors (Jensen 2008). On the other hand, competitive political environments may lead to institutional improvements, fostering investor confidence (Rodrik 2000). In Zambia, the alternation of power between the Patriotic Front (PF) and the United Party for National Development (UPND) has raised questions about how shifting policy priorities affect foreign investments in mining. Despite growing literature on the subject, research focusing on Zambia often overlooks how different forms of political contestation, electoral violence, legal disputes over resource control, and policy reversals uniquely influence FDI. Additionally, there is limited research on how foreign investors navigate these challenges by leveraging political networks or adjusting investment strategies. Zambia's mining sector has witnessed frequent policy changes, including fluctuating tax regimes, nationalization debates, and shifts in mining royalties (Fraser and Lungu 2007; Lungu 2008). Such unpredictability affects investor confidence. For instance, during the 2019 tax reform, the Zambian government increased mineral royalties, prompting threats of capital withdrawal from major investors such as First Quantum Minerals (International Council on Mining and Metals 2022).

Existing literature primarily focuses on how these policy changes impact investment decisions but fails to consider the role of political contestation in driving such changes (Abdul-Salam 2012). Furthermore, studies have not adequately addressed how investors respond to such uncertainties, whether by lobbying, engaging in corporate social responsibility to gain political favour, or diversifying risk across multiple jurisdictions.

2.4 Empirical Studies on Political Stability, Contestation, and FDI

Research on Africa's political stability and its connection to FDI highlights a multifaceted relationship, with various elements collectively shaping investor decisions across the continent (Faruq 2023; Podobnik et al. 2007; Rana and Kebewar 2014). Political stability is often posited as a critical determinant for attracting FDI because it reduces risks associated with investment and creates a conducive environment for economic activities. Studies have demonstrated that countries exhibiting higher levels of political stability tend to attract more FDI. For instance, research indicates that political stability positively correlates with increased FDI inflows because stable political environments provide predictability and security for investors (Faruq 2023). In contrast, political turmoil (e.g., recurrent government shifts, protests, or warfare) creates greater uncertainty and risk, making FDI less appealing.

Corruption, a facet often intertwined with political instability, has been shown to adversely affect FDI. High levels of corruption increase the cost of doing business and create an unpredictable investment climate, thereby discouraging foreign investment (Podobnik et al. 2007). Empirical analyses reveal a negative correlation between corruption and FDI inflows, suggesting that efforts to combat corruption could enhance a country's attractiveness to foreign investors. The quality of institutions also plays a pivotal role in mediating the relationship between political stability and FDI. Robust legal frameworks, transparent regulatory systems, and effective governance are instrumental in fostering investor confidence. Countries that have implemented reforms to strengthen institutional quality have observed increased FDI inflows because these reforms mitigate risks associated with political instability (Rana and Kebewar 2014).

However, the relationship between political stability and FDI is not unidimensional. Some resource-rich African countries have attracted significant FDI despite political challenges, primarily due to the lucrative opportunities in extractive industries. This phenomenon aligns with the "resource curse" theory, where countries rich in natural resources may experience less economic growth and worse development outcomes due to governance challenges (Bigsten and Tengstam 2015). Thus, while political stability is a significant factor, sector-specific opportunities can also influence FDI patterns. Empirical research has largely confirmed that political stability is a key determinant of FDI. Asiedu (2006) found that African countries with stable political environments attracted more FDI compared to those with frequent political upheavals. Similarly, Anyanwu (2012) demonstrated that coups, civil unrest, and weak governance structures significantly reduce FDI inflows. For Zambia, studies (Bigsten and Tengstam 2015) highlight that the mining sector, despite being a significant FDI destination, has been affected by policy inconsistency and government interventions during political turmoil.

However, existing empirical studies tend to treat political stability as a binary variable, either stable or unstable, without considering degrees of contestation within democratic settings. The impact of closely contested elections, shifts in ruling parties, and opposition strength on investor confidence remains understudied in the context of Zambia's mining sector.

2.5 Overall Critique of Existing Literature

While substantial research exists on the determinants of FDI, political stability, and mining policies, a review of the empirical literature reveals important gaps that warrant further investigation. In particular, although many studies examine political stability, few explore the role of political contestation and its impact on FDI. This gap highlights the need for a study such as this one. Additionally, there is insufficient research on how foreign investors strategically respond to political contestation and policy uncertainty in Zambia's mining sector. Extant literature (Bigsten and Tengstam 2015; Faruq 2023; Rana and Kebewar 2014) also lacks sufficient industry-specific studies. While general FDI studies exist, few focus specifically on Zambia's mining sector, which has unique challenges such as resource nationalism and government intervention.

3 Methodology

This research adopts a mixed-methods approach using a convergent parallel design. Quantitative (QUAN) and qualitative (QUAL) data are collected and analysed simultaneously (Edmonds and Kennedy 2017). In this concurrent triangulation design, the data is collected independently. The results are later compared and integrated into a cohesive framework. Qualitative data was analysed using NVivo. Quantitative data was stored and analysed using EViews and Stata, facilitating concurrent analysis.

3.1 Quantitative Data Analysis

To examine the impact of political stability and contestation on FDI inflows, part of this research employed a quantitative research design to analyse the relationship between political stability, contestation, and inward FDI, using an Autoregressive Distributed Lag (ARDL) model (Assefa 2020; Okeke and Kalu 2022; Pesaran et al. 1998).

Quantitative secondary data on political stability and contestation were obtained from the World Bank's World Governance Indicators (WGI) database, while data on inward FDI were sourced from the World Bank's World Development Indicators (WDI), in accordance with existing literature. The period of 1965–2023 was selected to capture long-term trends in inward FDI and political contexts. The period was also selected to obtain the latest available, reliable data at the time of the study. Further,

Table 1 Variables in the study

Variable	Description	Source
POS	Political Stability	World Bank—WGI
POC	Political Contestation	World Bank—WGI
FDI	Foreign Direct Investment in the mining sector, net inflows (% of GDP)	World Bank—WDI

Source Developed by the authors

the period covers data during which all regimes ruled the country, starting with the United Independence Party (UNIP), the Movement for Multiparty Democracy (MMD), the Patriotic Front (PF), and part of the time for the United Party for National Development (UPND). This ensured that the results could be generalized regardless of the party in power and its policies. Table 1 provides a summary of the data sources.

This research employed secondary time series data, which is particularly beneficial for identifying trends and patterns. Time series data allows researchers to analyze past events and forecast future trends. Recent advancements have also enabled causal analysis using time series data. Given that this research aims to examine the impact of political stability and contestation on inward FDI, considering short-term and long-term causal relationships, time series data was deemed the most appropriate. The analysis employed the Autoregressive Distributed Lag (ARDL) model, which is particularly well-suited for time series data that exhibit unit roots and potential cointegration. This model is perfect for capturing short-term dynamics and long-term equilibrium relationships between political stability, political contestation, and inward FDI (Shrestha and Batha 2018).

3.2 Qualitative Data Analysis

This research primarily gathered qualitative data through focus group discussions. According to recommendations in literature, effective focus groups should comprise six to twelve participants with diverse perspectives and knowledge (Lazar et al. 2017; Wilson 2015). The collected data was analysed using a combination of thematic analysis, content analysis, and grounded analysis within the NVivo software environment. Thematic analysis focused on identifying recurring themes and patterns (Braun and Clarke 2006; Lester and Lochmiller 2017; Morgan 2022) related to the influence of political stability and contestation on inward FDI. This method has been adapted from existing literature (Saad 2014). Content analysis was applied to examine the frequency and distribution of specific concepts and ideas (Bengtsson 2016; Elo et al. 2014; Erlingsson and Brysiewicz 2017). This method offered a deeper understanding of the factors affecting the link between political stability, contestation, and inward foreign direct investment (FDI), along with the underlying mechanisms and explanations for these effects.

Using grounded analysis, the research generated novel theoretical perspectives (Charmaz and Thornberg 2021; Edgington 1967; Timmermans and Tavory 2012) on the connections between these variables. This approach prioritizes developing theory based on empirical data. The mixed-methods strategy, employing a convergent parallel design, was particularly effective for this research because it enabled the simultaneous gathering and synthesis of qualitative and quantitative data (Tashakkori and Newman 2010). Integrating these two data types allowed for a richer, more nuanced examination of the intricate relationship between political stability, contestation, and inward FDI (Schoonenboom and Johnson 2017).

3.3 *Measurement of the Variables and Justification*

Measuring variables is critical to any research study (Table 2). In this research, the independent variables (political contestation) are measured as voice and accountability estimate using a scale from − 2.5 to + 2.5. In turn, political stability is measured as a proxy of political stability and absence of violence and terrorism measured in the index of − 2.5 to + 2.5 in line with extant literature (Ali et al. 2022; Mkonyi et al. 2022; Rehman et al. 2020). The dependent variable (FDI inflows) is measured as FDI net inflows expressed as a percentage of GDP. This approach aligns with existing literature on tax revenue (Castro and Camarillo 2014; Gwaindepi 2021; Ofori et al. 2018).

Table 2 Measurement of variables

Type of variable	Variable	Proxy	Code	Unit of measurement	References	Source of data
Dependant	FDI	FDI net inflow (% of GDP)	FDI	Percentage	Mahmood and Chaudhary (2013), Fraser and Lungu (2007)	World Bank
Independent	Political stability	Political stability and the absence of violence and terrorism	POS	Index of − 2.5 to 2.5	Mkonyi et al. (2022)	World Bank
Independent	Political contestation	Voice and accountability estimate	POC	Index of − 2.5 to 2.5	Aisen and Veiga (2013), Mkonyi et al. (2022)	World Bank

Source Compiled by the authors

The quantitative data in this study were analyzed using the procedures detailed in the following subsections. The analysis began with descriptive statistics, followed by unit root tests to assess the stationarity of the data. Optimal lag lengths were determined using various information criteria. Subsequently, the bounds testing approach was applied to examine long-run relationships. Short-run dynamics were then analyzed using the Error Correction (EC) representation of the ARDL model. Finally, diagnostic tests were conducted to validate the robustness of the model.

3.4 Unit Root Tests and Optimal Lag Selection

Unit root tests were conducted using the Augmented Dickey-Fuller (ADF) and Phillips-Perron (PP) tests. The Akaike Information Criterion (AIC) was employed to select the optimal lag in the ADF test. The Schwarz Information Criterion (SIC) was used for the PP test. The null hypothesis assumed that the series contained a unit root or was non-stationary. The ARDL model was applied to variables with a mixed order of integration, meaning some variables were integrated of order 1 (I(1)), requiring first differencing to become stationary, while others were integrated of order 0 (I(0)), indicating stationarity at the level.

3.5 Model Specification, Short and Long Run Dynamics

The primary model employed in this research is the ARDL model, which offers several advantages. It is particularly useful for analysing time series data with a mixed order of integration and non-stationary time series (Shrestha and Batha 2018). The model is applicable when some variables are integrated of order 1 (I(1)) and others of order 0 (I(0)), provided that none are integrated of order 2 (I(2)) or higher. Therefore, the ARDL model is well-suited for the analysis conducted. The model to investigate the influence of POS and POC on FDI in 1965–2023 can be specified as follows:

$$FDI_t = (POS_t, POC_t) \tag{1}$$

In Eq. 1, FDI refers to FDI inflows measured as a percentage of GDP. POS is political stability measured as the absence of violence and terrorism, measured in the index of − 2.5 to 2.5. POC refers to political contestations measured as voice and accountability estimate with an index of − 2.5 to 2.5. Equation 1 can also be written as follows:

$$FDI_t = \lambda_0 + \lambda_1 POS_t + POCt\ +\ \mu_t \tag{2}$$

The logs of each variable were taken to minimize the volatility and multi-collinearity of the time series data. The following log-linear model is obtained by applying logs in Eq. 2:

$$\log FDI_t = \lambda_0 + \lambda_1 \log POS_t + \lambda_2 \log POC_t + \mu_t \tag{3}$$

Analysis in the ARDL model can be done in two steps. Step one looks at long-run associations in the model. Step two looks at the short run. Equation 4 represents the ARDL model specification for this research.

$$\begin{aligned}\Delta \log FDI_t = \alpha_0 \sum_{i=1}^{p} \beta_{1j} \Delta \log FDI_{t-k} + \sum_{i=0}^{p} \beta_{2j} \Delta \log POS_{t-k} \\ + \sum_{i=0}^{p} \beta_{3j} \Delta \log POC_{t-k} + \lambda_1 \log FDI_{t-1} \\ + \log POS_{t-1} + \lambda_3 \log POC_{t-1} + \varepsilon_t \end{aligned} \tag{4}$$

In this equation, α_0 is the intercept, Δ is the first difference operator, p is the lag order, and Δ_t is the error term. The bounds test was used to check for long-run equilibrium in the relationships amongst POS, POC, and FDI. In this test, the null hypothesis (Ho) is $\delta_1 = \delta_2 = \delta_3 = 0$ (i.e., there is no cointegration). The alternative hypothesis (H1) is $\delta_1 \neq \delta_2 \neq \delta_3 \neq 0$ (i.e., there is cointegration). If the calculated F statistic or absolute t-statistic is greater than the upper level bound or absolute upper level bound, respectively, H0 is rejected (Pinto and Zhu 2016). The conclusion is that there is cointegration in the relationship among POS, POC, and FDI. If the statistics are below the lower bound, there is no cointegration. If the calculated statistics are between the upper and lower bounds, the result is inconclusive.

The JJ test (Jensen 2008), "cumulative sum recursive residuals (CUSUM) and cumulative of square of recursive residuals (CUSUMSQ)" can be used to check the robustness of the cointegration (Brown et al. 1975; Chandio et al. 2020).

The short-run relationships between POS, POC, and FDI were assessed using the following ECM form of the ARDL model:

$$\begin{aligned}\Delta \log FDI_t = \alpha_0 \sum_{i=1}^{p} \beta_{1j} \Delta FDI_{t-k} + \sum_{i=0}^{p} \beta_{2j} \Delta POS_{t-k} \\ + \sum_{i=0}^{p} \beta_{3j} \Delta POC_{t-k} + \alpha ECM_{t-1} + \varepsilon_t \end{aligned} \tag{5}$$

Model fit was checked using the R^2. This value ranges from 0 to 1. The higher the value, the better the fit (with 1 implying a perfect fit and 0 implying no fit at all). The Breusch-Godfrey test was used to check serial correlation. Additional diagnostic tests were conducted to evaluate various model characteristics. These included the Jarque–Bera test for normality, the ARCH and White tests for heteroscedasticity, the RESET test for linearity, and the CUSUM of squares test for model stability.

4 Results

4.1 Qualitative Findings of the Research

The findings from qualitative data analysis suggest that political stability plays a crucial role in determining the inflow of FDI into Zambia's mining sector. Participants, including industry experts, government officials, and investors, stressed that a stable political environment boosts investor confidence, encouraging long-term investments and capital inflows. Additionally, qualitative interviews highlighted that times of political instability, marked by electoral uncertainties, shifts in government policies, and social unrest, have historically resulted in capital flight and decreased investor interest. Several participants pointed to the fluctuating tax regimes in Zambia's mining sector as a significant concern, noting that frequent alterations in fiscal policies discourage long-term investments. Investors require a predictable operating environment. Any political turbulence that threatens policy consistency tends to deter FDI inflows.

Another key topic that emerged from the research is the role of government institutions in maintaining stability. Respondents indicated that weak institutional frameworks often lead to corruption, inefficiencies, and delays in project approvals, which negatively impact FDI. A strong institutional environment, coupled with political stability, was cited as a major determinant of investor confidence. Additionally, geopolitical risks and regional stability were cited as important factors. Some respondents pointed out that although Zambia is relatively stable compared to its neighboring countries, regional conflicts and economic crises in Southern Africa still have indirect impacts on investor sentiment. This underscores the interconnectedness of political stability and economic appeal in the region.

Overall, the qualitative findings emphasise that political stability plays a crucial role in determining FDI inflows into Zambia's mining sector. Stable governance structures, policy predictability, and institutional efficiency are essential for sustaining investor confidence. Conversely, political uncertainty, policy inconsistencies, and weak governance structures create an unfavourable investment climate, leading to reduced FDI inflows. The research reinforces the need for sound governance and policy continuity to maintain Zambia's attractiveness as a mining investment destination.

The research found that political contestation significantly influences foreign investors' decisions in the mining sector, primarily through heightened uncertainty, regulatory instability, and risk perception. Interviewees, including foreign investors, policy analysts, and industry experts, indicated that political disputes, particularly those involving elections, policy reversals, and nationalist rhetoric, create an unpredictable investment environment. Investors often perceive political contestation as a sign of potential policy shifts, which can lead to increased taxation, renegotiation of mining agreements, or even expropriation. Similar to the findings of Jensen (2012) and Vivoda (2009), the research established that this uncertainty discourages long-term capital commitments because investors seek stable and predictable regulatory frameworks to ensure returns on investment.

Moreover, respondents emphasised that political contestation influences the enforcement of mining laws and regulations. In politically contested environments, changes in leadership can result in inconsistent regulatory enforcement, favouring some investors over others. This aligns with Wilson (2015), who found that selective enforcement leads to concerns about fairness and transparency, pushing investors to either delay their investments or divert them to more politically stable jurisdictions. Additionally, interviewees noted that political contestation often fuels public debates about resource nationalism, wherein political actors advocate for greater state control over natural resources. Such rhetoric, even if not immediately translated into policy, creates apprehension among investors, who fear future restrictions on profit repatriation or increased government intervention.

Foreign investors also highlighted that political contestation can result in social unrest, which exacerbates investment risks. Protests, labor strikes, and community resistance often arise in politically contested settings, particularly when opposition groups mobilize around resource control issues. Several respondents pointed to instances where opposition parties, seeking political leverage, have championed anti-foreign investment narratives, further complicating investor decision-making. Consequently, many investors employ risk-mitigation strategies, such as lobbying for favorable policies, participating in corporate social responsibility activities to gain the support of local communities, or securing political risk insurance. However, in politically unstable environments, some investors may favor short-term extractive approaches instead of long-term sustainable investments to minimize their exposure to political risks. The findings indicate that political contestation creates significant uncertainty in the mining sector, which affects investor confidence and long-term decision-making. Investors tend to prioritize political stability and clear regulatory frameworks when making investment decisions because volatile political situations increase financial and operational risks. Therefore, host governments aiming to attract and retain foreign direct investment in mining must focus on mitigating the effects of political contestation by ensuring policy consistency, transparency, and strong legal protections for investors.

Drawing on qualitative data from in-depth focus group interviews with policymakers, economic analysts, and business leaders, alongside a review of relevant secondary sources, this study identifies several key policy interventions aimed at mitigating the adverse effects of political uncertainty on FDI inflows. One of the primary interventions emphasised by respondents is the strengthening of democratic governance and the rule of law. Political instability in Zambia is often linked to electoral disputes, contested leadership transitions, and policy unpredictability. Ensuring free, fair, and transparent elections, coupled with mechanisms for peaceful dispute resolution, was consistently highlighted as a critical factor in restoring investor confidence. Business leaders interviewed emphasized that investors are more likely to commit capital when they perceive political processes as stable and predictable.

Another important policy suggestion from respondents is the depoliticization of economic policies. Investors highlighted that uncertainty caused by sudden policy changes, especially in industries like mining and agriculture, is a significant concern. Establishing independent economic institutions (e.g., a non-partisan investment promotion agency and regulatory bodies with clear and stable mandates) can help insulate economic policies from political fluctuations. Participants also stressed the importance of long-term policy frameworks that are legally binding and difficult to alter arbitrarily. Additionally, strengthening public–private dialogue emerged as a recurring topic. Regular engagement between the government and the business community can help build trust and ensure that investor concerns are addressed promptly. Respondents emphasized the importance of institutionalizing stakeholder consultations in the policymaking process, particularly in key FDI-attracting sectors such as mining, energy, and infrastructure. Transparent communication of policy changes, coupled with a stable and predictable regulatory environment, was frequently identified as essential for maintaining investor confidence.

Judicial and legal reforms were also highlighted as crucial. Many investors are deterred by weak contract enforcement and a judiciary perceived as susceptible to political influence. This research found that strengthening the independence of the judiciary and ensuring swift legal recourse for investment disputes can enhance Zambia's attractiveness as an FDI destination. Some respondents recommended the establishment of specialized commercial courts to handle investment-related cases efficiently. Lastly, macroeconomic stability was identified as an indispensable factor in mitigating the negative effects of political contestation on FDI. Political instability often coincides with exchange rate volatility, inflationary pressures, and fiscal mismanagement, which further erode investor confidence. Experts advocated for policies that promote sound fiscal management, inflation control, and exchange rate stability. Maintaining Zambia's creditworthiness through responsible debt management was also emphasized as a necessary step to ensure investor interest.

In conclusion, qualitative insights suggest that Zambia can mitigate the adverse effects of political instability on FDI through a combination of governance reforms, policy consistency, institutional strengthening, legal protections, and macroeconomic stability. Implementing these measures can create a more predictable investment environment, fostering long-term economic growth and development.

4.2 *Quantitative Findings of the Research*

This part presents the findings based on quantitative data analysis, particularly the ARDL model. It starts with descriptive statistics, followed by inferential statistical results. Next, the authors discussed the outcomes of stationarity tests. The findings related to the long-run hypothesis are then examined, followed by testing of the short-run hypothesis. Finally, the authors provided a summary of the diagnostic test results.

Table 3 Descriptive statistics

Measure	FDI inflows	POC	POS
Mean	3.255	− 0.249	0.213
Median	2.351	− 0.300	0.159
Maximum	8.633	− 0.048	0.661
Minimum	− 0.958	− 0.49	− 0.1599
Std. Dev	2.634	0.136	0.245
Skewness	0.414	− 0.00042	0.12
Kurtosis	0.123	− 1.284	2.5
Jarque–Bera	1.789	1.307	1.2
Observations	27	19	18

Source Developed by the authors

4.3 Descriptive Statistics

The summary statistics are presented in Table 3, which primarily includes the measures of central tendency and dispersion for FDI, POS, and POC. The average POS is 0.213, with a maximum value of 0.661 and a minimum of − 0.1599. The standard deviation of 0.245 indicates that the FDI inflow values are relatively close to the mean. The positive skewness of 0.414 in FDI inflows suggests that most of the data points are clustered near the lower end, with fewer outliers closer to the upper end. Similarly, the positive skewness of 0.12 in POS implies that while most years displayed lower political stability, a few years had significantly higher stability. In contrast, the negative skewness of − 0.00042 in POC suggests that while most years showed relatively high political contestation (with minimal leftward skew), a few years had notably lower levels of contestation.

The FDI inflow was found to be approximately mesokurtic, indicating it has a tail behavior similar to a normal distribution. The Kurtosis value of 0.123 is close to 0, signifying that the FDI data is approximately normally distributed, with no significant outliers. This supports the reliability of the dataset for further statistical analysis or modeling (Kallner 2018; Kim 2013). Furthermore, the Jarque–Bera test statistic showed a p-value greater than 0.05, indicating that the FDI inflow follows a normal distribution.

4.4 Correlation Among POS, POC, and FDI

Table 4 illustrates the correlations between POS, POC, and FDI. The results indicate a significant negative correlation between POC and FDI, implying that as the number of political contestations in a country increases, FDI inflows decrease, and vice versa. Additionally, a strong positive correlation was found between POS and FDI,

Table 4 Correlations among POC, POS, and FDI

	Log FDI Inflow	Log POC	Log POS
Log FDI	1.0000		
Log POC	− 0.7320	1.0000	
Log POS	0.7032	0.0441	1.000

Source Developed by the authors

suggesting that greater political stability in a country is associated with higher FDI inflows, and vice versa.

4.5 *Stationarity Tests*

Prior to conducting time series analysis, it is necessary to perform stationarity tests. This is especially important for ARDL models, which require that none of the variables be integrated of order I(2). However, the variables can be I(0), I(1), or a combination of both. The ADF test was applied alongside the PP test for robustness. The results of the stationarity tests (Table 5) show that the variables are all I(1) and I(0). This result allows for the use of the ARDL model.

4.6 *Optimal Lag Selection*

To evaluate the long-term and short-term relationships between POC, POS, and FDI inflow, the ARDL method is employed. This requires determining the optimal lag structure. According to all information criteria, the optimal lag length is found to be lag one (Table 6).

4.7 *Bounds Test*

The results of the bounds test are presented in Table 7. For FDI, POC, and POS as outcome variables, the F statistics are 5.8545, 5.9356, and 4.9765, respectively. Two of these values exceed the upper bound critical value at the 5% significance level, indicating the existence of two cointegrating relationships.

The Johansen cointegration test was also run for robustness (Table 8). These indicate the presence of at least one cointegration equation. This suggests that a long-run association exists between FDI inflow and the studied antecedents (political contestations and political stability).

Table 5 Stationarity tests

	With a constant T-statistic	Prob.	With constant and trend T-statistic	Prob.	Without constant and trend T-statistic	Prob.
Panel 1: ADF						
Log FDI	− 2.370	0.281	− 2.049	0.546	− 0.243	0.516
Log POC	− 2.805**	0.063	− 2.991	0.174	− 0.652	0.399
Log POS	− 0.985	0.884	− 1.776	0.568	0.765	1.940
ΔLog FDI	− 6.246***	0.000	− 3.608*	0.073	− 4.832***	0.000
ΔLog POC	− 8.124***	0.000	− 8.470***	0.000	− 6.965***	0.000
ΔLog POS	− 4.108**	0.034	− 4.032**	0.048	− 3.089***	0.043
Panel 2: PP						
Log FDI	− 2.261	0.3674	− 2.0426	0.6902	− 0.2851	0.6231
Log POC	− 3.2453*	0.0701	− 2.7190	0.0986	− 0.3553	0.3895
Log POS	− 1.1436	0.7320	− 0.0699	0.6487	0.6745	0.8723
ΔLog FDI	− 5.8741***	0.0000	− 1.4623***	0.0000	6.2678**	0.0000
ΔLog POC	− 7.9231***	0.0000	− 2.6875***	0.0000	8.5324**	0.0000
ΔLog POS	− 2.9801**	0.0443	− 2.979**	0.0741	− 3.682**	0.0021

Note Augmented Dickey-Fuller (ADF) and Phillips-Perron (PP) tests with 10% (*), 5% (**), and 1% (***) significance levels. *Source* Compiled by the authors

Table 6 VAR optimal lag selection

Lag	LogL	LR	FPE	AIC	SC	HQ
0	− 25.1357	NA	0.00001	2.3217	2.8745	2.3424
1	44.3876	107.6542	1.36715	− 2.09542*	− 1.12894*	− 1.97856*
2	59.4668	16.94574	0.00000	− 1.98456	− 0.23484	− 1.45673

Note Lag chosen by criteria. *Source* Compiled by the authors
The asterisk indicates the values used to select the lag using the AIC and SC (i.e where lag 1 produced lower values compared to lag 0 and lag 2) and hence selected as the optimal lag

Table 7 Bounds test results

Variable	log FDI	log POC	log POS
F statistic	5.8545*	5.9356*	4.97651
Optimal lags	(1,0,1,0)	(1,0,1,0)	(1,1,1,1)
Best trend specifications	Constant and trend	Constant and trend	Constant and trend
Critical values	10%	5%	1%
Lower bound (0)	3.47	3.99	5.09
Upper bound (1)	4.49	4.98	5.97
Diagnostics			
R^2	0.9021	0.3904	0.9767
Adj. R^2	0.5874	0.3461	0.9451

Note 5% (*) significant levels. *Source* Developed by the authors

Table 8 Johansen cointegration test

Cointegration equations	Eigenvalue	Statistic	Critical value at 5%	Prob.
Trace statistic				
None	0.963004	53.89765	49.78634	0.0451
At most 1	0.421970	17.9747	28.4762	0.8135
At most 2	0.287689	4.56733	14.8995	0.6984
At most 3	0.142352	0.86542	4.0231	0.4032
Maximum eigenvalue				
None	0.85644	32.8973	26.8527	0.0079
At most 1	0.39563	13.2895	21.9894	0.5326
At most 2	0.24786	4.0546	13.4326	0.8958
At most 3	0.15479	0.7952	4.0542	0.8754

Source Developed by the authors

4.8 *The Long-Run Model Results*

The results of the long and short run estimates are shown in Table 9. The findings show that in the long run, political contestations have a negative relationship with FDI at the 5% level ($\beta = -0.1793$, $p = 0.0014$). In turn, political stability was found to have a positive relationship with FDI inflow at the 5% level ($\beta = 0.1680$, $p = 0.0023$).

Table 9 ARDL (1, 0, 1, 0) Regressing determinants on FDI using AIC

Variable	Coefficient	Se	T-Statistic	P-value
Panel 1: Long run estimates				
Log POC	– 0.1793	0.0608	– 3.2435	0.0014
Log POS	0.1680	0.0978	2.1765	0.0023
Panel 2: Short-run estimates				
ΔLog POC	– 0.1962	0.0395	– 3.967	0.0006
ΔLog POS	0.1342	0.0370	2.7500	0.0180
ECM (– 1)	– 0.8147	0.2134	5.1238	0.0001
Diagnostic tests (p-value in brackets)				
Durbin-Whiteson statistic	1.7342			
Adjusted R^2	0.7059			
R^2	0.7562			
X^2 SERIAL Breusch-Godfrey LM test	1.3087 (0.3124)			
X^2 White	24 (0.34056)			
X^2 NORMAL	3.047689(0.23217)			
X^2 ARCH	1.965432(0.1872)			
X^2 RESET	4.4612(0.04736)			
F statistic	12.0765 (0.0001)			

Source Developed by the authors

4.9 Short Run Dynamics

The short-run results are also shown in Table 9. Similar to the long run results, political contestations were found to have a negative short run relationship with FDI inflow at the 5% level. The estimated coefficient for political contestations was – 0.1962, while that of political stability was 0.1342. The cointegration coefficient was found to be – 0.8147 ($p = 0.0001$). Thus, in the short run, when there is a shock in the model, there is an 81.47% speed of adjustment to equilibrium. Overall, this model can explain 75.62% (R^2) of changes in FDI inflow.

4.10 Results of Hypothesis Testing

Table 10 summarizes the results of the hypothesis testing, all of which were supported. The supported hypotheses are as follows:

- Political stability has a short-run relationship with inward FDI;
- Political stability has a long-run relationship with inward FDI;
- Political contestation has a short-run relationship with inward FDI;

Table 10 Results of hypothesis tests

Hypotheses	Outcome
H1$_a$: Political stability has a short-run relationship with inward FDI	Supported
H2$_a$: Political stability has a long-run relationship with inward FDI	Supported
H3$_a$: Political contestation has a short-run relationship with inward FDI	Supported
H4$_a$: Political contestation has a long-run relationship with inward FDI	Supported

Source Developed by the authors

- Political contestation has a long-run relationship with inward FDI;

4.11 Diagnostic Tests

Diagnostic tests are presented in Table 9. The model successfully passed several diagnostic checks, including the Breusch-Godfrey test for serial correlation, the Jarque–Bera test for normality, the ARCH and White tests for heteroscedasticity, the RESET test for linearity, and the CUSUM and CUSUM of squares tests for model stability. Regarding model stability, the results from the CUSUM and CUSUM of squares tests showed that all plots remained within the 5% critical boundaries, confirming the stability of the estimated model parameters throughout the estimation period.

5 Discussion

The research results indicate that political stability plays a crucial role in attracting FDI into Zambia's mining sector. Stability creates a favorable investment environment by minimizing political risks and uncertainties. A stable political climate ensures consistent policies, predictable legal frameworks, and enhanced investor confidence, all of which are essential for long-term, capital-intensive investments like mining. These findings align with those of Assefa (2020), Busse and Hefeker (2007), Okeke and Kalu (2022), who found that political stability reduces the perceived risk of investing in a country, leading to increased levels of FDI. Their analysis demonstrated that countries with higher political stability scores attract more FDI compared to those with instability. Similarly, a study by Sabir et al. (2019) also emphasizes the importance of strong institutions and political stability as determinants of FDI flows. They found that countries with better governance and stable political environments tend to attract more foreign investments.

In Zambia, periods of political stability have been linked to increased FDI inflows, as investors are reassured by a secure environment, regulatory transparency, and fair business practices. In contrast, political instability marked by policy inconsistencies, governance challenges, and electoral uncertainties tends to deter foreign investment due to heightened risks of expropriation, contract renegotiation, and operational disruptions. These findings align with the findings of Asiedu (2006) and Bénassy-Quéré et al. (2007), who found a positive relationship between political stability and FDI inflows. Besides, in their assessment of the impact of conflicts and political instability on FDI, Aisen and Veiga (2013), Buthe and Milner (2008) reported that areas affected by civil unrest or political strife experience significant declines in FDI inflows, highlighting the deterrent effect of instability. As such, empirical evidence suggests that Zambia's mining sector has experienced fluctuating FDI trends in response to changes in governance structures, fiscal policies, and regulatory frameworks. For instance, shifts in mining tax regimes and nationalisation debates have influenced investor perceptions, sometimes leading to capital flight or reduced reinvestment. Therefore, maintaining political stability through sound governance, policy consistency, and institutional integrity is essential for sustaining and enhancing FDI inflows in Zambia's mining sector.

The findings suggest a significant correlation between changes in political contestation and fluctuations in FDI inflows in Zambia's mining sector. The impact of political contestation on FDI inflows in Zambia's mining sector is characterized by a complex interplay of political stability, regulatory consistency, and investor confidence. Higher levels of political contestation, often marked by electoral uncertainties, policy shifts, and governance challenges, tend to deter FDI inflows due to the increased risk perception among investors. Political contestation can result in policy inconsistencies, such as sudden changes in mining taxation, royalty rates, or ownership structures, which create an unpredictable business environment that deters long-term capital investments. On the other hand, when contestation occurs within a framework of democratic accountability and strong institutions, it can lead to policy reforms that boost investor confidence. In Zambia, historical trends indicate that periods of intense political contestation, particularly around national elections or transitions in government leadership, have often led to regulatory instability in the mining sector, contributing to declines in FDI. However, when contestation is coupled with transparent governance and constructive policy discussions, it can drive reforms that gradually improve investment conditions. Therefore, the overall impact of political contestation on FDI inflows in Zambia's mining sector depends on the degree to which political competition contributes to instability or institutional strengthening.

6 Conclusion and Recommendations

The research examined the impact of political stability and contestation on FDI in Zambia's mining sector. It has been shown that FDI plays a vital role in driving economic growth. However, its inflows are significantly influenced by political instability and contestation. The research indicates that political instability, marked by frequent changes in government, policy uncertainty, violent conflicts, or weak institutional frameworks, creates an unpredictable investment environment that discourages foreign investors. Nevertheless, effective policy measures can help counteract these negative effects, promoting a more stable and appealing investment climate.

According to the findings, a crucial policy measure is the strengthening of institutions, which includes reinforcing the rule of law, increasing governance transparency, and ensuring the independence of regulatory bodies. Stable and transparent institutions increase investor confidence by reducing risks related to policy unpredictability and contract enforcement issues. Moreover, political risk insurance mechanisms and investment guarantees, often facilitated by multilateral agencies such as the World Bank's Multilateral Investment Guarantee Agency (MIGA), can reassure investors by offering compensation in the event of political disruptions. Moreover, investment-friendly legal and regulatory frameworks, characterized by stable tax policies, secure property rights, and effective dispute resolution mechanisms, contribute to a more predictable and attractive business environment. Additionally, governments can strengthen investor confidence by entering into bilateral and multilateral investment treaties, which provide foreign investors with added protection against abrupt policy shifts.

Macroeconomic stability is vital in minimizing the adverse impacts of political instability. By ensuring sound fiscal and monetary policies, controlling inflation, and maintaining stable exchange rates, governments can offer reassurance to investors, even amidst political uncertainties. Additionally, engaging in public–private dialogue and involving stakeholders in policy development can reduce conflicts by promoting inclusiveness and aligning the interests of investors with national development objectives. In politically unstable regions, regional integration frameworks and economic blocs (e.g., the African Continental Free Trade Area, ASEAN, or the EU) can serve as stabilizing forces, offering investors wider market access and institutional protections. Ultimately, although political instability presents significant challenges to foreign direct investment (FDI), proactive policy measures focused on institutional reforms, risk mitigation, regulatory consistency, and macroeconomic stability can effectively lower investor uncertainty and maintain foreign investment, even in politically volatile settings.

Meanwhile, this research faced several limitations. One limitation of assessing FDI inflow and political contexts specifically for the mining sector is that it is a challenging task because several governments have refrained from publishing such figures for years. Consequently, many scholars rely on aggregated FDI in the affected countries. While disaggregated FDI data could provide deeper analytical insights, this study utilized aggregated FDI figures, given that the mining sector—Zambia's primary recipient of foreign investment—accounts for a substantial share of annual FDI inflows. Although FDI encompasses various forms, including inflows, outflows, Greenfield, and brownfield investments, FDI inflows are most relevant in the Zambian context. In practice, discussions of FDI in Zambia almost exclusively focus on inflows, as other forms are generally considered negligible. For instance, FDI outflows are seldomly discussed due to the limited number of stakeholders involved in them. Therefore, this research focused specifically on FDI inflows.

References

Abdul-Salam Y (2012) Socio-political instability and foreign direct investments in Ghana: An ARDL approach. J Econ Sustain Dev 3(13). Retrieved from https://www.iiste.org/Journals/index.php/JEDS/article/view/3491/0. Accessed 8 Apr 2025

Aisen A, Veiga FJ (2013) How does political instability affect economic growth? Eur J Polit Econ 29:151–167. https://doi.org/10.1016/j.ejpoleco.2012.11.001

Ali S, Yusop Z, Kaliappan SR, Chin L, Nazar R (2022) Impact of trade openness, human capital, and institutional performance on economic growth: evidence from Organization of Islamic Cooperation countries. J Public Aff 22(4):e2654. https://doi.org/10.1002/pa.2654

Anyanwu JC (2012) Why does foreign direct investment go where it goes? New evidence from African countries. Ann Econ Financ 13(2):433–470

Asiedu E (2006) Foreign direct investment in Africa: the role of natural resources, market size, government policy, institutions and political instability. World Econ 29(1):63–77. https://doi.org/10.2139/ssrn.717361

Assefa G (2020) Determinants of foreign direct investment in Ethiopia: the ARDL bounds testing approach (Master's thesis, Addis Ababa University). Addis Ababa University Institutional Repository, Addis Ababa. Retrieved from https://etd.aau.edu.et/items/c1b12566-bf96-48cc-94df-f1edb99156f6. Accessed 8 Apr 2025

Bellak C, Leibrecht M, Damijan J (2009) Infrastructure endowment and corporate income taxes as determinants of Foreign Direct Investment in Central- and Eastern European countries. World Econ 32(2):267–290. https://doi.org/10.1111/j.1467-9701.2008.01144.x

Bénassy-Quéré A, Coupet M, Mayer T (2007) Institutional determinants of foreign direct investment. World Econ 30(5):764–782. https://doi.org/10.1111/j.1467-9701.2007.01022.x

Bengtsson M (2016) How to plan and perform a qualitative study using content analysis. Nurs plus Open 2:8–14. https://doi.org/10.1016/j.npls.2016.01.001

Bigsten A, Tengstam S (2015) International coordination and the effectiveness of aid. World Dev 69:75–85. https://doi.org/10.1016/j.worlddev.2013.12.021

Bodea C, Ye F (2020) Investor rights versus human rights: do bilateral investment treaties tilt the scale? Br J Polit Sci 50(3):955–977. https://doi.org/10.1017/S0007123418000042

Bratton M, van de Walle N (1997) Democratic experiments in Africa: regime transitions in comparative perspective. Cambridge University Press, Cambridge

Braun V, Clarke V (2006) Using thematic analysis in psychology. Qual Res Psychol 3(2):77–101. https://doi.org/10.1191/1478088706qp063oa

Brown RL, Durbin J, Evans JM (1975) Techniques for testing the constancy of regression relationships over time. J Roy Stat Soc: Ser B (Methodol) 37(2):149–163. https://doi.org/10.1111/j.2517-6161.1975.tb01532.x

Buckley PJ, Clegg LJ, Cross AR, Liu X, Voss H, Zheng P (2007) The determinants of Chinese outward foreign direct investment. J Int Bus Stud 38(4):499–518. https://doi.org/10.1057/palgrave.jibs.8400277

Burnell PJ (2005) Political strategies of external support for democratization. Foreign Policy Anal 1(3):361–384. https://doi.org/10.1111/j.1743-8594.2005.00016.x

Busse M, Hefeker C (2007) Political risk, institutions and foreign direct investment. Eur J Polit Econ 23(2):397–415. https://doi.org/10.1016/j.ejpoleco.2006.02.003

Buthe T, Milner HV (2008) The politics of foreign direct investment into developing countries: increasing FDI through international trade agreements? Am J Polit Sci 52(4):741–762. https://doi.org/10.1111/j.1540-5907.2008.00340.x

Castro GA, Camarillo DBR (2014) Determinants of tax revenue in OECD countries over the period 2001–2011. Contaduría y Administración 59(3):35–59. https://doi.org/10.1016/S0186-1042(14)71265-3

Chandio AA, Jiang Y, Rehman A, Rauf A (2020) Short and long-run impacts of climate change on agriculture: an empirical evidence from China. Int J Clim Change Strateg Manage 12(2):201–221. https://doi.org/10.1108/IJCCSM-05-2019-0026

Charmaz K, Thornberg R (2021) The pursuit of quality in grounded theory. Qual Res Psychol 18(3):305–327. https://doi.org/10.1080/14780887.2020.1780357

Cheeseman N (2022) How Zambia's 2021 elections changed the political landscape. J Democr 33(1):79–93

Edgington ES (1967) Review of the discovery of grounded theory: strategies for qualitative research. Canadian Pschologist/Psychologie Canadienne 8a(4):360. https://doi.org/10.1037/h0083159

Edmonds WA, Kennedy TD (2017) An applied guide to research designs: Quantitative, qualitative, and mixed methods, 2nd edn. SAGE Publications, Thousand Oaks. https://doi.org/10.4135/9781071802779

Elo S, Kääriäinen M, Kanste O, Pölkki T, Utriainen K, Kyngäs H (2014) Qualitative content analysis: a focus on trustworthiness. SAGE Open 4(1):215824401452263. https://doi.org/10.1177/2158244014522633

Erlingsson C, Brysiewicz P (2017) A hands-on guide to doing content analysis. Afr J Emerg Med 7(3):93–99. https://doi.org/10.1016/j.afjem.2017.08.001

Faruq ATM (2023) The determinants of Foreign Direct Investment (FDI) A panel data analysis for the emerging Asian economies. arXiv preprint arXiv:2307.07037. https://doi.org/10.48550/arXiv.2307.07037

Fraser A, Lungu J (2007) For whom the windfalls? Winners & losers in the privatisation of Zambia's copper mines. Civil Society Trade Network of Zambia, Lusaka

Freedom House (2023) Zambia: freedom in the World 2023 report. Retrieved from https://freedomhouse.org/country/zambia/freedom-world/2023. Accessed 8 Apr 2025

Goldring E, Wahman M (2016) Democracy in reverse: the 2016 general election in Zambia. Afr Spectr 51(3):107–121. https://doi.org/10.1177/000203971605100306

Gwaindepi A (2021) Domestic revenue mobilisation in developing countries: an exploratory analysis of sub-Saharan Africa and Latin America. J Int Dev 33(2):396–421. https://doi.org/10.1002/jid.3528

International Council on Mining and Metals (2022, December 13) Role of mining in national economies: mining contribution index, 6th edn. ICMM, London, UK. Retrieved from https://www.icmm.com/en-gb/research/social-performance/2022/role-of-mining-in-national-economies. Accessed 8 Apr 2025

Jensen N (2008) Political risk, democratic institutions, and foreign direct investment. J Polit 70(4):1040–1052. https://doi.org/10.1017/S0022381608081048

Jensen NM (2012) Politics and foreign direct investment. University of Michigan Press, Ann Arbor

Johansen S, Juselius K (1990) Maximum likelihood estimation and inferenceon cointegration-with application to demand for money. Oxford Bull Econ Stat 52(2):169–210

Kallner A (2018) Laboratory statistics: handbook of formulas and terms: methods in chemistry and health sciences, 2nd edn. Elsevier, San Diego. https://doi.org/10.1016/B978-0-12-814348-3.00001-0

Kaulu B, Haabazoka L (2023) External Debt, FDI and copper prices and their relationship with tax revenue mobilization: the case of Zambia. In: Proceedings of the 4th African international conference on industrial engineering and operations management. IEOM Society International, Lusaka, Zambia, pp 69–81

Kim HY (2013) Statistical notes for clinical researchers: assessing normal distribution (2) using skewness and kurtosis. Restorat Dent Endond 38(1):52–54. https://doi.org/10.5395/rde.2013.38.1.52

Kolstad I, Wiig A (2011) Better the devil you know? Chinese foreign direct investment in Africa. J Afr Bus 12(1):31–50. https://doi.org/10.1080/1536710X.2011.555259

Larmer M (2023) Opposition politics and democratic change in Zambia. Routledge, London

Lazar J, Feng JH, Hochheiser H (2017) Research methods in human-computer interaction. Morgan Kaufmann, Waltham

Lester JN, Lochmiller CR (2017) An introduction to educational research: connecting methods to practice. SAGE Publications, Thousand Oaks

Lungu J (2008) Copper mining agreements in Zambia: Renegotiation or law reform? Rev Afr Polit Econ 35(117):403–415. https://doi.org/10.1080/03056240802411032
Mahmood H, Chaudhary AR (2013) Impact of FDI on tax revenue in Pakistan. Pakistan J Commerce Soc Sci 7(1):59–69. Retrieved from https://www.econstor.eu/handle/10419/188073. Accessed 8 Apr 2025
Mkonyi G, Macheru J, Ayako A (2022) Socio-economic determinants of FDI flow into the mining sector in Tanzania. Int J Econ 7(2):1–36. https://doi.org/10.47604/ijecon.1706
Morgan H (2022) Conducting a qualitative document analysis. Qualitat Rep 27(1):64–77. https://doi.org/10.46743/2160-3715/2022.5044
Ofori IK, Obeng CK, Armah MK (2018) Exchange rate volatility and tax revenue: evidence from Ghana. Cogent Econ Financ 6(1):1537822. https://doi.org/10.1080/23322039.2018.1537822
Okeke CT, Kalu CU (2022) Impact of political stability on foreign direct investment: evidence from Nigeria. Soc Sci Res 6(1). Retrieved from https://journals.aphriapub.com/index.php/SSR/article/view/1411. Accessed 8 Apr 2025
Pesaran MH, Shin Y (1998) An autoregressive distributed lag modelling approach to cointegration analysis. In: Storm S (ed) Econometrics and economic theory in the 20th century: the Ragnar Frisch Centennial symposium. Cambridge University Press, Cambridge, pp 371–413
Pinto PM, Zhu B (2016) Fortune or evil? The effect of inward foreign direct investment on corruption. Int Stud Quart 60(4):693–705. https://doi.org/10.1093/isq/sqw025
Podobnik B, Shao J, Njavro D, Ivanov PC, Stanley HE (2007) Influence of corruption on economic growth rate and foreign investments. arXiv preprint arXiv:0710.1995. https://doi.org/10.48550/arXiv.0710.1995
Rakner L (2017) Tax bargains in unlikely places: the politics of Zambian mining taxes. Extract Indus Soc 4(3):525–538. https://doi.org/10.1016/j.exis.2017.04.005
Rana AT, Kebewar M (2014) The political economy of FDI flows into developing countries: does the depth of international trade agreements matter? arXiv preprint arXiv:1402.0139. https://doi.org/10.48550/arXiv.1402.0139
Rehman FU, Ding Y, Noman AA, Khan MA (2020) The nexus between infrastructure and export: an empirical evidence from Pakistan. Glob J Emerg Mark Econ 12(2):141–157. https://doi.org/10.1177/0974910120930529
Resnick D (2011) Do electoral coalitions facilitate democratic consolidation in Africa? Party Polit 19(5):735–757. https://doi.org/10.1177/1354068811410369
Rodrik D (2000) Institutions for high-quality growth: what they are and how to acquire them. Stud Comp Int Dev 35(3):3–31. https://doi.org/10.1007/BF02699764
Saad N (2014) Tax knowledge, tax complexity and tax compliance: Taxpayers' view. Procedia Soc Behav Sci 109:1069–1075. https://doi.org/10.1016/j.sbspro.2013.12.590
Sabir S, Rafique A, Abbas K (2019) Institutions and FDI: evidence from developed and developing countries. Financ Innov 5:8. https://doi.org/10.1186/s40854-019-0123-7
Schoonenboom J, Johnson RB (2017) How to construct a mixed methods research design. Kölner Zeitschrift Für Soziologie und Sozialpsychologie 69(2):107–131. https://doi.org/10.1007/s11577-017-0454-1
Shrestha MB, Batha GR (2018) Selecting appropriate methodological framework for time series data analysis. J Financ Data Sci 4(2):71–89. https://doi.org/10.1016/j.jfds.2017.11.001
Siachiwena H (2021) A silent revolution: Zambia's 2021 general election. J Afr Elect 20(2):32–56. https://doi.org/10.20940/jae/2021/v20i2a3
Simutanyi N (1996) The politics of structural adjustment in Zambia. Third World Q 17(4):825–839. https://doi.org/10.1080/01436599615407
Tashakkori A, Newman I (2010) Mixed methods: integrating quantitative and qualitative approaches to research. In: Peterson PP, Baker E, McGaw B (eds) International encyclopedia of education, 3rd edn. Elsevier, Amsterdam; Boston, pp 514–520. https://doi.org/10.1016/B978-0-08-044894-7.00287-6
Timmermans S, Tavory I (2012) Theory construction in qualitative research: from grounded theory to abductive analysis. Soc Theory 30(3):167–186. https://doi.org/10.1177/0735275112457914

UNCTAD (2021) World Investment Report 2021: investing in sustainable recovery. United Nations Publications, New York

United Nations Conference on Trade and Development (UNCTAD) (2023) World Investment Report 2023. United Nations Publications, New York

Vivoda V (2009) Resource nationalism, bargaining and international oil companies: challenges and change in the new millennium. New Polit Econ 14(4):517–534. https://doi.org/10.1080/13563460903287322

Wilson JD (2015) Understanding resource nationalism: economic dynamics and political institutions. Contemp Polit 21(4):399–416. https://doi.org/10.1080/13569775.2015.1013293

World Bank (2019) Wealth beyond mining: Leveraging renewable natural capital. 12th Zambia Economic Brief. World Bank, Washington

Zambia Chamber of Mines, Association of Zambian Mineral Exploration Companies (2024, August 7) Zambia's proposed minerals law may hurt copper investment, mines lobby says. Reuters. Retrieved from https://www.reuters.com/markets/commodities/zambias-proposed-mining-law-bad-investment-miners-say-2024-08-07/. Accessed 8 Apr 2025

The Analysis of the Effects of Corporate Governance Practices on Agricultural Cooperatives: A Case of Lusaka District

Lubinda Haabazoka, Deborah Taulo, and Oscar Kaonga

Abstract This research analyzed the impact of corporate governance practices on the performance of agricultural cooperatives in the Lusaka district, guided by the objectives to assess the effects of transparency and accountability, policies and procedures, and monitoring and internal controls. The research used a descriptive quantitative approach, focusing on five multi-purpose agricultural cooperatives registered with the Patent and Company Registration Agency (PACRA) and managed by the District Agricultural Office (DAO). A sample of 323 respondents provided feedback via structured questionnaires. Data was analyzed using correlation and descriptive statistics through SPSS25. Findings revealed a positive relationship between corporate governance practices and cooperative performance. Proper implementation of governance mechanisms, particularly transparency, accountability, internal controls, monitoring, and clear policies, enhances cooperative sustainability and efficiency. The research provides several recommendations. First, policymakers allocate funding to implement research findings at various levels. Second, capacity-building and performance management systems should be strengthened for provincial and district cooperative offices. Third, a structured supervision and monitoring plan should be established for camp and block officers to ensure effective mentorship and trickle-down adoption of best practices. Additionally, future research should evaluate cooperatives' actual financial performance (profits and annual returns) and assess how political pronouncements influence administrative management. These measures aim to improve governance structures, ensuring agricultural cooperatives operate more effectively and sustainably.

L. Haabazoka · D. Taulo · O. Kaonga (✉)
Graduate School of Business, University of Zambia, Lusaka, Zambia
e-mail: oscarkaonga@gmail.com

L. Haabazoka
e-mail: lubinda.haabazoka@unza.zm

D. Taulo
e-mail: deborahtaulo@gmail.com

A. B. Karbekova (ed.), *Sustainable Development of Rural Territories Based on Climate-Smart Technologies*, Climate Change Management,
https://doi.org/10.1007/978-3-032-20106-5_31

Keywords Corporate governance · Agriculture cooperatives · Transparency · Accountability · Policies · Monitoring · Internal controls

JEL Codes A1 · G30 · L2 · Q13

1 Introduction

Since the eighteenth century, cooperatives have been acknowledged as the engine for socio-economic growth in most developed countries of the world. This is attributed to their entrenched entrepreneurial tenets that are anchored on internal and external dynamics. Internally, the propensity by cooperatives to enterprise hinges on factors such as vision, working capital, internal controls, accountability, and good decisions on investment portfolios (Merrett and Walzer 2004; Yubai 1999). Externally, the prospects by cooperatives to the enterprise are premised on matters pertaining to the legal framework and policies (Khumalo 2014). In recent years, cooperatives are increasingly viewed as a means to promote improved agricultural technologies and alleviate food insecurity and poverty (Chambo 2009; Verhofstadt and Maertens 2014). Cooperative membership is associated with higher crop yields, increased household income and assets, and lower transaction costs for accessing inputs and markets (Abebaw and Haile 2013; Ma et al. 2017; Mojo et al. 2017; Ortmann and King 2007). In most cases, cooperatives are associated with collective action and social capital. Thus, they are thought to be better placed in reducing poverty than other types of institutional innovations such as contract farming (Mojo et al. 2017; Verhofstadt and Maertens 2014; Yubai 1999). Similarly, a commonly accepted topic—despite conceptual differences among researchers—is that business models, when combined with strong corporate governance practices, can be leveraged to create, deliver, and capture both commercial and social value (Zott et al. 2010). Several case studies and surveys have demonstrated a clear relationship between good corporate governance practices and firm success (Casadesus-Masanell and Ricart 2010; Thurlow et al. 2009; Yubai. 1999; Zott et al. 2010), and suggest that such practices can even alter the economics of industry (Magretta 2002; Mitchell and Coles 2003). One study found that the most successful companies in maintaining a competitive position were those consistently making fundamental improvements across multiple dimensions of their business model to better serve customers, end users, and other key stakeholders.

Similarly, studies have confirmed the importance of good corporate governance in organizational performance and access to finance. On the other hand, the failure of companies has also provided a basis for studying bad corporate governance practices. The failure of companies, both internationally and locally, has provided valuable case studies from which lessons have been learned to strengthen corporate governance practices. For example, Finger (2016) reports that the loss of over $40 billion by the Bank of America in 2008 was a consequence of failing to separate the positions of board chairman and chief executive officer of the bank. From this failure, good corporate governance, as advocated by Cadbury (1992), emphasizes the need to separate the roles of board chair and chief executive officer, or at the very least, to ensure stronger oversight of the CEO by the board. Ignoring this principle can lead to serious consequences. In this regard, cooperatives that perform well are typically those that implement sound corporate governance practices and empower their members and communities by providing affordable goods and services (International Cooperative Alliance (ICA) 2007).

In Zambia, cooperatives have been part of each of the successful political administrations since gaining independence in 1964, with the most common being agricultural cooperatives (Ma et al. 2017; Mtonga 2012). In the early years, cooperatives were largely viewed as a mechanism for stimulating rural development, not necessarily as institutions for meeting the economic needs of their members (Becker et al. 2013). With an estimated number of over 2022 registered cooperatives as of 2020, the current objectives of cooperatives have evolved to better impact and contribute to development, such as playing an increased role in food production, distribution, and support of long-term food security (Mtonga 2012). With the formation of the Fertilizer Support Programme (FSP) in 2002, cooperatives have become more important because farmers would only access inputs through approved farmer cooperatives or other registered farmer groups.

However, about 70% of cooperatives in developing countries have generally been characterized by entrepreneurial stagnation and corporate failure. This is evident for cooperatives that are found in the Lusaka district in Zambia (Lolojih 2009). Additionally, the question of whether cooperatives have been effective or not in speeding up economic development is yet to be answered. Additionally, the contribution of cooperatives to employment creation has not been successful in Zambia (Lolojih 2009). According to research, this problem dates back to the first republic (1964–1991), where cooperatives were generally nurtured by the government. The nurturing of cooperatives by the government was mainly in the form of giving them handouts and preferential treatment on the market. This situation hindered cooperatives from sustainably developing their enterprises, expanding economic activities, and achieving self-reliance. Their tendency to prioritize investment in consumption-related projects over productive ones contributed to the development of a dependency syndrome (Moonga and Mgemezulu 2006).

Moreover, the mandate to manage cooperatives was transferred from the Ministry of Agriculture to the Ministry of Commerce, Trade, and Industry. It is currently under the Ministry of Small and Medium Enterprise Development. Additionally, the decentralized management of cooperatives through district cooperative units was introduced to enhance their enterprise capacity. However, these efforts have not led to significant improvements in cooperative productivity or sustainability (Lolojih 2009). Ministry of Agriculture and Livestock (MAL) (2015) asserted that periodical inspection reports from the Department of Cooperatives attributed the problem of non-enterprising cooperatives to their failure to adhere to good cooperate governance practices. Based on the above account, this research analyzed the effect of corporate governance practices on the performance of agricultural cooperatives in the Lusaka district. This analysis considered practices such as transparency and accountability, policies and procedures, and monitoring and internal controls using quantitative research methods.

1.1 Statement of the Problem

The number of agricultural cooperatives formed and registered in Zambia has increased over the past decade. As of 2020, the Lusaka district alone had an estimated total of over 2022 cooperatives (PACRA 2022). The rationale behind the formation of agricultural cooperatives is that, if effective and sustainable, they can strengthen the capacity of small-scale farmers to improve productivity, increase sales, and ultimately reduce poverty levels in the country. However, research indicates that most of these cooperatives are either defunct or non-performing, limiting their impact. As a result, the Zambia Cooperative Federation (ZCF) has struggled to influence government policy effectively. According to Lolojih (Lolojih 2009), this is due to how the cooperatives were nurtured in the past, which created a dependency syndrome. Compounding this challenging situation is the fact that the contribution of cooperatives to employment creation remains unknown, largely due to the absence of comprehensive records at the district agricultural cooperative offices. These offices often lack the funding and capacity necessary to support and monitor cooperative operations effectively. A study conducted in the Sesheke District of Western Province, using the routine annual inspection form, which assesses performance based on benchmarks such as the presence of a vision statement, a business plan, and adherence to acceptable business practices, found that the cooperatives in the area were not enterprising (Simammba 2018). Against this background, the need to empirically analyze the effects of corporate governance practices on the performance of agricultural cooperatives arose with a particular focus on three practices: transparency and accountability, policies and procedures, and monitoring and internal controls.

The main hypotheses of this research are as follows:

- $\mathbf{H_0}$: There is no causal relationship between corporate governance practices and agricultural cooperatives;
- $\mathbf{H_a}$: There is a causal relationship between corporate governance practices and agricultural cooperatives.

The research findings hold significant value for policymakers, implementers, and beneficiaries of agricultural cooperatives. First, policymakers are provided with evidence-based solutions to enhance the performance and sustainability of cooperatives. Since the research was conducted in the Lusaka District and the identified challenges are likely to be similar across other districts, the findings can be reasonably generalized to inform national policy. Second, implementers gain a practical framework of benchmarks to ensure that agricultural cooperatives become more enterprising and sustainable. Finally, if the recommendations are adopted, cooperative members and beneficiaries can achieve increased incomes and improved livelihoods.

2 Empirical Literature Review

N. A. Mrope defined Corporate Governance from the perspective of the investor as:

"... both the promise to repay a fair return on capital invested and the commitment to operate a firm efficiently given the investment. Corporate governance also refers to the way in which companies are governed and for what purpose. It identifies who has power and accountability and who makes decisions. It is, in essence, a toolkit that enables management and the board to deal more effectively with the challenges of running a company. Corporate governance ensures that businesses have appropriate decision-making processes and controls in place to balance the interests of all stakeholders (shareholders, employees, suppliers, customers, and the community). Governance at a corporate level includes the processes through which a company's objectives are set and pursued in the context of the social, regulatory, and market environment. It is concerned with practices and procedures for ensuring that a company is run in such a way that it achieves its objectives while ensuring that stakeholders can have confidence that their trust in that company is well founded" (Mrope 2011).

In the Zambian Context, cooperative societies are formal organizations that are registered by the Registrar of Cooperatives at the Patents and Companies Registration Agency (PACRA) and managed by the district agricultural cooperative office present in all districts in the country. The agricultural cooperatives are being managed by the District Agricultural Cooperatives office (DACO), which falls under the Ministry of Small and Medium Enterprise Development with support from the Ministry of Agriculture. The legal status of cooperatives is supported by the Cooperative Societies Act No. 20 of 1998 and the respective Cooperative By-laws. Each cooperative's By-laws are derived from the principal Act and reflect the legal and organizational character of the cooperative as a corporate entity. These By-laws define the relationships between the various organs of the cooperative and serve as a guiding framework for its day-to-day operations. This legal framework is crucial to developing enterprising cooperatives because it recognizes them as autonomous entities capable of conducting business similarly to enterprises and companies. However, the current framework falls short of creating an enabling environment for cooperatives to thrive. Particularly, it does not offer preferential treatment or market incentives that would support their competitiveness and enterprise growth.

Research related to the management of cooperatives and their performance has been conducted locally and in various countries and sectors of economies (Chambo 2009; Lolojih 2009; Simammba 2018). Upon observing high levels of inefficiency in the cooperative movement in Zambia, Lolojih (2009) concluded that this was due to the way the cooperatives were nurtured in the past, which created a dependency syndrome. In his study on why cooperatives in Zambia are not enterprising, Simammba (2018) attributes this partly to the limited interventions to foster enterprise development. He notes that the government and its cooperating partners have yet to provide lasting solutions, with most recommendations ultimately being shelved rather than implemented.

3 Theoretical Literature Review

3.1 The Stakeholder Theory

Stakeholders could be any groups or individuals in an organization that are affected by a company's operations. Stakeholder theory aims to find the best interests based on different stakeholder groups. Stakeholder theory promotes a practical, efficient, effective, and ethical way to manage organizations in a highly complex and turbulent environment (MBA Knowledge Base n.d.). Given the significant losses faced by cooperatives in Zambia and their limited impact on poverty reduction, applying corporate governance principles by all stakeholders could enhance performance. A review of theoretical literature suggests that sound corporate governance practices have a substantial influence on the effectiveness and sustainability of cooperatives.

3.2 The Cooperative Theory

There is now a renewed interest in a new type or "new generation cooperative" that addresses the weaknesses of traditional cooperatives by strengthening the assignments of property rights to its individual members and reducing the incentives for opportunistic behavior (Cook and Iliopoulos 2000). Since new innovative cooperatives require substantial capital investment beyond farmers' capacity, outside investors are allowed by statutes to buy tradable equity and have control by voting and are eligible for election as board of director members and distribution of benefits according to investment rather than in relation to use as well as the opportunity to capture increased cooperative value at the equity's disposition.

The stakeholder theory is perfect for this research because it highlights the importance of the organizational structures accountability to all stakeholders involved and the influence of leadership on cooperative performance. The cooperative theory is equally important to this research because it highlights a departure from theoretical ways of resolving cooperative challenges to practical ways of involving members of the cooperatives in improving performance. Thus, these two theories are suitable for this research because they help situate the research within the ongoing debate on how cooperatives' performance is influenced by corporate governance practices.

4 Methodology

4.1 Research Design

To achieve the research objectives and outline the philosophical assumptions guiding this inquiry, the researcher used a research matrix as a heuristic tool. This matrix outlined the nature of the research questions, linking them with an appropriate logical framework (Table 1).

Table 1 Research design

Research question	Research objective	Population and sampling	Data collection techniques	Data analysis techniques
1. How do transparency and accountability affect the performance of agricultural cooperatives?	1. To establish if transparency and accountability affect the performance of agricultural cooperatives	A maximum of 64 respondents were selected from each of the five agricultural cooperatives using maximum heterogeneity sampling	Structured interviews	Quantitative content analysis
2. How do monitoring and internal controls affect the performance of agricultural cooperatives?	2. To establish if Monitoring and Internal Controls affect the Performance of Agricultural Cooperatives			
3. How do policies and procedures affect the performance of agricultural cooperatives?	3. To establish if Policies and Procedures affect the Performance of Agricultural Cooperatives			
4. What measures can help enhance the performance/ sustainability of agricultural cooperatives in the Lusaka district?	4. To determine measures that would help to enhance the performance/ sustainability of cooperatives in the Lusaka district			Open-ended questions were used to obtain this information from the respondents

Source Developed by the authors

4.2 Study Population

The research targeted a total number of 323 cooperative members in the five sampled multi-purpose agricultural cooperatives in the Lusaka district. The selected cooperatives included the following: "Seeds of Hope," "Ngombe Agricultural Cooperative," "Lilanda Christian Coop," "Tilipamodzi," and "Batukashi Women."

4.3 Sample and Sampling Procedure

The research employed a maximum variation sampling technique to make the sample heterogeneous. Creswell (2008) argued that this technique is useful in selecting participants because it reflects variations of attributes along a continuum.

4.4 Data Collection

The research used a structured questionnaire to get feedback from 323 respondents in five cooperatives. The questionnaire had closed-ended questions to collect quantitative data, and these were administered to all the respondents in person. The sample size of 323 was realized from the Raosoft software, which derived the sample size from the total number of cooperatives registered in the Lusaka district (Raosoft 2022). Questionnaires were used because they are a quick way of getting feedback from respondents in a structured and standardized manner, which makes analysis very easy. Questionnaires also give respondents an opportunity to ask questions and have quality discussions around the topic.

4.5 Data Analysis

To derive meaning from the raw data that was collected from the participants, this research processed and analyzed the data using Microsoft Excel and Statistical Package for Social Sciences (SPSS). The research also used descriptive and correlational analysis to determine the mean, standard deviation, and the relationships among the variables.

5 Results and Analysis

The demographic profile of the sampled cooperatives is presented in Table 2. Table 3 presents the demographic profile of the respondents.

5.1 Descriptive Statistics

Table 4 presents a statistical summary of the variable "Transparency and accountability," which was the first construct in the questionnaire. Transparency and accountability were measured using five questions on a five-point Likert scale, where a score of "1" indicated "Strongly disagree" and "5" indicated "Strongly Agree." As shown in the univariate analysis in Table 4, most respondents gave high scores (mostly 4 and 5) to questions related to this topic. Notably, Questions 1 and 3 received relatively higher mean scores of 4.17 and 4.37, respectively. Overall, the mean scores ranged from 3.97 to 4.37, indicating a general tendency among participants to agree with the statements, suggesting a positive perception of transparency and accountability within the cooperatives.

Table 5 presents a statistical summary of the variable "Policies and procedures." The results revealed that most respondents gave high scores (either "4" or "5"), indicating general agreement with the statements. The highest mean scores were 4.01 and 3.89, respectively.

"Monitoring and internal controls" was the final construct of the questionnaire (Table 6). Respondents generally indicated agreement, with most selecting "Agree" or "Strongly Agree." The highest mean score was 4.5 for Question 3, followed by mean scores of 4.19 and 4.12 for subsequent questions.

Table 2 Demographic profile of the sampled agriculture (cooperatives)

Cooperative's name	Year of registration	Type	Core business	Members	
				Males	Females
Seeds of hope	2005	Multi-purpose	Agriculture	90	140
Ngombe agricultural cooperative	2012	Multi-purpose	Agriculture	32	46
Lilanda Christian coop	2008	Multi-purpose	Agriculture	80	150
Tilipamodzi	2007	Multi-purposc	Agriculture	65	115
Batukashi womens group	2019	Multi-purpose	Agriculture	20	20

Source Developed by the authors

Table 3 Demographic profile of the sampled population (respondents)

Variable		Frequency (n)	Percent (%)
Gender			
	Male	111	34.4
	Female	212	65.6
	Total	323	100
Age			
	25 years and below	9	2.8
	26–30 years	88	27.2
	31–35 years	128	39.6
	36–39 years	57	17.6
	Above 40 years	41	12.7
	Total	323	100
Academic qualifications			
	Grade 7	49	15.2
	Grade 12 certificate	73	22.6
	Craft certificate	133	41.2
	Diploma	65	20.1
	Degree	3	0.9
	Total	323	100
How long have you been a member of the cooperative?	Less than one year	29	9
	2–3 years	144	44.6
	4–5 years	114	35.3
	Above five years	36	11.1
	Total	323	100
Who pays for your contribution to the agriculture cooperative?	Myself	153	47.4
	Paid by others (relative/friend; employer; the state /exempted)	170	52.6
	Total	323	100

Source Developed by the authors

5.2 Correlation Analysis

The correlation matrix in Table 7, produced by SPSS version 26 software, depicts the relationships of the considered research variables. Gujarati et al. (2008) note that multicollinearity can render the coefficients of explanatory variables indeterminate and that it is a common issue in panel and time series data. Similarly, Pallant (2011) suggests that multicollinearity becomes problematic when pairwise correlation coefficients exceed 0.9. In this study, since all correlation coefficients among the

Table 4 Descriptive statistics

Questionnaire variables	Strongly disagree (1)	Disagree (2)	Neutral (3)	Agree (4)	Strongly agree (5)	Mean	Standard deviation
1. Transparency and accountability have a positive effect on the performance of agricultural cooperatives	0%	0%	11.8% (38)	47.1% (152)	41.2% (133)	4.17	1.175
2. Selecting leaders in a transparent manner has a positive effect on the performance of agricultural cooperatives	0%	5% (16)	7.1% (23)	24.2% (78)	63.8% (206)	3.97	1.051
3. Accountability from leaders has a positive effect on agricultural cooperatives performance	0%	0%	20.4% (66)	34.1% (110)	45.5% (142)	4.37	1.147
4. Holding regular meetings and documenting the minutes has a positive effect on agricultural cooperatives performance	0%	5.9% (19)	23.2% (75)	47.7% (154)	23.2% (75)	3.95	1.181
5. Transparency and accountability increase sales in agricultural cooperatives	4% (13)	5.6% (18)	15.5% (50)	31.9% (103)	43% (139)	4.04	1.083

Source Developed by the authors

Table 5 Statistical summary of the variable "policies and procedures"

Questionnaire variables	Strongly disagree (1)	Disagree (2)	Neutral (3)	Agree (4)	Strongly agree (5)	Mean	Standard deviation
1. Policies and procedures have a positive effect on the performance of agricultural cooperatives	5.3% (17)	8.45% (27)	14.9% (48)	29.1% (94)	42.4% (137)	4.01	1.176
2. Reviewing policies and procedures has a positive effect on the performance of agricultural cooperatives	9.6% (31)	14.6% (47)	23.2% (75)	28.5% (92)	24.1% (78)	3.43	1.265
3. Awareness of policies and procedures by members has a positive effect on the performance of agricultural cooperatives	6.2% (20)	10.8% (35)	12.4% (40)	42.1% (136)	28.5% (92)	3.76	1.16
4. Understanding policies and procedures by members have a positive effect on the performance of agricultural cooperatives	7.1% (23)	17.3% (56)	23.2% (75)	30% (97)	22.3% (72)	3.43	1.212
5. Having good policies and procedures increases sales in agricultural cooperatives	0%	7.7% (25)	20.2% (65)	36.5% (118)	35.6% (115)	3.89	1.142

Source Developed by the authors

Table 6 Statistical summary of other variables

Questionnaire variables	Strongly disagree (1)	Disagree (2)	Neutral (3)	Agree (4)	Strongly agree (5)	Mean	Standard deviation
1. Monitoring and internal controls have a positive effect on the performance of agricultural cooperatives	7.1% (23)	12.7% (41)	28.5% (92)	31% (100)	20.7% (67)	3.46	1.161
3. Good business plans and budgets have a positive effect on the performance of agricultural cooperatives	0%	0%	16.1% (54)	36.2% (117)	47.1% (152)	4.19	1.224
4. Performance evaluation at all levels (self and for the cooperative) has a positive effect on the performance of agricultural cooperatives	0%	7.1% (23)	14.2% (46)	34.4% (111)	44.3% (143)	4.12	1.129
4. Training on agricultural management and financial literacy has a positive effect on the performance of agricultural cooperatives	7.1% (23)	17.3% (56)	23.2% (75)	30% (97)	22.3% (72)	4.12	1.009

(continued)

Table 6 (continued)

Questionnaire variables	Strongly disagree (1)	Disagree (2)	Neutral (3)	Agree (4)	Strongly agree (5)	Mean	Standard deviation
5. Sufficient facilities and resources to support members have a positive effect on the performance of agricultural cooperatives	0%	0%	13% (42)	40.5% (133)	46.6% (150)	4.1	1.074
6. Convenient operating hours and the location of cooperatives have a positive effect on the performance of agricultural cooperatives	0%	0%	9% (29)	31.6% (102)	59.4% (192)	4.5	1.329
7. Regular monitoring and internal control of cooperative activities would increase sales for agricultural cooperatives	0%	6.8% (22)	11.8% (38)	35.3% (114)	46.15% (149)	4.1	1.288

Source Developed by the authors

variables are below 0.8, multicollinearity is not a concern. Furthermore, the correlation table indicates that all research constructs exhibit positive and statistically significant relationships.

Table 7 Correlation matrix

Variables	1	2	3	4
1. Cooperative performance	–			
2. Member participation	0.564**	–		
3. Sustainability	0.505**	0.506**	–	
4. Perceptions on monitoring and internal	0.600**	0.677**	0.408**	–
5. Perception on policies and procedures	0.679**	0.656**	0.506**	0.725**

Source Developed by the authors
** Correlation is significant at the 0.01 level (2-tailed)

6 Conclusion

6.1 Summary of Findings

The research aimed to analyze the effects of corporate governance practices (transparency and accountability, monitoring and internal controls, and policies and procedures) on the performance of agricultural cooperatives. A quantitative research approach was employed, with data collected using maximum variation sampling techniques. The analysis was guided by relevant theoretical frameworks, namely stakeholder theory and cooperative theory. The findings indicate that corporate governance practices significantly affect cooperative performance. Therefore, it is essential for agricultural cooperatives to implement these practices to enhance their effectiveness and sustainability.

6.2 Conclusions

This research sought to understand why agricultural cooperatives in the Lusaka District were underperforming. Specifically, it examined the impact of corporate governance practices on cooperative performance as a means of addressing this challenge. Previous studies have shown that many agricultural cooperatives in Zambia are either defunct or non-performing. As a result, the Zambia Cooperative Federation (ZCF) has struggled to effectively influence government policy (Lolojih 2009). In the context of this research, corporate governance was restricted to three main practices: transparency and accountability, policies and procedures, and monitoring and internal controls.

Therefore, to explore this topic, the research was anchored on the following three questions:

- Does transparency and accountability affect the performance of agricultural cooperatives?
- Do policies and procedures affect the performance of agricultural cooperatives?

- Does monitoring and internal controls affect the performance of agricultural cooperatives?

Thus, the research employed an action approach (i.e., the use of questionnaires and focus group discussions method). The following findings depicted the status of the research in relation to the problem statement as follows:

(a) Transparency and accountability were the first independent variable in this research. The research found a strong and positive relationship between corporate governance practices and cooperative performance (specifically in terms of sustainability and member participation) with correlation coefficients of $r = 0.505$ and $r = 0.564$. These findings are consistent with the study by Birchall and Richard (2004), which explored what motivates members to participate in cooperative and mutual businesses. Their research highlighted that poor legal frameworks, fraudulent transactions, and exploitation significantly demotivate members from engaging in the economic activities of cooperatives or similar enterprises;
(b) Policies and procedures were the second independent variable examined in this research. The results indicated a positive and significant relationship with cooperative performance, specifically in terms of sustainability and member participation, with correlation coefficients of $r = 0.506$ and $r = 0.656$. These findings align with the conclusions of (Garnevska et al. 2011) a study on the factors contributing to the successful development of cooperatives in Northwest China. Liu emphasized that cooperative success is strongly influenced by sound policies, effective regulations, strong leadership, good management, and the capacity building of members;
(c) Monitoring and internal controls were the third independent variable in this research. It was found that it had a positive and significant relationship with cooperative performance (sustainability and members participation) with a correlation coefficient ($r = 0.408$, $r = 0.677$).

The research revealed that corporate governance practices affect cooperative performance. Therefore, agriculture cooperatives need to implement these measures to perform as required.

6.3 Recommendations

According to the research, the effective implementation of corporate governance practices (particularly in the areas of transparency and accountability, policies and procedures, and monitoring and internal controls) can significantly enhance cooperative performance and promote long-term sustainability. The research makes the following main recommendations:

(a) There is a need for policymakers to increase funding to the agricultural sector to support the implementation and operationalization of research findings related to cooperatives. According to the International Fund for Agricultural Development (International Fund for Agricultural Development (IFAD) 2014), agriculture accounts for approximately 20% of Zambia's GDP and has the potential to drive significant economic growth and reduce rural poverty, given the country's abundant fertile land and favorable rainfall. Although the sector employs nearly half of the workforce, its contribution to GDP has declined, reaching just 12.2% by 2008. This trend underscores the urgent need for increased investment, including accessible loans to support the commercialization of smallholder agriculture, strengthened policy dialogue on rural financial services, and the promotion of sound corporate governance practices within cooperatives;

(b) There is a need to provide capacity-building and performance management systems for staff at the provincial and district agricultural cooperative offices. Initiatives such as those by the Common Market for Eastern and Southern Africa (COMESA) Business Council (CBC), which aim to provide technical support and promote local sourcing and consumption of homegrown products, should be localized and actively encouraged. Additionally, the finalization and implementation of the National Small and Medium Enterprises (SMEs) Development Policy would further empower implementers to strengthen the cooperative sector and drive sustainable growth;

(c) A well-structured supervision and monitoring plan should be established for camp and block officers, extending from the district to the provincial level. This would facilitate effective mentorship and ensure a trickle-down effect of best practices throughout the cooperative system. Furthermore, the SME Development Policy should include practical guidelines for supervising and monitoring staff within the sector to ensure that the policy's objectives are effectively realized and that the efforts of all stakeholders yield measurable outcomes.

References

Abebaw D, Haile MG (2013) The impact of cooperatives on agricultural technology adoption: empirical evidence from Ethiopia. Food Policy 38:82–91. https://doi.org/10.1016/j.foodpol.2012.10.003

Becker MS, Watson FGR, Droge E, Leigh K, Carlson RS, Carlson AA (2013) Estimating past and future male loss in three Zambian lion populations. J Wildl Manag 77(1):128–142

Birchall J, Richard AS (2004) What motivates members to participate in co-operative and mutual businesses? Ann Public Coop Econ 75(3):465–495. https://doi.org/10.1111/j.1467-8292.2004.00259.x

Cadbury A (1992) Report of the committee on the financial aspects of corporate governance. Gee Publishing Limited, London, UK

Casadesus-Masanell R, Ricart JE (2010) From strategy to business models and onto tactics. Long Range Plan 43(2–3):195–215. https://doi.org/10.1016/j.lrp.2010.01.004

Chambo SA (2009) Agricultural co-operatives: role in food security and rural development. Paper presented to expert group meeting on cooperatives. UN, New York, NY

Cook ML, Iliopoulos C (2000) Ill-defined property rights in collective action: the case of US agricultural cooperatives. In: Ménard C (ed) Institutions, contracts and organizations: perspectives

from new institutional economics. Edward Elgar Publishing, Cheltenham, UK, pp 335–348. https://doi.org/10.4337/9781781952764.00036
Creswell JW (2008) Research design: qualitative and quantitative approach. Sage Publications, New Delhi, India
Finger R (2016) Bank of America: a lesson in poor corporate governance. Huffpost. Retrieved from http://www.huffingtonpost.com/richard-finger/bank-of-america-a-lesson-_b_7257364.html. Accessed 4 April 2025
Garnevska E, Liu G, Shadbolt NM (2011) Factors for successful development of farmer cooperatives in Northwest China. Int Food Agribus Manag Rev 14(4):69–84. https://doi.org/10.22004/ag.econ.117603
Gujarati DN, Porter DC, Gunasekar S (2008) Basic econometrics, 5th edn. McGraw-Hill Education, New York, NY
International Cooperative Alliance (ICA) (2007) Review of International Co-operation, vol. 100, no 1. ICA, Geneva, Switzerland. Retrieved from https://ica.coop/sites/default/files/2021-11/Review%20of%20International%20Co-operation%202007.pdf. Accessed 4 April 2025
International Fund for Agricultural Development (IFAD) (2014) New directions for smallholder agriculture. Oxford University Press, Oxford, UK
Khumalo P (2014) Improving the contribution of cooperatives as vehicles for local economic development in South Africa. Afr Stud Q 14(4):61–79
Lolojih PK (2009) Bearing the brunt of a liberalized economy: a performance review of the cooperative movement in Zambia. CoopAFRICA working paper no. 16. ILO, Nairobi, Kenya. https://library.uniteddiversity.coop/Cooperatives/CoopAFRICA/Cooperative_Movement_in_Zambia.pdf. Accessed 4 April 2025
Ma W, Abdulai A, Goetz R (2017) Agricultural cooperatives and investment in organic soil amendments and chemical fertilizer in China. Am J Agr Econ 100(2):502–520. https://doi.org/10.1093/ajae/aax079
Magretta J (2002) Why business models matter. Harv Bus Rev 80(5):86–92. Retrieved from https://www.matizyasociados.com/wp-content/uploads/2014/02/Why-Business-Models-Matter-Magretta.pdf. Accessed 4 April 2025
MBA Knowledge Base (n.d.) Business model—definition, components, role and importance. Retrieved from https://www.mbaknol.com/general-business-articles/business-model/. Accessed 4 April 2025
Merrett CD, Walzer N (2004) Cooperatives and local development: Theory and applications for the 21st century. M.E. Sharpe, New York, NY
Ministry of Agriculture and Livestock (MAL) (2015) Provincial annual report. MAL, Mongu, Zambia
Mitchell D, Coles C (2003) The ultimate competitive advantage of continuing business model innovation. J Bus Strateg 24(5):15–21. https://doi.org/10.1108/02756660310504924
Mojo D, Fischer C, Degefa T (2017) The determinants and economic impacts of membership in coffee farmer cooperatives: recent evidence from rural Ethiopia. J Rural Stud 50:84–94. https://doi.org/10.1016/j.jrurstud.2016.12.010
Moonga LP, Mgemezulu O (2006) New challenges facing cooperatives and cooperative development in the Eastern Province of Zambia. ZAMSIF/Ministry of Finance and National Planning, Lusaka, Zambia
Mrope NA (2011) The role of agricultural marketing cooperatives in improving smallholder farmers' income: a case study of Tanzania (Master's thesis). University of Agder, Kristiansand, Norway
Mtonga E (2012) The role of agricultural cooperatives in promoting food security: a case study of Zambia (Doctoral dissertation). University of Zambia, Lusaka, Zambia
Ortmann GF, King RP (2007) Agricultural cooperatives I: history, theory and problems. Agrekon 46(1):18–46. https://doi.org/10.1080/03031853.2007.9523760
PACRA (2022) Types of companies obtainable in Zambia. Retrieved from https://www.pacra.org.zm/#/html/LCORegistrationHtml/. Accessed 4 April 2025

Pallant J (2011) SPSS survival manual: a step-by-step guide to data analysis using IBM SPSS, 4th edn. Open University Press, Maidenhead, UK

Raosoft (2022) Sample size calculator. Retrieved from http://www.raosoft.com/samplesize.html. Accessed 4 April 2025

Simammba H (2018) The role of cooperatives in economic development: a case of agricultural cooperatives in Zambia (Master's Thesis). University of Zambia, Lusaka, Zambia

Thurlow J, Zhu T, Diao X (2009) The impact of climate variability and change on economic growth and poverty in Zambia. IFPRI discussion paper 00890. International Food Policy Research Institute, Washington, DC

Verhofstadt E, Maertens M (2014) Smallholder cooperatives and agricultural performance in Rwanda: do organizational differences matter? Agric Econ 45(S1):39–52. https://doi.org/10.1111/agec.12128

Yubai (1999) Agricultural cooperatives and rural development: a case study of Mkushi District (Master of business administration thesis). Copperbelt University, Kitwe, Zambia

Zott C, Amit R, Massa L (2010) The business model: Recent developments and future research. J Manag 37(4):1019–1042. https://doi.org/10.1177/0149206311406265

Has Amalgamation Increased the Independence of Armenian Communities?

Hamazasp S. Galstyan, Kristine M. Tatosyan, and Gagik Z. Badadyan

Abstract This research aims to assess the impact of administrative-territorial reforms in Armenia, with a focus on how amalgamation has affected the relationship between community revenues and expenditures. It also evaluates the effects of fiscal decentralization on local economic development. Regression analysis conducted as part of the research indicates that, following amalgamation, communities in Armenia have become increasingly dependent on state budgetary support relative to their own income sources. This trend has contributed to a notable decline in local tax autonomy. Simultaneously, a ten-year analysis of Armenia's community budgets reveals a consistent increase in own-source revenues. However, this growth does not always correspond with a higher level of income decentralization. This discrepancy can be attributed to the fact that, despite the administrative-territorial reforms, communities in Armenia remain heavily reliant on state grants. As a result of this continued dependence, substantial state-funded investments (e.g., road construction projects and the development of preschool educational facilities) are still implemented directly through central government support. The analysis presents a nuanced and somewhat contradictory view of the effects of amalgamation. On the one hand, it underscores the importance of capital investment in amalgamated communities, made possible by state funding. On the other hand, it points to a decline in fiscal decentralization and the weakening of local financial independence.

Keywords Amalgamation · Revenues · Expenditures · Dependence · Communities · Financial decentralization

JEL Classification H72 · H75 · H76

H. S. Galstyan · K. M. Tatosyan · G. Z. Badadyan (✉)
Armenian State University of Economics, Yerevan, Armenia
e-mail: gagikbadadyan@gmail.com

A. B. Karbekova (ed.), *Sustainable Development of Rural Territories Based on Climate-Smart Technologies*, Climate Change Management,
https://doi.org/10.1007/978-3-032-20106-5_32

1 Introduction

The process of administrative-territorial reforms in the Republic of Armenia started in 2011, when the concept of "Amalgamation of communities and establishment of inter-communal associations" was adopted, which justified the significance and requirement of executing administrative-territorial reforms in Armenia. Moreover, from 2015 onwards, the process of community amalgamation in the Republic of Armenia started in a sequential manner. Initially, administrative-territorial reforms were guided by the principle of "cluster amalgamation," whereby a relatively stable community with adequate financial and human resources was merged with several smaller, less resilient neighboring communities. Since 2017, the process of community amalgamation has gradually begun to be carried out based on the notion of revitalizing the prior provinces of the Soviet era. As of January 1, 2025, the number of Armenia's communities was reduced from 915 to 71 (including the recently established 64 amalgamated communities, capital city Yerevan, Gyumri, and five communities populated by ethnic minorities that remain non-amalgamated) (Republic of Armenia 1995, 2002). During administrative-territorial reforms, it is essential to assess the extent to which the independence of community decision-making processes on local issues can be enhanced in the Republic of Armenia.

Due to the administrative-territorial reforms in the Republic of Armenia, there is currently a great need to develop indicators characterizing the level of fiscal decentralization and conduct appropriate analysis based on them, including assessments of the impact of fiscal policy on local economic development.

The following tasks were identified in the research:

- To analyze the level of fiscal decentralization in RA amalgamated communities;
- To conduct a regression analysis to assess the impact of fiscal decentralization on the own income or provincial GDP indicator that characterizes the state of local economic development (before and after amalgamation).

Within the scope of this research, the following key questions were identified:

- To what extent has amalgamation contributed to increasing the independence of communities in the Republic of Armenia (RA)?
- How has the degree of income decentralization affected the indicators reflecting the state of local economic development in RA communities?

2 Methodology

Within the scope of research conducted in Armenia, the impact of per capita community budget expenditures on the socio-economic indicators of specific territorial units was examined across Yerevan and the ten provinces. The study concludes that fiscal decentralization in Armenia has yielded positive outcomes in terms of improving access to and consumption of social services by the population. However, from the standpoint of addressing broader social challenges, community budgets have proven ineffective. Local self-governance institutions have been unable to reduce poverty and unemployment through increased budgetary spending. Furthermore, the analysis indicates that community budgets have had little to no impact on improving overall economic performance (Karapetyan 2019).

The analysis of the results drawn from evaluating the specialization of Armenia's regional economies has enabled domestic researchers to identify the comparative advantages of the territorial units of Armenia, which may prove beneficial for delivering targeted support to each Armenian province regarding the development of a territorial development policy while prioritizing their development opportunities (Movsisyan et al. 2022).

In November 2023, the Prime Minister of the Republic of Armenia ratified the "Concept of Decentralization of Authorities in the Republic of Armenia" by signing the corresponding decision, thereby initiating the transfer of various responsibilities to local communities across several sectors. While a considerable number of studies in Armenia have examined fiscal decentralization, they primarily focus on regional-level indicators. There remains a lack of research that specifically evaluates the impact of amalgamation on fiscal decentralization at the community level. The present analysis seeks to address this gap by assessing the effects of amalgamation. Its findings aim to inform future opportunities for decentralizing authority in Armenia and clarify the relationship between fiscal decentralization and local economic development indicators before and after amalgamation.

To achieve the primary objective of this research, assessing the efficiency of community amalgamation and the relationship between fiscal decentralization and selected indicators of local economic development, the research reviewed a wide range of scientific publications and works by international scholars.

The analysis of the Baltic states suggests that fiscal decentralization has a positive but moderate effect on economic growth, after which higher degrees of decentralization can have a negative impact on economic development (Slavinskaite et al. 2022).

The analyses conducted within the scope of the research certify that the administrative-territorial reforms implemented in different countries worldwide are primarily accompanied by the amalgamation of communities.

An analysis conducted in the Indonesian province of Papua indicated that the influence of local income and capital expenditures on local economic development has a negligible negative effect; the effect of capital expenditure expansion has a positive, though insignificant, effect (Ariadi et al. 2022). The findings of the analysis conducted in Jambi State indicate that government spending and investment significantly influenced the degree of fiscal decentralization, which in turn had a positive impact on the development of the state's industrial sector (Hidayat and Herlin 2019).

Moreover, to evaluate the effects of decentralization on the economic development of US urban and non-urban regions, the growth indicators of population, employment rate, and real per capita income were thoroughly analyzed and evaluated. The analysis included dependent variables (e.g., unemployment and employment levels) and a range of indicators related to the education sector. In non-urban areas, it was revealed that there is no significant correlation between income growth and the dependent variables included in the analysis (Hammond and Tosun 2009).

The analysis of the Baltic states suggests that fiscal decentralization has a positive but moderate effect on economic growth, after which higher degrees of decentralization can have a negative impact on economic development (Statistical Committee of the Republic of Armenia (Armstat) n.d.).

According to the data provided by the International Monetary Fund for different countries, Martinez-Vasqueso and Timfeev highlight the importance of assessing and interpreting indicators that evaluate the degree of decentralization considered independently. They consider that the income decentralization in the community depends on the shares of income and expenses and distinct political factors (Kim et al. 2013).

The research employed quantitative evaluation methods, including comparative analysis, data integration techniques, and regression analysis.

The justification for the chosen methods is due to the requirement to perform parallel comparisons and comparative analyses across the variables within data from Armenian communities, related to the period before and after the community amalgamation. The chosen methods will make it possible to identify correlations, patterns, and causal relationships between the indicators of fiscal decentralization of the amalgamated communities of the Republic of Armenia.

A set of indicators used to assess the level of fiscal decentralization and local economic development, described in detail in the research and analyzed through regression analysis, was applied to nine amalgamated communities, conditionally selected based on their early participation in the amalgamation process. The assessment was based on the following indicators:

- The share of local taxes in the total expenditures of the community budget (taxation autonomy);
- The share of state grants in the total expenditures of the community budget (dependence on the state budget);
- The share of community preschool education expenses within the total community budget expenditures;
- The share of the community's own income in the GDP indicator of the respective province;
- The share of the expenses of the community in the GDP indicator of the respective province.

Within the framework of the analysis, the ratio of community income to provincial GDP was selected as the dependent variable. This indicator was chosen to represent the level of local economic development because it reflects the relative weight of a community's income in relation to the GDP generated in the respective province. The independent variables included the ratio of local taxes to total expenses (indicating tax autonomy), the ratio of state grants to community expenses (indicating dependence on the state budget), the ratio of community expenses to provincial GDP, and the share of preschool education expenses within total community expenditures. The inclusion of preschool education expenses as an independent variable is based on the fact that preschool education falls entirely within the responsibilities of local governments. Moreover, community budgets have shown a consistent annual increase in expenditures, particularly in administrative and funding costs, driven by the expansion of subvention programs in recent years. Therefore, incorporating this factor into the model allows for an evaluation of the impact of investments in the preschool education system on local economic development.

The above-mentioned indicators were compiled and calculated using data from the annual official reports on the execution of budgets in the amalgamated communities, as well as from publications and findings provided by the RA Ministry of Territorial Administration and Infrastructure, the RA Ministry of Finance, the RA Statistical Committee, and various international institutions.

Drawing on international best practices and recognizing the need to evaluate the effectiveness of amalgamation in Armenia, our analysis focused on the following key areas:

1. Estimation of indicators characterizing the level of fiscal decentralization of the amalgamated communities of Armenia and the analysis of such data;
2. The assessment of the interaction of indicators characterizing the level of fiscal decentralization of the amalgamated communities of Armenia and local economic development (e.g., own income/provincial GDP).

3 Results

As a key component of administrative-territorial reforms, the amalgamation of communities is not the ultimate goal. One of the primary objectives of community amalgamation is directed towards optimizing the efficiency of financial and human resource management in small communities.

Prior to amalgamation, most communities in the Republic of Armenia were unable to fully provide the public services under their jurisdiction due to limited human and financial resources. A significant portion of their already insufficient budgets was allocated to maintaining community administrative structures, further constraining service delivery (Ministry of Territorial Administration and Infrastructure of RA n.d.).

In general, the key indicators for assessing the impact of community amalgamation include the levels of income and expenditure decentralization, as well as the community's own income and expenditure figures weighted against the GDP of the respective province.

In this context, a comprehensive set of indicators reflecting the state of fiscal decentralization and local economic development was calculated for communities that underwent amalgamation at comparatively early stages. Using these indicators, the study aimed to assess the impact of amalgamation on the ratio of communities' own income to the GDP of their respective provinces.

In the second regression model developed in this research, a statistically significant relationship was observed only between the community expenses/provincial GDP and preschool education expenses/community expenses indicators. Additionally, the increase in the state grants/community expenses ratio following amalgamation was found to contribute to a decline in the communities' own income/provincial GDP ratio. Notably, the state grants/community expenses variable, which was statistically insignificant in the first regression model (P-value = 0.204), became significant in the second model (P-value = 0.006), underscoring its increased influence after amalgamation.

In the first regression model, the index of local taxes/total expenses (tax autonomy) is significant (P-value = 0.007). In the second regression model, this independent variable was insignificant (P-value = 0.964).

The results of the regression analysis indicate that amalgamation has led to increased dependence on the state budget and a corresponding decline in tax autonomy among communities. Prior to amalgamation, a rise in tax autonomy was associated with an increase in the own income/provincial GDP ratio, and this relationship was statistically significant (P-value = 0.007). However, after amalgamation, this indicator lost its significance (P-value = 0.964). In contrast, the state grants/community expenses variable, reflecting dependence on the state budget, gained statistical significance and demonstrated an inverse relationship with local income generation. This shift confirms that, following amalgamation, communities became substantially more reliant on state financial support.

4 Conclusions

The results of the regression analysis demonstrate that, following amalgamation, the negative correlation between dependence on the state budget and the own income/provincial GDP ratio became statistically significant. This confirms that after amalgamation, communities' reliance on state funding increased at a faster rate than the growth of their own income sources. In contrast, prior to amalgamation, this relationship was not statistically significant. Instead, indicators of tax autonomy and the own income/provincial GDP ratio showed significance, suggesting that local fiscal independence played a more meaningful role before amalgamation.

These findings support the conclusion that tax autonomy has been eroded in the amalgamated communities of the Republic of Armenia as reliance on state budget transfers has grown. The regression analysis assessing the impact of fiscal decentralization on local economic development further underscores this trend: after amalgamation, there is a clear increase in dependence on the state budget and a simultaneous decline in tax autonomy. The analysis also reveals that while increased tax autonomy prior to amalgamation was associated with growth in own income/provincial GDP, this relationship lost its significance after amalgamation. Instead, heightened dependence on state transfers correlated with a reduction in locally generated revenues relative to provincial GDP.

These outcomes are particularly concerning in the context of ongoing efforts to deepen the decentralization of authority in Armenian communities. The growing reliance on central government funding raises serious concerns about the feasibility and effectiveness of both administrative and fiscal decentralization initiatives in the future.

Data Availability The results of the research conducted in the article are presented in three tables that are available at https://figshare.com/ with the identifier https://doi.org/https://doi.org/10.6084/m9.figshare.28784741.

References

Ariadi W, Anas M, Fedayyen MY, Evi T, Sidanti H (2022) Financial performance and economic growth of government in the province of Papua in the pandemic. Adv Econ Bus Manag Res 205:93–96. https://doi.org/10.2991/aebmr.k.220107.019

Hammond GW, Tosun MS (2009) The impact of local decentralization on economic growth: evidence from U.S. counties. Discussion Paper No. 4574. Bonn, Germany: Institute for the Study of Labor (IZA). Retrieved from https://docs.iza.org/dp4574.pdf. Accessed 16 April 2025

Hidayat MS, Herlin F (2019) Fiscal decentralization and its impact on industrial development in Jambi province. Jurnal Perspektif Pembiayaan Dan Pembangunan Daerah 7(3):297–304. https://doi.org/10.22437/ppd.v7i3.8493

Karapetyan N (2019) Government spending is efficient in central, or in local level? The impact of fiscal decentralization on the development of provinces of Armenia. Region World 10(4):108–115

Kim J, Lotz J, Blöchliger H (eds) (2013) Measuring fiscal decentralisation: concepts and policies. In: OECD Fiscal Federalism Studies. OECD Publishing, Paris, France. https://doi.org/10.1787/9789264174849-en

Ministry of Territorial Administration and Infrastructure of RA (n.d.) Budgetary revenues of RA communities. Retrieved from https://www.mtad.am/pages/budgetary-revenues-ra-communities. Accessed 16 April 2025

Movsisyan M, Karapetyan T, Ohanyan T, Minasyan A (2022) Assessment and analysis of the RA territorial disparity. Banber ASUE 4:59. https://doi.org/10.52174/1829-0280_2022.4-41

Republic of Armenia (1995) Appendix 2. In: The Law of "On administrative-territorial division" (Adopted by the National Assembly on November 7, 1995 No. ZR-18). Yerevan, Armenia. Retrieved from https://www.arlis.am/documentview.aspx?docid=165211. Accessed 16 April 2025

Republic of Armenia (2002) Article 8, Part 1, Point 2. In The Law of "On local self-government" (Adopted by the National Assembly on May 7, 2002 No. ZR-337). Yerevan, Armenia. Retrieved from https://www.arlis.am/documentview.aspx?docid=188985. Accessed 16 April 2025

Slavinskaite N, Lapinskiene G, Hlawiczka R, Vasa L (2022) Financial innovation management: impact of fiscal decentralization on economic growth of the Baltic countries. Market Manag Innov 1:257–271. https://doi.org/10.21272/mmi.2022.1-19

Statistical Committee of the Republic of Armenia (Armstat) (n.d.) Publications, regional statistics. Retrieved from https://armstat.am/en/?nid=111. Accessed 16 April 2025

Business Development Challenges in the Energy Sector of the Republic of Armenia

Mikayel S. Melkumyan, Taguhi A. Barseghyan, and Emil A. Grigoryan

Abstract The energy sector is of vital importance for the economic development of the Republic of Armenia because it is one of the main infrastructure components ensuring the country's socio-economic development. The primary goal of this research is to develop approaches for improving the quality of business processes in the energy sector in Armenia. The research presents the specifics of the tariff policy regulation in the energy sector. The energy liberalization has also made the sectors' business environment more attractive to companies. The regulation has created serious conditions for traders to organize their business activities. For the first time, Armenia has created necessary conditions for the activities of numerous business companies in the electricity sales or export markets, which will improve the competitive environment and increase export volumes. The authors outlined the licensing procedure for operating in the sector, along with the legal framework governing relevant legal relations. The research also discusses tariff policy, its alignment with global prices, and the principles behind establishing fair and transparent pricing. Regulating electricity distribution and consumption prices is a key factor in ensuring the efficiency, public satisfaction, and stability of the energy system. The combination of various economic, social, and environmental factors requires a well-founded and justified policy that must align the interests of consumers and energy companies while maintaining the financial stability of the system and the security of supply.

Keywords Business environment · Investment environment · Business attractiveness · Energy sector · Electricity tariff · Tariff policy

JEL Classifications O13

M. S. Melkumyan · T. A. Barseghyan (✉) · E. A. Grigoryan
Armenian State University of Economics, Yerevan, Armenia
e-mail: t.barseghyan.a@gmail.com

T. A. Barseghyan
National University of Architecture and Construction of Armenia, Yerevan, Armenia

A. B. Karbekova (ed.), *Sustainable Development of Rural Territories Based on Climate-Smart Technologies*, Climate Change Management,
https://doi.org/10.1007/978-3-032-20106-5_33

1 Introduction

The energy sector covers electricity generation, transmission, distribution, and consumption. The efficient development of this sector affects the performance of other sectors, from industry to agriculture and the service sector. The most important factors for business development, service organization, and management quality improvement in the energy sector are related to technological, organizational, and human resources. Investment in human capital, particularly education, may lead to decreased economic activity and consumption in the short term. However, in the long term, human capital development leads to a greater increase in economic activity and consumption (Melkumyan et al. 2023, p. 41). In recent years, during Armenia's independence, conditions have been established to support business organization and development. These conditions have simultaneously encouraged entrepreneurial activity, safeguarded state interests, and, from another perspective, created potential risks for corruption. Thus, a system that promotes the business environment in a way that balances the interests of all three key stakeholders (i.e., the state, businesses, and society) has yet to be fully established. While the problem remains unresolved, there is growing awareness and optimism that a solution may be found in the near future. In some countries, authorities actively support initiatives such as small and medium-sized enterprises (SMEs), which also serve social purposes, and green energy, which contributes to high ecological efficiency.

With regard to green energy opportunities in Armenia, it is worth noting that the Law of the Republic of Armenia "On energy" has included a provision for energy market liberalization for several years. Under this provision, any economic entity is permitted to purchase electricity directly from generating stations and either export it abroad or supply it to other domestic consumers. The entire sales process is governed by established legal and contractual frameworks.

The liberalization regulation has created serious conditions for the organization of business activities of traders. For the first time, Armenia has created necessary conditions for the activities of numerous business companies in the electricity sales or export markets, which resulted in an improved competitive environment and increased export volumes. The electricity transmission organization, the Electric Networks of Armenia CJSC, has to transfer electricity to consuming organizations without any hassles and for a certain set fee. This entire chain should be based on appropriate legal and contractual relationships. In terms of the attractiveness of the business environment, it is necessary to assess the barriers to entry and exit in a given market. To enter the electricity market, traders in the Republic of Armenia must be licensed by the state. The question is whether the licensing process is easy or there are various bureaucratic hassles. Therefore, it is important to analyze how businesses apply for such licenses, how many have been granted, and whether the requirements are reasonably attainable. From the perspective of creating clear and predictable rules for businesses, the presence of stable and consistent tax legislation and administration is crucial.

For example, the Tax Code has been in effect in Armenia for about ten years. However, amendments and additions are made to some documents at least several times a year. These changes mainly have a "worsening" effect on business (e.g., increase the tax burden). The latest changes, which came into effect on January 1, 2025, were to increase the turnover tax from 5 to 10%, which applied to SMEs. According to government sources, it aims to reduce shadow circulation and create equal conditions with companies operating in the general taxation field. This increases the tax burden if there are no documents for the purchase of all the goods. However, it is not clear whether it is possible to obtain all documents.

2 Methods

The development and implementation of intelligent solutions to the identified problems require the use of systemic, comparative, and statistical methods. In this research, the authors analyzed and compared the electricity output of various generating companies. A range of analytical tools was employed, including statistical, factorial, comparative, and systemic analysis, as well as grouping and tabular methods. By applying these approaches, the authors managed to formulate policy recommendations for the electricity market. Using the method of induction, the authors derived general conclusions from specific cases and facts—such as identifying legislative gaps and proposing targeted solutions to address them.

3 Results

Creating a competitive environment in the Armenian energy market plays an important role in the process of tariff regulation. There are certain tariffs in Armenia's energy sector that are eventually subject to review for the electricity supply system to continue to operate efficiently. Competition and the implementation of free market regulations can lead to lower prices, promote production optimization, and ultimately enhance the overall business environment.

Tariffs of electricity system companies are calculated and set based on the following:

- In accordance with the principles enshrined in Article 21 of the Energy Law of Armenia;
- Based on the tariff calculation methodologies and rules approved by the decisions of the Public Services Regulatory Commission of the Republic of Armenia;
- Considering the requirements of agreements concluded within the framework of public–private partnership programs.

In Armenia, energy regulation until 2022 was based solely on the "cost-based" mechanism. Starting from February 1, 2022, a gradual transition to market-based mechanisms was initiated while still maintaining the use of the "cost-based" approach. This cost-based mechanism incorporates several pricing and settlement methods, including the rate-of-return regulation method, the revenue cap regulation method, and the price cap regulation method.

The above methods are used in Armenia depending on the specifics of the companies' activities. "Since 2022, under the new liberalized electricity market model, "Electric Networks of Armenia" CJSC is no longer the sole buyer and seller of electricity. According to the Energy Law of Armenia, the company retains only the exclusive right to distribute electric energy, operate the distribution network, and serve as the guaranteed supplier—rights that remain in effect until July 1, 2025 (Republic of Armenia 2001a).

In accordance with this regulation, tariffs for the distribution of electricity have also been set for "Electric Networks of Armenia" CJSC. After July 1, 2025, if the function of the guaranteed supplier is no longer assigned to "Electric Networks of Armenia" CJSC, the latter's necessary revenue will be provided only from the provision of electricity distribution services. Table 1 reflects the voltage levels for connection to the distribution network and their corresponding tariffs.

The participants of the wholesale electricity market are as follows (Public Services Regulatory Commission of the Republic of Armenia 2019a):

- Electric energy (power) producers;
- Wholesale license holders;
- Suppliers;
- Guaranteed supplier;
- Transmission license holder;
- Distribution license holder;
- Micro-energy producers;
- Qualified consumers.

The participants in the retail electricity market are as follows (Public Services Regulatory Commission of the Republic of Armenia 2019b):

- Suppliers;
- Guaranteed supplier;
- Autonomous energy producers;

Table 1 Tariffs for the provision of electricity distribution services by "Electric Networks of Armenia" CJSC (effective from February 1, 2023)

No.	Voltage level of connection to the distribution network	AMD/kWh (including VAT)
1	110 kW	2.016
2	35 kW	3.389
3	6(10) kW	9.276
4	0.38 kW and below	21.102

Source Compiled by the authors based on Public Services Regulatory Commission of the Republic of Armenia (n.d.)

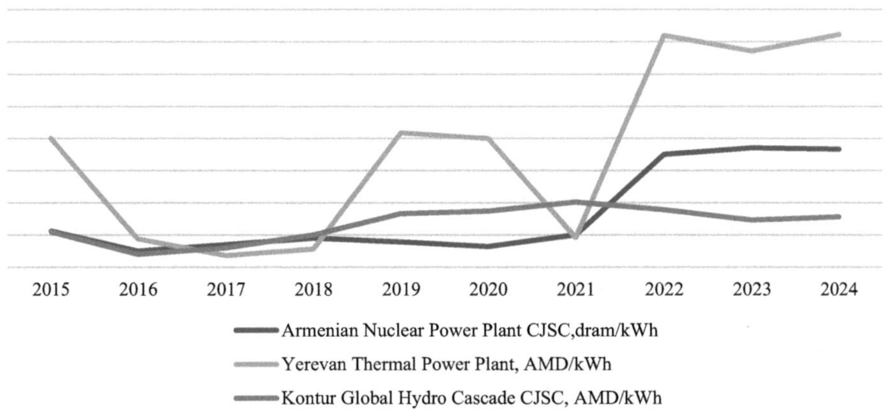

Fig. 1 Changes in electricity tariffs for companies in 2015–2024. *Source* Compiled by the authors based on Public Services Regulatory Commission of the Republic of Armenia (n.d.)

- Consumers (except qualified consumers).

Figure 1 shows that there has been a sharp increase in tariffs since 2022, with the exception of Contour Global Hydro Cascade CJSC, which maintains a relatively stable price.

Licensing is the granting of permission to perform certain activities, which is confirmed by a document of the same name. Some types of activities are licensed, meaning they can only be carried out after obtaining the appropriate permission (license) from the authorities (Republic of Armenia 2001b). Given that the power system is a highly strategic sector tied to the interests of broad segments of society—and to ensure effective state oversight—its activities have been defined by law as subject to licensing.

The types of activities subject to licensing in the electric power system are: production of electric energy (power) (including combined production of electric and thermal energy), transmission, distribution, supply and guaranteed supply of electric energy, service of the operator of the electric energy system, service of the operator of the electric energy market, and wholesale trade in electric energy (Public Services Regulatory Commission of the Republic of Armenia n.d.). The purchase and sale of electrical energy may only be carried out by persons holding a relevant activity license unless otherwise provided by law. Import or export of electrical energy may be carried out only within the framework of a type of activity license issued by the Commission that includes the right to import or export electrical energy. Import of electrical energy may also be carried out by a qualified consumer.

Licenses for electricity systems are issued by the PSRC. The PSRC defines the list of documents submitted to the PSRC for obtaining activity licenses, the forms, the procedure, and the conditions for considering the license application and issuing the license (Republic of Armenia 2001a).

The legislation provides for state fees for issuing licenses to carry out activities subject to licensing in the electricity system (Public Services Regulatory Commission of the Republic of Armenia 2019c). The rates of state duties for electricity generation are as follows:

- During construction of new production capacities: $250 per year;
- During electricity generation: $250 per year on plants with a capacity of up to 6 MW, $2500 per year on plants with a capacity of 6–15 MW, $6300 per year on plants with a capacity of 15–25 MW, $12,500 per year on plants with a capacity of 25–50 MW, and $37,500 per year on plants with a capacity of over 50 MW;
- For wholesale trade in electricity: $2500 per year;
- For transmission of electricity: $37,500 per year;
- For distribution of electricity: $65,000 per year;
- For supply of electricity (including guaranteed): $2500 per year;
- For service of the electricity system operator: $12,500 per year;
- For the electricity market for operator services: $12,500 per year (Public Services Regulatory Commission of the Republic of Armenia n.d.).

A person holding an electricity distribution license has the exclusive right to distribute electricity to consumers within a designated territory of the Republic of Armenia. This license also grants the authority to operate and regulate the distribution network, construct or reconstruct distribution infrastructure, and purchase electricity on the wholesale market to cover losses and meet internal needs (Republic of Armenia 2001a). A person holding an electricity distribution license is prohibited from engaging in the licensed activity of electricity production, as well as supply (except for guaranteed electricity supply). A person holding an electricity supply license has the right to purchase or import electricity (capacity) on the wholesale market in accordance with the procedures and conditions established by market rules. The license holder is obligated to supply electricity to consumers with whom it has concluded supply contracts, and to sell any surplus to other participants in the wholesale electricity market, in line with the terms of the license and relevant contractual agreements. A person holding a guaranteed electricity supplier license is entitled and obligated to provide electricity supply services to all consumers who do not use the services of another supplier or whose chosen supplier does not supply electricity for reasons beyond the consumer's control. A person holding a guaranteed electricity supplier license has the right to purchase or import electricity (capacity) from the wholesale market, as well as from autonomous energy producers, in accordance with the procedures and conditions established by market rules. The license holder is also permitted to sell any surplus electricity to other participants in the wholesale market. Supply contracts between the guaranteed supplier and consumers must comply with the mandatory terms and model forms approved by the PSRC.

Electricity generation licenses are issued to the following companies:

- "Armenian Nuclear Power Plant" CJSC;
- "Yerevan Thermal Power Plant" CJSC;
- "Armpower" CJSC (new thermal power plant built in Yerevan);

- "ContourGlobal Hydro Cascade" CJSC (Voratan Hydropower Complex);
- "International Energy Corporation" CJSC (Sevan–Hrazdan Cascade);
- "Hrazdan Energy Organization (HrazDPP)" OJSC;
- "Hrazdan-5" station of "Gazprom Armenia" CJSC;
- About 250 small power plants that are in the production stage (as of January 2023);
- About 30 small power plants that are in the construction stage (as of January 2023).

The electricity market is served by the following companies:

- "Electricity System Operator" CJSC;
- "High Voltage Electric Networks" CJSC;
- "Calculation Center" CJSC (market operator);
- "Electric Networks of Armenia" CJSC (distribution service).

The electricity supply licenses are issued to the following companies:

- "Electric Networks of Armenia" CJSC (guaranteed supplier);
- 13 supply licensees (as of January 2023).

The licenses for wholesale trade of electric energy are issued to 11 legal entities with wholesale trade licenses (as of January 2023).

To extend the term specified in the license, the licensee must submit an application to the PSRC no earlier than three months and no later than fifteen working days prior to the license's expiration, specifying the requested extension period.

The PSRC decides to extend the license validity period or reject the application within 25 working days from the date of the application's submission but no later than before the expiration of the license validity period. The expenses of other economic activities of licensed persons are not included in the calculation of tariffs. Additional income generated from the use of fixed assets involved in licensed activities for other purposes and additional income received from the implementation of electricity trading in the electricity market are subject to deduction from the necessary revenue of these persons in the cases and in the manner prescribed by this methodology.

Thus, the entire system operates on a foundation of legal and contractual relationships. However, the barriers to entry and exit in this market remain significant, raising questions about the accessibility of the licensing process and the extent of bureaucratic obstacles. Therefore, we propose a methodological approach to assess the impact of various factors on the investment environment and, consequently, on the overall attractiveness of doing business in the sector.

Suppose we have the following complex evaluation formula:

$$y = f(x_1, x_2, x_3 \ldots x_n) \quad (1)$$

where:

y is the level of attractiveness of the investment and business environment;
x_1 is the impact of changes in tax legislation;

x_2 is the impact of tax administration;
x_3 is the impact of credit interest rates;
x_4 is exchange rate fluctuations;
x_5 is the accessibility of logistics routes;
x_6 is the quality of competitive relations;
x_7 is the accessibility of licensing procedures;
$x_{(n-1)}$ is the impact of inflation;
x_n is the impact of external security risks.

In this regard, we can determine the three largest factors affecting the magnitude of y and, through their correlation analysis, assess the positive or negative expectations of the investment environment.

4 Conclusion

The Armenian electricity market continues to operate in a partially liberalized market structure, where the only company (i.e., "Electric Networks of Armenia" CJSC) holds the function of distribution (Republic of Armenia 2001a). This liberalization regulation has created serious conditions for the organization of business activities of traders. For the first time, Armenia has managed to create necessary conditions for the activities of numerous business companies in the electricity sales or export markets, which resulted in an improved competitive environment and increased export volumes. However, since the process has only just begun, it is necessary for state government bodies to avoid creating bureaucratic hassles regarding the licensing of traders and other documents. Simultaneously, the electricity transmission organization (i.e., "Electric Networks of Armenia" CJSC) should transfer electricity to consuming organizations without any hassles and for a certain set fee. This entire chain must be based on appropriate legal and contractual relationships. When evaluating the attractiveness of business organization conditions, it is essential to assess the barriers to market entry and exit. As our research has shown, these processes are often lengthy and procedurally complex. In Armenia, electricity market traders must obtain a state-issued license to operate. However, the licensing process involves various bureaucratic hurdles. This research provides a detailed overview of the application procedure, the number of licenses issued, and an assessment of how accessible the licensing requirements are for prospective market participants.

References

Melkumyan M, Igityan H, Sahakyan M, Baharyan F (2023) Introducing educational reforms in neoclassical model. Econ Stud J 32(6):41–53

Public Services Regulatory Commission of the Republic of Armenia (2019) Decision "On the approval of Decision No. 344 of August 9, 2017 of the Commission for Regulating Public

Services of the Republic of Armenia 'On approval of the trading rules of the wholesale electricity market of the Republic of Armenia and on recognizing the trading rules of the Republic of Armenia as void'" (December 25, 2019 No. 516-N). Yerevan, Armenia. Retrieved from https://www.arlis.am/DocumentView.aspx?docid=173571. Accessed 10 May 2025
Public Services Regulatory Commission of the Republic of Armenia (2019) Decision "On recognizing as void the trade rules of the Republic of Armenia operating based on the 'Decision by the Commission for Regulating Public Services of the Republic of Armenia on the approval of the trade rules of the retail electricity market of the Republic of Armenia'" adopted on December 27, 2006 No. 358 (adopted on December 25, 2019 No. 517-N). Yerevan, Armenia. Retrieved from https://www.arlis.am/DocumentView.aspx?docid=172646. Accessed 10 May 2025
Public Services Regulatory Commission of the Republic of Armenia (2019) Decision "On granting a license for guaranteed electricity supply to 'Armenian Electricity Networks' CJSC" (December 21, 2019 No. 748-A). Yerevan, Armenia. Retrieved from https://psrc.am/contents/document/8929. Accessed 10 May 2025
Public Services Regulatory Commission of the Republic of Armenia (n.d.) Official website. Retrieved from https://www.psrc.am/. Accessed 10 May 2025
Republic of Armenia (2001) The Law on energy (adopted by The National Assembly of Armenia on March 7, 2001). Yerevan, Armenia. Retrieved from https://www.arlis.am/documentview.aspx?docID=67721. Accessed 10 May 2025
Republic of Armenia (2001) The Law on licensing (adopted by The National Assembly of Armenia on May 30, 2001). Yerevan, Armenia. Retrieved from https://www.arlis.am/documentview.aspx?docid=73268. Accessed 10 May 2025

Comparative Assessment of the Investment Attractiveness of Territorial Units with the Example of Ten Regions of Armenia

Hamazasp S. Galstyan, Gagik Z. Badadyan, Kristine M. Tatosyan, and Mher S. Shahinyan

Abstract Assessment of the investment attractiveness of territorial units is a necessity for public administration bodies and individual investors. Public administration bodies assessing the investment attractiveness of territorial units can develop effective investment policies, while investors gain a solid information base for making informed investment decisions. In this context, creating a methodology for evaluating the investment attractiveness of individual regions is a critical challenge. This research addresses that challenge through detailed analysis and demonstrates the practical applicability of the proposed methodology using Armenia as a case study. The developed methodology is universal and, with some adaptation, can be applied to any country or territorial unit. The implementation of the developed methodology with software can provide the opportunity to obtain automatically generated data on the investment attractiveness of territorial units at any time.

Keywords Assessment methodology · Investment attractiveness · Investment risks · Investment opportunities · Territorial development

JEL Codes R11 · R53 · R58

1 Introduction

Investments in a country are the foundation of sustainable economic growth and long-term development within its economic system. The primary factors influencing the sustainable growth of investments are the investment environment and the tools

H. S. Galstyan · G. Z. Badadyan (✉) · K. M. Tatosyan · M. S. Shahinyan
Armenian State University of Economics, Yerevan, Armenia
e-mail: gagikbadadyan@gmail.com

H. S. Galstyan
e-mail: hamazasp.s.galstyan@gmail.com

A. B. Karbekova (ed.), *Sustainable Development of Rural Territories Based on Climate-Smart Technologies*, Climate Change Management,
https://doi.org/10.1007/978-3-032-20106-5_34

and established institutions that promote investment. An investor's ability to make effective investment decisions is closely tied to the concept of "investment attractiveness." This concept can be applied at various levels: the national economy, individual regions, specific economic sectors, and individual organizations. A comprehensive assessment of investment attractiveness across these levels is crucial because only such a multifaceted approach can provide a reliable information base for sound investment decisions. Within this context, evaluating the investment attractiveness of a country's regions becomes particularly important. This research focuses on developing a methodology for such an assessment and demonstrates its application using Armenia as a case study.

2 Methodology

Assessment of the investment attractiveness of territorial units is a complex process that implies a variety of assessment models. Depending on the objectives, the availability and depth of statistical data, the number of territorial units, and some other factors, the models for assessing the investment attractiveness of territorial units differ significantly. In this context, a literature review is necessary.

(Hryhoruk et al. 2022) developed a model for assessing the investment attractiveness of territorial units, which is based on the idea of clustering. By means of clustering, the territorial units are classified according to their distances from the centers of the clusters. Clustering is based on two groups of indicators: the first reflects the investment activity of business entities in the regions, while the second captures the development levels of the regional units. Before clustering, all indicators are standardized to ensure comparability. To validate and compare the results, a principal component analysis (PCA) was also conducted, which identified two main groups of principal components.

To evaluate the investment attractiveness of territorial units, (Niekrasova and Stepanenko 2019) developed a model based on six groups of indicators, namely property, innovation potential, financial capacity, productivity, social and labor potential, and socio-economic and environmental safety evaluations. The model is based on this group of indicators and their weighted average score calculated using the weights obtained by expert assessments. The model standardizes indicators according to the maximum and minimum indicators available in the country; the final result is the classification of territories according to comparisons of attractiveness.

The model developed by (Kharlamova 2014) for assessing the investment attractiveness of territorial units is based on several relative indicators. However, the indicators are weighted simultaneously by the area of territorial units and the population. For this purpose, the geometric mean of these two indicators is calculated. The indicators used for evaluations refer to the following groups: production potential, innovation capacity, institutional potential, infrastructural capabilities, human capital, natural geographical conditions, consumption opportunities, financial potential, social risks, economic risks, environmental risks, and crime risks. As can be seen, the first eight groups of indicators relate to assessing potential and opportunities, while the last four focus on risk evaluation. Clustering is used to classify territorial units based on these two groups of indicators. This classification yields an opportunity-risk matrix, where each region is positioned according to its unique characteristics.

Varavin et al. (2023) developed a model for assessing the investment attractiveness of territories, which is based on the volume of previous investments in the unit, existing fixed assets, and the volume of atmospheric emissions. The model evaluates investment attractiveness according to the areas of activity.

Konstantinova and Bychkova (2019) developed a model to evaluate the investment attractiveness of territorial units based on the use of statistical tools, expert evaluations, and the approach of classifications. The foundation of investment attractiveness should be based on sector profitability, prospects, and investment risks. The proposed model suggests that investment attractiveness is influenced by both strong and weak factors. Strong factors include elements such as location and natural resources, while weak factors encompass policies, market expectations, financial incentives, and similar variables.

To evaluate the investment attractiveness of territories, Myakshin and Grishina (2021) suggest a model that assesses several key dimensions: production and financial factors, natural resource endowment, socio-economic and political conditions, and development prospects. Each dimension is evaluated using a set of standardized indicators. The result is a network map for each perspective, visualizing the indicators that characterize it. In other words, for each territorial unit, a network map is created across these four dimensions to comprehensively assess its investment attractiveness.

Mustafakulov (2017) proposes a model for assessing the investment attractiveness of territorial units based on a comprehensive assessment. The weighing tools in the model are the profitability levels in the territorial units and the inverse indicators of the risk. Calculations are based on indicators that characterize the investment capacity, investment risks, and investment activity in the territorial unit.

The model developed by Avetisyan (2020) for assessing investment attractiveness primarily defines it based on the volume of foreign direct investment (FDI) in a given territorial unit. In this framework, the evaluation is based on factors such as property rights, road infrastructure, inflation, number of graduates, availability of skilled labor, market dominance, sectoral competition, tariff complexity, border conditions, and the costs associated with starting a business. The analysis and estimates were carried out using a statistical software package.

The model developed by Skrynnikova and Skrynnikov (2023) bases the assessment on the opportunities and risks. The assessment of investment opportunities is based on measuring indicators such as production capacity, labor force, financial and consumption potential, natural resources, infrastructure development, innovation potential, and tourism appeal. In turn, investment risk assessment focuses on financial, infrastructural, and environmental risks. The model employs clustering, where the significance coefficients derived from the clustering process are used as weights for the indicators. All indicators are standardized using their maximum and minimum values, with adjustments made based on whether their relationship with investment attractiveness is direct or inverse.

The results of the conducted research are summarized in two tables and thirteen figures presented in the “Data Availability” section. The calculations in the tables and figures have been made by the authors based on the publications of the Armstat (Statistical Committee of Armenia (Armstat) n.d.). Based on the analysis of the aforementioned models, the authors developed a methodology for assessing the investment attractiveness of Armenia’s regions. The first step in this process involves identifying a set of relevant indicators and normalizing their values. These indicators are divided into two groups: those reflecting investment attractiveness and those reflecting investment risks. The specific indicators within each group are listed. The selection of indicators presented mostly based on the reviewed literature. However, considering the statistics maintained and published by the statistical committee of Armenia, the series has been adjusted. The presented indicators are essentially in a direct comparative relationship with investment attractiveness and investment risks, which implies that a high level of each indicator characterizing investment attractiveness leads to the improvement in investment attractiveness (and vice versa), and a high level of each of the indicators characterizing investment risks leads to the increase investment risks (and vice versa Each of the 41 indicators is relative because they are weighted according to various factors such as population, geographic area, the number of organizations, and so on, depending on the nature of each indicator. Given that these indicators differ significantly in scale and units, standardization is necessary to ensure comparability across the dataset. Standardization will be carried out according to the following formula:

$$\boldsymbol{stXi} = \boldsymbol{Xi}/\max(\boldsymbol{Xi}) \tag{1}$$

where:

stXi is the standardized indicator X of the ith region (standardized values will range from 0 to 1);

Xi is the indicator X of the ith region (relative, weighted);

max(Xi) is the maximum value of the indicator X (relative, weighted).

In the next step, the authors carried out a separate assessment of the investment attractiveness and investment risks of Armenia’s regions. The assessment was carried out using standardized values and their weights. The weights were derived from a survey conducted among 305 individuals with a background in economics. Participants assessed the significance of each of the 10 groups of indicators in shaping the investment attractiveness of regions, using a 3-point scale: 1, 2, and 3. The “Analytic Hierarchy Process” (AHP) method was applied to analyze the survey results. Based on this analysis, the weights for six groups of indicators related to investment attractiveness (with a total sum of 10) and four groups of indicators related to investment risks (also summing to 10) were calculated separately. Within the ten groups, each indicator has equal importance. As a result, the investment attractiveness and risks of regions were evaluated according to the following formulas:

$$\boldsymbol{IAi} = \sum \boldsymbol{Wj}^{*}\boldsymbol{Mean}(\boldsymbol{stXij}) \tag{2}$$

where:

IAi is the assessment mark of the investment attractiveness of the region i;

Mean (stXij) is the average of the standardized indicators included in the group j of the region i;

Wj is the weight of the group j of the indicators.

The assessment marks of the investment attractiveness of the region range from 0 to 10.

$$\boldsymbol{IRi} = \sum \boldsymbol{Wj}^{*}\boldsymbol{Mean}(\boldsymbol{stXij}) \quad (3)$$

where:

IRi is the assessment mark of the investment risks of the region i;

Mean (stXij) is the average of the standardized indicators included in the group j of the region i;

Wj is the weight of the group j of the indicators.

The assessment marks of the investment risks of the region range from 0 to 10.

With the investment attractiveness and investment risk assessments of the regions in hand, a classification of Armenia's regions was conducted based on these evaluations. To enable qualitative insights from the quantitative data, clustering was performed using a method based on defining clustering centers, following the logic of a management network. Each region was assigned to one of five clusters, determined by its distance from the respective clustering centers. This approach allowed for both quantitative and qualitative analysis of the regions, grounded in their assessed levels of investment attractiveness and investment risk.

3 Results

With the methodological basis, the necessary weights, and a standardized information base, the authors assessed the investment attractiveness and risks for the regions of Armenia. However, before moving on to the analysis of those two summary assessments, it is necessary to address the situation in the groups:

1. Assessment of production potential: The assessment of the production potential mainly focuses on the regions' output opportunities. Syunik consistently held the leading position in all the observed years. This leadership is primarily attributed to the region's high volume of industrial production. However, it is important to note that Syunik's dominance extends beyond industry, reflecting strong performance in other areas as well. Notably, from 2017 to 2022, Syunik's lead became increasingly pronounced because the gap between its production potential score and that of the second-ranking region widened over time. According to 2022 data, Syunik outperformed Kotayk (i.e., the region with the second highest production potential) by 3.9 points. In contrast, Kotayk was only 2.5 points ahead of Gegharkunik, the region with the lowest score. Throughout

2017–2022, Tavush, Shirak, Lori, and Gegharkunik consistently ranked at the lower end in terms of production potential.

2. Evaluation of the volume of foreign investments: As a result of the analysis of the second group of indicators characterizing the investment opportunities of regions (total net foreign investments and direct net foreign investments), it can be recorded that during 2017–2022, net foreign investments in different regions were irregular, being positive in some years and negative in others. In 2022, in all Armenia's regions, except for Syunik, foreign net investments either had a negative value or had a positive and insignificant value (compared to Syunik). An almost similar situation was recorded in 2017–2021. Syunik recorded the highest indicators of the investment inflow, significantly surpassing other regions. These data show that Syunik has been the main destination for foreign investments in Armenia.
3. Assessment of infrastructural capabilities: Among Armenia's regions, Armavir stood out with the highest level of infrastructure development throughout the 2017–2022 period, consistently leading in this category. Ararat and Kotayk also recorded relatively high infrastructure scores during the same years. The strong performance of these three regions is largely attributed to their proximity to Yerevan. Notably, Shirak and Lori also demonstrated comparatively strong infrastructure indicators. In contrast, most of the remaining regions scored below 4 in nearly all observed years. Vayots Dzor, Gegharkunik, and Tavush were particularly marked by low levels of infrastructure development.
4. Evaluation of consumption opportunities: Based on the level of salaries, consumption expenditures of households, as well as the volume of retail trade turnover, it can be stated that the highest consumption opportunities in Armenia are characteristic of Syunik. This progress is largely driven by the high wages recorded in the region's mining sector, which, in turn, contributes to increased consumption volumes. Kotayk, Armavir, and Ararat (i.e., regions bordering Yerevan) also stand out for their relatively high consumption potential. Interestingly, Tavush shows a positive deviation in this regard, with comparatively high volumes of retail trade turnover. On the other hand, Gegharkunik consistently ranked among the lowest in all observed years, indicating limited consumption opportunities. This weak consumer market reduces the region's appeal to investors and negatively affects its overall investment attractiveness.
5. Assessment of institutional and financial potential: The assessment of the institutional and financial potential of the Armenian regions is based on broad inclusion characteristics, such as the volumes of financial and insurance activities, the total budget revenues in communities, the volumes of production and costs of commercial organizations, the number of entrepreneurs, and others. As with infrastructure, the assessment of institutional and financial potential plays a significant role in determining the investment attractiveness of regions. In this area, Syunik stands out as the leader, with a score of 9.4, indicating strong performance across all indicators that define this potential. Notably, Gegharkunik once again ranks at the bottom, trailing Syunik by 4.5 points, highlighting a consistent gap in key development factors between the two regions.

6. Evaluation of human capital: Syunik is the leader in Armenia in terms of human capital development, which has high enough results in terms of all indicators included in the assessment. Shirak, Lori, and Tavush also stand out with relatively high results. High evaluations of human capital of these regional units are largely due to the relatively high number of preprofessional, secondary professional, and higher education institutions and their graduates. The lowest estimate was recorded in Aragatsotn, where this estimate is about 2.8 times lower than the same estimate in Syunik. Such a situation in this region is mainly due to the relatively low number of preprofessional, secondary, and higher education institutions and their graduates.
7. Assessment of social risks: Tavush ranks as the most vulnerable region in Armenia in terms of unemployment, poverty levels, and security risks, with a social risk score of 8.8 out of a possible 10. In this region, all four indicators used to assess social risks received standardized scores above 0.75 (on a scale of 0 to 1), clearly highlighting Tavush as the highest-risk and most problematic area in this context. Shirak, Gegharkunik, Vayots Dzor, and Lori also exhibit relatively high social risks, each with scores exceeding 6 out of 10. In contrast, Aragatsotn and Kotayk show the lowest levels of social risk, with scores below 5, indicating comparatively more favorable conditions.
8. Assessment of risks related to crimes: Similar to the previous case, Tavush ranks highest in terms of crime risk assessment, with a risk score of 8 out of 10. This score is based on two indicators: the number of registered crimes and violations of environmental legislation. While Tavush does not lead in the number of crimes per 1000 people, ranking behind Aragatsotn, Ararat, Armavir, Lori, Kotayk, Shirak, and Vayots Dzor, it has the highest number of environmental violations per 1000 people in Armenia. This makes Tavush the most at-risk region in this category, primarily due to widespread illegal deforestation. It is also worth noting that the remaining nine regions show relatively similar levels of crime risk, with scores ranging from 0.35 to 0.55. The lowest risk level in the Republic of Armenia was recorded in Shirak.
9. Assessment of environmental risks: The highest level of environmental risks in Armenia (7 points out of 10 possible) was recorded in Syunik, which is due to the structure of the region's economy (the mining industry is developed in the region). The region is a leading producer of waste, which significantly contributes to the magnitude of environmental risks. Kotayk also stands out due to a high level of environmental risks, with a risk assessment score of 6.7 out of 10 possible. Such a size in that region is due to the volume of gas and water consumption per capita. In the case of the remaining regions, the marks of environmental risks do not exceed 4 points out of a possible 10, which implies that the environmental risks in 8 regions are significantly smaller compared to Syunik and Kotayk. In this regard, the lowest risk mark was recorded in Tavush (2 points).
10. Assessment of economic risks: According to 2022 data, Vayots Dzor recorded the highest level of economic risks in Armenia. This is primarily due to the high amount of mandatory taxes and fees per organization, as well as the significant

number of organizations that cease operations or have short life cycles. Ararat also exhibited a relatively high level of economic risk, mainly driven by the high turnover of organizations. It is important to note that economic risk levels are elevated across all regions of Armenia, with all scores exceeding 5.5 out of a possible 10. This consistently high level of economic risk poses a serious challenge to the country's overall investment attractiveness.

Having analyzed the calculated marks of the regions received from ten groups of indicators determining the investment opportunities and investment risks of Aremnia's regions, it is necessary to refer to the summary evaluations of investment opportunities and investment risks of the regions. From 2017 to 2022, Syunik consistently held the highest investment opportunity score among Armenia's regions. In 2022, Syunik achieved its peak score of 8.3 out of 10—the highest recorded during the observed period. This significantly exceeds the score of Kotayk, which ranked second with 5.3 points, creating a 3-point gap. Notably, the difference between Kotayk and Gegharkunik—the region with the lowest investment opportunity score of 3.2—is only 2.1 points. This means that Kotayk is closer to the lowest-ranked region than to Syunik, underscoring Syunik's dominant position. In contrast, most other regions cluster more closely together. In 2022, the lowest investment opportunity scores were recorded in Gegharkunik (3.2), Aragatsotn (3.8), and Tavush (4.2), further highlighting the wide gap between Syunik and the rest. The situation is somewhat different regarding investment risks. Tavush consistently recorded the highest investment risk score throughout 2017–2022, reaching 6.3 points in 2022. However, unlike investment opportunities, investment risk scores across the regions are more evenly distributed. In 2022, the range of scores spanned just 1.6 points. Vayots Dzor held the second-highest risk score at 6.1, followed by Syunik with 5.7. The lowest investment risk levels were observed in Aragatsotn and Armavir, both scoring 4.7 out of 10.

4 Conclusions

As a result, the clustering of Armenia's regions based on their investment opportunity and risk assessments is developed (with the data of 2022). According to the clustering results, none of Armenia's regions fall into the first quartile, characterized by above-average investment opportunities and below-average investment risks. This indicates that no region in the country has yet developed a fully favorable and attractive investment environment. Even more concerning is the fact that 6 out of 10 regions are located in the third quartile, which reflects a combination of below-average investment opportunities and above-average investment risks. This suggests a generally unfavorable investment climate across most of Armenia, where the risks are high but the opportunities remain limited. Syunik and Kotayk are positioned in the second quartile, where above-average investment opportunities are accompanied by above-average investment risks. While this scenario is also suboptimal, Syunik's position is relatively more favorable, as its investment opportunities outweigh its risks. Aragatsotn and Armavir are in the fourth quartile, marked by low levels of both investment risks and investment opportunities. Among them, Aragatsotn presents a particularly notable case: while it has the second-lowest investment opportunities in the country, it also boasts the lowest investment risks. This unique balance may offer potential for strategic, risk-averse investments, provided targeted efforts are made to develop its economic potential.

Data Availability The results of the conducted research are presented in two tables and thirteen figures, which are available at https://figshare.com/ with the identifier https://doi.org/10.6084/m9.figshare.28783532.v1.

References

Avetisyan AG (2020) Country attractiveness: analysis of the main factors. Finance: Theory and Practice 24(4):58–74. https://doi.org/10.26794/2587-5671-2020-24-4-58-74

Hryhoruk PM, Khrushch NA, Grygoruk SS, Ovchynnikova OR (2022) Multidimensional statistical analysis of investment attractiveness and regional changes in the COVID-19 pandemic. In: M^3E^2-MLPEED 2022: the 10th international conference on monitoring, Modeling & Management of emergent economy. CEUR Workshop Proceedings, Odessa, Ukraine, pp 150–167

Kharlamova G (2014) Investment attractiveness of Ukrainian regions: rating assessment and marketing promotion. J Int Stud 7(1):9–26. https://doi.org/10.14254/2071-8330.2014/7-1/1

Konstantinova M, Bychkova N (2019) Methods of assessing the country's investment attractiveness. Scientific Herald of Uzhhorod National University 27(1):66–73. https://doi.org/10.32782/2413-9971/2019-27-13

Mustafakulov S (2017) Investment attractiveness of regions: Methodic aspects of the definition and classification of impacting factors. Eur Sci J 13(10):433–449. https://doi.org/10.19044/esj.2017.v13n10p433

Myakshin VN, Grishina IV (2021) Assessment of the region investment attractiveness in complex relationship with sustainable development on the basis of the balanced scorecard. In: Proceedings of ISSDRI 2021: international scientific and practical conference on sustainable development of regional infrastructure. SciTePress, Setúbal, Portugal, pp 338–395. https://doi.org/10.5220/0010591103880395

Niekrasova L, Stepanenko M (2019) Assessment of regional investment attractiveness as a strategic factor of enterprise development. The Scientific Journals of the University of Technology in Katowice 11:89–102

Skrynnikova I, Skrynnikov N (2023) Mathematical methods for assessing the investment attractiveness of territories. E3S Web of Conferences 371:01065. https://doi.org/10.1051/e3sconf/202337101065

Statistical Committee of Armenia (Armstat) (n.d.) Publications, regional statistics. Retrieved from https://armstat.am/am/?nid=130. Accessed 19 Aug 2024

Varavin YV, Kozlova MV, Kuur OV, Pestunova GB (2023) Assessment of investment attractiveness of regional industries in the context of green development. Economy of Regions 19(2):494–510. https://doi.org/10.17059/ekon.reg.2023-2-15

Innovative Climate Smart Integration Projects in Sustainable Agricultural Policy of CARICOM

Vera V. Ozhigina

Abstract The research aims to identify the features and directions of sustainable agricultural integration policy of CARICOM in the context of the implementation of climate-smart agricultural projects (CSAIPs). The authors studied the texts of integration agreements and other official documents of CARICOM, information from CARICOM institutions and authorities, and data from the World Bank and UNCTAD. The research proposes a meso-level approach to understanding the role of the integration system in the unification of national economies and the integration of the national and global levels. As one of the structural policies, the agricultural integration policy refers to the meso-level of the integration system and is formed by various stakeholders based on the project principle and social partnership. The features and problems of agricultural development, strategic and programming documents, successful CSAIPs, sources of their financing, and coordinating integration institutions and authorities of CARICOM are identified. Based on the analysis of trends in 2000–2023, the authors conclude about the effectiveness of CSAIPs within the framework of a sustainable agricultural policy of CARICOM in increasing the food sector's share in GDP and mutual trade, developing (agricultural) tourism, employing youth and women, and increasing import substitution and food security.

Keywords International integration · Integration policy · Integration project · Sustainable development · Climate-smart agriculture · Innovation · CARICOM

JEL Classification F13 · F15 · N56 · O13 · Q16 · Q17 · Q18 · Q54 · Q55 · Q56 · Q58

V. V. Ozhigina (✉)
Belarus State Economic University, Minsk, Belarus
e-mail: kme@bseu.by

A. B. Karbekova (ed.), *Sustainable Development of Rural Territories Based on Climate-Smart Technologies*, Climate Change Management,
https://doi.org/10.1007/978-3-032-20106-5_35

1 Introduction

The relevance of this research lies in the importance of agriculture and the food security challenges in CARICOM. In 2023, agriculture accounted for 4% of GDP, 3.5% of land area (from 1 to 39% by country), 0.3% of permanent cropland (from 0.1 to 23% by country), 12% of employment (rural population is 53%—from 16 to 81% by country), 13% of exports, and 20% of imports (UNCTAD 2025; World Bank 2025).

Caribbean agriculture is negatively affected by topographic complexity, land use diversity (Alfaro et al. 2025), climate variability, high temperatures, irregular rainfall, natural disasters (floods, hurricanes, droughts, tropical cyclones, pests and diseases), land degradation, aging population, falling crop yields, lack of technologies (Caribbean Agricultural Research and Development Institute (CARDI) 2025), supply shortages, rising prices, political discord (Forgenie et al. 2024), low production levels, food imports dependence, and vulnerability to external shocks and climate change (Isaac et al. 2020).

The following solutions are proposed to address the problems:

- Risk management by climate forecasts (Alfaro et al. 2025);
- Irrigation technologies monitoring crops and environment (Balbontín et al. 2024);
- ICT and agro-tech start-ups under the assistance of international high-tech agrifood companies, funds, and organizations (Belkina 2024);
- Nature-based, climate-smart, resource-saving technologies and local agricultural practices, training, subsidies, and investment (De Caires et al. 2024);
- Sustainable agricultural GVCs (Forgenie et al. 2024);
- Pest and disease forecasting simulation models (Hall-Hanson and Sinha 2012);
- Certification and audit scheme to determine CSA green projects and justify the amount of a grant or loan (Maximay 2016);
- Ecosystems of carbon sequestration in the soil (Sierra et al. 2025);
- Development assistance distribution based on factor analysis and a linear programming model (Smolińska-Bryza and Kiryluk-Dryjska 2025);
- Microorganisms in low-yield or natural soils (Arellano-Wattenbarger et al. 2025).

This research aims to identify the features and directions of CARICOM's sustainable agricultural integration policy in the context of innovative climate-smart agricultural integration projects (CSAIPs).

The research tasks are as follows:

- To study the features and problems of agricultural development and food security in CARICOM;
- To characterize the primary strategic and program documents, CARICOM integration authorities, and institutions coördinating cooperation and integration in sustainable and climate-smart agriculture (CSA);
- To identify successful CSAIPs and available sources of their funding;
- Based on the analysis of statistical data, to identify trends and evaluate the results of the CSAIPs and the sustainable agricultural integration policy.

2 Methodology

The research used the texts of CARICOM integration agreements and other official documents, information from the websites of CARICOM institutions and authorities (Caribbean Agricultural Research and Development Institute (CARDI) 2016; Caribbean Agricultural Research and Development Institute (CARDI) 2025; Caribbean Institute for Meteorology and Hydrology (CIMH) 2010; CARICOM 2023), and statistical data from the World Bank and UNCTAD for 2000–2023 (UNCTAD 2025; World Bank 2025).

As the analysis of publications showed, CSA publications mainly concern the study of the features and problems of the Caribbean region (Alfaro et al. 2025; Caribbean Agricultural Research and Development Institute (CARDI) 2025; Forgenie et al. 2024; Isaac et al. 2020) and ways to solve them at the national (Alfaro et al. 2025; Arellano-Wattenbarger et al. 2025; Balbontín et al. 2024; De Caires et al. 2024; Forgenie et al. 2024; Hall-Hanson and Sinha 2012; Sierra et al. 2025) and global levels (Belkina 2024; Maximay 2016; Smolińska-Bryza and Kiryluk-Dryjska 2025). However, regulation is multi-level; it includes a regional (integration) level. In this regard, the existing publications focus more on regulation at the level of Caribbean countries (Alfaro et al. 2025; Belkina 2024; Forgenie et al. 2024; Hall-Hanson and Sinha 2012; Isaac et al. 2020), rather than on coordinated actions, cooperation, and integration within the CARICOM common agricultural integration policy. Therefore, this research focused on identifying the features and directions of the CARICOM sustainable agricultural integration policy and CSAIPs. The novelty of this methodology lies in its approach to studying integration not as the creation of a regional organization but as a process of self-organization and co-evolution among a group of countries connected by integration agreements. This process occurs at the meso-level of the global economy, intermediate between the micro-level (national) and the macro-level (global). Integration systems, as meso-groups of states, emerge through cooperation, integration, and competition. They establish regional institutions that bridge the gap between global and national institutions, account for national and regional interests, represent these interests at the global level, and localize global rules through adaptation. Agricultural integration policy, as a form of structural policy, is shaped by a range of stakeholders, including national governments, CARICOM bodies, rural communities, MSMEs, farmers, women, and youth, through project-based initiatives and social partnerships.

3 Results

During the CARICOM economic integration process, several measures have been undertaken to develop sustainable agriculture. These include the liberalization of mutual food trade, harmonization of standards, promotion of food self-sufficiency, and encouragement of investment in food systems. Efforts also support the development of rural and indigenous communities and the introduction of climate-resilient agricultural strategies, methods, and technologies aimed at enhancing food and nutrition security, increasing productivity, and establishing climate-resilient agricultural global value chains (GVCs). Priority is given to market orientation, commercialization, competitiveness, environmental friendliness, productivity, adaptation to climate change, infrastructure development, and the use of ICT in CSA practices, as well as job creation, human resource development, and research. The CSAIPs are implemented using financial resources of external donors, member countries' budgets, and the CARICOM Development Fund.

Since 2014, CARICOM has been implementing the CSA and Climate-Smart Fisheries programs under the auspices of the Technical Center for Agricultural and Rural Cooperation (CTA) of the Organization of ACP States (OACPS) and the EU.

In 2021, a Special Ministerial Taskforce on Food Production and Food Security (MTF) was established to coordinate the agrifood systems program, achieve food security and nutrition, and promote economic growth and rural resilience. The MTF removes technical barriers to mutual trade, implements the CARICOM Cross-Border AgriFood Investment Strategy to support production corridors and the E-Agriculture Strategy, develops risk mitigation measures in rural areas, and identifies priority sectors for commercialization.

The "Vision 25 by 2025" initiative aims to ensure food security and nutrition while reducing food imports by 25% by the year 2025. This goal is to be achieved through the liberalization and facilitation of mutual trade, promotion of cross-border investments in climate-smart agriculture (CSA) innovations, increased production and productivity, human capacity building, and the development of transport and logistics infrastructure in rural areas. It is implemented by MTF based on a social partnership between countries, integration bodies, the CARICOM Private Sector Organization, civil society, rural areas, farmers, and international donors.

The Emergency Response Strategy and Action Plan for the Agriculture Sector in the Caribbean aims to improve emergency preparedness, response, recovery, and rehabilitation, as well as to strengthen agricultural resilience (CARICOM 2023).

The Caribbean Agricultural Research and Development Institute (CARDI) promotes sustainable rural development by conducting research and introducing farmers to new climate-smart food production, processing, storage, and distribution technologies, developing rural infrastructure, increasing the adaptive capacity of agriculture to climate change, and mitigating its impacts.

The CARDI Strategic Plan 2024–2028, titled "Using Science and Innovation to Improve Caribbean AgriFood Systems," is aimed at the following projects:

- Eliminating gaps in production and developing GVCs of priority products for import substitution and commercialization: production and supply of planting material, increasing production, productivity, and climate change resistance of plant varieties and livestock breeds;
- The dissemination of innovative technological agricultural solutions: transition from subsistence to commercial agriculture; development of reliable, stable,

and progressive agrifood enterprises; application of new or improved (post-) production technologies to rise product quality; competitive and sustainable food supplies; increase in production of nutritious foods to combat malnutrition; involvement of rural MSMEs, youth, and women; and removal of mutual barriers;
- Public education, dissemination of knowledge and research results, brand renewal, communication strategy, and expansion of the CARDI knowledge product portal;
- Digital transformation: implementation of cloud solutions and computerization of agrifood systems; increasing agribusiness efficiency and productivity by digital tools; using integrated strategies and computer models for managing soil, water resources, pests, and diseases; stimulation of R&D; ICT infrastructure modernization; business processes automation and optimization; creation of digital agriculture architecture and climate-smart farming; digital maturity of rural areas;
- Institutional strengthening: mobilization of own resources, increase of traditional and non-traditional grant funding; restructuring of the management; human capital development; infrastructure modernization; social partnership in CSAIPs (Caribbean Agricultural Research and Development Institute (CARDI) 2025).

CARDI, in collaboration with the EDF, implemented the CSAIP by advancing germplasm management and strengthening in vitro facilities. The initiative also included the development of agro-ecological zone maps and assessments of flood, salinity, and drought risks. Additionally, it involved the identification and establishment of a climate-resilient germplasm bank for sweet potato, maize, cassava, pigeon pea, and beans (Caribbean Agricultural Research and Development Institute (CARDI) 2016).

Successful cases of country CSAIPs include the following:

- ICT solutions in St. Lucia;
- CSA technologies and food production methods in schools of the Bahamas;
- Self-sufficient rural communities in Grenada;
- Agricultural greenhouse gas inventories, vulnerability assessment, and adaptation of water resources in Jamaica;
- Crop and spatial modeling in Guyana;
- Resilience to climate change in the Belize sugar industry;
- Reconstruction of seed storage and training at the Jamaica Agricultural Research Station;
- Research on production and farming methods resilient to climate change in St. Kitts and Nevis, etc.

The Caribbean Institute for Meteorology and Hydrology (CIMH) is the research and climate, calibration, and maintenance center for the Caribbean Meteorological Organization. The Caribbean Society for Agricultural Meteorology (CariSAM) provides an online platform for forums and information on weather and climate.

CARDI and CIMH, with support from the EU, WMO, and CARICOM countries, implemented the interdisciplinary Caribbean Agrometeorological Initiative (CAMI).

This initiative contributed to increased farm productivity as part of climate-smart agriculture (CSA) through several key actions: the coordinated collection and dissemination of weather and climate information; forecasting of pests, diseases, cyclones, floods, droughts, and hurricanes; and mitigation of the impacts of climate change and natural disasters. CAMI also fostered networking in applied research and development of advanced weather service software tools, including mesoscale models, weather forecasting systems, earth remote sensing models, and geographic information systems. Additionally, it promoted the use of digital decision support platforms, such as business and meteorological models, and farming systems, to improve crop performance, irrigation, pest and disease management, bio- and quality control, fertilizer application, farm revenue, and export planning (Caribbean Institute for Meteorology and Hydrology (CIMH) 2010).

To coordinate CSAIPs, disseminate knowledge and technologies, increase understanding and awareness, and obtain feedback in multi-stakeholder social partnerships, the Caribbean Agriculture Week, the CARICOM Farmer of the Year competition, the Forum with farmers and agricultural agencies, and the Caribbean Food Systems Summit are held annually.

The Caribbean Community Climate Change Centre coordinates adaptation to climate change by effective solutions and projects to combat its impact on the rural environment, provides advice and recommendations on CARICOM climate policy, and acts as an intermediary in attracting donor CSA finance and information repository. The examples of CSAIPs include the following:

- The Green Climate Fund (GCF) funded a project aimed at strengthening the resilience of the sugarcane sector to climate change by addressing key agronomic, technological, and institutional barriers that intensify its impacts. The project focused on climate change adaptation and improved agronomic practices, sustainable water and land management, training and capacity building, as well as enhanced information sharing;
- CSA project funded by the Inter-American Institute for Cooperation on Agriculture (IICA) to increase productivity and food security, adapting to climate change, reducing greenhouse gas emissions in agriculture with an emphasis on intergenerational and gender approaches;
- Integrated Climate Adaptation Strategies (ICCAS), funded by Germany within the framework of the International Climate Initiative (IKI), aimed at helping local communities survive and thrive in an era of rapid global environmental change;
- A GCF-funded project to mainstream climate change adaptation strategies into national policies, supporting national planning, implementing CSAIPs to strengthen regional resilience to the impacts of climate change, promoting sustainable agricultural practices, and strengthening food security.

4 Conclusion

Since the implementation of CSAIPs from 2014 to 2023, the food sector's share of GDP increased slightly from 3.8 to 4.0%, while its share in exports declined from 15 to 13%, and food imports rose from 16 to 20%. Despite a decrease in agricultural employment, from 15 to 12% overall (with male employment falling from 11 to 8% and female from 5 to 3%), due to population aging and the adoption of new technologies, fertilizer use decreased significantly from 113 to 69 kg per hectare of arable land. Simultaneously, domestic food production increased by 72%. Food exports as a share of total food production decreased from 42 to 25%, despite an increase in the absolute volume of food exports of 4%. Meanwhile, food imports as a share of total consumer products dropped from 46 to 41%, although import volumes rose by 40%. Between 2000 and 2023, following the entry into force of the Treaty of Chaguaramas in 2001, which established the CARICOM Single Market and Economy (CSME), mutual food exports increased by 239% (including a 25% increase from 2014 to 2023), while exports to third countries rose by only 60% (0.8% during 2014–2023). The food sector's share in mutual CARICOM exports increased from 17 to 22% between 2000 and 2014, and further to 29% in 2023. The share of mutual food exports in total exports also grew, from 11 to 17% in 2000–2014, reaching 21% in 2023. For comparison, the share of mutual trade in manufactured goods reached just 8% in 2000, 10% in 2014, and 11% in 2023. For fuels, the figures were 16%, 19%, and 7%, respectively. Mutual food imports within CARICOM also grew by 223% overall (15% between 2014 and 2023), while food imports from third countries increased slightly more—by 265% overall, including a 43% increase from 2014 to 2023 (UNCTAD 2025; World Bank 2025).

These trends indicate positive results of CSAIPs and sustainable agricultural integration policy: growing importance of agriculture in GDP and mutual trade, increased consumption of local products, improved import substitution, and food security. In the long term, the development of fresh produce production, the promotion of unique CARICOM gastronomy, the use of local and indigenous products, reducing the carbon footprint, food loss, waste in rural areas, and the protection of the region's rich natural capital and biodiversity will positively affect the (agricultural) tourism GVCs of CARICOM. This will contribute to the growth of youth employment (while the population is ageing) and women's employment (while it is still lower than men's employment).

References

Alfaro E, Hidalgo H, Pérez-Briceño PM, Calderón-Solera B (2025) Climate change scenarios in the southern Caribbean region of Central America. Revista de Biologia Tropical 73(S1):e64044

Arellano-Wattenbarger GL, Córdoba-Agudelo M, Rocha J (2025) Ancestral roots: exploring microbial communities in traditional agroecosystems for sustainable agriculture. Geoderma Reg 41(1):e00960. https://doi.org/10.1016/j.geodrs.2025.e00960

Balbontín C, Rubio JG, Calera A, Bavestrello-Riquelme C, Garcia C, Otero A et al (2024) Modernization of technological tools for crop monitoring and water use efficiency in LAC agriculture. In: Proceedings of SPIE: remote sensing for agriculture, ecosystems, and hydrology XXVI, vol 13191. SPIE, Edinburgh, UK, pp 190–199. https://doi.org/10.1117/12.3033827

Belkina SI (2024) Introduction of digital technologies in the food sector: the experience of Latin America and the Caribbean. Ekonomika i upravlenie problemy i resheniya [Economics and Management: Problems and Solutions] 4-6(145):88–94. https://doi.org/10.36871/ek.up.p.r.2024.04.06.011

Caribbean Agricultural Research and Development Institute (CARDI) (2016) Improving Caribbean food security in the context of climate change [Blog]. Retrieved from https://www.cardi.org/blog/improving-caribbean-food-security-in-the-context-of-climate-change/. Accessed 1 July 2025

Caribbean Agricultural Research and Development Institute (CARDI) (2025) CARDI strategic plan, 2024–2028: using science and innovation to improve Caribbean agrifood systems. St. Augustine, Trinidad and Tobago: CARDI; The University of the West Indies. Retrieved from https://www.cardi.org/download/23213/?tmstv=1742503737. Accessed 1 July 2025

Caribbean Institute for Meteorology and Hydrology (CIMH) (2010) Caribbean agrometeorological initiative (CAMI). Funded by the European Union under the African, Caribbean and Pacific Group of States (ACP) Science and Technology Programme. Retrieved from https://cimh.edu.bb/cami/index.html. Accessed 1 July 2025

CARICOM (2023) Emergency Response Strategy and Action Plan for the Agriculture Sector in the Caribbean. In: Climate Smart Agriculture, Georgetown, vol II - Action Plan, Guyana: Caribbean Agribusiness. Retrieved from https://agricarib.org/climate-smart-agriculture/. Accessed 1 July 2025

De Caires S, Smith BAM, Atwell MA, Wuddivira MN (2024) Technological adoption framework for mitigating food insecurity in Caribbean small Island developing states: a bibliometric approach. Trop Agric 101(3):340–368

Forgenie D, Hutchinson SD, Muhammad A (2024) Welfare implications of rising imported food costs in the Caribbean. J Agric Food Res 18:101489. https://doi.org/10.1016/j.jafr.2024.101489

Hall-Hanson RA, Sinha A (2012) Climate sustainable agriculture: training in pest and disease modelling to increase crop protection and productivity in the Caribbean. In: Caribbean Agricultural Research and Development Institute (CARDI). Retrieved from https://www.cardi.org/wp-content/uploads/2012/05/Climate-Sustainable-Agriculture-Pest-and-Disease-Modelling-rhallhanson-and-a.sinha_.pdf. Accessed 1 July 2025

Isaac W-AP, Ganpat WG, Bridgemohan P, Attzs M (2020) Defining a policy nexus for sustainable agriculture and food security in the Caribbean region. In: Venkatramanan V, Shah S, Prasad R (eds) Global climate change: resilient and smart agriculture. Springer, Singapore, pp 1–13. https://doi.org/10.1007/978-981-32-9856-9_1

Maximay S (2016) Determining the degree of climate-smart agriculture compliance, with respect to projects, processes and products. In: Academia. Retrieved from https://www.academia.edu/32715368/Determining_the_degree_of_Climate_Smart_Agriculture_compliance_with_respect_to_projects_processes_and_products. Accessed 1 July 2025

Sierra J, Stark F, Fanchone A (2025) Modeling the effect of land use and manure management on soil carbon sequestration in tropical mixed crop-livestock systems: a case study in Guadeloupe (Caribbean). Geoderma Reg 41(1):e00968. https://doi.org/10.1016/j.geodrs.2025.e00968

Smolińska-Bryza K, Kiryluk-Dryjska E (2025) The development aid for the agricultural sector of African, Caribbean and Pacific countries—determinants and allocation. Agriculture 15(2):190. https://doi.org/10.3390/agriculture15020190

UNCTAD (2025) UNCTADstat Data centre. Retrieved from https://unctadstat.unctad.org/datacentre/. Accessed 1 July 2025

World Bank (2025) World Bank Open Data. Retrieved from https://data.worldbank.org/. Accessed 1 July 2025

Prospects for Implementing ESG Concepts in China with the Development of Wooden Housing Construction

Sergey D. Tsymbalov, Olga S. Smirnova, Elena P. Kozlova, Evgeny Y. Denisov, and Ekaterina P. Garina

Abstract The Chinese market is oversaturated with housing, with a population of 1.41 billion and a family of three people requiring 400–500 million houses. China is one of the world's largest contributors to air pollution, yet it is actively working to reduce greenhouse gas emissions and decrease its reliance on fossil fuels. Concrete high-rise buildings characterize the country, while emissions from various industries heavily pollute the air. In the 1980s, a ban on the use of wood in construction was introduced to help restore the nation's forest resources. However, the high level of greenhouse gas emissions linked to residential buildings has prompted China to actively promote wooden housing construction as a means of reducing carbon output. Forests naturally absorb carbon dioxide from the atmosphere, storing it in trees as they grow. The use of wood in construction is considered more environmentally friendly than traditional materials such as concrete, brick, and steel, as it helps reduce greenhouse gas emissions and lowers the carbon footprint of the construction industry. Moreover, wooden buildings can be more energy-efficient and cost-effective to operate. China actively conducts research and develops innovative technologies and methods for utilizing wood in construction.

Keywords Emissions · Carbon trace · Greenhouse gases · Energy efficiency · Wooden houses

JEL Code R31

S. D. Tsymbalov (✉) · O. S. Smirnova · E. P. Kozlova · E. Y. Denisov · E. P. Garina
Minin Nizhny Novgorod State Pedagogical University, Nizhny Novgorod, Russia
e-mail: sergey.cymbalov@mail.ru

O. S. Smirnova
e-mail: volga0077@bk.ru

A. B. Karbekova (ed.), *Sustainable Development of Rural Territories Based on Climate-Smart Technologies*, Climate Change Management,
https://doi.org/10.1007/978-3-032-20106-5_36

1 Introduction

Over the past century, the average global temperature of the Earth's atmosphere has increased by 0.48 °C. According to several scientists, the main reason for the universal warming is the widespread use of fossil hydrocarbon fuel, which produces carbon dioxide, which belongs to the category of GG. In the twentieth century, the average temperature rose approximately 0.6 °C. In the Northern Hemisphere, the period of ice and snow melt now begins about nine days earlier than it did 150 years ago, while autumn frosts start roughly ten days later.

Amid worsening environmental challenges, 175 states officially signed the Paris Climate Agreement on April 22, 2016. By November 1 of the same year, the agreement had been ratified by 92 countries, whose combined greenhouse gas emissions accounted for 65.82% of the global total, surpassing the required threshold values. The Paris Agreement highlights the following key indicators and goals:

1. Make sure that the average temperature of the Earth's atmosphere does not increase by more than two degrees Celsius relative to the industrial period;
2. The Paris Agreement mentions that the global emissions of the GG should reach their peak as soon as possible. The most important thing is to achieve pure zero emissions of the GG around the world to the second half of the current century.

According to Xi Jinping, the Secretary General of the Central Committee of the Communist Party of China, the Paris Agreement was a starting point for a new round of international interaction in the field of environmental protection. China supports the ideas of introducing innovations, cooperation, environmental friendliness, openness, and joint use of resources. The country actively contributes to the development of a green and low-carbon economy. It is also taking decisive measures to combat climate change.

2 Methodology

Many experts identify the following as the main causes of climate change:

1. Rapid population increase, which threatens the balance of natural ecosystems;
2. The rapid increase in environmental pollution;
3. Warming of the marine environment. A large amount of toxic chemical waste and solid household waste produced as a result of human activity is constantly dumped into the ocean, leading to the destruction of the marine environment;
4. Factors of land destruction. The main cause of soil erosion and desertification is unsustainable agricultural production. As we all know, healthy vegetation can prevent soil erosion;
5. Factors of a sharp reduction in forest resources. All over the world, the area of forests is rapidly decreasing due to natural and human-made factors.

Some experts believe that global warming is associated with a solar cycle and has little in common with human activity (Netease 2025).

The contribution of China to the global change in climate from the moment of the industrial revolution in 1750–2016 was 10 (± 4) % (Encyclopedia by Baidu 2023). Nowadays, China is the largest source of emissions of PG. In this regard, many developed countries require a quantitative reduction in emissions. However, if we consider the situation from the point of view of per capita emissions and historical data, China cannot take on obligations that developed countries require. Since 2005, China has been responsible for more than 20% of global carbon dioxide emissions. According to research, the total volume of carbon dioxide emissions in China has increased in recent years, but its contribution to global warming has remained relatively stable at around 8%–9%, with a maximum of no more than 12% (Encyclopedia by Baidu 2023).

3 Results

China continues to help reduce GG emissions. It was expected that in 2023, energy consumption from renewable energy sources would account for 17.9% of energy consumption, while coal use would decrease from 67.4% in 2013 to 55.3%. In this case, the absorption of carbon dioxide reaches 19.493 billion cubic meters, which is 6.5 billion cubic meters more than in 2005 (Chinese Government Network 2024). As time has shown, mainly these results were achieved. The Ministry of Ecology and the Environment held a press conference on the topic "Active response to climate change and the joint construction of a clean and beautiful world." Xia Insyan, Director of the Department of Climate Response at the Ministry of Ecology and Environment, stated that as of the end of July 2024, the total installed capacity of wind and solar power had reached 1.206 billion kilowatts, which is 2.25 times higher than at the end of 2020. This means that the 2030 target for installed capacity will be achieved more than six years ahead of schedule (Chinese Government Network 2024; Zibao 2016).

The Ministry of Ecology and Natural Resources will continue to work on improving the system of voluntary reduction of PG emissions. In the future, the department will actively participate in market mechanisms to monitor and reduce greenhouse gas emissions. According to the annual report, the national climate change adaptation strategy for 2035 is being successfully implemented by various departments, while local communities are actively taking measures. Most provinces across the country have also developed and are carrying out their own regional action plans. The Ministry of Ecology and Natural Resources has selected 39 cities nationwide to implement pilot projects aimed at creating climate-resilient cities. Specialists in the department are actively studying possible models and approaches to constructing such cities. As part of its work, the Ministry of Ecology and Natural Resources published data on the average level of carbon dioxide emissions in the national energy system for 2021. The research also considers key sources of energy, such as coal, electricity, oil, gas, and transport. This will help determine the factors affecting the carbon trace of products. Local authorities support cooperation with regulatory authorities for the pilot certification of products based on carbon trace. This will allow the first labeling of products and evaluate its effectiveness. Additionally, China plans to actively exchange information on the carbon footprint of products at the international level, monitor and analyze developments in international trade policies related to carbon emissions, and track the evolving rules on product carbon footprints.

Addressing the challenges of climate change and pursuing green, low-carbon development requires the active participation of everyone, from energy production to daily consumption. It is essential to reduce overall consumption, minimize losses and pollutant emissions, eliminate waste, and ensure the efficient and rational use of energy (General Office of Hubei Provincial People's Government 2024).

On October 27, in the city of Jinman, a meeting on environmentally friendly and low-carbon urban renewal was held, organized by the Architectural Journal of the Ministry of Housing and Urban–Rural Development together with the Jinman municipal government. As one of the country's first demonstration cities for rapid building development, and the only one in Hubei Province, Jinman has become a pioneer in smart construction and large-scale urban renewal. In its parks, business districts, and industrial zones, numerous multifunctional green buildings have been built, covering a total area of more than 4 million square meters. This wave of sustainable construction meets modern standards and contributes to improving the city's environmental conditions (Encyclopedia by Baidu 2011).

In China, there are more than 600 million residential buildings. About 70% of urban residents live in their own homes, 20% rent housing, and the remaining 10% live in dormitories (Chinese Government Network 2025). According to the report of the UN program on the environment for 2022, published in 2023, the construction industry and related sectors consume 36% of all energy in the world and produce 36% of carbon dioxide emissions (Fig. 1). Of these, 6% falls on emissions in the production of concrete, steel, and aluminum used in construction. Another 6% emissions from other materials (e.g., bricks). According to expert estimates, emissions from glass production account for about 2%–4% of total emissions, while non-residential buildings contribute 11% and residential buildings 16%. The construction industry is the largest source of carbon dioxide emissions worldwide, and the share of buildings within this sector remains particularly high. Therefore, the active promotion of green and low-carbon buildings is an essential path toward achieving global sustainable development.

In 2014, the "Guide Instructions of the Ministry of Housing and Urban–Rural Development on the Active Implementation of Innovative Construction Technologies Using Wooden Structures" proposed that by 2020, wooden buildings should account for 8% of total construction. In February 2016, during discussions on the State Council's opinions on further strengthening urban planning and construction management, it was suggested that over the following ten years, 30% of new houses would be built while actively promoting the development of modern wooden buildings (Fig. 2) (Encyclopedia by Baidu 2025). On July 31, 2024, the Central Committee

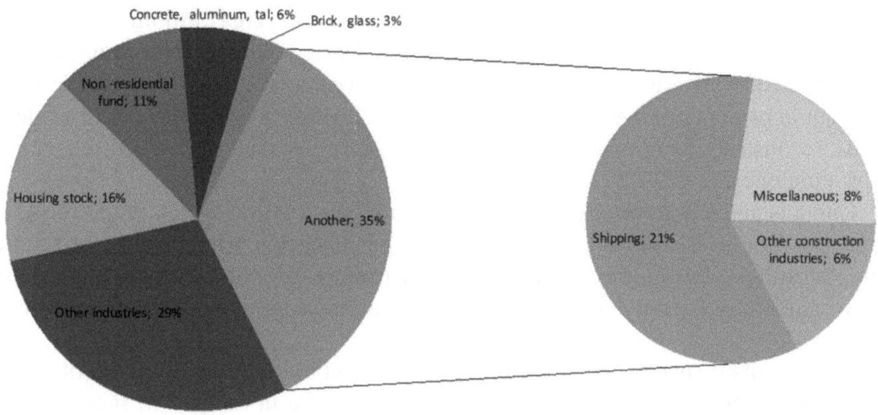

Fig. 1 Carbon dioxide emissions in the construction field. *Source* Developed and compiled by the authors

of the Communist Party of China and the State Council announced their recommendations to accelerate the comprehensive green transformation of economic and social development. They emphasized the need to adopt an environmentally friendly, low-carbon, and high-quality development model, while actively promoting the green and low-carbon transformation of the energy sector (Yetyan 2024).

Wooden structures are a type of rapid-construction building that aligns with the principles of environmental protection and supports the active development of sustainable building practices. They open up new prospects for growth. China is actively restoring its forest resources and developing prefabricated wooden houses for export to other countries. To achieve this, high-quality materials from abroad (e.g., Russian larch) are supplied to the domestic market at affordable prices. Chinese experts emphasize that the development of wooden structures should focus on the

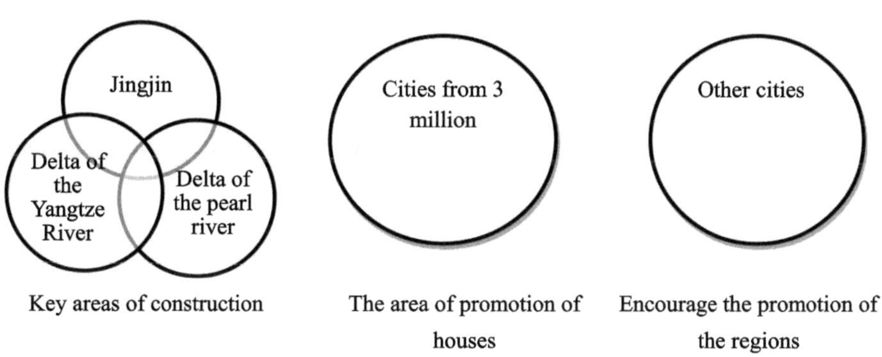

Fig. 2 Recommendation No. 71 of 2016, issued by the General Office of the State Council, addressed the development of rapid-construction buildings, including wooden structures. *Source* Developed by the authors

construction of new villages, as well as housing in seismic areas. Additionally, there is a need to construct multi-storey mixed-use wooden buildings in both urban and rural areas, as well as temporary facilities for ecotourism. This approach is especially relevant in rural regions, where most residents own single-storey houses and estates. Thus, reconstruction not only preserves existing land use patterns but also increases their efficiency.

China is actively developing wooden housing construction as one of the ways to reduce carbon emissions. A tree is a natural source of carbon, which is absorbed from the atmosphere during growth. Thus, the construction of wood is considered more environmentally friendly compared to traditional building materials (e.g., concrete and steel).

China actively investigates and introduces new technologies and methods of using wood in construction. Engineering solutions are being developed to increase the strength and durability of wooden structures, as well as special methods of wood processing for protection against fire and decay.

Thanks to the development of wooden housing construction in China, the construction industry seeks to reduce its impact on the environment and decrease GG emissions. Additionally, such houses can be more energy-efficient and economical in operation. At the beginning of 2021, the Chinese government announced plans to develop wooden housing construction in the country to reduce environmental pollution. To implement this project, China began actively purchasing wooden houses from Russia, where this type of construction has a long and successful history. Such projects support the development of the wooden housing industry in both countries and strengthen friendly relations between China and Russia. The fight against climate change is a global endeavor that necessitates international cooperation and the rigorous enforcement of global agreements. Every country must contribute to preserving the planet for future generations. Therefore, China is developing wooden housing construction to reduce carbon emissions.

In the early 1980s, China faced serious problems caused by the depletion of forest resources. These difficulties forced the government to take several measures to restrict the use of wooden building materials. Simultaneously, the state began to actively stimulate the transition to alternative materials, such as aluminum, steel, and plastic. The industrialization in the country contributed to the gradual replacement of traditional wooden structures by modern building materials, such as bricks, reinforced concrete, and steel. Wooden houses began to give way to new architectural solutions, marking the end of an era. Restrictive measures on the use of wood had pushed the development of modern wooden architecture into a dead end. However, the situation changed dramatically after China's accession to the World Trade Organization (WTO) (Web Platform SOHU 2019). The status of the use of wooden structures in construction has improved significantly. Over the past few years, interest in wooden construction has been revived against the backdrop of deteriorating environmental conditions and an increasing focus on energy conservation and low-carbon architecture. Simultaneously, wood production specifically tailored to the needs of the construction industry began to grow. This gave new momentum to the development of modern wooden structures that combine environmental sustainability with advanced technologies. Such buildings are expected to secure a strong position in China's real estate market.

On January 19, 2011, at 12:19 p.m. local time, the city of Anqing in Anhui Province experienced a 4.8-magnitude earthquake. The disaster affected three administrative districts with a combined population of about 15.25 million people. In Yancey City, 90 elderly residents were relocated to other nursing homes with a five-year living guarantee. The earthquake caused the collapse of four private houses and 11 apartment buildings, and damaged 8,964 additional structures. The total direct economic loss amounted to 36.245 million yuan, including 25.985 million in personal property losses and 10.2 million in damages to state social protection institutions.

In 2024, mainland China experienced 25 earthquakes of magnitude 5.0 or higher on the Richter scale, most of which struck Xinjiang Province. One of the most significant events occurred in the city of Ice, where a 7.1-magnitude earthquake caused tragic consequences: three people lost their lives, and more than 900 houses were destroyed (UN n.d.).

Fortunately, apart from the incident in the city of Ice, earthquakes in other regions (e.g., Sichuan, Anhui, Tibet, and Yunnan) did not result in casualties or major economic losses. However, the overall impact was still severe: 114,000 people were affected, three lives were lost, and 24,000 residents had to be urgently evacuated to safe zones. More than 900 buildings were completely destroyed; another 63,000 sustained varying degrees of damage. Direct economic losses exceeded 3.79 billion yuan (Encyclopedia by Baidu 2018). Wooden structures have unique characteristics of strength and stability. Due to its box-shaped form, a wooden house can evenly distribute the loads that occur during operation. The lightness and elasticity of the structure make it resistant to sharp blows and repeated vibrations. Therefore, even with strong earthquakes, seismic influences are minimized; the foundation of a wooden building prevents destruction. When the soil is displaced, the wooden house retains its shape due to the high elasticity of the material, not succumbing to destruction. An example of such resistance is seen in cases in the cities of Kobe (Japan) and Los Angeles (USA), where, during powerful earthquakes, wooden buildings were only slightly deformed but did not collapse. Even with the most intense seismic influences, when the building could have shifted slightly from its foundation, its structure remained intact. This phenomenon clearly demonstrates the seismic resistance and reliability of wooden buildings in extreme loads. Following the devastating earthquake in Kobe, the Japanese government decided to reconstruct civilian residences using North American technologies for wooden houses, which later led to the introduction of new JAS construction standards. Based on the above, it can be argued that wooden buildings are the most resistant to seismic activity among all existing types of structures. Additionally, the use of special protective coatings can prevent corrosion and extend the life of wooden elements, ensuring their durability and safety.

4 Conclusion

The economic effect of wooden housing construction is manifested in several key aspects:

1. Reducing construction costs: Wooden structures often require lower material and labor costs compared to buildings made of traditional materials such as concrete and brick. The lightness of wood simplifies both transportation and installation, further reducing overall construction expenses;
2. Energy efficiency: Houses made of wood have high thermal insulation properties, which reduces the need for heating in winter and air conditioning in the summer. This leads to energy saving and reducing utility costs;
3. Environmental stability: The use of wood instead of other materials helps reduce carbon dioxide emissions, as trees absorb CO_2 during their growth. Therefore, a wooden house can play a role in climate strategy while also providing long-term economic benefits;
4. The speed of construction: The construction of wooden houses is much faster than traditional building methods. This accelerates project completion and allows income to be generated from the building's use more quickly;
5. Resistance to natural disasters: The tree has high strength and flexibility, which makes wooden houses resistant to earthquakes and other natural phenomena. This reduces the risk of destruction of buildings and minimizes the cost of recovery after natural disasters;
6. Durability and maintainability: With proper care, wooden buildings can serve for a long time, requiring only periodic repairs and updating. This ensures the stable operation of the building without the need for a large investment;
7. Secondary processing market: Wood is easily processed and reused, creating additional opportunities for profit through the sale of production waste or recycled materials;
8. Raising the value of real estate: The environmental friendliness and energy efficiency of wooden houses make them attractive to buyers and tenants, which increases the market value of such objects.

These aspects together form a positive economic effect of wooden housing construction, making it advantageous for developers and end users.

The analysis shows that China has significant potential for the further development of wooden housing construction. The use of wood in residential and non-residential buildings offers several advantages, most notably ensuring environmental safety. Additionally, this approach reduces energy consumption, which is especially important given rising global energy prices. Another key benefit is that wooden structures lower the costs of addressing the consequences of natural disasters and technological accidents, a factor particularly relevant in regions with high seismic activity. Taken together, these advantages can positively influence the country's economic development by stabilizing financial indicators and enhancing the competitiveness of the Chinese economy in the international arena.

References

Chinese Government Network (2024) Acceleration of green transformation of economic and social development. Beijing ICP No. 05070218. Retrieved from https://www.gov.cn/zhengce/202408/Content_6967663.htm. Accessed 28 Jan 2025

Chinese Government Network (2025) The main situation with natural disasters in the country in 2024. Beijing ICP No. 05070218. Retrieved from https://www.gov.cn/lianbo/bumen/202501/content_6999765.htm. Accessed 28 Jan 2025

Encyclopedia by Baidu (2011) Earthquake in Ancin 1 19. Beijing ICP No. 030173. Retrieved from https://baike.baidu.com/item/1·19安庆地震/17190734. Accessed 28 Jan 2025

Encyclopedia by Baidu (2018) The state of development of modern wooden architecture of China. Beijing ICP No. 030173. Retrieved from https://tieba.baidu.com/p/5511402341. Accessed 28 Jan 2025

Encyclopedia by Baidu (2023) How many houses are there in the country? Retrieved from https://baijiahao.baidu.com/s?id=1758619529995240088&wfr=spider&for=pc. Accessed 28 Jan 2025

Encyclopedia by Baidu (2025) Global warming. Beijing ICP No. 30173. Retrieved from https://baike.baidu.com/item/全球气候变暖/1034504. Accessed 27 Jan 2025

General Office of Hubei Provincial People's Government (2024) Jingmen "builds blocks" to build 4 million square meters of green buildings. Hubei ICP No. 05011090. Retrieved from https://www.hubei.gov.cn/hbfb/szsm/202410/t20241029_5390824.shtml. Accessed 27 Jan 2025

Netease (2025) National party media information public platform. License B2-20090058. Retrieved from https://www.163.com/dy/article/jgbapa890530Mkmj.html. Accessed 27 Jan 2025

UN (n.d.) Climate change. Retrieved from https://www.un.org/en/global-issues/climate-change. Accessed 28 Jan 2025

Web Platform SOHU (2019) Review of the current state of modern wooden buildings in China. Beijing ICP No. 030367. Retrieved from https://www.sohu.com/a/289768647_822803. Accessed 27 Jan 2025

Yetyan H (2024) The study of carbon dioxide emissions in the construction industry. Beijing ICP No. 030173. Retrieved from https://baijiahao.baidu.com/s?id=1799628371417544070&WFR=Spider&For=PC. Accessed 27 Jan 2025

Zibao J (2016) After the industrial revolution, the contribution of China to global warming amounted to about 10%. Shanghai ICP No. 14003370. Retrieved from https://www.thepaper.cn/newsdetail_forward_1460091. Accessed 27 Jan 2025

The European Union's Food Security Experience in the Context of Globalization

Murat Zh. Abdiev, Tulkin Kh. Farmanov, Elmurat A. Obdunov, Aigul O. Okenova, and Bayana T. Mametova

Abstract By examining the experiences of EU nations in achieving integration and developing a unified food market, this research aims to identify theoretical and practical approaches for strengthening food security and managing food imports in the Kyrgyz Republic. Based on the analysis and evaluation of food security in the EU countries, the authors highlight their key advantages in planning and implementing food security initiatives, assess their level of development, and underscore the relevance of applying their experience to the Kyrgyz context. Within the broader framework of economic integration, the research focuses on economic mechanisms that ensure food security as its main subject and object. The findings of this research provide a solid theoretical and methodological foundation for advancing food security and offer practical recommendations for implementation. The analysis of food security in the Kyrgyz Republic is grounded in an in-depth examination of theoretical sources and the development of applied recommendations addressing existing challenges in this field. The proposed recommendations serve as a basis for further research on food security by graduate students, doctoral candidates, and other

M. Zh. Abdiev (✉) · B. T. Mametova
Osh Technological University Named After M.M. Adyshev, Osh, Kyrgyzstan
e-mail: mabdiev1977@mail.ru

B. T. Mametova
e-mail: bayana310597@list.ru

M. Zh. Abdiev
Fergana State Technical University, Fergana, Uzbekistan

T. Kh. Farmanov
Tashkent State University of Economics, Tashkent, Uzbekistan

Tashkent State Agrarian University, Tashkent, Uzbekistan

E. A. Obdunov
Osh State University, Osh, Kyrgyzstan
e-mail: obdunov@oshsu.kg

A. O. Okenova
Research Institute of Economics at the Scientific Research University "Kyrgyz Economic University Named After M. Ryskulbekov", Bishkek, Kyrgyzstan

A. B. Karbekova (ed.), *Sustainable Development of Rural Territories Based on Climate-Smart Technologies*, Climate Change Management,
https://doi.org/10.1007/978-3-032-20106-5_37

interested scholars. The research employs a range of research methods, including comparative analysis, economic and statistical techniques, systems analysis, and monographic approaches.

Keywords Food security · National food security · Agri-food sector · Economic integration · Foreign trade · Agri-food market · Support · Regulation · Reform

JEL Classification D20 · E23 · F15 · Q12 · R11 · R58

1 Introduction

Food security forms the foundation of a nation's overall security and serves as a key priority in addressing social and economic challenges. It is closely linked to global processes of globalization and integration, encompassing areas such as the development of a green economy, environmental sustainability, the production of organic products, and the promotion of public health.

The historical aspects of food security have been particularly significant since the twentieth century. However, since the dawn of human civilization, issues related to food production, quality, storage, accessibility, and safety have always been among the state's primary responsibilities. With population growth and the advancement of science and technology, global civilizations, national governments, and international organizations have increasingly focused on addressing the challenges associated with food security (Balabanov and Borisenko 2002; Momunov 2017).

During the economic crisis and the application of sanctions by other countries, the problem of ensuring food security has become one of the pressing national tasks. Examples of this include the spread of COVID-19, the conflict between Russia and Ukraine, and the situation in the Middle East. To address the related problems of food security, it is necessary to study and practically apply the experiences of foreign countries, considering the state's internal policies and the country's regional characteristics, as some developed countries have extensive experience in creating and improving mechanisms for ensuring food security (Abdiev 2020; Balabanov and Borisenko 2002).

In this research, the authors examined the experiences of EU countries in addressing food security issues. In developed EU countries, the state provides material and financial support in every possible way to achieve the efficiency and competitiveness of agriculture and the processing industry to solve food security problems. This support is enhanced in light of the state of the food market.

2 Methodology

The authors applied sources from Kyrgyz and foreign scholars, textbooks, teaching aids, monographs, and internet materials on food security issues to form the theoretical and methodological foundation of this research.

To identify potentially effective directions for developing food security, the research aims to examine the basic and advanced forms and methods of state support and regulation across interrelated sectors, particularly agriculture, processing, and service industries. It also seeks to analyze the main approaches used at the national level and systematize and generalize the experiences of other countries.

During the research, the authors used various methods, including monographic, comparative, statistical, economic, and mathematical, as well as calculation and design methods.

3 Results

Political, environmental, social, and economic conditions and factors are considered when evaluating food security indicators. These factors are identified according to the extent of their influence on solving food security problems. Issues related to ensuring food security are determined primarily by state policies aimed at providing the population with food products, mechanisms of regulation and support, as well as by natural and climatic conditions (Abdiev et al. 2024, 2023).

According to statistical data, the criteria and indicators of food security in Western countries are higher than in other regions. Based on the results of this research, special publications are being prepared (FAO, Ifad, and WFP 2014, 2015).

Countries with high ratings indicate a positive assessment of food security, which is largely the result of state support and regulation of the food market (Table 1).

The EU experience represents one of the most successful examples of coordinated decision-making and implementation of a common agricultural policy aimed at maintaining food security The Economist Intelligence Unit (The Economist Intelligence Unit (EIU) 2023). The EU's effective food security policy has been achieved through the intensive development of the agricultural sector, increasing its efficiency and competitiveness, and ensuring dynamic growth. This success is also attributed to the integration of agricultural processing, logistics, and inter-industry economic sectors, as well as the digitalization of production processes, the introduction and widespread adoption of modern innovative technologies, and the attraction of domestic and foreign investments.

The EU countries solve the problems mentioned above by implementing and carrying out the Common Agricultural Policy of the EU, which was established in 1957 in accordance with the Treaty of Rome. The primary objective of the Common Agricultural Policy (CAP) is to support rural producers in achieving stable development and maximizing their income, while ensuring that the population has guaranteed

Table 1 Ranking of countries by level of food security in the world

Position in the ranking	Country rating	Overall rating	Economic availability of food resources	Physical availability of food for the population	Quality and safety of food resources	Sustainable development and adaptation of agriculture
1	Finland	83.7	91.9	70.5	88.4	82.6
2	Ireland	81.7	92.6	70.5	86.1	75.1
3	Norway	80.5	87.2	60.4	86.8	87.4
4	France	80.2	91.3	69	87.7	70.3
6	Japan	79.5	89.8	81.2	77.4	66.1
7	Canada	79.1	88.3	75.7	89.5	60.1
9	UK	78.8	91.5	71.6	77.6	71.1
13	USA	78	87.1	65.1	88.8	69.4
19	Germany	77	87.9	67	79.9	70.8
25	China	74.2	86.4	79.2	72	54.5
32	Kazakhstan	72.1	78	67.2	76.3	65.4
43	Russia	69.1	77.8	61.4	78.7	56.6
55	Belarus	64.5	67.8	61.9	69	58.5
68	India Algeria	58.9	59.3	62.3	62.1	51.2
71	Ukraine	57.9	66.6	48.1	71.3	43.5
Next come the countries of the African continent						

Source Developed by the authors based on The Economist Intelligence Unit's Global Food Security Index 2023 The Economist Intelligence Unit (The Economist Intelligence Unit (EIU) 2023)

access to high-quality food products. By employing subsidy mechanisms and stabilizing food prices, including setting minimum price levels, the CAP seeks to enhance food security and improve the efficiency of agricultural production within the EU. This unified policy aims to enhance the efficiency and competitiveness of agricultural production, while ensuring food security through subsidies for rural producers and the regulation of food price standards.

In EU countries, agricultural exports account for 18% of global trade turnover. The Common Agri-Food Policy serves as an effective mechanism for promoting the sale of agricultural products and ensuring food security. The European Union is the world's third-largest exporter of grain and poultry. Moreover, it is the second-largest exporter of dairy products and pork (European Commission 2011; Uzenbaev et al. 2019).

Creating advantageous conditions for domestic farmers is one of the tasks of the unification process in the agriculture of the EU countries.

The EU countries strive to sell their products at a low cost compared to other developed countries, sometimes by 35% or more, thanks to the adoption of a coordinated agricultural policy. The united strategy guarantees that buying agricultural products from EU farmers will always take precedence over imports. Reducing agricultural product overproduction and assisting small farmers have been the main focuses of this program in recent years (Dzhumabaev 2004; Kupuev and Obdunov 2009; Musakozhoev et al. 2016).

During the analyzed period, EU countries achieved significant success through the creation of value chains, deeper integration, and the establishment of cooperatives and clusters, developments that should be taken into consideration. In the 1980s, many unresolved issues existed regarding the supply of food products. At that time, due to the low efficiency of the food industry, shortages of manufactured food affected between 8 and 23% of the total population. With the proper organization and implementation of the CAP within the EU, a single food market was established, effectively addressing supply and export challenges and contributing to the stable development of the industry.

It was possible to achieve self-sufficiency in all socially significant food products within the EU, even though the agricultural sector's share of national economies in the EU countries has drastically decreased to around 1.4% in 2023 due to a sharp increase in its efficiency and competitiveness. Wheat accounts for 123.0% of self-sufficiency, rice for 92.0%, vegetable oils for 136.0%, sugar for 111.0%, potatoes for 108.0%, milk for 106.0%, meat for 100.-%, and eggs for 100.0%. These figures are significantly higher compared to those of the EAEU countries, particularly from Russia and Belarus, by 24.3%, 2.7%, 34.0%, 43.0%, 39.0%, and 36.0%, respectively (Krylatykh 2015). The policies and measures implemented to ensure food supply in EU countries are primarily aimed at achieving stability and dynamic growth in agricultural production, promoting expanded reproduction, and maintaining the steady development of the food market.

It is particularly noteworthy that pricing remains a significant issue in EU countries, as national agricultural markets operate under a unified principle and strive to maintain a consistent price level for agricultural products. This is due to the fact that the transition to a common pricing system required significant harmonization of national agricultural policies among EU countries. In this context, the following fundamental principles were established as the foundation for the implementation of the CAP within the EU:

- Improving the market for the production, sale, and consumption of agricultural and processed goods;
- Replacing imported products with domestic ones;
- Ensuring the provision of necessary financial resources, including credit, leasing, and the rational use of working capital.

Prior to the establishment of the EU, each European country had its own set of regulations for managing agricultural supply and demand, with the goal of ensuring the long-term development of agriculture. Within the scope of the single agricultural market, these laws were replaced with a single European system for managing supply and demand, which eliminated price disparities.

In EU countries, fixed prices are applied in the food market to maintain price stability in transactions. Simultaneously, minimum prices are established for agricultural products. These processes are managed, planned, and continuously monitored by the relevant EU authorities.

Various types of pricing mechanisms are used in the EU food market. Planned prices are applied to support and further improve food markets, enabling the purchase or sale of surplus agricultural products as needed. Foreign trade (dumping) prices are used to stimulate agricultural exports and promote goods in foreign markets. The adoption of universal pricing regulation standards has further ensured the stability of farmers' incomes.

A central principle of the CAP is the strategy of import substitution and the preferential consumption of food products produced within EU countries in their domestic markets. Simultaneously, the responsible EU bodies manage the reduction of food imports, support exports, establish quotas, provide loans, enforce standards, and certify products through organizational and economic measures, thus ensuring the stability of the agri-food market.

The EU has a system of support and protection for farmers from competitors who have cheaper prices for their products. For this purpose, the compensation price is used in accordance with EU regulations and directives. Those economic entities involved in the export of products are helped with additional types of assistance to obtain profit and solve socio-economic problems. EU bodies also conduct direct negotiations with supplying organizations so that they comply with the price according to the standards established in the EU countries.

Additionally, the significance of collective solidarity and financial support for rural producers is growing, particularly in light of the heightened competition in the global agro-food markets. By implementing the CAP, all EU countries are required to provide equitable financial assistance for agricultural policy. Within the parameters of the common agricultural strategy that the CAP is pursuing, financial assistance for agriculture is provided in two ways. One system of payments per unit of farmed area is part of the first direction, which regulates assistance measures within the single market framework. The requirement to adhere to regulations to guarantee food and environmental security determines the extent of public assistance.

More than half of the EU budget goes to the European Agricultural Guarantee Fund (EAGF), which provides this support. Another important direction of state support includes measures related to increasing profitability, achieving efficiency in the agricultural sector, conquering the market, and, as a result, raising the standard of living in rural areas. The examples of this include export assistance, additional payments for low-income rural areas, and restrictions on the import of food.

Increasing agricultural production and achieving food self-sufficiency are the primary goals of the protectionist policies outlined in the EU CAP. Favorable economic conditions are created for the industry's competitive development by the instruments of coordinated regulation and development support that have been put in place. Within the CAP framework, significant attention is given to promoting high-value agricultural production, the food industry, and logistics, which together contribute to the creation of final added value. At the same time, a separate component has been established to support larger economic entities, facilitating the development of agricultural cooperation and the consolidation of farms.

The EU places significant emphasis on providing financial support to farmers in its member states, both at the national level and within the framework of international initiatives. Currently, the EU allocates approximately 55 billion euros annually to the implementation of agricultural and food policies, which represents about 35% of the total EU budget. However, the agricultural sector accounts for only about 1.5% of the EU's gross domestic product. To maintain stable prices and incomes, a significant portion of this funding (around 78%) is distributed as subsidies to agricultural producers.

The remaining amount of this capital is allocated for the development of rural areas. In general, within the structure of agricultural production in the EU, 37% of funds are allocated as assistance to rural producers, and $850 is invested per hectare to improve agricultural land (FAO, Ifad, and WFP 2015; Krylatykh 2014).

In EU countries, financial assistance is provided through the consolidated EU budget and the national budgets of individual states. Together, these two funding sources form a supranational level of financial support. The departments operating under the EU Fund possess substantial financial resources aimed at addressing issues within the agricultural sector, ensuring food supply, supporting the food market, and providing loans and subsidies. Their activities directly influence the stability and dynamism of the industry, further reinforcing the implementation of the CAP.

In the EU, financial assistance is provided from several sources: direct support through the budgets of EU member states (0.78% of each country's budget), 0.80% of each country's value-added tax, and import duties and customs fees on products imported into the EU. The amount of these duties ranges from 1.20% to 1.27% of the gross domestic product of EU countries.

In EU countries, the agricultural sector receives significant support at the state level. State assistance for the agro-food sector in the EU depends on each country's financial capacity, budget, and domestic policy priorities. Notably, in countries such as the Netherlands, Finland, France, and Germany, the level of financial support provided to agriculture is considerably higher than in other EU member states, as the sector is regarded as a national economic priority and plays a vital role in promoting economic growth.

Within the framework of this policy, the following priority areas for development should be considered based on the study of EU countries' experiences in ensuring food security:

- Ensuring food independence through the diversification and dynamic development of the domestic agro-food sector and its key industries, grounded in innovative modernization, as well as promoting the comprehensive development of rural areas;

- Developing and implementing a national program for import substitution aimed at maximizing specialization in agricultural production, moving away from inefficient models focused solely on grain independence, and reviving and accelerating the growth of traditional industries such as animal husbandry and the cultivation of industrial crops (e.g., sugar beet, raw cotton, etc.);
- Reforming state monetary policy to significantly increase targeted financial assistance for the agricultural sector to correct existing fiscal imbalances. To ensure sustainable economic growth, the agricultural sector should also prioritize the planning and implementation of major import-substitution projects.

In recent years, businesses operating in a competitive environment have expanded production as a result of improved management practices and the modernization of manufacturing processes, with a strong focus on advanced technologies in the country's food industry. The gradual increase in the market share of their products abroad has contributed to a reduction in import volumes. Examples of such enterprises include "Bishkek-Sut" OJSC, "Coca-Cola Bishkek Bottlers" CJSC, "Riha" LLC, "Bear Beer" CJSC, "Kulikovsky Confectionary House," "Shoro," and others.

In addition to improving product quality and creating new employment opportunities, the establishment of food industry clusters provides the Kyrgyz Republic with specific trade advantages through its membership in the WTO, EAEU, and CIS. Furthermore, increased exports of processed goods and the adoption of new technical standards for product quality and processing can help expand the Kyrgyz Republic's presence in international markets.

4 Conclusion

The results of this research indicate that foreign experience in ensuring food security demonstrates that, in EU countries, great attention is given to state support and regulation of agriculture, the implementation of the CAP, the development of the processing industry, the integration of components within the agro-industrial sector, and the enhancement of cooperation among related industries. This approach is considered essential for the effective implementation of a unified agricultural and food policy across EU countries. All forms of material and financial support for farmers are provided through both national and supranational budgets.

The level of financial assistance depends directly on the condition of national and municipal budgets. In the least developed member states, the amount of government support is significantly influenced by supranational financial resources, while in more industrialized countries with greater fiscal capacity, domestic agriculture receives stronger support from national funds rather than supranational ones.

As the research further reveals, in the context of globalization, food security at the national, regional, household, and individual levels directly depends on the overall level of food security in the global community, as it affects such fundamental needs as nutrition, national stability, and personal well-being. Food shortages can lead to social tensions, deteriorating relationships, conflicts, and humanitarian crises.

References

Abdiev MZ, Toktorov KK, Abdrakhmanov OE, Momunov UN, Ergeshov KA (2024) Ways to improve the innovative development of Kyrgyz agriculture. In: Popkova EG, Bogoviz AV, Sergi BS, Kaurova OV, Maloletko AN (eds) Sustainable development of the agrarian economy based on digital technologies and smart innovations. Springer, Cham, Switzerland, pp 51–54. https://doi.org/10.1007/978-3-031-51272-8_9

Abdiev MZ (2020) Improving the competitiveness of the food industry of the Kyrgyz Republic in the context of economic integration (Synopsis of dissertation of doctor of economic sciences). Kyrgyz Economic University named after M. Ryskulbekov, Bishkek, Kyrgyzstan

Abdiev MZ, Abdrakhmanov OE, Zulpukarova T (2023) Criteria and indicators of food security of the Kyrgyz Republic. Economy and Business: Theory and Practice 2(96):7–10. https://doi.org/10.24412/2411-0450-2023-2-7-10

Balabanov VS, Borisenko EN (2002) Food security: international and internal aspects. URSS, Moscow, Russia.

Dzhumabaev KZ (2004) Economic problems of formation of the food market in the Kyrgyz Republic (Synopsis of Dissertation of Doctor of Economics). Almaty, Kazakhstan

European Commission (2011) Agriculture in the EU: statistical and economic information report 2010. Publications Office of the European Union, Luxemburg. Retrieved from https://op.europa.eu/en/publication-detail/-/publication/bfb03c92-1e93-436d-8bf4-46360a1f2b25/language-en. Accessed 23 Sept 2025

FAO, IFAD, and WFP (2014) The state of food insecurity in the world 2014. Strengthening the enabling environment for food security and nutrition. FAO, Rome, Italy. Retrieved from http://www.fao.org/3/a-i4030e.pdf. Accessed 23 Sept 2025

FAO, IFAD, and WFP (2015) The state of food insecurity in the world 2015. Meeting the 2015 international hunger targets: taking stock of uneven progress. FAO, Rome, Italy. Retrieved from https://openknowledge.fao.org/server/api/core/bitstreams/63863832-4cb5-4e05-9040-4b22d9a92324/content. Accessed 23 Sept 2025

Krylatykh EN (2014) Concepts and methodological foundations of studying food security. In: Theoretical and methodological problems of measuring, forecasting and managing food security. VIAPI named after A. A. Nikonov, Moscow, Russia, pp 3–5

Krylatykh EN (ed) (2015) Agrarian Europe in the 21st century. Summer Garden, Moscow, Russia

Kupuev PK, Obdunov E (2009) Food security: realities and prospects in the Kyrgyz Republic. Ilim, Bishkek, Kyrgyzstan

Momunov UN (2017) Mechanisms for ensuring food security in the context of deepening integration (based on the materials of the Kyrgyz Republic) (Synopsis of dissertation of candidate of economic sciences). Osh State University, Osh, Kyrgyzstan

Musakozhoev SM, Kamchybekov TK, Japarov AU (2016) Economy of Kyrgyzstan for 25 years: problems and development prospects. Publishing House Turar, Bishkek, Kyrgyzstan

The Economist Intelligence Unit (EIU) (2023) Global food security index 2023 (unpublished). EIU, London, UK

Uzenbaev RA, Mardalieva LA, Abdiev MZ, Umarov ST, Ergeshov KA (2019) Prospects for development of Kyrgyzstan's food market in the conditions of integration in the EAEU. In: Popkova EG (ed) Ubiquitous computing and the internet of things: prerequisites for the development of ICT. Springer, Cham, Switzerland, pp 859–869. https://doi.org/10.1007/978-3-030-13397-9_88

The System of State Regulation of the Insurance Market and Its Role in the Agrarian and Credit Policy of Kyrgyzstan

Elnura I. Murzalieva, Umut M. Omurzakova, Gulnara T. Pulatova, Ainura K. Askarova, and Gulchehra B. Abdyrakhmanova

Abstract Drawing upon a systemic-functional lens, this research examines state governance of the insurance market. It outlines critical features that constitute the framework for governmental administration within the sector, alongside fundamental guidelines shaping its operation. The research aims to dissect the nuances of state regulation through this perspective, determine constituent elements and underlying principles, and clarify the function of governmental entities in maintaining an efficient national insurance system and harmonizing the objectives of all involved parties. The research utilized a systemic-functional approach, which allowed for the examination of state regulation as a holistic system. Moreover, the authors employed a comparative legal examination to contrast how doctrines and legislation define "state regulation of the insurance market and the system of state regulation of the insurance market. The analytical method was used to examine the functions and powers of the Financial Market Regulation and Supervision Service of the Kyrgyz Republic and their impact on the development of the insurance sector. The research clarifies the notion and organization of state regulation within the insurance market, pinpointing its vital features and outlining primary operational standards.

E. I. Murzalieva · U. M. Omurzakova
Kyrgyz State Technical University Named After I. Razzakov, Bishkek, Kyrgyzstan
e-mail: elnura.murzalieva@kstu.kg

U. M. Omurzakova
e-mail: umut.abdyldaeva@kstu.kg

G. T. Pulatova
Education-Research and Industrial, Complex International University Named After Kantoro Sharipovich Toktomamatov, Manas, Kyrgyzstan
e-mail: gulnarapulatova.74@mail.ru

A. K. Askarova (✉) · G. B. Abdyrakhmanova
Jalal-Abad State University Named After B. Osmonov, Manas, Kyrgyzstan
e-mail: aynura.7474@mail.ru

G. B. Abdyrakhmanova
e-mail: Gulchehra1860@mail.ru

A. B. Karbekova (ed.), *Sustainable Development of Rural Territories Based on Climate-Smart Technologies*, Climate Change Management,
https://doi.org/10.1007/978-3-032-20106-5_38

Keywords Insurance · Insurance market · Social structure · Regulation · Development · Agricultural policy · Credit system

JEL Classification G22 · O10 · Q14 · R10

1 Introduction

Digitalization of Kyrgyz insurance hinges on rapid ICT adoption and digital environment development. Kyrgyz insurers adopt digital solutions, aligning with global trends and customer needs to boost efficiency.

Public and entrepreneurial financial literacy, effective use of insurance strategies for managing risk and capital, and supportive government policies also play major roles in shaping outcomes. Regulatory authorities must remain adaptable as conditions evolve. For example, in an age of increasing globalization and expanding cross-border trade, regulators need to be prepared to respond quickly to shifts and emerging risks originating outside their own markets. This readiness requires continuous monitoring and analysis of market trends, as well as ongoing communication and coordination among regulatory agencies and stakeholders across the insurance industry.

The need to study this topic arises from the requirement for a comprehensive understanding of the mechanisms of state regulation of the insurance market and their interaction with digitalization processes. Such understanding makes it possible to develop effective models for the development of the insurance sector adapted to the context of the Kyrgyz Republic, where digital infrastructure and insurance culture are still in the active stage of formation. The study of this issue is significant for formulating state policy and improving the professional competencies of specialists in the financial and insurance industries, enhancing supervisory mechanisms, creating digital platforms, and strengthening public trust in insurance institutions.

Thus, research on state regulation and digitalization of the insurance market represents a critically important area for developing strategic decisions aimed at the sustainable growth of the national financial system, boosting investment activity, and increasing the competitiveness of the national economy.

The system of state regulation of the insurance market in Kyrgyzstan plays a key role in maintaining the stability of the national economy and ensuring the effective functioning of the agricultural and credit sectors. In the agricultural sector, government oversight of insurance activities supports rural development by providing mechanisms for risk mitigation associated with natural disasters, climate fluctuations, and market instability. Through subsidized agricultural insurance programs, the state helps farmers safeguard their incomes and maintain production continuity even under adverse circumstances. This approach enhances food security and promotes sustainable development in rural areas by integrating insurance into the broader framework of agrarian policy.

In the credit sector, the system of state-regulated insurance increases the reliability of financial institutions and expands access to credit resources. Insurance of loans and financial guarantees reduces the risk of default for banks and credit unions, encouraging them to provide financing to small and medium-sized enterprises and agricultural producers. This creates a multiplier effect: insured credit risk stimulates investment activity and economic growth while maintaining overall financial stability.

Thus, state regulation of the insurance market in Kyrgyzstan operates as a strategic component of agrarian and credit policies. It forms a controlled and transparent environment for managing financial risks, balancing the interests of the state, businesses, and citizens, and supporting the sustainable development of the national socio-economic system.

2 Materials and Methods

According to the Dictionary of Financial and Economic Terms, compiled by Professor M. A. Eskindarov, state regulation is defined as government actions intended to guide the overall economy and private enterprise toward equitable societal progress. It aims to avert or lessen the impact of economic and social disruptions while harmonizing the needs of diverse communities. Such oversight utilizes methods including direct intervention via legislation and enforcement, fiscal levers through taxation, and adjustments in monetary policy to manage capital movement (Sharkova et al. 2015).

Yu. D. Subbotina defines economic governance as guiding business practices and impacting market forces, aiming to bolster the overall market while addressing ecological and social challenges (Subbotina 2022).

According to A. G. Karandaev, state regulation involves governmental actions designed to structure social interactions, encouraging, aiding, and safeguarding participants within them (Karandaev 2023).

According to A. S. Markov, financial sector supervision comprises the regulations and laws that entities like banks, credit unions, insurers, investment brokers, asset managers, and others in finance must comply with. Moreover, Markov emphasizes that financial governance is not simply about having those regulations in place; it additionally demands ongoing monitoring to confirm their observance (Markov 2024).

In the context of the insurance sector, some scholars define state regulation of insurance activities as the establishment of framework conditions by the state that allow the participants in the insurance market to make their own decisions within these boundaries (Polyakova et al. 2018).

From the perspective of G. A. Nasyrova, state regulation plays a critical role in ensuring economic security, particularly in insurance (Nasyrova 2016).

In her doctoral dissertation, E. L. Prokopyeva emphasizes that state regulation is one of the factors affecting insurance because it serves as the primary mechanism for managing insurance risk. This regulation is a constraining (due to constant changes in legislation) and stimulating factor (through strengthened state oversight) (Askarova et al. 2024; Prokopyeva 2021).

Building upon the foundational perspectives of system theory elaborated by V. N. Sadovsky and E. G. Yudin, Russell L. Ackoff, A. Tsyganov, and N. P. Sakhirova, it becomes evident that the insurance market, as an institution, functions as a complex, self-organizing system characterized by hierarchical interconnections and dynamic interactions among its elements. According to V. N. Sadovsky and E. G. Yudin, a system represents an integrated totality whose components are interrelated and simultaneously form part of broader, more complex frameworks. This notion underscores the interdependence of the insurance market with the national economy, illustrating that regulatory and market mechanisms operate within a multilayered structure where macroeconomic, institutional, and social subsystems interact within a unified whole (Sadovsky and Yudin 1969).

In line with Russell L. Ackoff's conception, a system is not merely a set of elements but an organized entity defined by the relationships that exist among its parts. Russell L. Ackoff posits that the significance of each element emerges only within the context of the system's overall purpose and structure, emphasizing feedback, adaptability, and strategic coherence as critical attributes. Translating this logic to the insurance market, regulation emerges as a mechanism of systemic alignment that supports the adaptive balance between state oversight, business autonomy, and consumer interests (Ackoff 2002).

Further extending this theoretical framework, A. A. Tsyganov and N. P. Sakhirova highlight the institutional nature of insurance markets, emphasizing their structured yet evolving character. They argue that the system's stability depends on the consistency and organization of interactions among subsystems such as regulatory bodies, insurers, and consumers. Within this construct, each subsystem maintains its own operational objectives, which occasionally diverge from the system's collective purpose but ultimately contribute to its broader equilibrium. Their institutional approach demonstrates that regulatory reform, digitalization, and innovation in the insurance sector are not isolated phenomena but elements of systemic evolution aimed at achieving resilience, efficiency, and sustainable growth (Tsyganov and Sakhirova 2013).

Thus, synthesizing the theoretical insights of V. N. Sadovsky and E. G. Yudin, Russell L. Ackoff, A. Tsyganov, and N. P. Sakhirova, state regulation of the insurance market can be conceptualized as a complex system of interconnected agents and mechanisms functioning within a dynamically adaptive institutional environment.

3 Results

The authors believe that having a shared objective, along with the willingness of the system's various components to work toward that objective, is a defining feature of how the government oversees the insurance sector, shaping agricultural and lending practices. In this context, the central goal is to safeguard the security and long-term sustainability of the national socio-economic system by maintaining balanced market conditions and protecting the interests of private businesses (e.g., insurers), consumers of insurance services (e.g., policyholders), and society and the state. Based on these considerations, the system of state regulation of the insurance market in the Kyrgyz Republic can be schematically illustrated as follows (Fig. 1).

Additionally, the advanced training of government officials overseeing insurance activities plays a key role, especially in the context of increasing digitalization. These tools collectively enable the state to regulate the insurance market efficiently, ensuring transparency in the insurance process and maintaining market stability.

This framework is part of the broader state influence over the insurance market and its participants. To safeguard the stability and strength of the Kyrgyz economy and society, the authors strive for an equitable market environment and the protection of key stakeholders, such as businesses offering insurance, people relying on it, and the governing body.

The most important principle is systematization, which implies a comprehensive approach to regulating insurance activities, considering the interconnections

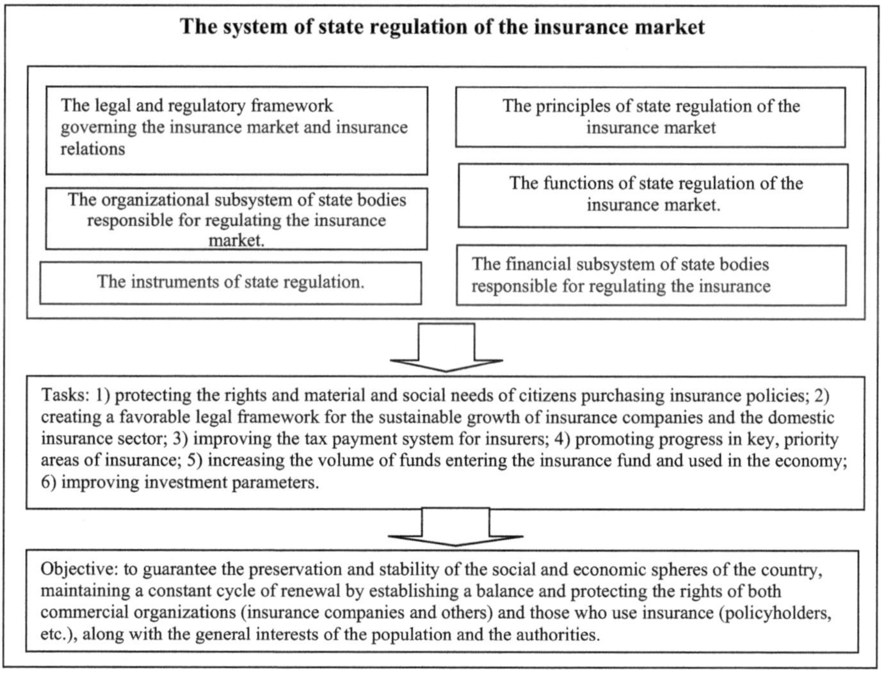

Fig. 1 The framework for government supervision concerning the insurance industry in Kyrgyzstan. *Source* Developed by the authors based on (Cabinet of Ministers of the Kyrgyz Republic 2021; Republic 1998, 2021; Abdyrahmanova 2019; Askarova et al. 2021; Kochumova and Mukhammetgeldieva 2024; Sayakbaeva et al. 2019; Sergeeva and Zhukovsky 2024; Tsyganov and Sakhirova 2013; Yurgens and Nichiporuk 2017)

between financial and socio-economic processes. State influence is directed toward maintaining a balance between control and business freedom, which is expressed in the principle of proportionality: supervisory measures must correspond to the scale and nature of the risks specific to each insurance organization.

The principle of transparency requires the openness of information regarding the financial condition of insurance companies, licensing procedures, and reporting, thereby ensuring trust among consumers and market participants. Another crucial principle is fair competition, which promotes innovation and improves the quality of insurance services (Fig. 2).

Additionally, under digitalization, the principles of adaptability and technological integration are gaining increasing importance, emphasizing the need for continuous improvement of regulatory tools, implementation of digital technologies (InsurTech and SupTech), and the use of analytical data for risk forecasting.

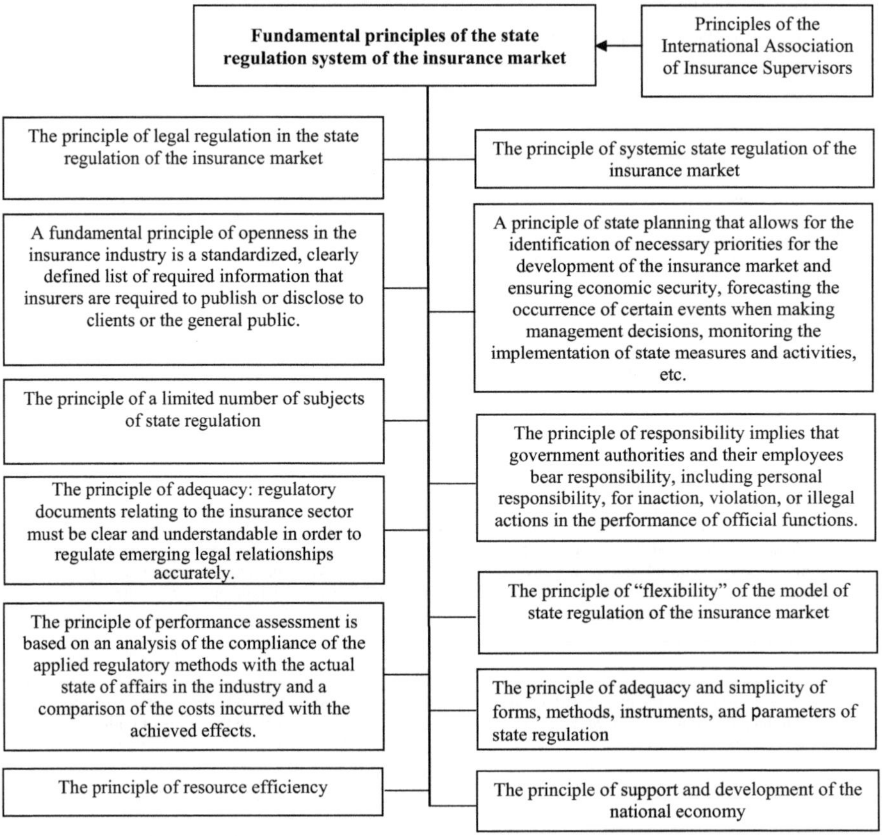

Fig. 2 Fundamental principles guiding state regulation in the insurance sector. *Source* Developed by the authors based on (Askarova et al. 2022; International Association of Insurance Supervisors (IAIS) 2019; Karandaev 2023; Markov 2024; Nasyrova 2016; Shkolik 2024; Subbotina 2022)

Thus, the principles of insurance market regulation form the foundation for the sustainable development of the industry, ensure financial security for participants, and provide effective protection of the economic interests of citizens and the state. State regulation and digitalization of Kyrgyzstan's insurance market represent interconnected elements of a unified system that ensures the country's financial stability and competitiveness. Innovations in insurance based on the adoption of digital solutions enable domestic insurers to improve efficiency and align with international trends. Simultaneously, key factors for successful development include the financial literacy of the population, entrepreneurial skills in risk management, and consistent government policy.

The activities of state supervisory bodies in the insurance sector require adaptability and continuous monitoring of global economic processes. In the context of international integration, the government must ensure a prompt response to external challenges, strengthen coordination among market participants, and enhance the transparency of insurance operations through digital control mechanisms. Improving the professional training of officials responsible for oversight becomes a crucial condition for effective governance.

Thus, state regulation of the insurance sector is a purposeful activity aimed at balancing the interests of the state, insurance companies, and service consumers, as well as maintaining the stability and security of the national socio-economic system.

4 Conclusion

Guided by foundational principles, state governance of the insurance sector largely assumes the shape of financial law. This feature is intimately connected with the specific nature of the insurance industry and the wide array of monetary tools forming an evolving framework of governmental oversight within the insurance marketplace, enabling authorities to impact this economic area. Through these instruments of state financial regulation, the key financial indicators of insurers (such as capital size, insurance reserves, profits, etc.) are controlled, along with the determination of the general directions and parameters of their investment activities, reinsurance policies, and relationships with policyholders in the event of insolvency of insurance companies (Askarova et al. 2021, 2024).

State regulation and digitalization of the insurance market represent complementary directions that ensure the stability, transparency, and efficiency of the insurance system. Within the framework of state regulation, legal, fiscal, financial, and institutional mechanisms are used to protect the interests of policyholders, maintain the solvency of insurance companies, and prevent systemic risks. These measures create the legal and economic foundation for the functioning of the industry, ensuring a balance between the interests of the state, business, and society.

Digitalization of the insurance market enhances the effectiveness of traditional regulatory tools by introducing new forms of interaction between market participants and supervisory authorities. The use of digital platforms, artificial intelligence, big data, and blockchain enables the automation of licensing, reporting, and control processes, makes insurance operations more transparent, and reduces fraud risks. The development of electronic supervisory systems provides for real-time analysis of market data and allows for a quicker response to potential threats to its stability.

Furthermore, digital transformation contributes to the creation of an open insurance environment where government agencies and insurance companies interact based on data exchange and transparent standards. As a result, a modern ecosystem of insurance regulation is formed, in which supervision becomes proactive and technologically advanced, while the insurance market gains additional opportunities for innovation, enhanced competitiveness, and strengthened consumer trust.

References

Abdyrahmanova GB (2019) Supervision and government regulation in insurance of Kyrgyzstan. Eurasian Sci Union 6–4(52):223–226

Ackoff RL (2002) Ackoff on management (Volkova LA, Transl. from English). Piter, St. Petersburg, Russia

Askarova AK, Orozova ChS, Abdeshimova GA (2021) Digitalization process in the Kyrgyz Republic. J Vector Econ 3(57):47

Askarova AK, Zholboldueva DS, Murzalieva EI (2022) Assessment of the implementation of projects for the digitalization of the economy of the Kyrgyz Republic. In: Shuvalov VP (ed) Infocommunication technologies: current issues of the digital economy. Collection of scientific papers of the II international scientific and practical conference. Yekaterinburg, Russia, pp 192–195

Askarova AK, Abdyldaeva UM, Murzalieva EI, Zholboldueva DS, Abdyrakhmanova GB (2024) Current trends in the development of the financial and credit system in the context of digitalization. In Sergi BS, Popkova EG, Ostrovskaya AA, Chursin AA, Ragulina YV (eds) Ecological footprint of the modern economy and the ways to reduce it. Springer, Cham, Switzerland, pp 213–217. https://doi.org/10.1007/978-3-031-49711-7_37

Cabinet of Ministers of the Kyrgyz Republic (2021) Decree on the issues of subordinate divisions and organizations of the ministry of economy and commerce of the Kyrgyz Republic (December 10, 2021 No. 300). Bishkek, Kyrgyzstan. Retrieved from https://cbd.minjust.gov.kg/158757/edition/6054/ru. Accessed 4 Nov 2025

International Association of Insurance Supervisors (IAIS) (2019) Insurance core principles and common framework for the supervision of internationally active insurance groups. IAIS, Basel, Switzerland. Retrieved from https://www.iaisweb.org/uploads/2022/01/191115-IAIS-ICPs-and-ComFrame-adopted-in-November-2019.pdf. Accessed 4 Nov 2025

Karandaev AG (2023) Comparative legal study of the category of state regulation in the legislation of Russia and Kazakhstan. Legal Sci 10:101–105. Retrieved from https://tinyurl.com/y3azreth. Accessed 4 Nov 2025

Kochumova AG, Mukhammetgeldieva T (2024) State regulation of insurance and its need for development of insurance sector. Int Sci J Sci Herald 9(78):20–23. Retrieved from https://tinyurl.com/muk7v94h. Accessed 4 Nov 2025)

Kyrgyz Republic (1998) Law on the organization of insurance in the Kyrgyz Republic (July 23, 1998 No. 96). Bishkek, Kyrgyzstan. Retrieved from https://cbd.minjust.gov.kg/105/edition/1179456/ru. Accessed 4 Nov 2025

Kyrgyz Republic (2021) Constitutional law on the cabinet of ministers of the Kyrgyz Republic (October 11, 2021 No. 122). Bishkek, Kyrgyzstan. Retrieved from http://cbd.minjust.gov.kg/act/view/ru-ru/112301. Accessed 4 Nov 2025

Markov AS (2024) State regulation of the financial market: problems and ways to increase efficiency. Mod Sci: Actual Probl Theor Pract. Ser: Econ Law 3:42–47. Retrieved from https://tinyurl.com/2rwd88sh. Accessed 4 Nov 2025

Nasyrova GA (2016) Institutional support of the insurance activity regulation system (Synopsis of dissertation of doctor of economic sciences). St. Petersburg State University of Economics, St. Petersburg, Russia

Polyakova AA, Kozhanchikova NY, Dudareva AB (2018) Modern trends in the functioning of the insurance market in Russia. Manage Consult 10:70–83

Prokopyeva EL (2021) Regional insurance market and its influence on the socio-economic development of the region (Dissertation of doctor of economic sciences). Abakan, Russia: Khakass Technical Institute—Branch of Siberian Federal University

Sadovsky VN, Yudin EG (eds) (1969) Studies in general systems theory: a collection of translations. Progress Publishing House, Moscow, USSR

Sayakbaeva AA, Shamyratova NS, Sartova RB (2019) Some aspects of the organizational and legal regulation of the insurance business in Kyrgyzstan. Eurasian Scientific Union 6–4(52):273–276

Sergeeva IA, Zhukovsky AV (2024) Application of the concept of a common insurance space in the EAEU: analysis of the regulatory framework and economic effects. Herald Bashkir Inst Soc Technol 1(62):33–42. https://doi.org/10.47598/2078-9025-2024-1-62-33-42

Sharkova AV, Kilyachkov AA, Markina EV, Soliannikova SP, Chaldaeva LA (2015) Financial and economic terms dictionary. Dashkov and Co, Moscow, Russia

Shkolik OA (2024) The main aspects of public law regulation of insurance in the Central European states. Legal Sci 6:82–87. Retrieved from https://tinyurl.com/y4jjkuxv. Accessed 4 Nov 2025

Subbotina YD (2022) State regulation of the national economy: educational and methodological guide. Perm State National Research University, Perm, Russia

Tsyganov AA, Sakhirova NP (2013) Institutional development of the insurance market: theoretical and methodological aspects. Sci Notes Russ Acad Entrepreneurship 36:261–277

Yurgens IY, Nichiporuk ND (2017) Regulation of insurance business in Russia. J Bus Soc Power 1(26):99–113. Retrieved from https://tinyurl.com/ej8xbryh. Accessed 4 Nov 2025

Applied Solutions for the Sustainable Development of Rural Areas Based on Climate-Smart Technologies

Toolkit for Stimulating the Transfer of Low-Carbon Technologies: International Practices and the Strategy for Structural Transformations of Enterprises

Siwen Tan

Abstract The research analyzes a multi-level toolkit for intensifying the transfer of low-carbon technologies through the lens of advanced international experience, enabling the identification of optimal configurations of stimulating mechanisms applicable to various economic, technological, and institutional contexts. The author proposes a typology of existing instruments based on different criteria, forming the conceptual basis for a comparative study of national practices. Building on this typology, the research offers an in-depth analysis of low-carbon technology transfer stimulation models implemented by leading low-carbon development countries (the EU, Japan, the Republic of Korea, and China), revealing common patterns and specific national nuances in the design of the corresponding toolkit. The author argues that the effectiveness of low-carbon technology transfer depends critically on macroeconomic policy parameters and the structural transformation strategies at the level of the direct actors in the innovation process—the companies receiving the technologies. In this context, the research examines directions for restructuring corporate governance models in implementing low-carbon technologies, including a systemic green reorientation of R&D and technological management, the greening of production and logistics processes, and the development of low-carbon competencies and technological training potential.

Keywords Low-carbon technologies · International transfer · Enterprises · Incentive instruments · Structural transformation strategies · Sustainable development

JEL Classification O32 · Q55 · Q58 · L52

S. Tan (✉)
RUDN University, Moscow, Russia
e-mail: TanSiwen@yandex.ru

A. B. Karbekova (ed.), *Sustainable Development of Rural Territories Based on Climate-Smart Technologies*, Climate Change Management,
https://doi.org/10.1007/978-3-032-20106-5_39

1 Introduction

Amid escalating anthropogenic pressure on the biosphere and accelerating climate change, achieving sustainable development (i.e., a harmonious balance of environmental, economic, and social dimensions) has become imperative for human existence and interaction. As emphasized by the UN Sustainable Development Goals (SDGs), preserving natural capital and maintaining ecosystem functions is possible only by integrating advanced technological, managerial, and institutional innovations into production and consumption processes (United Nations 2015). Low-carbon technologies, which reduce greenhouse gas emissions while maintaining economic efficiency, are one of the essential tools for this integration. The importance of low-carbon technologies in facilitating a low-carbon transformation is increasingly recognized in the climate agendas of many countries. However, the practical implementation of their large-scale transfer faces a complex array of interrelated economic, technological, institutional, and socio-cultural barriers. Simona Pedde notes that "there is a gap between high-level climate adaptation policies, adaptation practices, and the production and transfer of knowledge" (Pedde et al. 2024). Global experience shows that overcoming these barriers requires a comprehensive toolkit of stimulation measures that spans the entire innovation cycle, from fundamental research to commercialization and the widespread adoption of green solutions. However, the effectiveness of this toolkit critically depends on its adaptation to the specific context of each country.

A comparative analysis of advanced international practices, which is the focus of this research, reveals the most effective approaches to accelerating low-carbon technologies under various economic, technological, and institutional conditions and evaluates the feasibility of adapting these approaches to national specifics. Among the leading countries in promoting low-carbon innovations are EU members such as Germany, Denmark, and the Netherlands, which implement long-term strategies aligned with ambitious climate neutrality goals, as well as the Republic of Korea and Japan, which have successfully integrated low-carbon technologies into their innovation growth models. Particularly noteworthy is China's approach, which combines robust state support for priority low-carbon sectors with mechanisms for technological transfer through global value chains.

Moreover, experts emphasize that effective implementation of advanced low-carbon technology stimulation measures requires their deployment at the macro level of national policy and the sectoral micro level within individual companies and production facilities (Fan et al. 2024). In the process of technological transfer, enterprises must undergo a profound restructuring of business models, organizational processes, and management approaches, necessitating the development of robust strategies for structural transformation. As V. Chen points out, companies are advised but must indeed adopt innovative business strategies if they are to mitigate adverse environmental impacts during low-carbon development (Chen 2022). Thus, the research aims to identify the main tools and best practices for intensifying the transfer of low-carbon technologies based on international experience and substantiate strategic approaches for structural transformation at the level of enterprises engaged in low-carbon innovation processes.

2 Materials and Methods

The empirical and theoretical analysis of the issues surrounding the stimulation of low-carbon technology transfer is based on statistical data, program documents, and scholarly publications of international and country-specific nature, including materials from the European Patent Office (EPO), research reports from the European Union, data from the World Intellectual Property Organization (WIPO), and statistical reports on low-carbon technology transfer from the World Bank. A structural–functional analysis is employed for the typology and systematization of the incentive toolkit for low-carbon technology transfer, allowing for identifying the directions and mechanisms of influence while considering the multifaceted and multi-level nature of this phenomenon.

3 Results and Discussion

Analyzing international practices for stimulating the transfer of low-carbon technologies involves understanding the conceptual foundations and developing a typology of the applied instruments on several bases (Kuznetsova and Sipailo 2017). Depending on the target phase of technology transfer, we can identify the instruments that focus on intensifying the following:

- Research and development (grants, concessional loans, and tax credits for R&D);
- Demonstration and implementation (pilot projects, public procurement, preferential tariffs);
- Dissemination and replication (environmental taxes, emissions trading, and green certificates).

Based on the criterion of obligatoriness, instruments are divided into voluntary (agreements, standards, and labeling) and mandatory (regulations, limits, and taxes). From the perspective of the innovation process participants, measures can be distinguished that support developers (innovation vouchers and venture financing), manufacturers (preferential loans, state guarantees, and accelerated depreciation), and consumers of low-carbon technologies (subsidies, tax credits, and information dissemination) (Elkerbout and Egenhofer 2018). One notable trend in the last decade has been the increasing role of green financing in the toolkit for stimulating low-carbon technology transfer. A wide range of innovative instruments is used to mobilize large-scale investments in low-carbon projects: green loans and bonds, climate funds, preferential credit lines provided by national and multilateral development banks, and public–private partnerships. A crucial factor in the effectiveness of green financing is the existence of universally recognized definitions and taxonomies that channel capital flows toward truly climate-significant technologies and prevent greenwashing (Amandeep and Sardana 2021).

Structurally, climate financing is characterized by a predominance of public funding (82%) (private financing mobilized by the public sector—15.7%; export credits—2.3%), as well as an asymmetric allocation of funds between mitigation (67%) and adaptation (24%) objectives, with a significant share of loans (72%), the majority of which are provided on market (non-preferential) terms (Jensen and Roniger 2023). The European Union and its member states are the largest donors of climate financing (28.5 billion euros in 2022) (Jensen and Roniger 2023). Summarizing the typological variations of instruments for stimulating the transfer of low-carbon technologies in the context of country-specific features, this distribution can be presented in tabular form (Table 1).

Effectively intensifying the transfer of low-carbon technologies requires establishing a comprehensive support ecosystem encompassing every stage of the innovation cycle—from fundamental science to commercialization and involving the state, businesses, and society. A distinctive feature of the European toolkit is the combination of ambitious climate goals with extensive government support for low-carbon innovations, focusing on technological aspects and organizational, marketing, and social dimensions. Particularly noteworthy is the "Horizon Europe" (2021–2027) framework program for research and innovation, aimed at advancing low-carbon technologies. Funding for climate research has doubled, and 35% of the total budget (95.5 billion euros) is allocated to sustainable development objectives.

The EU Emissions Trading System (EU ETS) plays a significant role in stimulating low-carbon investments, although its effectiveness is noted to be limited by covering only 40% of greenhouse gas emissions in the energy and industrial sectors (Elkerbout and Egenhofer 2018). This raises the question of extending the system to other sectors, such as automotive transport and building heating, which could further discourage carbon-intensive production. Moreover, in the methodological context of studying mechanisms for transferring low-carbon technologies, patent activity serves as a meaningful indicator of how effectively technological innovations diffuse into EU production systems, while 2012 saw relative symmetry in the intensity of transferring electric (400 patent applications) versus traditional technological solutions (312 applications), by 2021, there was a substantial divergence in technology transfer trajectories (patent applications related to electric vehicles (743) greatly exceeded those related to traditional vehicles (63)). The exponential growth in patent applications in the electric transport segment (European Patent Office (EPO) 2022) (+85.8%), coupled with a reduction in activity for traditional technologies (−79.8%), indicates the institutionalization of effective mechanisms to stimulate low-carbon technology transfer, reflected in the shift of technological development priorities in the automotive industry toward decarbonization. It is interesting to note the similarity between the approaches of the EU and Japan to promoting low-carbon technologies, confirmed by similar targets for achieving carbon neutrality by

Table 1 Instruments to stimulate NUT transfer

Types		Examples from world practice
By target phase	R&D	Research grants (USA, EU, Japan, China, and Russia); R&D tax credits (UK, France, China, Russia)
	Demonstration and implementation	"Green" public procurement (USA, Japan, Korea); Preferential tariffs (Germany, Spain, China); Pilot projects (Denmark, the Netherlands, UAE)
	Dissemination	Carbon taxes (Sweden, France, Canada); Emissions trading system (EU, China, Korea); "Green" certificates (UK, Italy, Sweden)
By obligation	Voluntary	Sectoral and corporate agreements (Japan, the Netherlands); Eco-labeling (EU, China); Voluntary "green" standards (LEED, BREEAM, EDGE)
	Mandatory	Energy efficiency and emissions regulations (USA, EU, China, Russia); Emissions limits and quotas (EU, Korea, New Zealand); Environmental taxes and fees
By subjects	Developers of low-carbon technology	Innovation and venture funds (USA, Israel, France); Technology transfer centers (EU, Korea, Saudi Arabia, China); Patent pools and consortia (USA, Japan, the Netherlands)
	Manufacturers of low-carbon technology	Accelerated depreciation of low-carbon assets (China, Malaysia); Investment and production subsidies (India); Concessional loans (Germany, Brazil)
	Consumers of low-carbon technology	Tax incentives for electric vehicles (Norway, the Netherlands, China); Subsidies for green housing (Japan, UAE); Eco-labeling (EU, USA, Australia)
By instruments	Debt	Green loans and bonds (China, USA, France, Russia); Green financing funds (UK, Australia); Credit lines from multilateral banks (EBRD, AIIB, NDB)
	Equity	"Green" stock indices (S&P, FTSE, MSCI) "Green" exchange-traded funds (BlackRock, Invesco, StateStreet)
	Guarantee	Green guarantee funds and programs (USA, EU, India); Climate risk insurance (Switzerland)
	Mixed	PPPs in green infrastructure (Canada, Japan); Project and mezzanine financing (USA, UAE)

Source Developed by the authors

2050 (Table 2) and bilateral cooperation (under the "Green Alliance" declaration) (D'Ambrogio 2021).

The Japanese experience in intensifying low-carbon technology transfer is particularly notable for its effective coordination and concentration of resources on breakthrough areas under the "Society 5.0" and "Carbon Neutrality by 2050" strategies. To focus limited budgetary resources on the most promising technological directions, the Japanese government employs the "Strategic Innovation Programs" (SIP) tool, which brings together various ministries, agencies, research institutions, and industries to

Table 2 Comparative analysis of EU and Japan carbon neutrality achievement rates, 2019

Indicators in 2019	Japan (%)	EU (%)
Energy self-sufficiency	12.1	39.3
Share of fossil fuels in TPES	88	71.2
Emission reduction target	46 (compared to 2013)	55 (compared to 2013)

Source Developed by the authors based on (D'Ambrogio 2021)

achieve specific, measurable medium- and long-term goals. Program documents provide for creating a specialized fund of 2 trillion yen (approximately 15.4 billion euros) to stimulate environmental entrepreneurship and innovative activity. Structurally, the Japanese model for stimulating low-carbon technologies is based on five cross-sectoral policy instruments and fourteen growth sectors. The expected economic impact of this strategy is estimated at 90 trillion yen (686 billion euros) by 2030, rising to 190 trillion yen (1450 billion euros) by 2050 (D'Ambrogio 2021).

In technological terms, Japan demonstrates clear leadership in developing hydrogen technologies, planning to increase annual hydrogen consumption from zero to 3 million tons by 2030 and 20 million tons by 2050. The target cost for hydrogen in generation and transportation is to be reduced from the current 170 yen to 30 yen per normal cubic meter by 2030 and to 20 yen by 2050. A crucial element of the Japanese model is the active involvement of local authorities and the private sector. As of August 2021, 444 local governments, representing 87.9% of the country's population, had committed to achieving net-zero emissions by 2050. Over 80 Japanese companies have implemented internal carbon pricing systems, and 50 companies have joined the international RE100 initiative, committing to 100% renewable energy (D'Ambrogio 2021).

In turn, China is charting a unique path to becoming a global leader in low-carbon technologies by combining robust state support with the effective borrowing and adaptation of foreign innovations. Studies note that just five countries (China, Germany, Japan, the Republic of Korea, and the USA) account for 75% of costly innovations in climate change mitigation, with China leading in the number of patent applications (World Intellectual Property Organization (WIPO) 2023, p. 8). The state acts as the most significant driver of low-carbon technology transfer, providing targeted support to national enterprises while relying on a dynamically growing domestic market. Among China's 70 most innovative companies, a significant share (20%) specializes in developing and implementing low-carbon technology solutions, indicating the high effectiveness of the state's technological modernization policy. It is important to note the dual nature of the mechanisms for stimulating technology transfer: on one hand, the state participates directly in the innovation process through state-owned enterprises that demonstrate high patent activity in strategically important sectors (such as energy infrastructure and distribution networks); on the other hand, a comprehensive policy is implemented to stimulate the private sector by creating favorable institutional and economic conditions for technological transformation in line with the imperatives of "green" development, as reflected in the dynamic growth of innovative activity among non-state enterprises in sectors such as electric vehicle manufacturing and battery technologies (Fig. 1) (Insights by Greyb 2024).

In the field of renewable energy investments, China holds a leading position by committing $24 billion to 33 major projects, reflecting the country's comprehensive approach to technological modernization: the state not only participates directly through state-owned enterprises but also creates attractive conditions for private businesses (Fig. 2).

The international practices discussed above provide a general framework of mechanisms and instruments for stimulating the transfer of low-carbon technologies; however, their practical implementation necessitates profound transformations at the level of individual companies and production processes. In executing technology transfer, companies must restructure their business models, organizational processes,

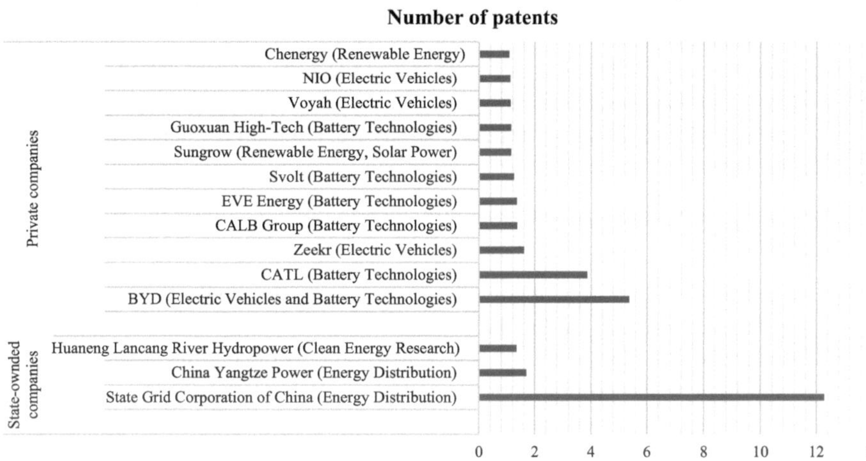

Fig. 1 Chinese companies specializing in the development and implementation of low-carbon technologies by ownership structure and number of registered patents, 2024. *Source* Developed by the authors based on (Insights by Greyb 2024)

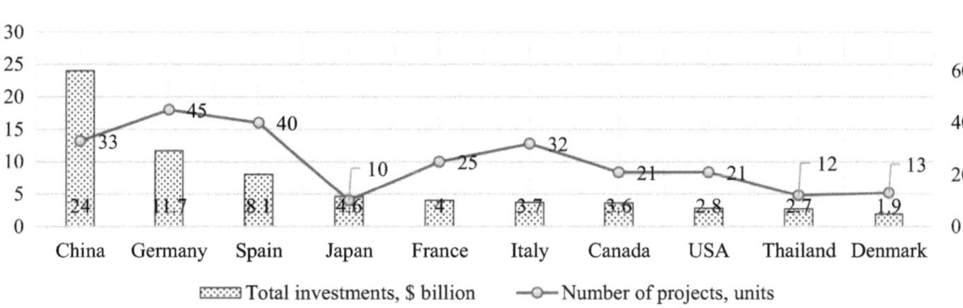

Fig. 2 Major renewable energy investor countries (2018). *Source* Developed by the authors based on (World Intellectual Property Organization (WIPO) 2023)

and management approaches in accordance with the imperatives of low-carbon development. First and foremost, implementing low-carbon technologies requires a systematic overhaul of R&D processes and technology management to integrate environmental priorities into every stage of innovation development and commercialization. This entails not only reorienting corporate R&D toward resource-saving and low-carbon technologies but also establishing a system for environmental monitoring, assessment, and selection of innovative projects that considers decarbonization effects alongside traditional techno-economic parameters. Simultaneously, it is essential to establish channels for technological vigilance and feedback to promptly identify promising low-carbon technology directions, diffuse best practices within the company, and engage with external sources of knowledge and expertise. Key areas include the following:

1. Reorienting corporate R&D toward low-carbon technologies:
   - Incorporating low-carbon technologies among the strategic priorities of corporate research;
   - Quantitatively evaluating and monitoring the outcomes of R&D projects in terms of emission reductions;
   - Increasing R&D funding for resource-saving, alternative energy, and "green" materials.
2. Integrating environmental criteria into technology development and transfer processes:
   - Including carbon footprint considerations in decision-making for technology implementation and product launches;
   - Conducting eco-audits and life cycle assessments during the development of technological and product innovations;
   - Establishing an expertise institute to evaluate the risks and opportunities associated with implementing low-carbon technologies;
   - Formulating a system of KPIs to monitor the progress and effectiveness of green innovations.
3. Establishing technological cooperation in the field of low-carbon technologies:
   - Actively scanning the external environment to identify breakthrough low-carbon technologies within current and related sectors;
   - Participating in industry consortia, technological platforms, and alliances, as well as engaging in international joint research;
   - Developing partnerships with universities and research institutions (including international ones) to access innovative research and talent.
4. Developing competencies and the capacity for technological training:
   - Implementing professional development and training programs in the field of low-carbon technologies;
   - Targeted recruitment and retention of specialists with expertise in low-carbon technology development;

- Appointing directors or managers responsible for low-carbon technology initiatives;
- Creating a robust knowledge management system.

Enhancing companies' receptivity to low-carbon technology implementation requires systematic efforts across these interrelated areas—from fine-tuning R&D management systems to establishing external and internal mechanisms for technological training and transfer. Advanced practices highlight the importance of a proactive, anticipatory approach that involves pre-adjusting a company's innovation processes in line with trends in green transformation across industries and the broader economy. Exemplary cases of this approach include technological leaders such as Siemens (Germany), Schneider Electric (France), Daimler (Germany), BASF (Germany), Vestas (Denmark), BYD (China), CATL (China), Tsinghua Energy (China), ABB (Switzerland), Mitsubishi Heavy Industries (Japan), LG Chem (South Korea), and Envision (China), which are actively reorienting their R&D and technology management models toward developing innovative solutions for a low-carbon economy.

Equally important is the ecological modernization of production and logistics processes as part of the structural transformation of companies in the context of low-carbon technology transfer. This involves optimizing resource usage, reducing the energy intensity and carbon footprint of production, minimizing waste, and closing material loops within circular business models. Typical practices in this area include the implementation of environmental management systems and green standards (ISO 14000, EMAS), the adoption of best available technologies (BAT) for low-carbon production, the use of advanced resource- and energy-saving technologies, the optimization of production processes and supply chains (for example, by establishing reverse logistics and infrastructure for the return, recycling, and reuse of products), and material recycling. This approach is not merely about isolated improvements in environmental metrics but represents a systematic transformation of the production model in line with the principles of low-carbon development. As demonstrated by the leading practices of sustainable production pioneers such as Unilever, Toyota, Cisco, and Maersk, success in this area hinges on combining technological innovations (through the implementation of low-carbon and best-available technologies) with organizational measures (creating business processes and incentives that support green changes) and collaborative strategies (building chains and ecosystems oriented toward circular exchanges).

4 Conclusion

Thus, the analysis of international practices demonstrates the diversity and multi-level nature of the toolkit for stimulating the transfer of low-carbon technologies. A systematic and synergistic application of various support mechanisms—from fiscal and regulatory incentives to partnership initiatives and corporate strategies—is a necessary factor for accelerating the so-called "creative destruction" of traditional production and consumption models and transitioning toward low-carbon growth trajectories. In this process, the complementarity and coherence of policies at the national, sectoral, and corporate levels are paramount, as is ensuring the compatibility of incentives for generating, transferring, and absorbing low-carbon technologies among all participants in the innovation ecosystem. The experience of developed and leading emerging economies shows that the greatest success is achieved by those who strike an optimal balance between targeted government support for priority areas and the creation of a favorable environment for dynamic low-carbon entrepreneurship through public–private partnerships, network collaboration, and technological platforms. It appears that the main direction for the future development of the low-carbon technology transfer toolkit should be its enhanced focus on achieving the goals of sustainable development in a broad sense, considering social and humanitarian effects. Only by adhering to the principles of a just transition and preventing further polarization can the green transformation realize its enormous potential to improve welfare and quality of life.

References

Amandeep A, Sardana S (2021) A conceptual study of green financing in promoting sustainable development. Int J Commer Bus Manag 15(1):50–58

Chen W (2022) Can low-carbon development force enterprises to make digital transformation? Bus Strateg Environ 32(4):1292–1307. https://doi.org/10.1002/bse.3189

D'Ambrogio E (2021) Japan's 2050 goal: a carbon-neutral society [Briefing]. European parliamentary research service. European Parliament, Brussels, Belgium. Retrieved from https://www.europarl.europa.eu/RegData/etudes/BRIE/2021/698023/EPRS_BRI(2021)698023_EN.pdf. Accessed 20 Dec 2024

Elkerbout M, Egenhofer C (2018) Tools to boost investment in low-carbon technologies: five possible ways to create low-carbon markets in the EU. CEPS Policy Insight No. 2018/11. Centre for European Policy Studies, Brussels, Belgium

European Patent Office (EPO) (2022) Insight into smart urban mobility. Retrieved from https://report-archive.epo.org/about-us/annual-reports-statistics/statistics/2021/insight-into-smart-urban-mobility.html. Accessed 10 Oct 2024

Fan L, Guo Y, Wang Y, Wang W (2024) Navigating green innovation in high-tech manufacturing: the roles of customer concentration and digital transformation. Sustainability 16(15):6358. https://doi.org/10.3390/su16156358

Insights by Greyb (2024) China most innovative companies (Updated 2024). Retrieved from https://insights.greyb.com/china-most-innovative-companies/. Accessed 20 Dec 2024

Jensen L, Roniger J (2023) International climate finance: status quo, challenges and policy perspectives [Briefing]. European parliamentary research service. European Parliament, Brussels, Belgium. Retrieved from https://www.europarl.europa.eu/RegData/etudes/BRIE/2023/754616/EPRS_BRI(2023)754616_EN.pdf. Accessed 20 Dec 2024

Kuznetsova T, Sipailo L (2017) State system of stimulation and regulation of innovative activities of enterprises in developed countries of the world. Int J New Econ Soc Sci 6(2):33. https://doi.org/10.5604/01.3001.0010.7638

Pedde S, Grendelman R, Cumiskey L, McCullagh, Vinke-de Kruijf J, Hölscher K (2024) Leveraging climate resilience capacities by (un)learning from transdisciplinary research projects. Clim Risk Manage 47:100675. https://doi.org/10.1016/j.crm.2024.100675

United Nations (2015) Sustainable development goals. Retrieved from https://www.un.org/sustainabledevelopment/sustainable-development-goals/. Accessed 20 Dec 2024

World Intellectual Property Organization (WIPO) (2023) Green technology book: solutions for climate change mitigation. WIPO, Geneva, Switzerland. https://doi.org/10.34667/tind.48717

A New Effective Method for the Steam-Thermal Processing of Rice Shali Grain in Kyrgyzstan

Eltar A. Smailov, Ruslanbek N. Arapbaev, Mansurbek O. Ergeshov, Bektur K. Omurzakov, and Kalysbek Koshmamat Uulu

Abstract The rice shali grain, harvested from the fields for rice production, is processed by two methods: (1) polishing, which employs a mechanical technique; and (2) hulling, a method used in Kyrgyzstan. The most significant difference between hulled and polished rice is that the former retains higher levels of protein, fiber, thiamine, and niacin. During hulling, the rice shali grain remains undamaged and preserves its original appearance and chemical composition. Under natural climatic conditions, the harvested rice plants—with their panicles—are bundled into heaps and allowed to warm naturally for two to twelve days or more, depending on the moisture content of the stems and the ambient temperature. In other words, a natural fermentation process occurs. During this process, the starch contained in the rice shali grain breaks down into micro- and macronutrients as well as amino acids beneficial to the human body. Additionally, certain physiological and biochemical changes occur within the grain, enhancing its quality—an effect that largely depends on the duration of fermentation. Simultaneously, the color of the rice shali grain changes to brown, and the rice acquires a glassy appearance with dark-brown streaks. Afterward, combine harvesters thresh the grain for subsequent sun drying and husk removal by hulling on special antique water-driven milling complexes known as "Ak-zhavaz."

Keywords Rice · Shali grain · Bran · Steam-thermal treatment · Proteins · Starch · Hemicellulose · Ash · Amino acid sum · Ak-uruk · Laser · Kara-kyltyryk

E. A. Smailov (✉)
International Uzgen Institute of Technology and Education Named After Academician B. Murzubraimov OshTU, Uzgen, Kyrgyzstan
e-mail: eltar_uito@mail.ru

R. N. Arapbaev · M. O. Ergeshov
Osh State University, Osh, Kyrgyzstan
e-mail: rarapbaev@oshsu.kg

B. K. Omurzakov · K. K. Uulu
Osh Technological University Named After M.M. Adyshev, Osh, Kyrgyzstan

A. B. Karbekova (ed.), *Sustainable Development of Rural Territories Based on Climate-Smart Technologies*, Climate Change Management,
https://doi.org/10.1007/978-3-032-20106-5_40

1 Introduction

It is well known that the grains of major cereal crops—rice, wheat, and rye—are primarily used for human consumption. Corn, sorghum, and millet serve as food crops in some countries but are generally cultivated for animal feed. Barley and oats are widely used in the production of food products and beverages; however, most of their grains are allocated for livestock feed, largely due to their composition. For much of the global population, rice is one of the key agricultural products. The U.S. Department of Agriculture (1963) presented the composition of rice and other cereal grains, making it possible to assess grain quality for their potentially effective use in food production. The data (U.S. Department of Agriculture 1963) indicate that, among all cereal products, hulled rice contains the lowest protein content (7.5%)—a very low figure given that rice quality is measured by its protein content—whereas rice cultivated in Kyrgyzstan contains up to 13% protein (Smailov et al. 2024). Its ash content is also low (1.2%), representing the non-combustible microelements, while rice grown in Kyrgyzstan can have an ash content of up to 4.8% (Smailov et al. 2011). Therefore, this research aims to thoroughly investigate the secret behind the quality of rice cultivated in Kyrgyzstan.

The harvested rice shali grain, intended for rice production, is processed by two methods: (1) polishing, which employs a mechanical technique, and (2) hulling, a method used in Kyrgyzstan. The degree of polishing refers to the extent to which the bran layer and germ are removed from the starchy endosperm. Polished rice is either completely devoid of the seed coat or retains only minimal traces on the surface of the endosperm. Consequently, mechanical polishing reduces the nutritional value of the rice because protein, fat, vitamins, and minerals are concentrated in the germ and the outer layer of the starchy endosperm, which are removed along with the husk during processing. Moreover, excessive polishing results in a lower overall grain yield as well as a decrease in the yield of whole rice, thereby diminishing the product's market value. The FAO Food and Agriculture Organization of the United Nations (1962) distinguishes three degrees of polishing: fully polished, moderately polished, and minimally polished rice, although the definitions of these degrees remain somewhat ambiguous.

2 Methodology

Amino acid analysis was performed using HPLC (high-performance liquid chromatography). Amino acids were determined by pre-derivatization with o-phthalaldehyde (Carl Roth, Karlsruhe, Germany), followed by HPLC-FD analysis according to the Lee and Dresher method.

3 Results

The most significant difference between hulled and polished rice is that the former contains higher amounts of protein, fiber, thiamine, and niacin. Rice grain has been studied less extensively than the grains of other cereals such as corn, barley, and wheat. It consists of an edible portion (the caryopsis) and protective coverings, or husks, with the husk making up 18%–28% of the grain's mass (Cagampang et al. 1966; Hinton 1948; Hinton and Shaw 1954; Primo et al. 1970).

It is well known that the cell walls of the aleuron layer are stained with dyes specific for protein, hemicellulose, and cellulose (Little and DawsonM 1960). In upland rice varieties, more aleuron layers form on the dorsal (back) and ventral (front) sides of the grain compared to varieties grown under irrigation (Hoshikawa 1967).

When cultivating rice under the natural climatic conditions of Kyrgyzstan, it is strongly recommended to adhere strictly to the scheduled agronomic practices (Smailov 2022; Smailov et al. 2020). Moreover, since predominantly early-ripening rice varieties are grown, their maturation occurs in the latter half of July and throughout August, when ambient temperatures peak, thereby contributing to producing high-quality rice.

Furthermore, in Kyrgyzstan (Uzgen district, Osh Region), where the famous Uzgen rice has been cultivated for generations, a method of natural steam-thermal processing of rice shali grain is employed, followed by the hulling process to produce finished rice using antique milling complexes known as "Ak-zhuvaz." General view of the working element of rice mill is presented at Data Availability section. The operation of the "Ak-zhuvaz" milling complex is driven by the blades of a water drum. In recent years, due to the increasing number of rice mills, an electric drum drive has gained popularity.

Prior to these steps, certain technological operations must be carried out. The rice, harvested with its panicles, is heaped into bundles (with the stalks interleaved) and allowed to warm under natural conditions for anywhere from 2 to 12 or more days, depending on that year's climatic conditions. During this period, various physiological and biochemical processes occur within the rice grain. Subsequently, combine harvesters thresh the grain, yielding the rice shali for further sun-drying and husk removal by hulling on specialized antique water-driven milling complexes.

From a single variety grown in Kyrgyzstan, rice of three different colors is produced: ranging from white (beige)—light brown to dark brown (developed by the authors, Patent No. 1961 dated June 30, 2017) (Smailov et al. 2017). Type and color of rice can be seen in the photo presented at Data Availability section. The chemical composition, as well as the content of micro- and macroelements and amino acids, varies according to the color of the rice. This is due to the duration of the process during which the starch contained in the rice grain gradually decomposes into proteins, carbohydrates, micro- and macroelements, and amino acids beneficial to human health. In the varieties cultivated in Kyrgyzstan, the starch content reaches up to 70%. Consequently, the protein content in rice grown in Kyrgyzstan ranges from 10.41% to 13%, which can be seen from the graph presented at Data Availability section. In contrast, rice varieties imported from Uzbekistan and cultivated in Kyrgyzstan exhibit significantly lower protein levels.

In the rice and the rice bran of locally cultivated varieties, the protein content is between 8.09% and 9.34%, whereas in varieties imported from Uzbekistan and grown in Kyrgyzstan, it is only 6.55%. Thus, by employing the method of natural steam-thermal processing, the microelements, including proteins, from the bran are transferred to the rice grain, resulting in rice that is light to dark brown and exhibits a glassy appearance.

The curves showing changes in the chemical composition of the "Kara-kyltyryk" rice variety as a function of the duration of natural steam-thermal processing are presented at the Data Availability section The results reveal increases in pectic substances by 0.7%, polysaccharides by 3.3%, hemicellulose by 3.98%, and ash by 1.3%, indicating a rise in microelements.

The results of the chemical composition analysis of rice varieties cultivated in Kyrgyzstan in comparison with Chinese rice available on the market are presented at the Data Availability section. Chinese rice contains the lowest level of polysaccharides, while the varieties grown in Kyrgyzstan have the highest. Conversely, hemicellulose and ash contents are highest in Chinese rice, at 26% and 32% respectively. In rice varieties grown in Kyrgyzstan, hemicellulose content ranges from 2 to 6%, with local varieties exhibiting a minimum of 3% and the Uzbek-imported variety "Laserniy" reaching 6%. Starch, primarily found in the starchy endosperm, is present in the highest concentration in Chinese rice (up to 32%), at 11% in rice imported from Uzbekistan and cultivated in Kyrgyzstan, and only up to 7% in local varieties. The ash content is high in local varieties such as "Kara-kyltyryk" and "Ak-uruk" (over 6%) and minimal in Chinese rice, indicating that Chinese rice contains the least amount of microelements. To conclusively validate the quality assessment of rice cultivated in Kyrgyzstan using a new technology of natural steam-thermal processing of rice shali, followed by sun drying and processing on antique milling complexes by hulling, special studies were conducted for the first time to determine the total content of amino acids essential for human health (histidine, leucine, arginine, phenylalanine, valine, isoleucine, methionine, lysine) according to rice variety and color. The research revealed that during the natural steam-thermal processing of rice shali, their amino acid content increases significantly (Fig. 1).

These studies showed that the renowned Uzbek variety "Laserniy" (which does not undergo steam-thermal processing) exhibits a total amino acid content of 17.6%–18.6%, regardless of the cultivation region. In contrast, regional varieties grown in Kyrgyzstan without natural steam-thermal processing have a total amino acid content ranging from 18.8% to 21.7%, indicating that, by nature, their amino acid levels are considerably higher than those of the famous "Laserniy" rice from Uzbekistan.

Overall, the rice shali cultivated in Kyrgyzstan that undergo natural steam-thermal processing exhibit total amino acid levels of 18.8% to 24.2% in the regional "Kara-kyltyryk" variety and 20.6% to 25.1% in the "Ak-uruk" variety, depending on the duration of processing and the resulting rice color. This indicates that additional changes in the chemical composition and quality of the rice occur during natural

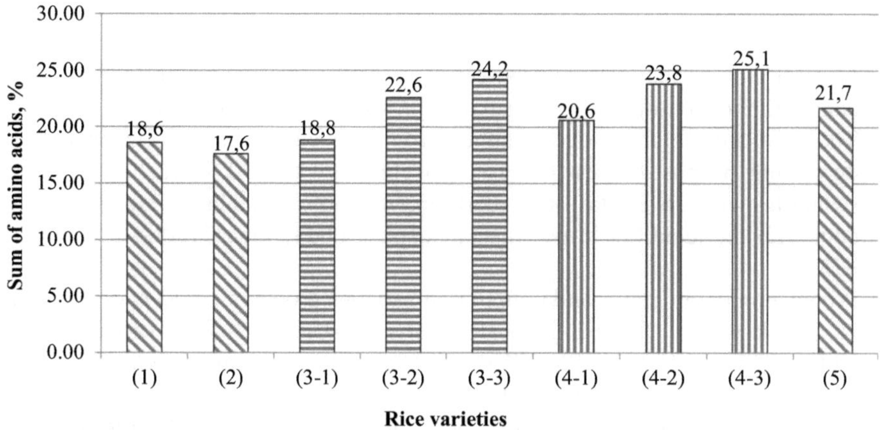

Fig. 1 Total amino acid content in different varieties and types (by color) of rice: (1) "Laserniy" (white, Uzbekistan); (2) "Laserniy" (white, Kyrgyzstan); (3–1) "Kara-kyltyryk" (white); (3–2) "Zarcha"; (3–3) "Dasta saryk"; (4–1) "Ak-uruk" (white); (4–2) "Zarcha"; (4–3) "Dasta saryk"; (5) "Musa" (white). *Source* Developed by the authors

steam-thermal processing, irrespective of the variety, as reflected in its amino acid content. It is also noteworthy that the increase in total amino acids in rice grown in Kyrgyzstan using this technology reaches 4.5% for "Ak-uruk" and 5.45% for "Kara-kyltyryk." Moreover, variations in rice color are accompanied by changes in the total amino acid content.

According to the authors' observations, this improvement is due to the breakdown of starch into more beneficial micro- and macroelements and amino acids for the human body. Therefore, we consider it important to test and implement the system we previously recommended (Smailov et al. 2011), which features automatic control of the steam-thermal processing of rice shali and monitors the biochemical changes occurring within the rice grain. This system will also help determine the optimal duration of steam-thermal processing to maximize both quality and the beneficial amino acid content.

4 Conclusion

1. The studies demonstrate the effectiveness of natural steam-thermal processing technology and rice production through hulling in enhancing rice quality;
2. The research found that during steam-thermal processing of rice shali, the starch decomposes into amino acids that are more beneficial to human health;
3. To investigate the changes in biochemical processes within rice shali during steam-thermal processing, further experimental studies must be conducted to identify factors that can further improve rice quality.

Data Availability

- General view of the working element of rice mill is available at https://figshare.com with the identifier https://figshare.com/s/f980c561b0efdbda6f58.
- The type and color of rice obtained from one variety in the Uzgen district depending on post-harvest processing technology (1a–Ak-Kuruch, 1b–Zarcha (white-brown), 1c–Dasta saryk (dark-brown)) are available at https://figshare.com with the identifier https://figshare.com/s/506d42066a477c0bb4fc?file=52867571.
- Protein content in grains and bran of rice varieties grown in the Kyrgyz Republic (1–Uzbek variety "Laser," 2–local variety "Ak-Uruk," 3–local variety "Kara-Kyltyryk") is available at https://figshare.com with the identifier https://figshare.com/s/cebff10e96b93b59c036?file=52867580.
- The curves showing changes in the chemical composition of the "Kara-kyltyryk" rice variety as a function of the duration of natural steam-thermal processing are available at https://figshare.com with the identifier https://figshare.com/s/bc36155ff242a8da70d1.
- The results of the chemical composition analysis of rice varieties cultivated in Kyrgyzstan in comparison with Chinese rice presented on the market are available at https://figshare.com with the identifier https://figshare.com/s/d15f1188efb975adcd62.

References

Cagampang GB, Cruz LJ, Espiritu SG, Santiago RG, Juliano BO (1966) Studies on the extraction and composition of rice proteins. Cereal Chem 43(2):145–155

Food and Agriculture Organization of the United Nations (FAO) (1962) The world rice economy: vol I. commodity bulletin series no. 36. FAO, Rome, Italy

Hinton JJC (1948) The distribution of vitamin B_1 in the rice grain. Br J Nutr 2(3):237–241. https://doi.org/10.1079/BJN19480041

Hinton JJC, Shaw B (1954) The distribution of nicotinic acid in the rice grain. Br J Nutr 8(1):65–71. https://doi.org/10.1079/BJN19540010

Hoshikawa K (1967) Studies on the development of endosperm in rice m 5 the number of aleurone cell layers its varietal difference and the influence of environmental factors. Proc Crop Sci Soc Jpn 36(3):221–227

Little RR, DawsonM EH (1960) Histology and histochemistry of raw -4nd cooked rice kernels. J Food Sci 25(5):611–622. https://doi.org/10.1111/j.1365-2621.1960.tb00005.x

Primo T, Barber S, Tortosa E, Camacho J, Ulidemolins J, Jimenez A et al (1970) Composition quimica del arroz. V. Subproductos obtenidos en las differentes etapas del diagrama de elaboration [Chemical composition of rice. V. By-products obtained in the different stages of the processing diagram]. Revista de Agroquimica y Tecnologia de Alimentos 10(2):244

Smailov EA (2022) Agrotechnics of rice cultivation in Kyrgyzstan. In: Proceedings of the republican scientific and practical conference. International Uzgen Institute of Technology and Education, Osh, Kyrgyzstan, pp 243–248

Smailov EA, Samiyeva ZT, Smailova KE (2011) Rice is a unique culture. Bishkek, Kyrgyzstan

Smailov EA, Smailova KE, Samiyeva ZT et al (2017) Patent no. 1961 of the Kyrgyz Republic, A01D 45/04 method of obtaining high quality Uzgen rice (published on May 31, 2017, bulletin no. 6). Phytotechnology Innovation Center of National Academy of Sciences of Kyrgyz Republic, Bishkek, Kyrgyzstan

Smailov EA, Tashmatova NK, Akmataliev AT (2020) Recommendations to production on agrotechnics and technology of rice cultivation in Batken Region. Ministry of Agriculture, Food Industry, and Land Reclamation of the Kyrgyz Republic, Bishkek, Kyrgyzstan

Smailov EA, Arapbaev RN, Kochkonbaeva AA, Samieva ZT, Tashmatova NK (2024) Formation of new mechanisms for sustainable development of the rice farming in Kyrgyzstan. In: Popkova EG, Bogoviz AV, Sergi BS, Kaurova OV, Maloletko AN (eds) Sustainable development of the agrarian economy based on digital technologies and smart innovations. Springer, Cham, Switzerland, pp 3–6. https://doi.org/10.1007/978-3-031-51272-8_1

U.S. Department of Agriculture (1963) Composition of foods: dairy and egg products. Raw. Processed. Prepared. Agriculture handbook No. 8–1. U.S. Department of Agriculture; Agricultural Research Service, Washington, DC

Improving the Economic Model for Organizing Integrated Vaccination Recordkeeping for Cattle-Farming Enterprises in an Automated Environment

Svetlana A. Chernyavskaya, Victoria G. Bush, Fatima Kh. Kirgueva, Sofia M. Dzidzoeva, and Egor G. Fedchenko

Abstract Ensuring that the population has access to high-quality food products, particularly cow's milk, requires a scientifically grounded approach for the proper and timely vaccination of animals (cattle), the effective planning and scheduling of vaccination programs by disease, with detailed breakdowns by age and gender for each individual head of cattle, and recording these activities in the financial and managerial accounting systems using automated software. Enhancing the control function of accounting in relation to animal vaccination involves creating primary documents to record the planning and the actual execution of vaccinations in accordance with regulatory standards. These regulatory parameters, when documenting the operational aspects of vaccination and revaccination, include the following: disease name, corresponding measures, the cattle's age and gender group, frequency (the normative interval—i.e., the age when the next vaccination should occur), the product name, its dosage, the method of vaccine administration, and the minimum interval between vaccinations. The need for proactive vaccination planning on the long-term and short-term (daily for the next seven days) basis with details down to the individual animal, including its placement within the appropriate age and gender group and its location on the farm, significantly increases the accountability of veterinary professionals at agricultural producers and simplifies oversight by the manager of the dairy and livestock farm regarding the actual administration of vaccinations.

S. A. Chernyavskaya (✉)
I.T. Trubilin Kuban State Agrarian University, Krasnodar, Russia
e-mail: docsveta17@gmail.com

V. G. Bush · E. G. Fedchenko
Russian University of Cooperation, Mytishchi, Russia
e-mail: v.g.bush@ruc.ru

F. Kh. Kirgueva · S. M. Dzidzoeva
North Ossetian State University Named After K.L. Khetagurov, Vladikavkaz, Russia

A. B. Karbekova (ed.), *Sustainable Development of Rural Territories Based on Climate-Smart Technologies*, Climate Change Management,
https://doi.org/10.1007/978-3-032-20106-5_41

Keywords Accounting · Control · Automation · Economic model · Managerial accounting · Vaccination

JEL Classification M410

1 Introduction

The development of dairy farming in the Krasnodar Territory is a priority, forming an essential component of the doctrine for ensuring food security for the region and the country.

Food production (specifically, the production of milk and beef is impossible without accounting for cattle vaccination activities and recording this information in paper-based "Vaccination Journals" or specialized zootechnical accounting programs such as "Seleks" and "Afifarm" but primarily by importing or exporting data into the automated program "1C Accounting" with detailed records for each animalBzhasso 2014; Cozzani and Zanelli 2007; Hodarinova 2017).

Recording cattle vaccination activities in accounting is mandatory, with costs allocated separately for breeding and non-breeding animals, especially for agricultural producers who hold the status of "Breeding Farm" or "Breeding Reproducer" and receive subsidies from the regional Ministry of Agriculture for breeding and milk production.

Moreover, the integrated accounting data recorded in the automated program "1C Accounting" forms the basis for making timely managerial decisions to strengthen control over the vaccination plan for each animal, ensuring compliance with mandatory vaccination requirements while considering the physiological condition of each individual animal (Stein et al. 2001).

With the constant rise in the cost of veterinary pharmaceuticals, the management of agricultural enterprises is forced to find ways to reduce production costs (milk yield, the number of calves per 100 cows, herd breeding efficiency). Thus, implementing vaccination activities and recording them in the accounts with detailed information on the source of funding (whether from the regional Ministry of Agriculture or from the enterprise's funds), by disease type, and for each animal (utilizing advanced information technologies) enables management to obtain timely information in this area.

Therefore, accounting and economics specialists are compelled to automate this segment of work, which is the focus of this scientific study and underscores its relevance.

2 Methodology

The need to enhance animal productivity (milk yield per fodder cow), increase the total milk production and weight gains of young cattle, and uncover reserves for reducing milk production costs and boosting profits from its sale imposes specific requirements for strengthening control over the accounting of vaccination activities (Chernyavskaya et al. 2024).

For this research, the authors consider the agricultural enterprise JSC "Viktoria-Agro" in the Kanevsky district of the Krasnodar Territory, which holds the status of "Breeding Reproducer." At the end of 2023, the cattle population numbered 2502 heads, divided into the following age and gender groups:

- Heifers, first half of the calving period—152 heads;
- Heifers, second half of calving period—73 heads;
- Heifers, 1 month before calving—37 heads;
- Fattening cows—136 heads;
- Fodder cows/1 to 30 days of lactation—112 heads;
- Fodder cows/111 to 305 days of lactation—449 heads;
- Fodder cows/31 to 110 days of lactation—283 heads;
- Fodder cows / dry period, first half—64 heads;
- Fodder cows / dry period, second half—64 heads;
- Young bulls from 2022, 12–16 months—3 heads;
- Young bulls from 2023, 0–2 months—46 heads;
- Young bulls from 2023, 2–4 months—26 heads;
- Young bulls from 2023, 4–6 months—45 heads;
- Young bulls from 2023, 6–12 months—28 heads;
- Heifer calves from 2022, 12–16 months—227 heads;
- Heifer calves from 2022, 16–18 months—35 heads;
- Heifer calves from 2023, 6–12 months—275 heads;
- Heifer calves from 2023, 0–2 months—158 heads;
- Heifer calves from 2023, 2–4 months—155 heads;
- Heifer calves from 2023, 4–6 months—134 heads.

The cattle are housed at two dairy and livestock farms.

3 Results

Figure 1 shows the livestock numbers for the main agricultural animals in the region by category of enterprise. In 2023, the cattle population across all types of agricultural enterprises in the Krasnodar Territory increased by 21,000 heads (3.90%) compared to 2015. However, the number of cows in 2023 decreased by 6000 heads (2.78%) relative to 2015. In agricultural organizations, the reduction in the cow population in 2023 compared to 2015 was even more pronounced at 7000 heads (5.19%). Moreover, the proportion of cows within the overall livestock structure of these enterprises declined from 63% in 2015 to 61% in 2023 (Department of the Federal State Statistics Service for the Krasnodar Territory and the Republic of Adygeya 2024).

Milk production in all types of enterprises in the Krasnodar Territory reached 1703.7 thousand tons in 2023, which is 376.1 thousand tons (28.33%) higher than in 2015. The growth in cow's milk production in agricultural organizations is even more significant: in 2023, production amounted to 1230.4 thousand tons, exceeding the 2015 level by 380.1 thousand tons (44.70%).

In 2023, the sale of milk and dairy products across all categories of enterprises in the Krasnodar Territory reached 1518 thousand tons, exceeding the 2015 level by 408 thousand tons, or 36.76%, even though milk production increased by only 28.33%. In agricultural organizations, milk and dairy product sales amounted to 1204 thousand tons, which is 394 thousand tons (48.6%) above the 2015 level; milk production grew by 44.70%.

The milk yield per cow in all categories of enterprises in the Krasnodar Territory in 2023 was 8594 kg, an increase of 2347 kg (37.57%) over 2015. In agricultural organizations, it was 9999 kg, up by 3375 kg (50.95%) compared to 2015.

Profit from sales in the livestock sector within the region's agricultural organizations reached 7204 million rubles, which is 1706 million rubles (29.86%) higher than in 2015, although the volume of milk sold grew by 48.6%. Meanwhile, the profitability in the livestock sector in these organizations in 2023 stood at 22.96%, which is 4.2 percentage points lower than in 2015 and 8.2% below that of the crop production sector in 2023.

Average prices and the price ratios for identical types of products at various stages of the supply chain are as follows:

- The price for raw cattle milk for agricultural producers in 2023 was 34.09 rubles per kilogram;
- The consumer price for whole, pasteurized drinking milk with 2.5–3.2% fat in retail chains in 2023 was 70.37 rubles per kilogram.

Thus, the price ratio for individual types of goods at different stages of the supply chain in the Krasnodar Territory in 2023 was 2.1 times, compared to 176.0% in 2021 and 189.7% in 2022, indicating a concentration of profits in the retail chains and a withdrawal of funds from agriculture amid rising prices for consumed resources.

The need to create an integrated production accounting system in the 1C "Accounting 8.3 (Corp.)" program for recording the operational aspects of cattle vaccination activities, is dictated by the labor-intensive process of preparing an up-to-date and highly detailed "Vaccination Schedule" for each animal, distributed across various periods (daily, weekly, monthly, yearly).

Moreover, errors in vaccination procedures (e.g., failing to observe the minimum intervals between vaccinations, especially in calves under 6 months) have led to animal losses and reduced herd survival. Therefore, the regulatory parameters for recording vaccination and revaccination activities include the following:

- Disease name (anthrax, foot-and-mouth disease, tuberculosis, etc.);

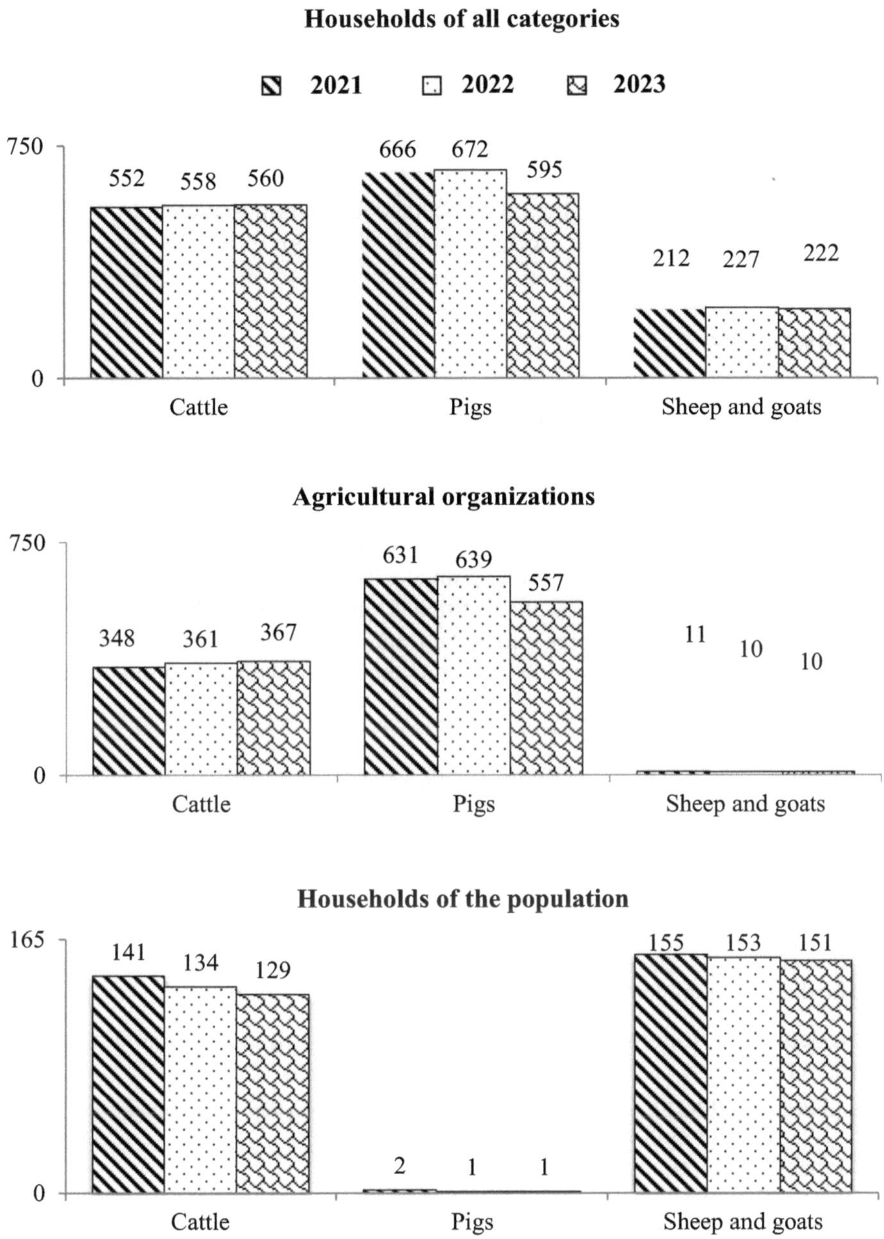

Fig. 1 Change in the number of livestock and poultry (end of year, thousand heads) in the Krasnodar Territory. *Source* Developed by the authors based on (Department of the Federal State Statistics Service for the Krasnodar Territory and the Republic of Adygeya 2024)

- Activities (vaccination, treatment, and testing);
- Age and gender group of the cattle;
- Frequency (the normative interval—i.e., the age at which the next vaccination should occur);
- Product name (e.g., foot-and-mouth vaccine, tuberculin, etc.);
- Dosage per animal;
- Method of vaccine administration (subcutaneous, intramuscular);
- Minimum interval between vaccinations (10–14 days).

4 Discussion

The need for proactive vaccination planning (long-term and short-term (daily for the next seven days)), with details down to each individual animal, including its placement within the appropriate age and gender group and its location on the farm, significantly increases the accountability of the veterinary staff at agricultural production enterprises and simplifies oversight by the manager of the dairy and livestock farm regarding the actual execution of animal vaccinations.

Subsequently, automated documents titled "Daily Act for the Write-Off of Veterinary Pharmaceuticals" were prepared in the Data Availability section, which include detailed records with the following indicators entered into the accounting registers:

- Disease (IRT IPG3);
- Activity (vaccination);
- Age and gender group of the cattle (fodder cows/111 to 305 days of lactation);
- Type of cattle (breeding);
- Treatment scheme (8: Vak.-Sules Premium Vaccine (5));
- Details for each specific animal with their identification numbers;
- Materially responsible person (E.V. Krutikov);
- Division (Farm No. 3).

The document contains the following columns:

- Product name from the "Nomenclature" directory (Vak.-Sules Premium Vaccine 100 ml);
- Unit of measure (ml);
- Standard consumption per animal (5 ml);
- Actual consumption per animal (1 ml);
- Write-off amount—the lesser of the above values (1 ml);
- Accounting entries with details by "subaccount" levels (D 20.01.2 K 10.16).

The developed format of the directory "Types and characteristics of vaccines for cattle" in the research object is presented in the Data Availability section, specifying the following parameters:

- Revaccination frequency for the animal's lifetime, with normative dates and ages for the next vaccination;

- Minimum interval between vaccinations (10–14 days);
- Notes (budget vaccine, do not use for heifers, sick animals, administer one month before calving, etc.).

The report form "Report on the Implementation of the Cattle Vaccination Schedule" at JSC "Viktoria-Agro" is also presented in the Data Availability section.

5 Conclusion

The construction of an integrated production accounting system using the 1C "Accounting 8.3 (Corp.)" program will address the following objectives:

1. Develop the following directories, primary documents, and reports: the directory "Types and Characteristics of Vaccines for Cattle"; the document "Daily Act for the Write-Off of Veterinary Pharmaceuticals"; the report "Vaccination Schedule for Cattle"; and the report "Vaccination Schedule Compliance Report for Cattle." Industry experts (chief zootechnician, chief veterinary officer), as well as mid-level specialists (farm manager, veterinary doctor), will enhance the control function of accounting by:
    - Having a clear overview of the planned volume of animal vaccinations for both the short-term (daily, weekly, and monthly) and the long-term (month-by-month until the end of the year), with detailed breakdowns by disease type, age, and gender group, vaccine (product) requirement, and the anticipated movement of the cattle herd within the specified time interval;
    - Monitoring the work of farm veterinarians in the timely execution of the "Vaccination Schedule" in compliance with regulatory standards, not only in the aggregate but also with details for each animal.
2. Improve herd preservation, reduce animal losses, extend the lifespan of cows, and enhance their productivity, ultimately contributing to lower milk and weight-gain production costs, increased profits, and, consequently, higher sales profitability.

Thus, these measures will contribute to enhancing the economic efficiency of dairy cattle farming, increasing the competitiveness and financial stability of agricultural producers, and meeting the requirements set for recipients of subsidies from the Ministry of Agriculture of the Russian Federation for breeding and milk production.

Data Availability

- The automated document "Daily Write-Off Act for Veterinary Drugs" is available at https://figshare.com with the identifier https://doi.org/10.6084/m9.figshare.28450571.v1.
- The recommended format for the reference guide "Types and Characteristics of Cattle Vaccines" in the "1C: Accounting" software is available at https://figshare.com with the identifier https://doi.org/10.6084/m9.figshare.28450655.v1.

- The recommended report format "Report on the Implementation of the Cattle Vaccination Schedule" in the "1C: Accounting" software is available at https://figshare.com with the identifier https://doi.org/10.6084/m9.figshare.28450679.v1.

References

Bzhasso AA (2014) Key objectives of anti-crisis regional economies management. Bull Adyghe State Univ 2(141):30–35

Chernyavskaya SA, Bush VG, Mnatsakanyan NV, Turchenko VV, Chechetka EL (2024) Analysis and modeling of the use of fixed assets in the digital economy. In: Popkova EG, Kaurova OV, Maloletko AN (eds) Sustainable cooperation for the creation of green supply chains based on environmental technologies and responsible innovations. Springer, Cham, Switzerland, pp 257–267. https://doi.org/10.1007/978-3-031-70962-3_29

Cozzani V, Zanelli S (2007) An approach to the assessment of domino accidents hazard in quantitative area risk analysis. J Hazard Mater 130:24–50

Department of the Federal State Statistics Service for the Krasnodar Territory and the Republic of Adygeya (2024) Agriculture of the Krasnodar region 2023: statistical collection. Book 1: the state of the agricultural sector of the Krasnodar region. Krasnodarstat, Krasnodar, Russia

Hodarinova NV (2017) Organization and evaluation of the effectiveness of internal and external control of modern agricultural organizations. J Econ Entrepreneurship 4–1(81):536–547

Stein J, Usher S, LaGatutta D, Youngen J (2001) A comparables approach to measuring cashflow-at-risk for non-financial firms. J Appl Corp Financ 13(4):100–109. https://doi.org/10.1111/j.1745-6622.2001.tb00430.x

Developing an Algorithm to Study the Vulnerability of Occupations to Artificial Intelligence in the Framework of Sustainable Development Research

Michail V. Firsov, Olga N. Filatova, Marina N. Bulaeva, Margarita I. Koldina, and Andrey M. Morozov

Abstract The global spread of artificial intelligence (AI) across all aspects of human life prompts deep philosophical questions: what roles will remain for people, which professions should we prepare the next generation for, and how should society adapt to changes with no clear answers. The research examines the problem of the vulnerability of professions to AI. In 2023, experts at Goldman Sachs projected that AI could displace up to 300 million jobs. However, they also noted that AI would simultaneously create new roles and boost productivity in those that remain. Meanwhile, the International Monetary Fund (IMF) forecasted a much sharper rise in costs associated with employing young, highly skilled, and well-paid workers, driven by the growing integration of AI technologies. One of the tools for monitoring the sustainable development of organizations is the implementation of the ESG (Environmental, Social, and Corporate Governance) approach. This approach, focused on addressing global environmental, social, and governance (ESG) challenges, enables the integration of relevant monitoring mechanisms directly into an enterprise's business processes. The potential of ESG ratings as tools for evaluating sustainable development has been explored, including the introduction of a model that accounts for positive and negative cumulative assessments. Additionally, an algorithm has been developed to assess the vulnerability of various professions to AI. This involves selecting projects of varying complexity aligned with students' skill levels. Key data elements and procedures necessary for designing training datasets have been identified. Furthermore, an automated service framework has been created to support the large-scale development of custom neural networks.

M. V. Firsov · O. N. Filatova (✉) · M. N. Bulaeva · M. I. Koldina
Minin Nizhny Novgorod State Pedagogical University, Nizhny Novgorod, Russia
e-mail: olgaf91074@yandex.ru

A. M. Morozov
Lobachevsky Nizhny Novgorod State University, Nizhny Novgorod, Russia

A. B. Karbekova (ed.), *Sustainable Development of Rural Territories Based on Climate-Smart Technologies*, Climate Change Management,
https://doi.org/10.1007/978-3-032-20106-5_42

Keywords ESG · Rating · Profession · Sustainable development · Business process · Machine learning · Information system · Monitoring system · Professional education

JEL Classification M15 · O14 · O32 · O33

1 Introduction

The global penetration of artificial intelligence (AI) into all areas of human activity prompts us to consider philosophical questions about what a person will do, which professions to teach the younger generation, and other questions that do not have a clear answer (Izilyaeva et al. 2024). The research examines the vulnerability of professions to the impact of artificial intelligence. Having analyzed this problem, we propose a practical solution—the use of the ESG (Environmental, Social, and Corporate Governance) approach. This approach, which involves participation in solving global environmental, social, and management problems, also allows creating and integrating appropriate monitoring processes into the business processes of the enterprise.

2 Literature Review

In the field of technological leadership policy, the problems and prospects for the application of technology are considered in the works of Russian and foreign authors.

L. O. Izilyaeva, Ya. K. Vasiliev, K. S. Mirokiyants, and A. I. Yasavieva (Bylieva 2023; Izilyaeva et al. 2024; Karasev and Klimachev 2024) determined that a wide range of opportunities and challenges represent the prospects for the development of the application of AI in politics and economics.

A. V. Gushchin, O. N. Filatova, M. N. Bulaeva, and others emphasize that changes associated with the use of AI should affect professional education, which should prepare specialists for the digital economy (Filatova et al. 2022; Firsov et al. 2020; Markova et al. 2021).

E. V. Slavutskaya, L. A. Slavutskii, and M. V. Firsov consider how machine learning methods can be applied to analyze data and processes (Firsov 2024; Slavutskaya and Slavutskii 2024).

3 Materials and Methods

3.1 Analysis of Trends in the Impact of IT and AI Development on the Sustainability of Economic Development

According to the International Monetary Fund (IMF), AI will affect up to 40% of jobs worldwide and worsen inequality in access to technology. According to IMF analysts, AI is expected to impact up to 60% of all jobs in developed economies. The positive impact will be 50% when AI is used jointly with traditional jobs. Accordingly, in the remaining 50% of cases, AI will reduce the work that people do, which will lead to a decrease in demand for the corresponding professions and a decrease in wages (Karasev and Klimachev 2024).

In less developed economies, AI is expected to affect 26% of jobs due to slow integration into existing jobs and the elimination of fewer jobs. In 2023, Goldman Sachs experts predicted that AI would replace 300 million jobs. Simultaneously, new jobs will be created, and productivity will increase in the remaining ones. However, many low-income countries lack the necessary infrastructure and a sufficiently skilled workforce to leverage these advancements, thereby reinforcing existing technological disparities between nations.

The International Monetary Fund (IMF) forecasts a rapid rise in labor costs for young, highly skilled, and well-paid employees as a result of AI integration. However, the case of Russia presents a contrasting picture: wages for manual laborers are growing faster than those in other sectors. This discrepancy highlights the uneven impact of AI across different economies and labor markets. To address these shifts, governments must implement large-scale initiatives to support and retrain workers in professions most vulnerable to technological disruption. Such efforts are essential to protect living standards and mitigate growing technological inequality. The ongoing technological revolution, unfolding over several decades, continues to eliminate certain professions while giving rise to new ones. However, the core concern is not the disappearance of jobs but the sharp decline in the number of available positions. In most fields, the nature of job responsibilities is undergoing significant transformation, requiring workers to adapt to evolving demands. Equally important is ensuring the safe implementation of AI technologies. Achieving this requires a fundamental shift in legal frameworks and regulatory approaches. While leading countries are striving to define standards for how algorithms should be developed and applied, the AI sector, like any other, resists strict regulation due to its complexity and rapid evolution. Nevertheless, aspects of the development process can and should be governed by legislation. The legal domain is being reshaped by the accelerating advancement of information technology and AI.

Measuring the vulnerability of professions to AI requires studying the tasks that workers perform and comparing them with the capabilities of algorithms. This is primarily done by surveying workers. The number of people employed in jobs that could potentially be replaced by AI, such as filling out spreadsheets or analyzing data, has grown by 5% in five years. The demand for neural network training specialists has grown 30-fold. Expenses for marketers and copywriters are rapidly decreasing as they reach an average level of quality. There has been a significant decline in employment across several traditional sectors, with the number of agricultural workers falling tenfold, alongside sharp reductions in roles such as machinists, foresters, fishermen, and market sellers. These trends highlight the uneven impact of AI on job creation and destruction.

While official unemployment rates remain historically low (6% in EU countries, 3.5% in the USA, and 2.8% in Russia) as of the beginning of 2025, these figures may not fully reflect reality. In Russia, many unemployed individuals do not register with employment services due to the minimal benefits they receive. Residents of remote areas report that the cost of traveling to claim these benefits often exceeds the value of the payments themselves, leading to widespread underreporting. The broader discourse around automation and AI's impact on employment remains contentious. Studies vary widely in their projections, with no unified picture emerging. Analysts often reference historical precedents (e.g., the disappearance of occupations like coachmen, telegraphists, and telephone operators) to argue that society has always adapted to technological progress. This narrative supports the belief that, despite job losses, humanity continues to evolve and thrive. These forecasts are not supported by comprehensive statistics by industry and type of activity. However, AI developers are not as optimistic and insistently advise governments to consider unconditional basic income and changes to employment. The salary will be an additional part of the unconditional basic income.

The following conclusions can be drawn:

- The rate of technology adoption has increased dramatically over the past 50 years;
- Nowadays, the IT sector is reducing traditional jobs faster than new ones are being created. We need to talk about new jobs with an adjustment for the time it takes to prepare workers for them (i.e., training). Constant retraining of employees in new skills over several months occurs only on paper. The majority of people struggle to quickly acquire new skills, especially when those skills do not spark personal interest. Learning tends to be significantly more effective when it aligns with a learner's curiosity or enthusiasm. For example, children often enjoy studying basic programming in school, but their engagement typically lasts only as long as the material remains interesting to them. Nevertheless, the younger generation holds a distinct advantage by being introduced to modern, in-demand skills from an early age (Firsov et al. 2020);
- The imbalance in automation between the introduction of robots and software development. Russia's experience shows that the introduction of robots is lagging behind (Merkushova and Filatova 2023). Accordingly, jobs in the service sector and among office workers are being eliminated more rapidly. There are relatively few robots in production, primarily due to the limited scale of localized manufacturing in Russia. The most rapid adoption of robots could be expected in the logistics sector, which overlaps with trade. However, in practice, we are

witnessing an expansion of mass employment in the form of courier work (Petrov et al. 2020);

- Less developed countries have always had less access to technology. More developed countries have tried to take control of their resources. With the increasing use of mobile technologies (including small machines, 3D printers, advanced equipment and devices, and drones), employment opportunities are becoming available in remote and less developed regions. Global and infrastructure technologies that require the concentration of capital will develop only in developed countries;
- Movement of labor. The choice remains between the influx of labor and the transfer of technology to less developed countries. How much less expensive is the influx of labor? It is necessary to consider the increased costs of social infrastructure and benefits, as well as the degradation of domestic production;
- The decrease in automation efficiency caused by cheaper labor results in a temporary economic benefit. Simultaneously, it undermines the scientific and technological foundation. Rebuilding and strengthening this base requires substantial, large-scale investments. These dynamics contribute to stress and instability for the economy and citizens. In some cases, it can even lead to military conflict due to the deterioration of national defense capabilities. Abrupt changes are rarely welcomed;
- Automation is developing faster in private companies compared to state ones under the condition of labor shortage or its high cost. With the mass import of labor, the pace of automation is sharply reduced. However, the burden on all social funds continues to increase. They are spread out among new arrivals with families and older people. In 2024, the share of pensions in relation to the average salary in Russia decreased to 26%. This marks a significant decline from the recent level of 36%, which was much closer to the internationally recommended benchmark;
- The lack of financial transparency and unwillingness to pay taxes sharply slows down automation;
- Too liberal an economic policy always goes hand in hand with corruption and betrayal. The accumulation of a national wealth fund, the placement of assets in international securities, the monthly purchase of currency to prevent the strengthening of the ruble against the backdrop of losses from sanctions and the confiscation of international currency reserves over time looks like an attempt to prevent investments in domestic technological infrastructure and an increase in the well-being of citizens with a strong ruble;
- The accelerated pace of automation in recent years has not resulted in a reduction of the working week. Historically, when Henry Ford introduced the progressive assembly line in his factories, he simultaneously reduced the workday to eight hours and raised workers' wages. This combination of increased income and more free time fueled economic growth in the USA. However, such developments are not being replicated today. Despite significant advancements in automation, we see no similar initiatives from large or medium-sized companies;
- In cases where a company lacks immediate financial resources to motivate employees, a mechanism can be applied that provides additional days off in exchange for achieving high productivity indicators and scientific results. This

mechanism will make it possible to launch the process of implementing a four-day work week;

- In Russia, the trend is quite the opposite. Due to the decline in the purchasing power of wages, more people are taking on second and even third jobs. With the average salary in many regions of the Russian Federation ranging between 60,000 and 70,000 rubles, a worker earns only two pension points out of a possible ten;
- In countries like China and India, automation has given many people free time. This is indirectly evidenced by the first two places for India and China in the World Chess Championship and the rapid development of science in China;
- It is worth considering the application of positive and negative point assessments within the ESG rating system. In this model, all positive and negative points would accumulate, resulting in an overall score that could shift in either direction. Such a system would carry corresponding consequences for companies, especially if supported by appropriate legislation that formally integrates ESG ratings into regulatory frameworks;
- Automatic algorithms for marking and training neural networks are being actively developed. Therefore, the initial surge in employment in this area will stabilize.

Thus, we live in an era of rapid and large-scale change. In this connection, the importance of ensuring sustainable development at the level of countries and the world increases.

3.2 Using ESG Ratings to Monitor Sustainable Development

One of the tools for monitoring the sustainable development of organizations is the implementation of the ESG approach. This approach, which focuses on addressing global environmental, social, and governance challenges, also enables the development and integration of relevant monitoring processes into an enterprise's core operations. Adherence to ESG principles is increasingly viewed as a key indicator of a company's social responsibility. This approach usually relies on rating assessments to evaluate performance in these areas for the benefit of investors and the public. However, in the future, it has strong potential to complement the established ISO 9000 standards, which guide the implementation of quality management processes. Under such integration, additional processes would be introduced to support sustainable development and the continuous monitoring of related activities. ESG is closely aligned with the concept of corporate social responsibility (CSR), though the two differ in scope and governance structure. While CSR spans all levels of organizational management, ESG is generally developed and implemented at the highest levels of corporate leadership, often with direct implications for financial and credit decision-making.

The main areas of ESG are as follows:

- Solving environmental problems: Control of harmful emissions, implementing sustainable production technologies, and ensuring efficient use of natural resources;
- Solving social issues of employees: Ensuring the physical and mental health of employees;
- Management criteria: Analysis of the performance of managers, ensuring transparency of reports, anti-corruption control, and regulation of relations with shareholders.

Other ESG areas include IT, investment, raw materials, finance, and marketing. Criteria for the ESG rating are developed for each area.

In general, statistics show that companies with a high ESG rating demonstrate more stable and even higher financial and economic indicators compared to companies that do not participate in the rating. This approach enhances the effectiveness of investors' decisions when allocating funds to a company's financial instruments. Independent research agencies (e.g., Bloomberg, S&P Dow Jones Indices, JUST Capital, MSCI, and Refinitiv) compile ESG ratings for organizations using a standardized 100-point scale.

In 2021, Raex-Europe, a rating agency, published the first ESG ratings of Russian companies, including MTS, Sberbank, Russian Railways, and Rosatom. To further encourage organizations toward sustainable development, the authors proposed a model that incorporates both positive and negative cumulative point assessments.

3.3 Designing an Algorithm to Study the Vulnerability of Professions to Artificial Intelligence

To assess the risks AI poses to various professions, it is important to analyze the specific tasks workers perform and compare them with the current capabilities of algorithms. This assessment is typically carried out through worker surveys (Markova et al. 2021). However, there is a need to move beyond surveys toward more meaningful data. To achieve this, research should be conducted within professional educational institutions, involving students who are closely connected to practical experience. Projects of varying complexity should be selected to match the students' skill levels, enabling a more accurate evaluation of how AI may impact different job functions (Firsov 2024).

This can be implemented through the following directions:

- Search for already implemented projects and their results;
- Selection of the most significant common sets of actions by profession and their implementation separately and in complex tasks by the most popular neural networks (e.g., ChatGPT and others). Simultaneously, it is possible to compare the capabilities of various neural networks and language models;

- Implementation of custom neural networks within the framework of educational and diploma projects (Slavutskaya and Slavutskii 2024).

When designing neural networks on a large scale, the challenge arises of automating typical user tasks (e.g., data preparation, model training, and result presentation). At this point, there is a decision to be made: whether to use existing services or to develop custom solutions.

The latter is a non-trivial task. Let us consider a possible architecture of such a service. The main approach to the architecture will be modular (Slavutskaya and Slavutskii 2024).

Through the web interface, the user uploads a dataset for training and can choose to train the model on the entire database or apply filters to select specific data. The interface also provides tools to monitor training statistics and view the results. Once training is complete, the model can be tested. The required datasets for training depend on the specific application area. Regardless of the domain, all data included in the dataset must be anonymized in advance, ensuring that any personal information is removed prior to processing. The data may look as follows:

- In trade and sales, it is necessary to save all information during the sales stages to select sales tools and channels;
- In medicine, diagnoses, symptoms, and results of various treatment options are stored;
- In the banking sector, when issuing loans, a set of data is saved that is used to assess the creditworthiness of clients;
- For regression forecasting of sales volumes, macroeconomic and other market and sales volume indicators are retained;
- For production, it is important to determine defects based on the processing of product images. Therefore, it is necessary to save large images;
- Automatic counting of the number of products and visitors will require storing images of products as well as images representing various visitor actions, especially if action identification is needed. This image data serves as the basis for training models to accurately detect and interpret both quantities and behaviors;
- Recognition of elements in drawings and generation of a description of the technological process for manufacturing parts will require saving images of drawing elements and compiling tables of correspondence between drawing elements and the necessary processing operations.

Initially, the user enters the following parameters: subtask type, training type, metric, and training sample.

An automated service may consist of the following components: a user interface on the frontend; a backend for executing and transferring frontend commands to various internal services; a typical, customizable process for training and applying models for classification and prediction; a service for visualizing results; and a data preprocessing and labeling service that automatically extracts data in parts from a running database (Bylieva 2023).

4 Results

The Entity Embeddings approach is used in machine learning model training. It transforms words and phrases, images, product categories, user characteristics, and other metrics into continuous vectors within a multidimensional space. During training, categorical variables with thousands of distinct categories are represented as low-dimensional vectors, typically ranging from 50 to 100 dimensions. In this vector space, similar entities are positioned near each other, which enhances the accuracy of predictions.

To start the training process, the parameters (configuration) of the model training are transferred from the front to the backend. If the model is constantly rewritten, we can use separate GIT branches in Gitlab or save separate files with the code for the models. Datasets are stored on Hadoop (or in a simple database). The processes are started from Gitlab.

Model images are built on Jenkins or Gitlab. Then, DAG (Directed Acyclic Graph) is launched in Airflow (a set of tasks that are executed in a strictly defined sequence according to the schedule). The model code starts in a Docker container and is managed on a pod in Kubernetes. We will also need a queue (e.g., RabbitMQ) to launch containers with limited memory and processor power.

Implementing an automated model training service enables a focus on exploring data and models, rather than preparing them for training.

5 Conclusion

The trends in the impact of IT and AI development on the sustainability of economic growth are analyzed. The analysis highlights the shortcomings of the liberal economic model in Russia, which fails to ensure sustainable development in terms of economic and demographic indicators. It also falls short in areas such as technological advancement, effective utilization of the labor force, and the implementation of robotics. Additionally, this model contributes to a decline in national defense capability and overall life satisfaction.

The potential of ESG ratings as a tool for assessing sustainable development has been examined. A model has been proposed that incorporates both positive and negative cumulative point assessments. Additionally, an algorithm has been developed to evaluate the vulnerability of professions to AI. This involves selecting projects of varying complexity that align with the skill levels of students. The core data and processes required for designing training datasets have been identified. Furthermore, a scheme for an automated service has been created to support the large-scale development of proprietary neural networks.

References

Bylieva DS (2023) Collective and artificial intelligence. Vestn Minin Univ 11(2):13

Filatova ON, Bulaeva MN, Gushchin AV (2022) Application of neural networks in professional education. Probl Mod Pedagog Educ 73(3):243–245

Firsov MV (2024) The concept of developing security components based on the development of the company's logistics business processes. Russ J Water Transp 79(2):164–175. https://doi.org/10.37890/jwt.vi79.481

Firsov MV, Filatova ON, Gushchin AV (2020) Advanced training in future skills through the development of computer-based simulators and digital assistants with artificial intelligence. Tidings Balt State Fish Fleet Acad Psychol Pedagog Sci 3(53):11–16

Izilyaeva LO, Vasiliev YK, Mironiyants KS, Yasavieva AI (2024) The possibilities and risks of using artificial intelligence in the field of political relations of The Russian Federation. Econ Manag Sci Pract J 1(175):136–139. https://doi.org/10.34773/EU.2024.1.24

Karasev NA, Klimachev TD (2024) Theoretical aspects of studying problems and prospects of artificial intelligence in the management of Russian companies. Creat Econ 18(2):337–356. https://doi.org/10.18334/CE.18.2.120490

Markova SM, Tsyplakova SA, Sedykx EP, Filatova ON, Khizhnaya AV (2021) University modernization in the conditions of industrialization of production and intelligent machines. In: Shakirova OG, Bashkov OV, Khusainov AA (eds) Current problems and ways of industry development: equipment and technologies. Springer, Cham, Switzerland, pp 940–947. https://doi.org/10.1007/978-3-030-69421-0_104

Merkushova ES, Filatova ON (2023) Application of robotics in the field of education. Bull MAkmulla Bashkir State Pedagog 3(S1–68):10–13

Petrov YN, Petrov AY, Petrova NS, Syrova NV, Filatova ON (2020) Change of the level of working capacity of a student in the system of professional education under mental load. In: Popkova E, Sergi B (eds) Scientific and technical revolution: yesterday, today and tomorrow. Springer, Cham, Switzerland, pp 925–931. https://doi.org/10.1007/978-3-030-47945-9_99

Slavutskaya EV, Slavutskii LA (2024) Error analysis of machine learning methods as the educational background for its use skills formation. Vestn Minin Univ 12(2):4

Sustainable Rural Development Through Climate-Smart Agricultural Technologies: A Vision for 2035

Aktam U. Burkhanov, Nuraiym Timur kyzy, Shuhrat Z. Muxitdinov, Kalybek Z. Abdykadyrov, and Elena G. Popkova

Abstract The research aims to identify the requirements of the digital economy in rural areas whose fulfillment fosters the development of climate-smart agriculture and encourages population inflows into these territories. To this end, it summarizes statistics from 20 countries with the highest levels of digital technology adoption in 2025. The authors constructed an economic-mathematical model to describe how the sustainability of rural development depends on the regulatory approaches to the digital economic development of these areas in 2025. The research concludes that the primary requirement for the digital economy of rural territories, one whose fulfillment promotes the growth of climate-smart agriculture and attracts people to rural areas, is the robustness of the legal and social institutions that support climate-smart agricultural practices in these regions. A regulation-enhanced forecast is presented, according to which, by 2035, the share of rural residents in the total population of the sampled countries is expected to rise to 28.23%, in contrast to the decline to 21.75% projected by an alternative regulation-neutral forecast. The resulting projections reveal the prospects for the sustainable development of rural territories depending on the regulatory environment governing climate-smart agricultural technologies through 2035.

A. U. Burkhanov
Alfraganus University, Tashkent, Uzbekistan

A. U. Burkhanov (✉)
National University of Uzbekistan named after Mirzo Ulugbek, Tashkent, Uzbekistan
e-mail: burxanov.a@afu.uz; a.burkhanov@nuu.uz

N. Timur kyzy
Osh State University, Osh, Kyrgyzstan

S. Z. Muxitdinov
Namangan State Technical University, Namangan, Uzbekistan

K. Z. Abdykadyrov
Kyrgyz State University named after I. Arabaev, Bishkek, Kyrgyzstan

E. G. Popkova
Armenian State University of Economics, Yerevan, Armenia

Pan-European University (PEU), Bratislava, Slovakia

A. B. Karbekova (ed.), *Sustainable Development of Rural Territories Based on Climate-Smart Technologies*, Climate Change Management,
https://doi.org/10.1007/978-3-032-20106-5_43

Keywords Climate resilience · Sustainability of rural areas · Climate-smart technologies · Agriculture · Outlook 2035

JEL Classification P25 · Q16 · Q18 · Q54

1 Introduction

The emergence of digital economies as an advanced organizational and technological form of economic activity at the national scale has accelerated urbanization. With the advent of the Fifth Industrial Revolution, urban areas have become especially attractive for living, thanks to superior telecommunications infrastructure, and employment, due to the concentration of knowledge-intensive jobs available to digital professionals.

By now, this has heightened the problem of rural depopulation to a critical level. Developing economic and managerial mechanisms that would encourage population flows back into rural areas has become urgent for several reasons. First, urban spaces have become overcrowded, reducing the comfort of living in them. Densely built-up and heavily populated urban environments can no longer guarantee the previous quality of life to new internal migrants arriving from rural areas and require alleviation.

Second, the urban environment is becoming increasingly unfavorable due to the significant volume of non-recycled household and industrial waste, including air pollution caused by motor vehicles. Therefore, supporters of green communities among city dwellers are interested in relocating to rural areas, where the environment is considerably more favorable.

Third, the current level of digital economic development is sufficiently high to provide not only urban but also rural territories with telecommunications infrastructure that enables the active use of smart technologies. This creates the technological groundwork for the widespread adoption of climate-smart agricultural practices in rural areas, allowing them to develop more sustainably.

The challenge lies in the lack of clarity regarding the requirements for the cyber-social environment of rural territories, requirements that must be met to facilitate the spread of climate-smart agriculture and ensure sustainable rural development. This uncertainty needs to be addressed because it reduces the effectiveness of the digital modernization of rural economies, whose contribution to stimulating sustainable development remains insufficient to halt the ongoing population outflow.

This has defined the research objective, which is to identify and scientifically justify the requirements for the digital economy of rural territories whose fulfillment promotes the development of climate-smart agriculture and stimulates the inflow of population into rural areas.

2 Literature Review

The concept of climate-smart agriculture as a source of sustainable economic development in rural areas is reflected in the works of Hakhverdyan and Badadyan (2023) and Litvinova et al. (2016), providing the theoretical foundation for the present research. According to this concept, the following serve as measures of sustainable rural development:

- The yield of grain crops, reflecting labor productivity in agriculture (Mkrtchyan et al. 2023);
- The gap between potential and actual grain yields, indicating the degree to which the potential for productivity growth in agriculture has been realized (Drobot et al. 2024);
- Performance in achieving the Sustainable Development Goals (SDGs) related to food security (SDG 2), wage levels and economic growth (SDG 8), the development of telecommunications and innovation activity (SDG 9), and the effectiveness of climate-change mitigation (SDG 13) (Sozinova et al. 2023);
- The share of rural residents in a country's total population (Bogoviz et al. 2016).

The regulatory mechanisms of digital development in rural economies that affect the conditions for implementing climate-smart agriculture include the following:

- Ensuring the availability of digital agricultural personnel in rural areas, specialists competent in the use of climate-smart technologies in agriculture (Galoyan et al. 2023a);
- Developing telecommunications infrastructure in rural areas to support the application of climate-smart agricultural technologies (Mkrtchyan and Navasardyan 2024);
- Strengthening the legal and social institutions that govern climate-smart agricultural practices in rural territories (Matevosyan et al. 2024);
- Expanding access for rural agribusinesses to climate-smart agricultural innovations (Benčič et al. 2020);
- Stimulating innovative agribusiness projects in rural areas aimed at adopting climate-smart technologies (Galoyan et al. 2023b);
- Public budget financing for the development of climate-smart agriculture in rural territories (Popkova and Abdurakhmanova 2024);
- Enhancing the investment attractiveness of innovative agribusiness projects in rural areas focused on climate-smart technologies (Konina 2024).

A review of the available literature on this topic reveals that it does not specify the consequences of employing the wide range of existing regulatory mechanisms for the digital development of rural economies with regard to the spread of climate-smart agriculture and the sustainability of rural development. This lack of clarity constitutes a gap in the literature and raises the research question: "Which approaches to digital development in rural economies facilitate the adoption and use of climate-smart agricultural technologies and encourage population inflow into rural territories?"

3 Materials and Methods

To answer the research question posed, the authors conducted a regression analysis to evaluate how the sustainability of rural development, measured by grain crop yield ($RtSd_1$); the ratio of actual to potential grain yield ($RtSd_2$); performance on SDG 2 ($RtSd_3$), SDG 8 ($RtSd_4$), SDG 9 ($RtSd_5$), and SDG 13 ($RtSd_6$) based on UNDP statistics (UNDP 2025); and the share of rural residents in a country's total population ($RtSd_7$) according to World Bank data (World Bank 2025), depends on the application of regulatory mechanisms for the digital development of rural economies, using Tortoise Media statistics (Media 2025), including the following:

- Availability of digital agricultural personnel (Rdd_1);
- Telecommunications infrastructure (Rdd_2);
- Strength of legal and social institutions (Rdd_3);
- Accessibility of smart agribusiness innovations (Rdd_4);
- Prevalence of innovative agribusiness projects (Rdd_5);
- Reliability of public budget financing for smart innovations (Rdd_6);
- Investment support for innovative agribusiness projects (Rdd_7).

The research sample consists of 20 countries with the highest levels of digital technology adoption, ranking as leaders in the artificial intelligence economy in 2025 according to Tortoise Media (2025). The empirical data used in the research are presented in Table 1 (outcome variables) and Table 2 (factor variables).

Drawing on the regression equations for $RtSd_{1-7}$, the research identifies those regulatory mechanisms of digital development in rural economies whose implementation enhances the sustainability of rural development, namely, the Rdd_{1-7} factors whose regression coefficients are positive in all equations. Using 2015 statistics in comparison with 2025 data from Table 1, the ten-year change in the share of rural residents in the total population of the sampled countries is assessed.

The prospects for sustainable rural development are then determined depending on the regulation of climate-smart technologies in agriculture through 2035. To this end, two alternative forecasts are constructed. The first, a regulation-neutral forecast, extrapolates the 2015–2025 trend into the period 2025–2035. The second, a regulation-enhanced forecast, is derived by substituting the maximum possible values of the selected factor variables from among $RtSd_{1-7}$ into the regression equations for Rdd_{1-7}.

4 Results

The results obtained through the regression analysis of the statistical data in Table 1 are reflected in the following regression equations, which show how the sustainability of rural development depends on the application of regulatory mechanisms for the digital development of rural economies in 2025:

Table 1 Sustainable development of rural areas in 2025

Country	Cereal yield (tons per hectare of harvested land)	Yield gap closure (% of potential yield)	Goal 2 score (0–100)	Goal 8 score (0–100)	Goal 9 score (0–100)	Goal 13 score (0–100)	Rural population (% of total population)
	$RtSd_1$	$RtSd_2$	$RtSd_3$	$RtSd_4$	$RtSd_5$	$RtSd_6$	$RtSd_7$
Austria	6.90	69.89	77.91	77.70	97.66	71.07	40.18
Belgium	8.60	77.25	74.94	78.24	97.98	67.73	1.78
Czech Republic	5.93	57.80	66.51	85.05	76.73	74.64	25.26
Denmark	7.24	66.53	77.57	81.02	97.22	73.53	11.37
Estonia	4.23	40.70	68.00	75.83	82.22	71.90	29.98
Finland	3.77	52.66	66.97	81.69	96.69	75.03	14.13
France	6.66	75.09	78.88	81.61	93.13	82.23	17.96
Germany	7.13	77.19	76.19	83.65	96.85	72.64	22.11
Greece	4.14	51.59	68.05	74.96	77.22	79.96	19.02
Hungary	4.03	63.28	69.35	77.95	74.09	84.17	26.82
Ireland	8.70	74.57	69.92	79.22	88.60	68.59	35.24
Italy	4.75	58.94	72.65	78.65	89.44	81.86	27.71
Luxembourg	6.15	65.01	67.37	76.21	89.51	48.77	7.73
Netherlands	8.94	76.18	73.32	81.94	94.74	68.39	6.55
Poland	4.86	48.07	69.32	82.18	77.84	80.48	39.67
Slovakia	4.76	48.19	71.08	75.31	70.47	79.47	45.83
Slovenia	5.57	65.56	68.06	79.67	82.70	76.99	43.56
Spain	3.31	44.52	69.85	81.72	83.90	83.64	18.20
Sweden	6.11	68.96	72.37	80.98	98.46	78.96	11.02
Switzerland	6.08	64.12	74.78	74.62	98.33	64.55	25.67

Source Developed by the authors based on the materials from the UNDP (2025) and World Bank (2025)

$$\begin{aligned} RtSd_1 &= 5.090 + 0.226Rdd_1 - 0.032Rdd_2 \\ &+ 0.008Rdd_3 - 0.237Rdd_4 + 0.101Rdd_5 \\ &- 0.036Rdd_6 - 0.024Rdd_7 \end{aligned} \tag{1}$$

In model (1), grain crop yield increases:

- By 0.226 tons per hectare of harvested land, with a one-point increase in the availability of digital agricultural personnel in rural areas;
- By 0.008 tons per hectare of harvested land with a one-point strengthening of the legal and social institutions governing climate-smart agriculture in rural areas;

Table 2 Regulation of climate-smart technologies in agriculture in 2025, score 0–100

Country	Talent,	Infrastructure	Operating environment	Research	Development	Government strategy	Commercial
	Rdd_1	Rdd_2	Rdd_3	Rdd_4	Rdd_5	Rdd_6	Rdd_7
Austria	15.00	22.00	61.00	9.00	3.00	33.00	6.00
Belgium	16.00	20.00	66.00	6.00	6.00	27.00	9.00
Czech Republic	10.00	19.00	47.00	4.00	3.00	45.00	4.00
Denmark	17.00	25.00	75.00	7.00	5.00	44.00	9.00
Estonia	9.00	19.00	59.00	4.00	0.00	29.00	13.00
Finland	17.00	33.00	81.00	9.00	13.00	39.00	13.00
France	25.00	31.00	70.00	18.00	31.00	59.00	19.00
Germany	35.00	32.00	83.00	16.00	14.00	59.00	17.00
Greece	10.00	20.00	27.00	5.00	3.00	17.00	7.00
Hungary	7.00	23.00	57.00	2.00	3.00	26.00	5.00
Ireland	13.00	26.00	69.00	5.00	14.00	31.00	12.00
Italy	16.00	23.00	100.00	7.00	2.00	53.00	5.00
Luxembourg	21.00	35.00	69.00	9.00	7.00	35.00	8.00
Netherlands	23.00	40.00	67.00	10.00	11.00	43.00	10.00
Poland	11.00	23.00	64.00	3.00	4.00	31.00	4.00
Slovakia	7.00	18.00	58.00	2.00	0.00	0.00	4.00
Slovenia	8.00	22.00	76.00	5.00	0.00	28.00	5.00
Spain	17.00	26.00	74.00	6.00	10.00	66.00	7.00
Sweden	17.00	26.00	88.00	8.00	5.00	23.00	12.00
Switzerland	30.00	34.00	50.00	18.00	11.00	18.00	12.00

Source Developed by the authors based on the materials from Tortoise Media (2025)

- By 0.101 tons per hectare of harvested land, with a one-point increase in the prevalence of innovative agribusiness projects.

$$\begin{aligned}\text{RtSd}_2 = 49.636 + 0.858\text{Rdd}_1 - 0.195\text{Rdd}_2 + 0.181\text{Rdd}_3 \\ - 0.117\text{Rdd}_4 + 0.736\text{Rdd}_5 - 0.223\text{Rdd}_6 - 0.542\text{Rdd}_7\end{aligned} \tag{2}$$

In model (2), the ratio of actual to potential grain yield rises:

- By 0.858% with a one-point increase in the availability of digital agricultural personnel;
- By 0.181% with a one-point strengthening of legal and social institutions supporting climate-smart agriculture;
- By 0.736% with a one-point increase in the prevalence of innovative agribusiness projects.

$$\begin{aligned}\text{RtSd}_3 = 72.870 + 0.331\text{Rdd}_1\text{-}0.433\text{Rdd}_2 + 0.062\text{Rdd}_3 \\ + 0.397\text{Rdd}_4 + 0.203\text{Rdd}_5\text{-}0.049\text{Rdd}_6\text{-}0.258\text{Rdd}_7\end{aligned} \tag{3}$$

In model (3), the SDG 2 indicator increases:

- By 0.331 points with a one-point rise in the availability of digital agricultural personnel;
- By 0.062 points with a one-point strengthening of legal and social institutions for climate-smart agriculture;
- By 0.397 points with a one-point expansion of agribusiness access to climate-smart innovations;
- By 0.203 points with a one-point increase in the prevalence of innovative agribusiness projects.

$$\begin{aligned}\text{RtSd}_4 = 74.732 + 0.119\text{Rdd}_1 - 0.008\text{Rdd}_2 + 0.021\text{Rdd}_3 - 0.363\text{Rdd}_4 \\ + 0.135\text{Rdd}_5 + 0.108\text{Rdd}_6 - 0.052\text{Rdd}_7\end{aligned} \tag{4}$$

In model (4), the SDG 8 indicator rises:

- By 0.119 points with a one-point increase in the availability of digital agricultural personnel;
- By 0.021 points with a one-point strengthening of legal and social institutions for climate-smart agriculture;
- By 0.135 points with a one-point increase in the prevalence of innovative agribusiness projects;
- By 0.108 points, with a one-point increase in public budget financing for the development of climate-smart agriculture in rural areas.

$$\begin{aligned}\text{RtSd}_5 = 60.496 + 0.233\text{Rdd}_1 + 0.123\text{Rdd}_2 + 0.194\text{Rdd}_3 \\ + 0.857\text{Rdd}_4 - 0.366\text{Rdd}_5 - 0.037\text{Rdd}_6 + 0.568\text{Rdd}_7\end{aligned} \tag{5}$$

In model (5), the SDG 9 indicator increases:

- By 0.233 points with a one-point rise in the availability of digital agricultural personnel;
- By 0.123 points with a one-point improvement in telecommunications infrastructure in rural areas;
- By 0.194 points with a one-point strengthening of legal and social institutions for climate-smart agriculture;
- By 0.857 points with a one-point expansion of agribusiness access to climate-smart innovations;
- By 0.568 points, with a one-point increase in the investment attractiveness of innovative agribusiness projects in rural areas.

$$\begin{aligned} RtSd_6 = 87.512 - 0.579Rdd_1 - 0.742Rdd_2 + 0.107Rdd_3 \\ + 0.523Rdd_4 + 0.506Rdd_5 + 0.099Rdd_6 - 0.335Rdd_7 \end{aligned} \tag{6}$$

In model (6), the SDG 13 indicator increases:

- By 0.107 points with a one-point strengthening of the legal and social institutions governing climate-smart agriculture in rural areas;
- By 0.523 points with a one-point expansion of agribusiness access to climate-smart innovations;
- By 0.506 points with a one-point increase in the prevalence of innovative agribusiness projects;
- By 0.099 points with a one-point increase in public budget financing for the development of climate-smart agriculture in rural areas.

$$\begin{aligned} RtSd_7 = 50.672 - 1.102Rdd_1 - 0.794Rdd_2 + 0.144Rdd_3 \\ + 2.156Rdd_4 + 0.199Rdd_5 - 0.162Rdd_6 - 1.144Rdd_7 \end{aligned} \tag{7}$$

In model (7), the share of rural residents in a country's total population increases:

- By 0.144% with a one-point strengthening of the legal and social institutions supporting climate-smart agriculture;
- By 2.156% with a one-point expansion of agribusiness access to climate-smart innovations;
- By 0.199% with a one-point increase in the prevalence of innovative agribusiness projects.

Summarizing the results of models (1–7), it is established that the only regulatory mechanism of digital development in rural economies whose implementation improves rural development sustainability across all its measured dimensions is the strengthening of the legal and social institutions governing climate-smart agriculture in rural areas (the regression coefficients are positive in every equation only for Rdd_3).

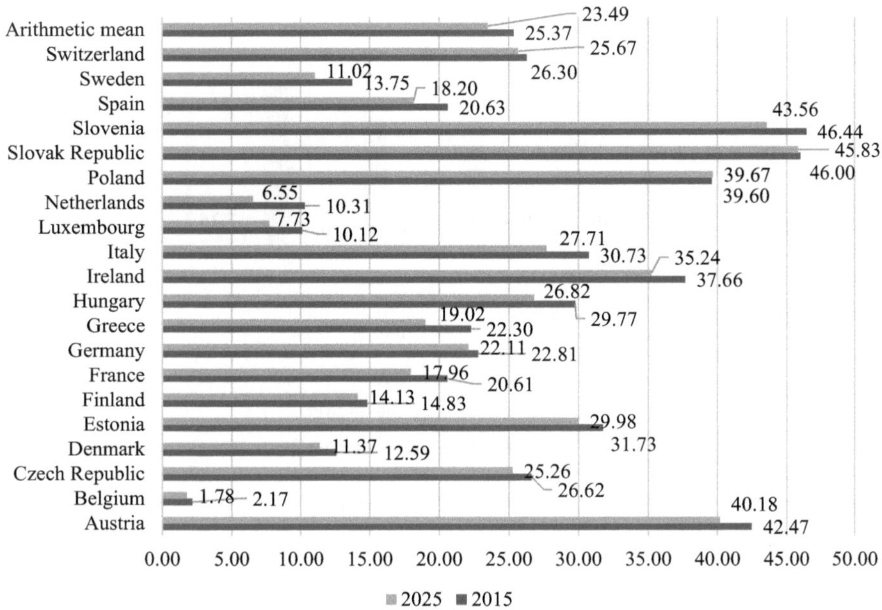

Fig. 1 Share of rural residents in the total population of the sample countries in 2015 and 2025, %. *Source* Developed by the authors

Using 2015 statistics illustrated in Fig. 1, compared with the 2025 data in Table 1, the ten-year change in the share of rural residents in the total population of the sampled countries has been assessed.

According to Fig. 1, the share of rural residents in the total population of the sampled countries decreased from 25.37% in 2015 to 23.49% in 2025, a decline by a factor of 0.93. Therefore, under the regulation-neutral forecast of sustainable rural development, which assumes that the 2015–2025 trend will continue through 2025–2035, the share of rural residents is expected to fall to 21.75% by 2035.

An alternative projection is offered by the regulation-enhanced forecast of sustainable rural development (Fig. 2), constructed by substituting the maximum possible value (100.00 points) of the selected factor variable (Rdd_3) into models (1–7).

As shown in Fig. 2, under the regulation-enhanced forecast, driven by a 49.14% strengthening of the legal and social institutions supporting climate-smart agriculture in rural areas relative to 2025, the following progress in sustainable rural development is expected by 2035:

- An increase in grain crop yield by 4.49%, from 5.89 tons per hectare of harvested land in 2025 to 6.16 tons per hectare by 2035;
- A rise in the ratio of actual to potential grain yield by 9.57%, from 62.30% in 2025 to 68.27% by 2035;
- An improvement in the SDG 2 indicator by 2.86%, from 71.65 points in 2025 to 73.70 points by 2035;

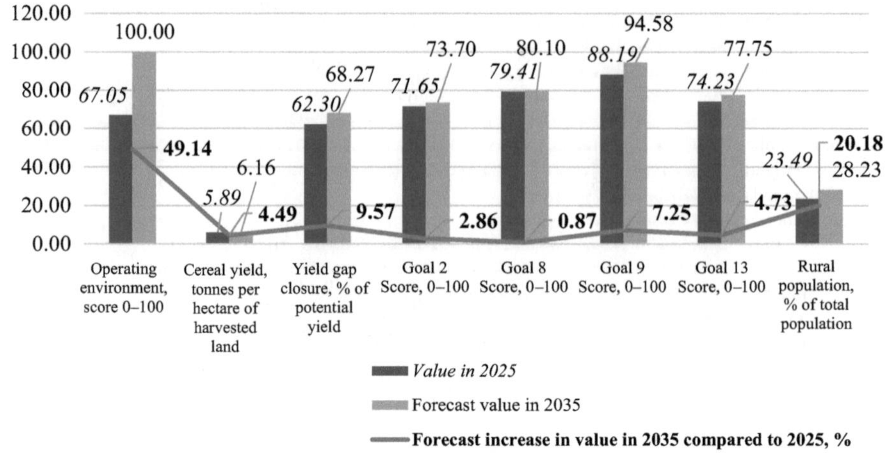

Fig. 2 Regulatory-enhanced forecast for sustainable development of rural areas until 2035. *Source* Developed by the authors

- An increase in the SDG 8 indicator by 0.87%, from 79.41 points in 2025 to 80.10 points by 2035;
- An increase in the SDG 9 indicator by 7.25%, from 88.19 points in 2025 to 94.58 points by 2035;
- An improvement in the SDG 13 indicator by 4.73%, from 74.23 points in 2025 to 77.75 points by 2035;
- An increase in the share of rural residents in the total population of the sampled countries by 20.18%, from 23.49% in 2025 to 28.23% by 2035.

5 Discussion

The findings in this research extend the literature of Bogoviz et al. (2016), Drobot et al. (2024), Hakhverdyan and Badadyan (2023), Litvinova et al. (2016), Mkrtchyan et al. (2023), and Sozinova et al. (2023), enriching the concept of climate-smart agriculture as a driver of sustainable economic development in rural areas by clarifying the consequences of applying existing regulatory mechanisms of digital development in rural economies for the diffusion of climate-smart agriculture and the sustainability of rural development.

Confirming the conclusions of Matevosyan et al. (2024), the analysis demonstrates that strengthening legal and social institutions comprehensively enhances rural sustainability across all its dimensions. Simultaneously, it is shown that other mechanisms of digital development in rural economies, such as providing digital agricultural personnel (contrary to the findings of Galoyan et al. (2023a)); expanding telecommunications infrastructure (contrary to the findings of Mkrtchyan and Navasardyan (2024)); increasing the availability of smart agribusiness innovations (contrary to the findings of Benčič et al. (2020)); stimulating innovative agribusiness projects (contrary to the findings of Galoyan et al. (2023b)); expanding public budget financing of smart innovations (contrary to the findings of Popkova and Abdurakhmanova (2024)); and strengthening investment support for agribusiness innovation (contrary to the findings of Konina (2024)), only partially support sustainable rural development, exerting a contradictory influence on the adoption of climate-smart technologies in agriculture.

6 Conclusion

The most significant result of this research lies in the development of an economic-mathematical model describing how the sustainability of rural development depends on the application of regulatory mechanisms for the digital development of rural economies in 2025. The conclusion derived from this model is that the key requirement of the digital economy in rural areas, one whose fulfillment promotes the development of climate-smart agriculture and encourages population inflows, is the strength of the legal and social institutions that support climate-smart agricultural practices in rural territories.

The model's practical value lies in its ability to determine the prospects for sustainable rural development depending on the regulation of climate-smart agricultural technologies through 2035. According to the regulation-enhanced forecast constructed in this study, by 2035, the share of rural residents in the total population of the sampled countries can be increased to 28.23%, instead of declining to 21.75% under the regulation-neutral forecast.

References

Benčič S, Kitsay YA, Karbekova AB, Giyazov A (2020) Specifics of building the digital economy in developed and developing countries. In: Popkova E, Sergi B (eds) Digital economy: Complexity and variety vs. rationality. Springer, Cham, Switzerland, pp 39–48. https://doi.org/10.1007/978-3-030-29586-8_5

Bogoviz AV, Ragulina YV, Kutukova ES (2016) Economic zones as a factor of increased economic competitiveness of the region. International Journal of Economics and Financial Issues 6(8SI):1–6

Drobot E, Petrenko Y, Hajduk L, Bencic S (2024) The concept of business ethics circles in the context of sustainable development. Journal of Infrastructure Policy and Development 8(7):5162. https://doi.org/10.24294/jipd.v8i7.5162

Galoyan DR, Mayilyan FN, Bazinyan YA, Movsisyan ME, Torosyan AA (2023a) Triple exploitation of human capital in developing countries. In: Popkova EG, Sergi BS (eds) ESG management of the development of the green economy in Central Asia. Springer, Cham, Switzerland, pp 375–383. https://doi.org/10.1007/978-3-031-46525-3_40

Galoyan D, Mkrtchyan T, Hrosul V, Buhrimenko R, Smirnova P, Balamut H (2023b) Formation of adaptation strategy for business entities in the context of digital transformation of the economy. Review of Economics and Finance 21(2023):616–621

Hakhverdyan D, Badadyan G (2023) Assessing disproportionate territorial development: Insights from 10 countries. Journal of Liberty and International Affairs 9(3):106–120. https://doi.org/10.47305/JLIA2393106h

Konina OV (2024) Improving business innovation management through artificial intelligence. Journal of Trends and Challenges in Artificial Intelligence 1(1):21–32. https://doi.org/10.61552/JAI.2024.01.003

Litvinova TN, Khmeleva GA, Ermolina LV, Alferova TV, Cheryomushkina IV (2016) Scenarios of business development in the agricultural machinery market under conditions of international trade integration. Contemporary Economics 10(4):323–332. https://doi.org/10.5709/ce.1897-9254.219

Matevosyan A, Grigoryan A, Khachatryan M, Matevosyan M, Israyelyan S, Hovakanyan L (2024) The recommended approach to EPS risk factor assesment. Journal of Lifestyle and SDGs Review 5(1):e03015. https://doi.org/10.47172/2965-730X.SDGsReview.v5.n01.pe03015

Mkrtchyan T, Navasardyan M (2024) Qualitative dimensions of economic growth: A new perspective. Proceedings on Engineering Sciences 06(2):863–874. https://doi.org/10.24874/PES06.02A.013

Mkrtchyan T, Bazinyan Y, Karapetyan N, Sargsyan H, Petrosyan N (2023) Overreactive to international markets? Pricing mechanisms and value chains in RA vegetable oil market. Proceedings on Engineering Sciences 05(4):601–608. https://doi.org/10.24874/PES05.04.002

Popkova EG, Abdurakhmanova GK (2024) The prospects for developing green supply chains based on implementing environmental technologies and responsible innovations in sustainable cooperation. In: Popkova EG, Kaurova OV, Maloletko AN (eds) Sustainable cooperation for the creation of green supply chains based on environmental technologies and responsible innovations. Springer, Cham, Switzerland, pp 3–12. https://doi.org/10.1007/978-3-031-70962-3_1

Sozinova AA, Shadiyeva AA, Giyazov AT, Litvinova SA (2023) Sustainable development goals as a milestone of strategic alliances: A viewpoint from the perspective of the game theory. Advances in Business Marketing and Purchasing 27:111–121. https://doi.org/10.1108/S1069-096420230000027012

Tortois Media (2025) The Global AI Index. Retrieved from https://www.tortoisemedia.com/data/global-ai (Accessed 5 December 2025)

UNDP (2025) The Sustainable Development Goals Report 2025. New York, NY: UN Publications. Retrieved from. https://desapublications.un.org/publications/sustainable-development-goals-report-2025 (Accessed 5 December 2025)

World Bank (2025) Rural population (% of total population). Retrieved from https://data.worldbank.org/indicator/SP.RUR.TOTL.ZS (Accessed 5 December 2025)

Results of Experimental Studies of New Biotechnological Processing of Rice Grains

Eltar A. Smailov, Ruslanbek N. Arapbaev, Zhyrgal T. Samieva, Bektur K. Omurzakov, and Mansurbek O. Ergeshov

Abstract In Kyrgyzstan, the technological process of post-harvest processing of rough rice differs significantly from that used in other rice-growing regions and countries. To enhance the nutritional value, quality, and taste of rice, a biotechnological fermentation process of rough rice is applied. During this process, complex biochemical reactions occur within the grain, which small-scale rural producers are unable to properly control. As a result, they rely on intuition, and the desired rice quality is not always achieved. The authors conducted field experiments using rice plant sheaves with panicles, as well as rough rice obtained through direct combine harvesting, processed with a new installation. The total duration of the experimental study with rice sheaves and panicles was 8 h and 48 min. During the experimental tests of rough rice harvested by direct combining on the new installation "A," the authors found that, due to the increased density of the rough rice, the steam generator operated at full power (75 kW) for 8 min, while the water temperature in the generator reached 94 °C. The total duration of the process was 17 h and 57 min. As a result, rice of the "Zarcha" type, with a light brown color, was obtained.

Keywords Rice · Rough rice grain · Steam-thermal treatment · Harvesting · Direct combining · Temperature · Moisture · Steam generator · Automation instruments

---

E. A. Smailov · Z. T. Samieva
Kyrgyz-Uzbek International University named after Batyraly Sydykov, Osh, Kyrgyzstan

E. A. Smailov (✉) · R. N. Arapbaev · M. O. Ergeshov
Osh State University, Osh, Kyrgyzstan
e-mail: eltar_uito@mail.ru; smailoveltar@gmail.com

R. N. Arapbaev
e-mail: rarapbaev@oshsu.kg

B. K. Omurzakov
Osh Technological University named after M.M. Adyshev, Osh, Kyrgyzstan

R. N. Arapbaev
Jalal-Abad State University, Manas, Kyrgyzstan

A. B. Karbekova (ed.), *Sustainable Development of Rural Territories Based on Climate-Smart Technologies*, Climate Change Management,
https://doi.org/10.1007/978-3-032-20106-5_44

JEL Classification Q160

1 Introduction

The primary technological method of rice harvesting applied worldwide is direct combining, which allows for the rapid collection of the crop from the field and the acquisition of rough rice for subsequent processing into finished rice products.

However, the technological process of post-harvest processing of rough rice in Kyrgyzstan differs significantly from that used in other rice-growing regions and countries. A biotechnological fermentation process of rough rice is applied to enhance the nutritional value, quality, and taste of rice. During this process, complex biochemical reactions occur within the grain, which small-scale rural producers perform by intuition and are unable to properly control. As a result, the required rice quality is not always achieved (Smailov et al. 2024).

The issue of determining the optimal duration of the process, during which it should be completed under natural conditions, remains unresolved and uncontrollable due to the absence of reliable monitoring and the inability to maintain proper natural steam-thermal treatment regimes. In some cases, this leads to the appearance of unpleasant odors of decay and spoilage of raw materials. Consequently, the maximum content of essential micro- and macroelements, as well as amino acids beneficial to human health, in the rice grain is reduced. This is particularly important in today's conditions, where the production of environmentally friendly, high-quality, and health-promoting products, as well as the preservation of the unique "brand" of rice produced in Kyrgyzstan, are of great significance.

Another drawback of the existing technology is the reliance on manual harvesting (cutting rice plants with panicles by hand), which reduces the efficiency of the harvesting process, increases labor costs, and prolongs the harvesting period, all of which are undesirable for local farmers. Therefore, the authors propose a technology for steam-thermal treatment of rough rice during harvesting by direct combining.

This research aims to develop and justify a technology for the biotechnological treatment of rough rice. The research task is to conduct field experimental studies using a new installation for the biotechnological processing of rough rice.

2 Materials and Methods

The object of the experimental research is an automatic steam-thermal installation, developed and manufactured with funding from the German GIZ project, based on patents (Shalpykov et al. 2017; Smailov et al. 2023), and subsequently refined according to the results of preliminary tests. The material used in the study consisted of rice varieties zoned for cultivation in the Uzgen district (Ministry of Agriculture, Food Industry and Melioration of the Kyrgyz Republic 2020). Field experimental studies were conducted in accordance with the established methodology (National Committee for Standardization (Gosstandart) 1980).

3 Results and Discussion

In general, rural producers cannot carry out high-quality natural steam-thermal treatment while maintaining the proper modes and process parameters, which determine the chemical composition of rice and the content of micro- and macroelements and amino acids beneficial to human health. Moreover, in such cases, it is impossible to state with certainty when the natural steam-thermal treatment process has been completed, since there is no way to monitor or confirm its conclusion.

Additionally, the globally accepted technology of harvesting rice plants by direct combining has only begun to be widely used in Kyrgyzstan over the past three years. It is developing rapidly. Currently, nearly 80% of rice fields are harvested using this method. However, the traditional Kyrgyz process of natural steam-thermal treatment of rough rice, essential for improving the quality and appearance of the renowned Uzgen rice, is not being applied at all under these new conditions. To compensate for this, farmers cultivating and producing rice attempt to carry out steam-thermal treatment by covering the rough rice with tarpaulins, layering straw on top, and leaving it for 25–30 days (Osh State University 2024).

Some producers even place layers of green plants up to 30 cm thick above and below the rice inside the tarpaulin. However, despite these efforts, fermentation of the rough rice does not occur, even after 30 days. At best, the rice obtained after threshing is beige in color, which significantly reduces its quality, consumer demand, and market value (by at least 25%). The main reason for this is the inability to create optimal conditions and regimes for proper natural steam-thermal treatment.

As a result, the total amino acid content beneficial to human health in such rice varieties is only about 18–20%, whereas in rough rice that has undergone proper natural steam-thermal processing, this figure is no less than 26%. This clearly demonstrates that, under direct combining conditions, it is impossible to perform the natural steam-thermal treatment process effectively in a natural environment. Therefore, it has become necessary to develop a specialized installation equipped with a steam generator to ensure the required processing conditions, along with automation instruments for controlling the technological process parameters. Solving this critical problem would significantly improve the quality of raw materials and increase demand for rice produced in Kyrgyzstan (Kochkonbaeva 2024).

Considering these circumstances, the authors developed an installation for automatic regulation of the entire steam-thermal treatment process of rough rice, for which a patent has been obtained (Smailov et al. 2023). This experimental installation performs the steam-thermal treatment process for both rice plant sheaves with panicles and rough rice obtained by direct combining. Based on this experimental design (Osh State University 2024; Smailov et al. 2023; Kochkonbaeva 2024) and theoretical calculations, a prototype installation was manufactured to conduct field experimental studies. The experimental investigations of the steam-thermal installation with automatic process-regime control are presented in the Data Availability section.

In the fall of 2024, production tests of the steam-thermal treatment installation for rough rice were carried out, featuring automatic control of technological process parameters. The process of loading rice sheaves and rough rice into the hopper for steam-thermal treatment is presented in the Data Availability section.

During experimental testing of the new installation for steam-thermal treatment of rice plant sheaves with panicles, it was established that after 4 min of steam generator operation at a total power of 75 kW, steam began to flow into the hopper of the steam-thermal treatment unit for rough rice. At that point, three heating elements (with a combined power of 45 kW) automatically switched off, leaving two elements (30 kW total) in operation. Steam began entering the hopper when the water temperature in the steam generator reached 88 °C. The optimal temperature inside the hopper (55–57 °C), necessary for steam-thermal treatment, was reached after 28 min. At this stage, steam penetrated not only the plant stems but also the rough rice grain.

Afterward, for 2 h and 40 min, all heating elements automatically shut off, while the temperature inside the hopper remained stable at 55 °C. The temperature inside both the hopper and the sheaves with panicles stabilized after 1 h and 43 min of operation. Subsequently, after 2 h and 40 min, as the temperature inside the hopper began to decrease, a single 15 kW heating element automatically switched on for 1 min and off for 3 min, repeating this cycle alternately until the end of fermentation. This alternating on-off cycle continued steadily until the steam-thermal treatment process was completed. The total duration of the steam-thermal treatment of rice sheaves with panicles was 7 h 05 min. The resulting rice had a pale brown color.

During experimental tests on the new installation using rough rice harvested by direct combining, it was found that, due to the higher density of the rough rice, the steam generator operated at full power (75 kW) for 8 min, twice as long as in the sheaf variant. Only when the water temperature in the steam generator reached 94 °C did the three 45 kW heating elements automatically shut off. Two heating elements (30 kW total) continued to operate for 1 h 16 min, until the steam generator temperature reached 100 °C, at which point all heating elements switched off.

Then, one 15 kW heating element operated in an alternating mode (2 min on, 2 min off) for 1 h 20 min, bringing the total steam generator operating time to 2 h 44 min. The optimal temperature of 55 °C in the hopper stabilized after 1 h and 55 min of operation. Afterward, the mode changed to 2 min on and 4 min off, maintaining the optimal 55 °C temperature inside the hopper. After 12 h 26 min, the steam generator began operating in 20 min on—20 min off cycles, continuing until 17 h 47 min of total treatment time. After that, all heating elements were switched off, and the installation continued to operate until 19 h 52 min. Since the process officially begins once the temperature inside the hopper and the rough rice stabilizes, the effective duration of the steam-thermal treatment was 17 h 57 min. The resulting rice was of the "Zarcha" variety, with a light brown color. Because the color of the rice could be determined immediately after treatment, before milling, further studies are planned to extend the effective processing time for biotechnological treatment of rough rice.

The conducted experimental studies indicate the necessity of continuing research on how the content of micro- and macroelements and amino acids changes depending on the set modes, efficiency, and duration of the steam-thermal treatment process of rough rice using the new installation. In other words, it is essential to determine the optimal regimes and process durations that yield high-quality rice enriched with components beneficial to human health.

In 2025, the authors plan to continue these studies, extending the process duration up to 30 h or until the quality indicators (e.g., the content of micro- and macroelements and amino acids) begin to decline. Further research will also include other varieties of rough rice and rice-growing regions, including those from Uzbekistan.

To monitor changes in the chemical composition of rice (specifically, its micro- and macroelements and amino acids) during the steam-thermal treatment of rough rice obtained by direct combining, samples were taken every four hours. The results of these analyses will be presented in our subsequent reports.

Since most farmers are now fully transitioning to the economically advantageous method of rice harvesting by direct combining, all research planned for 2025 will focus on rough rice harvested using this technology.

4 Conclusion

1. The total duration of the experimental study of rice plant sheaves with panicles was 8 h 48 min (starting at 16:12 and ending at 01:00). The temperature inside the hopper, within the rice sheaves with panicles, stabilized after 1 h 43 min. Subsequently, over the next hour, the 15 kW heating elements operated for a total of 15 min. The steam-thermal treatment process of the rice sheaves with panicles lasted 7 h and 5 min. The resulting rice had a pale brown color;
2. During experimental testing of the new installation using rough rice harvested directly from the field by a combine, it was established that, due to the higher density of the rough rice, the steam generator operated at full power (75 kW) for 8 min. The water temperature in the steam generator reached 94 °C, after which three heating elements with a total power of 45 kW automatically switched off. The remaining two heating elements (30 kW) continued to operate for 1 h 16 min, and when the steam generator temperature reached 100 °C, all heating elements were switched off. Afterward, one 15 kW heating element operated alternately (2 min on and 2 min off) until 17 h 47 min. The optimal temperature of 55 °C in the hopper stabilized after 1 h and 55 min of steam generator operation. All heating elements were then switched off until the end of the process (i.e., up to 19 h 52 min of total installation time). The total duration of the process was 17 h 57 min. The resulting rice was of the "Zarcha" type, with a light brown color;
3. The moisture content of the rice plant stems entering steam-thermal treatment was 39.6%, while the moisture content of the rough rice grain was 12.0%. After steam-thermal treatment, the moisture content of the rough rice decreased to 8.2%.

Data Availability The experimental investigations of the steam-thermal installation with automatic process-regime control, are available at https://figshare.com/ with the identifier https://doi.org/10.6084/m9.figshare.30205804.

The process of loading rice sheaves and rough rice into the hopper for steam-thermal treatment, is available at https://figshare.com/ with the identifier https://doi.org/10.6084/m9.figshare.30205825.

References

Ministry of Agriculture, Food Industry and Melioration of the Kyrgyz Republic (2020) State Register of Plant Varieties and Hybrids Approved for Use in the Kyrgyz Republic: Official Publication. Bishkek, Kyrgyzstan. Retrieved from https://tinyurl.com/5ey783z8 (Accessed 12 October 2025)

National Committee for Standardization (Gosstandart) (1980) GOST 24058–80. Agricultural machinery. Methods of operational-technological evaluation of machine complexes, special and universal machines. Stage of testing (Approved by Order No. 893 on March 29, 1988). Moscow, USSR: Gosstandart

Osh State University (2024) Scientific report under contract No. 43 dated April 24, 2024 on the project: Development and practical application of a new technology for post-harvest processing of rice in southern Kyrgyzstan, German project of the GIZ Applied Research Fund within the framework of the project "Vocational education for economic growth sectors in Central Asia." Osh, Kyrgyzstan, Osh State University

Shalpykov KT, Kaarashova AA, Akmatov AO, Samiev ZT, Smailova KE, Smailov EA et al. (2017) Patent of the Kyrgyz Republic No. 1961 Method for obtaining high-quality Uzgen rice (application No. 20160039.1, published in bulletin No. 6 of May 31, 2017). Bishkek, Kyrgyzstan: State Agency for Intellectual Property and Innovation under the Government of the Kyrgyz Republic (Kyrgyzpatent)

Smailov EA, Osmonov YJ, Smailov AE, Arapbaev RN, Akmataliyev AT Kochkonbaeva AA et al (2023) Patent of the Kyrgyz Republic No. 2328 "Mobile device for steam-thermal treatment of rice sheaves" (application No. 20220005.1, published in bulletin No. 3 of February 28, 2023). Bishkek, Kyrgyzstan: State Agency for Intellectual Property and Innovation under the Government of the Kyrgyz Republic (Kyrgyzpatent)

Kochkonbaeva AA (2024) Development of technology and justification of the parameters of working components of the installation for rough rice fermentation (abstract of dissertation of candidate of technical sciences), Bishkek, Kyrgyzstan

Smailov EA, Arapbaev RN, Kochkonbaeva AA, Samieva ZT, Tashmatova NK (2024) Formation of new mechanisms for sustainable development of the rice farming in Kyrgyzstan. In: Popkova EG, Bogoviz AV, Sergi BS, Kaurova OV, Maloletko AN (eds) Sustainable development of the agrarian economy based on digital technologies and smart innovations. Springer, Cham, Switzerland, pp 3–6. https://doi.org/10.1007/978-3-031-51272-8_1

Ecological Features of the Growth of the Malus Niedzwetzkyana Dieck in the Conditions of the Sary-Chelek State Biosphere Reserve

Zulfiia Sakbaeva, Sovetbek Kenzhebaev, Gulum Kalaeva, Mairamkan Asanova, and Altynai Mamyrbaeva

Abstract The research examines the impact of climate change on rare and endangered plant species within the walnut-fruit forests of southern Kyrgyzstan and the ecosystem of the Sary-Chelek Biosphere Reserve. Forests cover approximately 31% of the Earth's land surface and harbor more than 80% of global biodiversity. However, anthropogenic pressures and unsystematic exploitation of plant resources lead to habitat degradation and population decline. Particular attention is devoted to *Malus niedzwetzkyana*, a critically endangered endemic species, with an assessment of its distribution, population structure, and reproductive biology. The research employed field observations, route-based surveys, GPS mapping, phenological monitoring, and soil analyses to evaluate habitat conditions and reproductive potential. Results indicate that *Malus niedzwetzkyana* occurs at elevations between 1241 and 2000 m, primarily within walnut-fruit forest zones. The number of seeds per fruit ranges from 4 to 15, with a germination rate of up to 96%. These findings provide essential data for the development of conservation strategies, propagation programs, and habitat management measures. The research underscores the urgent need to preserve biologically valuable species and maintain ecological balance under ongoing climate change.

Keywords *Malus niedzwetzkyana* · Rare plants · Endemic species · Sary-chelek state biosphere reserve · Climate change · Soil moisture

JEL Classification M13 · O54

Z. Sakbaeva (✉) · G. Kalaeva · M. Asanova · A. Mamyrbaeva
Jalal-Abad State University named after B. Osmonov, Manas, Kyrgyzstan

S. Kenzhebaev
Jalal-Abad Scientific Center of the National Academy of Sciences of the Kyrgyz Republic, Manas, Kyrgyzstan

A. B. Karbekova (ed.), *Sustainable Development of Rural Territories Based on Climate-Smart Technologies*, Climate Change Management,
https://doi.org/10.1007/978-3-032-20106-5_45

1 Introduction

Forests occupy a total area of 4.06 billion hectares worldwide, accounting for approximately 31% of the Earth's terrestrial surface. They harbor more than 80% of global biodiversity (Mamadjanov 2004).

Kyrgyzstan is a mountainous country with a relatively small forested area, nearly 90% of which is located at elevations around 1500 m above sea level. Currently, the restoration of forest ecosystems, the conservation, and sustainable utilization of plant resources are of paramount importance. The increasing anthropogenic pressure on the environment and unsystemic exploitation of plant resources are major causes of bioceonotic disruption, resulting in changes to plant communities and their components, with some species facing the risk of extinction (Republic 1999; Ayupov and Zhunusov 2001).

Scientific evidence indicates that entire biological systems are disappearing, leading to the disruption of ecological balance. According to the Food and Agriculture Organization of the United Nations (FAO), over the 30 years from 1990 to 2020, the global forest area decreased by approximately 178 million hectares, while thousands of species are annually lost without monitoring or assessment. Species extinction and deforestation increase the likelihood of climate change and pose serious threats to human life (Food and Agriculture Organization of the United Nations (FAO) 2020).

Populations of certain endemic and rare species are declining. The loss of a single species results in an irreplaceable reduction of biodiversity, as each plant species represents a unique component of the gene pool.

Among the key conservation areas in Kyrgyzstan, the Sary-Chelek State Biosphere Reserve is particularly significant due to its unique natural environment. Currently, global climate change and demographic growth in areas surrounding the reserve negatively impact its ecosystem. Therefore, studying the bioecological characteristics of rare and endangered species, developing propagation methods, conducting comprehensive monitoring, evaluating the effects of anthropogenic pressures, updating information technologies, and enriching scientific data are of critical importance.

2 Methods

Field research was conducted through surveys of residents and reserve staff. During the research, the authors documented the habitat conditions of rare species. Methods for studying plant cover included field surveys, route-based inventory of tree and shrub species (area share, density, biodiversity), morphological description, phenological observations, and GPS-based coordinate mapping.

Comprehensive data on species distribution were collected; geobotanical descriptions were prepared. Each recorded species (tree or shrub) was documented; coordinates were determined using a Garmin eTrex 10 GPS device (Standards and Committee 2024).

Soil studies were conducted to evaluate the growth conditions of *Malus niedzwetzkyana* (Niedzwetzky's Apple). Soil samples were taken, genetic horizons identified, and soil moisture content measured (GOST 28268-89 1989).

3 Results

The Sary-Chelek State Biosphere Reserve is one of the most unique and picturesque regions in the world. According to botanical-geographical zoning, it belongs to the Chatkal floral region (Kenzhebaev et al. 2006). Its natural ecosystem plays a significant role in maintaining the ecological stability of the area. The reserve is located on the southern slopes of the Chatkal Range, in the Arkyt Gorge, covering an area of 23,868 ha at elevations ranging from 1200 to 4247 m above sea level. Established in 1959, it was designated as a State Biosphere Reserve in 1978 and has been part of the UNESCO World Network of Biosphere Reserves since 1979 (Sary-Chelek Nature Reserve 2019; Sary-Chelek Nature Reserve 2022).

The flora of Sary-Chelek was first catalogued by H. U. Borlakov, listing 62 families, 376 genera, and 969 species, including rare taxa (Borlakov and Golovkova 1971; Velasco et al. 2010). A list of rare plant species in the reserve is presented in Table 1.

Malus niedzwetzkyana is a highly rare and endangered species, occurring individually on the slopes of the Chatkal and Fergana ranges. It grows in forests, which

Table 1 Rare plant species of Sary-Chelek biosphere reserve

Species	IUCN classification	Rare category in Kyrgyzstan	Rare category in Central Asia	Population status and distribution
Abies semenovii	Rare endemic	Red book of Kyrgyzstan, 3.7 k ha	Not found outside Kyrgyzstan	Occurs in Toktogul and Aksy districts
Malus sieversii	Valuable gene pool element	Red book of Kyrgyzstan	Rare species	Seedlings prone to degradation
Malus niedzwetzkyana	Critically rare endemic	Red book of Kyrgyzstan	Critically rare species	Only 80 out of 113 trees remain in the reserve
Persian rowan	Highly endangered, restricted range	Red book of Kyrgyzstan	Critically rare	Single trees in coniferous zones
Uzun-Akhmat grape	Rare endemic	Red book of Kyrgyzstan	Extremely rare endemic	Occurs in the southern part of the reserve

Source developed by the authors

limits cross-pollination and may reduce the expression of its unique biological characteristics. Conservation measures include protection against grazing and propagation efforts. Nursery experience in the reserve exists, and forest management should establish separate plots for artificial plantations.

The bioecological characteristics, distribution, and propagation methods of some rare species, such as *Malus niedzwetzkyana*, remain insufficiently studied.

Malus niedzwetzkyana is included in the IUCN Red List as Endangered (EN). In the Red Book of Central Asia, it is classified as EN B2ab (IUCN Red List of Threatened Species 2007b). In the Kyrgyz Red Book, it is listed as Vulnerable (VU) (IUCN Red List of Threatened Species 2007a; State Agency for Environmental Protection and Forestry under the Government of the Kyrgyz Republic 2006). The species grows individually or in small groups, 3–5 m tall, with red-brown branches, thick grey-green leaves, and light-red flowers. Fruits are spherical, 3–7 mm in diameter, and dark red.

The rare *Malus niedzwetzkyana* occurs in the reserve at elevations ranging from 1241 to 2000 m above sea level (on the southern shore of Lake Kyla). During the study period, individuals of *Malus niedzwetzkyana* were found as solitary trees and in groups ranging from 3 to 16 trees (Kenzhebaev et al. 2011; Kenzhebaev et al. 2007). The locations of the recorded trees, whose map can be found in the Data Availability Section, were mapped using Google Earth.

Malus niedzwetzkyana is rarely found growing as a solitary tree at elevations of 1200–1300 m above sea level. Most recorded trees are located at elevations between 1400 and 1700 m, within the range of walnut-fruit forests.

During the research, the authors recorded 64 *Malus niedzwetzkyana* trees on various slopes within the reserve. As shown in the diagram (Fig. 1), more than half of the trees (60%) were found on southeast-facing slopes; only 2% were located on southern-facing slopes.

Fruit maturation depends on elevation: at 1255 m, fruits ripen in early September, while at 2000 m they mature by late September. Each fruit contains between 4 and 15 seeds. The thousand-seed weight averages 23.3 g, and the seeds show an 82%

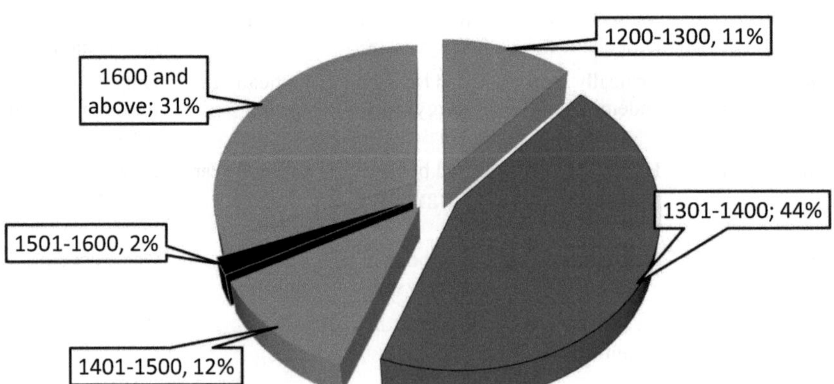

Fig. 1 Distribution of trees by elevation above sea level. *Source* developed by the authors

germination rate. Laboratory stratification experiments identified optimal dormancy-breaking conditions at 2–4 °C, with germination occurring 80 days after stratification. Such variation in the number of seeds per fruit can be explained by uneven pollination during the flowering period and the subsequent development of ovules. After pollination, some ovules are eliminated early, while others gradually cease development after fertilization. This process accounts for the observed variability in seed number per fruit.

Based on morphometric analyses of *Malus niedzwetzkyana* seeds, the weight of 1000 seeds was determined to range from 23.3 to 33.0 g. Germination capacity was assessed using an indigo carmine staining method, revealing a germination rate of 96%.

The favorable soil conditions of Sary-Chelek support the growth and natural regeneration of *Malus niedzwetzkyana*. The characteristics of the soils in the studied area were described in the studies of I. G. Roychenko. In the walnut-fruit forests of the Sary-Chelek State Biosphere Reserve, where *Malus niedzwetzkyana* occurs, mountain-forest brown soils are predominant, with humus content reaching 7–11% (Roychenko 1960). According to the research results, the soil moisture content in areas where *Malus niedzwetzkyana*, a rare and endemic species of the Sary-Chelek State Biosphere Reserve, grows is relatively high, which is an important factor in the context of ongoing climate change (Table 2).

The study of soil moisture distribution within the 0–60 cm profile revealed a pronounced decrease with depth. The highest moisture content was recorded in the 0–10 cm layer (42.27%), reflecting active surface horizon wetting. Between 10 and 40 cm, moisture gradually declined to 24%–28%, indicating reduced water availability. Minimum values were observed at 40–50 cm (25.70%). In the 50–60 cm layer, a slight increase in moisture (19.82%) was noted, likely due to subsoil water influence. These findings are essential for assessing the soil water regime and planning measures for soil moisture conservation.

Table 2 Soil moisture in walnut-fruit forests where *Malus niedzwetzkyana* is found

Depth (cm)	Cup No	Weight (g)					Soil moisture, % of dry weight
		Empty cup	Cup with moist soil	Cup with dry soil	Evaporated water	Dry soil	
0–10	447	28.2	79.97	64.59	15.38	36.39	42.27
10–20	449	30.01	81.74	67.62	14.12	37.52	37.64
20–30	409	29.89	87.21	74.39	12.82	44.50	28.81
30–40	439	22.02	87.24	74.24	12.99	52.22	24.87
40–50	415	30.0	91.17	78.66	12.505	48.66	25.70
50–60	431	23.01	105.7	92.18	13.69	69.08	19.82

Source developed by the authors

4 Conclusions

The research was written based on the results of a scientific project entitled "Study of the impact of climate change on rare and endangered species in nut-fruit forests and the ecosystem of the Sary-Chelek State Biosphere Reserve." The current increase in the pressure of anthropogenic and environmental factors is leading to the threat of extinction of some species. In turn, the disappearance of one species affects the ecosystem and leads to a disruption of its ecological balance. This threat also affects the *Malus niedzwetzkyana*, which is found in the Sary-Chelek State Biosphere Reserve.

Malus niedzwetzkyana occurs at elevations ranging from 1241 to 2000 m above sea level, primarily within walnut–fruit forests. The species is found in small groups of 3–16 trees per hectare, indicating significant habitat fragmentation and underscoring the need for conservation measures to prevent further population decline.

Morphometric analysis of seeds revealed that the weight of 1000 seeds ranges from 23.3 to 33.0 g, while laboratory studies confirmed a high germination rate (up to 96%), providing a basis for the artificial propagation of the species.

The results of this research have practical significance for the development of conservation, restoration, and sustainable management programs for walnut-fruit forests in Kyrgyzstan and may serve as a foundation for subsequent monitoring and introduction efforts.

Data Availability The map of *Malus niedzwetzkyana* found within the reserve area is shown at https://figshare.com/ with the ID https://doi.org/10.6084/m9.figshare.30674936.v1.

References

Ayupov FG, Zhunusov NS (2001) Ecology of nut-bearing forests of southern Kyrgyzstan (condition factors). Bishkek, Kyrgyzstan

Borlakov KU, Golovkova AG (1971) Issue 4: vegetation of the Sary-Chelek biosphere reserve. In: Proceedings of the Sary-Chelek reserve. State Committee for Forestry under the Council of Ministers. Kirghiz SSR, Frunze, Kirghiz SSR

Food and Agriculture Organization of the United Nations (FAO) (2020) Global Forest Resources Assessment 2020: Main report. FAO, Rome, Italy. https://doi.org/10.4060/ca9825en

GOST 28268-89 (1989) Soils. Methods of determination of moisture, maximum hygroscopic moisture and moisture of steady plant fading (approved by the Resolution of the USSR State Committee on Standards dated on no. 2924. Moscow, USSR Publishing House of Standards

IUCN Red List of Threatened Species. (2007a). *Malus niedzwetzkyana*. Retrieved from https://www.iucnredlist.org/species/63477/12681555 Accessed 14 Nov 2025

IUCN Red List of Threatened Species. (2007b). *Malus sieversii*. Retrieved from https://www.iucnredlist.org/species/32363/9693009 Accessed 14 Nov 2025

IUCN Standards and Petitions Committee (2024) Guidelines for using the IUCN red list categories and criteria. Version 16. IUCN Red List. Retrieved from https://nc.iucnredlist.org/redlist/content/attachment_files/RedListGuidelines.pdf, Cambridge, UK. Accessed 14 November 2025

Kenzhebaev SK, Aidarbekov AA, Aitikeev NZ (2006) Nut-bearing forests of southern Kyrgyzstan: condition, problems, and prospects. Bulletin of ZhAGU 55

Kenzhebaev SK, Karasheva BG, Nurmanbaev MZ (2007) Manual on growing walnut forests, protection from pests and diseases. Jalal-Abad, Kyrgyzstan
Kenzhebaev SK, Burkhanov AM, Mamadzhanov DK, Shamshiev BN (2011) Manual on establishment of walnut plantations and improvement of dense forests. Jalal-Abad, Kyrgyzstan
Kyrgyz Republic (1999) Forest code of the Kyrgyz Republic 66. Bishkek, Kyrgyzstan. Retrieved from https://cbd.minjust.gov.kg/10/edition/2815/ru Accessed 14 Nov 2025
Mamadjanov DK (2004) Walnut fruit forests and diversity of walnut tree (*Juglans regia* L.) in Kyrgyzstan. ISHS Acta Horticulturae 705:173–176. https://doi.org/10.17660/ActaHortic.2005.705.20
Roychenko GI (1960) Soils of southern Kyrgyzstan. Publishing House of the Academy of Sciences of the Kirghiz SSR, Frunze, Kirghiz SSR
Sary-Chelek Nature Reserve (2019) Chronicle of Nature of the Sary-Chelek state biosphere reserve. Arkyt, Kyrgyzstan
Sary-Chelek Nature Reserve (2022) Chronicle of Nature of the Sary-Chelek state biosphere reserve. Arkyt, Kyrgyzstan
State Agency for Environmental Protection and Forestry under the Government of the Kyrgyz Republic (2006) Red data book of the Kyrgyz Republic, 2nd edn. Bishkek, Kyrgyzstan
Velasco R, Zharkikh A, Affourtit J, Dhingra A, Cestaro A, Kalyanaraman A et al (2010) The genome of the domesticated apple (Malus × domestica Borkh). Nat Genet 42:833–839. https://doi.org/10.1038/ng.654

Innovative Biotechnology for Unconventional Tobacco Cultivation

Eltar A. Smailov, Keneshbai Turdumambetov, Kanybek K. Isakov, Zarema I. Kalchaeva, and Nurgazy Erkinbay Uulu

Abstract In the past, Kyrgyzstan produced up to 76,000 tons of high-quality Oriental aromatic tobacco of the Dubek type per year, which was used for natural aromatization in cigarette manufacturing. Globally, Oriental aromatic tobacco accounts for only about 8% of total production. The exceptional quality of Kyrgyz tobacco was supported by favorable high-altitude natural and climatic conditions. In Kyrgyzstan, tobacco is cultivated on small areas in the traditional tobacco-growing districts of the Osh Region (Nookat and Uzgen districts). In the Batken Region, a special variety of makhorka is grown for the production of the smokeless tobacco product nasvai. Nasvai is a type of smokeless tobacco traditionally used in Central Asia. The text presents solutions to certain challenges in mechanized post-harvest processing and fermentation technologies, using 100% solar energy, as well as approaches to unconventional uses of tobacco and tobacco waste for producing materials for medicine and industry. As part of this research, the authors conducted exploratory studies to select suitable extractants, examine the kinetics of extraction processes, assess factors influencing diffusion kinetics, and separate group fractions of tobacco raw materials and waste into individual components. A conceptual process diagram has been developed for the primary processing of freshly harvested tobacco stems, along with a proposed scheme for the integrated processing of tobacco raw materials and a technological line for obtaining protein and processed tobacco stems.

E. A. Smailov (✉)
Kyrgyz-Uzbek International University named after Batyraly Sydykov, Osh, Kyrgyzstan

K. Turdumambetov
Institute of Chemistry and Phytotechnology of the National Academy of Sciences of the Kyrgyz Republic, Bishkek, Kyrgyzstan

K. K. Isakov · N. E. Uulu
Osh Technological University named after M.M. Adyshev, Osh, Kyrgyzstan

Z. I. Kalchaeva
International Uzgen Institute of Technology and Education, Uzgen, Kyrgyzstan

A. B. Karbekova (ed.), *Sustainable Development of Rural Territories Based on Climate-Smart Technologies*, Climate Change Management,
https://doi.org/10.1007/978-3-032-20106-5_46

Keywords Tobacco · Nicotine · Resin · Essential oil · Plant sap · Raw protein mass · Purified protein · Nicotine · Solenasol · Tobacco oil · Polysaccharide · Tobacco meal

JEL Classification Q16

1 Introduction

Global tobacco production amounts to 6.3 million tons, with 76% consisting of two varieties (Virginia and Burley), which readily absorb fillers, allowing manufacturers to produce cigarettes of any desired flavor, according to consumer demand, without regard for their harm to health.

The history of tobacco cultivation in Kyrgyzstan dates back to 1896. In the past (during the 1980s–1990s), Kyrgyzstan produced up to 76,000 tons of high-quality Oriental aromatic tobacco of the Dubek type annually, used for natural aromatization in cigarette manufacturing. Simultaneously, global production of Oriental aromatic tobacco accounts for only 8%. The exceptional quality of Kyrgyz tobacco was largely due to favorable high-altitude natural and climatic conditions; therefore, Kyrgyzstan consistently attracted attention as a producer of premium tobacco raw materials.

Currently, tobacco in Kyrgyzstan is cultivated on small areas in the traditional tobacco-growing regions of the Osh Province (Nookat and Uzgen districts). In addition, a special variety of makhorka is grown in the Batken Province for the production of the smokeless tobacco product *nasvai*. Nasvai is a type of smokeless tobacco traditionally used in Central Asia. Despite these changes, tobacco cultivation remains an important contributor to Kyrgyzstan's economy. For this reason, the government continues to take active measures to support the development of the tobacco industry. However, further progress, aligned with global trends, faces new challenges shaped by the evolving mindset of rural producers, such as the widespread adoption of mechanization and automation in agricultural production, minimizing or eliminating manual labor (Smailov et al. 2025). This includes the use of solar energy in post-harvest tobacco processing (Smailov et al. 2024), increasing yields, producing environmentally friendly and waste-free products, and achieving higher profitability per unit area through unconventional uses of tobacco. One of the key conditions for expanding tobacco cultivation and production in Kyrgyzstan is increasing the purchasing price of tobacco raw materials. This goal could be achieved by creating a cluster based on zero-waste cultivation and processing technologies. Thus, further growth in tobacco production is closely tied to addressing these issues.

The research aims to identify pathways for the further development of tobacco cultivation in Kyrgyzstan and its unconventional applications in the national economy. Research objectives include expanding tobacco cultivation and production and developing primary and integrated technological schemes for processing tobacco and its waste.

2 Methodology

Field experiments were conducted on the lands of the scientific-production seed cooperative "Kyrgyzstan Dyubek." The soils of the experimental site are classified as long-irrigated typical sierozems. All measurements and observations of tobacco growth and development were carried out in accordance with established methods for field agrotechnical experiments with tobacco and makhorka. Exploratory work on selecting suitable extractants, studying the kinetics of extraction processes to evaluate factors of diffusion kinetics, and separating group fractions of tobacco raw materials and their waste into individual components was performed under laboratory conditions using standard chemical equipment and instruments in the Laboratory of Plant Chemistry of the Institute of Chemistry and Phytotechnology of the National Academy of Sciences of the Kyrgyz Republic. A gravimetric method was used to determine the total extractable substances. Ash content was measured according to standard procedures. The extractive content in the initial raw material was determined gravimetrically in accordance with GOST 24027 (GOST 1980a; GOST 1980b). Parameters of diffusion kinetics were calculated using established formulas. Solanasol was isolated through extractive crystallization, and the protein was obtained via acid and thermal coagulation.

3 Results

The analysis of global tobacco cultivation shows that while production amounted to 5.6 million tons in the 2000s, it has now reached 6.3 million tons, a significant increase that has proven difficult to curb, even though tobacco smoking is widely considered harmful. Speaking at the World Forum on Pharmacology and Medicinal Chemistry (Netherlands, Amsterdam) on April 3, 2024, Professor E. A. Smailov directly asked whether it was still worth conducting scientific research on tobacco-related issues. The forum moderator, a professor from the United States, openly responded that every plant has both positive and negative effects on human health; the key lies in responsible consumption. He also emphasized a second point: the annual global income from tobacco cultivation is around $150 billion. This is a tremendous sum; no one is willing to forgo it. Therefore, research in this field must continue, especially regarding the non-smoking uses of tobacco and its waste products (Smailov et al. 2023).

For this reason, we focus on the mechanized technology of post-harvest processing and fermentation, using 100% solar energy, as well as on unconventional uses of tobacco and its by-products for producing materials for medicine and industry.

In recent years, the production of tobacco raw materials in Kyrgyzstan has declined due to both objective and subjective factors. Despite this, in 2025, with the involvement of foreign investment, a new tobacco factory for manufacturing smoking products was built in the city of Tokmok. Additionally, the Ministry of Water Resources,

Agriculture, and Processing Industry of the Kyrgyz Republic has allocated seeds of promising tobacco varieties for testing in 2025 on the fields of the scientific-production seed cooperative “Kyrgyzstan Dyubek,” which will help increase the volume of tobacco production in the country.

This is further supported by the high economic efficiency and profitability of tobacco cultivation. The average selling price of 1 kg of tobacco raw material in Kyrgyzstan is $1.27, whereas in EU countries and Turkey, it ranges from $3 to $6.50 per kg (Smailov et al. 2023). As a result, revenues from exports in these countries amount to hundreds of millions of dollars, even with comparable export potential.

It is also important to emphasize that Greece exports 3.5 times more tobacco than Kyrgyzstan (35,000 tons), generating 300–330 million euros (i.e., 90% of its total revenue). In contrast, Kyrgyzstan produces 10,000 tons of tobacco and exports 95% of it, earning only $13 million. Considering the difference between the dollar and the euro, Kyrgyzstan receives seven times less income from exports.

After Kyrgyzstan gained independence, its supply chains for delivering tobacco raw materials to Russia were disrupted. This happened because Russia’s tobacco factories were privatized by five global companies (Philip Morris, JTI, BAT, Gallaher, and Reemtsma), which together control more than 61.8% of global cigarette production. Intermediary companies such as “Daiman International,” “Stan-Sun,” and others purchased our tobacco at artificially low prices. By 2000, the situation had become so dire that farmers could not sell tobacco for even 3 soms per kilogram, forcing many into bankruptcy and causing them to lose interest in tobacco cultivation. Meanwhile, purchasing companies in Kyrgyzstan buy tobacco at $1.27 per kg, resell it, or supply it to Russian tobacco factories at $7–$8 per kg. In 2024, the purchasing price for tobacco raw material reached 130 soms/kg, or about $1.5, essentially the same level as before.

To increase tobacco production in Kyrgyzstan, it is necessary to create a cluster that brings together tobacco growers (through a growers’ association), fermentation plants, and the Tokmok tobacco factory. Such a cluster would renew interest in tobacco cultivation by ensuring higher purchasing prices and guaranteed procurement of harvested tobacco. If the purchasing price were increased by 2–3 times, a realistic goal within a cluster, the economic output of Kyrgyzstan’s agro-industrial sector would multiply significantly.

Considering these circumstances, and in line with current global trends toward a green economy, we have developed a mechanized technological line for post-harvest processing and fermentation of tobacco using 100% solar energy for drying (Smailov et al. 2025), fermentation, and moisturizing of tobacco leaves (patents No. 1527, 155, and 2360) (Smailov et al. 2013a; Smailov et al. 2013b; Smailov et al. 2005). Designed to process the harvest from 40 hectares, this system includes a detailed technological map and labor organization plan. Most importantly, under this technology, large fermentation plants become unnecessary. In the mechanized complex, tobacco fermentation takes place in a loose mass, within the technological flow, simultaneously with leaf drying. As a result, the financial profit normally earned by fermentation plants is retained by tobacco producers themselves, namely, the

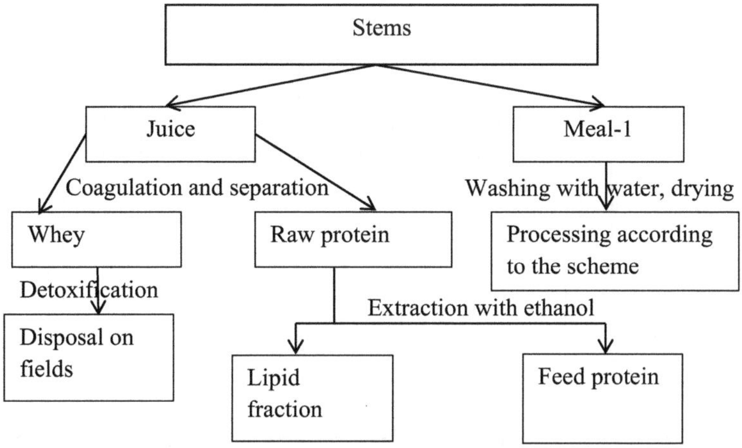

Fig. 1 Conceptual flowchart of primary processing technologies for freshly harvested tobacco stems. *Source* Developed by the authors

farming households. Thus, all issues related to cultivation, drying, and fermentation are resolved in favor of rural producers of tobacco raw materials.

Another important direction in tobacco use involves its nontraditional applications for producing raw materials for medicine and industry (Smailov 2025; Smailov and Abdullaeva 2018; Smailov and Abdullaeva 2019); (Smailov et al. 2019a; Smailov et al. 2019b).

Based on these studies, a conceptual scheme has been developed for the primary processing of freshly harvested tobacco stems (Fig. 1), along with an integrated processing scheme for tobacco raw materials (Fig. 2). A technological line for obtaining protein from tobacco stems has also been proposed.

These investigations show that, in the near future, it will be advisable to organize a large-scale experiment to grow tobacco biomass specifically for chemical and technological processing.

Additional research [(Smailov et al. 2007); (Smailov and Abdullaeva 2018; Smailov and Abdullaeva 2019); (Smailov et al. 2019a; Smailov et al. 2019b)] and patents [(Smailov et al. 2015a; Smailov et al. 2015b; Smailov et al. 2006); (Smailov et al. 2013c)] have demonstrated the feasibility of cultivating tobacco simultaneously for both smoking products and seed production to obtain oil, as well as methods for increasing nicotine yield from various parts of the tobacco plant.

Despite this, as noted above, the equipment required for biotechnological processing of the tobacco plant demands substantial financial resources and investment. Therefore, we are seeking interested partners for joint scientific research and investment collaboration. We would be pleased to work together.

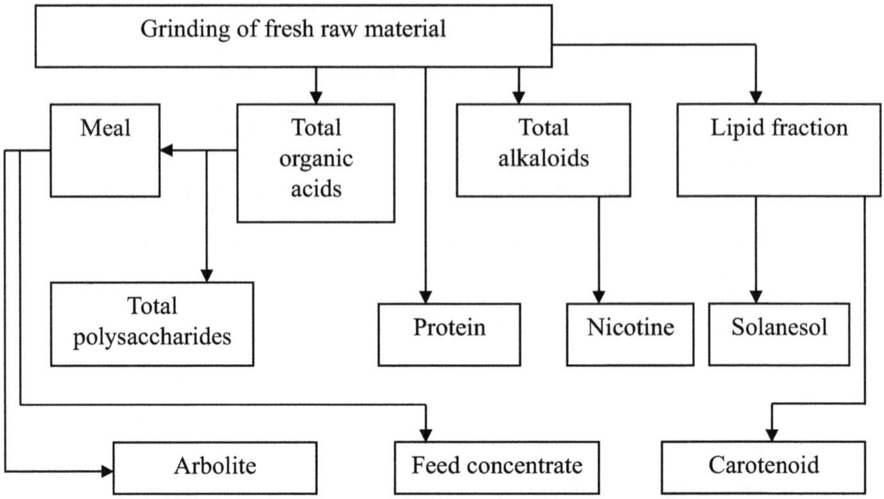

Fig. 2 Integrated processing scheme for tobacco raw materials. *Source* Developed by the authors

4 Conclusion

The tobacco plant is a valuable raw material for biotechnology, used to produce pharmaceuticals, food-grade protein, solanesol, and other products essential for medicine and industry.

References

GOST 24027.0–80 (1980a) Vegetal medicine raw stuff. Acceptance rules and sampling (adopted by resolution of the state committee of the USSR for standards dated March 6, 1980 No. 1038), Gosstandart, Moscow, USSR

GOST 24027.2–80 (1980b) Methods for determination of moisture, ash content, extractive and tannin materials, essential oil (adopted by Resolution of the State Committee of the USSR for Standards dated March 6, 1980 No. 1037), Gosstandart, Moscow, USSR

Smailov EA, Atamkulova MT, Zulpuev ZB et al (2013a) Device for separating dry tobacco leaves from a cord (Patent No. 1527; No. 20120005.1; declared January 27, 2012; published March 30, 2013, bulletin No. 3, 7 p. IPC A24B 3/07), State Patent Office of the Kyrgyz Republic, Bishkek, Kyrgyzstan

Smailov EA, Atamkulova MT, Zulpuev ZB et al (2013b) Device for cleaning tobacco leaves from resinous coatings of aphids (Patent No. 155; No. 20130003.2; declared February 20, 2012; published May 30, 2013, bulletin No. 5, 7 p. IPC A24B 3/18), State Patent Office of the Kyrgyz Republic, Bishkek, Kyrgyzstan

Smailov EA, Turdumambetov K, Murzubraimov BM et al (2015a) Method for obtaining nicotine and tar from the residues of tobacco raw materials (Patent No. 1721; No. 20140031.1; declared March 11, 2014; published April 30, 2015, bulletin No. 4, 6 p. IPC A24B 15/00), State Patent Office of the Kyrgyz Republic, Bishkek, Kyrgyzstan

Smailov EA, Turdumambetov K, Murzubraimov BM et al (2015b) Method for producing food protein (Patent No. 1750; No. 20140059.1; declared June 6, 2014; published July 31, 2015, bulletin No. 7, 6 p. IPC A23J 1/14, A61K 9/14) State Patent Office of the Kyrgyz Republic, Bishkek, Kyrgyzstan

Smailov EA, Turdumambetov K, Samyeva ZhT et al (2006) Antiseptic agent "Kord" (Patent No. 917; No. 20050030.1; declared April 7, 2005; published December 30, 2006, bulletin No. 12, 6 p. IPC A61K 31/4704), State Patent Office of the Kyrgyz Republic, Bishkek, Kyrgyzstan

Smailov EA, Turdumambetov K, Samyeva ZhT et al (2007) Method for producing tobacco oil (Patent No. 666; No. 2003018.1; declared March 14, 2003; published July 30, 2007, bulletin No. 7, 6 p. IPC A23L 1/36), State Patent Office of the Kyrgyz Republic, Bishkek, Kyrgyzstan

Smailov EA, Turdumambetov K, Samyeva ZhT et al (2013c) Method for producing cellulose (Patent No. 1592; No. 20120073.1; declared July 25, 2012; published November 30, 2013, bulletin No. 11, 6 p. IPC D21C 3/02), State Patent Office of the Kyrgyz Republic, Bishkek, Kyrgyzstan

Smailov EA, Tursunbaev Zh, Zulpuev ZB et al (2023) Mechanized technological line for drying tobacco (Patent No. 2360; No. 20230005.1; declared January 30, 2023; published October 31, 2023, bulletin No. 10, 8 p. IPC A01D 45/04), State Patent Office of the Kyrgyz Republic, Bishkek, Kyrgyzstan

Smailov E (2025) Tobacco plant of Kyrgyzstan: a valuable crop for biotechnology in medicine. In: 6th global summit on advances in medicinal chemistry and pharmacology. Peers Alley Media, Amsterdam, Netherlands, pp 43–44

Smailov EA, Abdullaeva RA (2018) Economic aspects of the cultivation and production of tobacco for nicotine. Izvestiya VUZov Kyrgyzstana 9:11–15

Smailov EA, Abdullaeva RA (2019) The effect of soil moisture and mineral fertilizers on the accumulation of nicotine in tobacco (Nicotiana T). Sci, New Technol Innov Kyrgyzstan 12:166–173

Smailov EA, Samyeva ZT, Abdullaeva RA (2005) Tobacco and makhorka material as raw material for obtaining organic acids. Izvestiya OshTU 1:136–140

Smailov EA, Samyeva ZT, Abdullaeva RA (2019a) The influence of soil type and its moisture content on nicotine accumulation dynamics in leaves of various tobacco varieties (Nicotiana Tabacum L.). Bull Altai State Agric Univ 6(176):36–46

Smailov EA, Samyeva ZT, Abdullaeva RA, Kaparov MK (2019b) The effect of soil moisture on nicotine accumulation dynamics in different parts of a tobacco plant (Nicotiana Tabacum L.). Bull Altai State Agric Univ 8(178):77–86

Smailov EA, Tursunbaev ZZ, Zulpuev ZB, Akmataliev AT, Matisakov TK (2025) Mechanized tobacco post-harvest processing complex for leaf curing using solar energy. In: Popkova EG (ed) Corporate social responsibility to the green growth of business and economy. Springer, Cham, Switzerland, pp 357–361. https://doi.org/10.1007/978-3-031-83041-9_59

Smailov EA, Tursunbaev ZZ, Zulpuev ZB, Atamkulova MT, Abdykadyrov AB, Kalchaeva ZI (2024) Effective methods and techniques for drying tobacco leaves using solar energy in Kyrgyzstan. Bull Altai State Agric Univ 4(234):31–37. https://doi.org/10.53083/1996-4277-2024-234-4-31-37

Phytoncide Activity of Woody Plants Used in Landscaping

Tazhimamat K. Erkebaev, Akim T. Tazhibaev, Zhypargul D. Abdullaeva, Yrysbubu T. Saryeva, and Toktomamat kyzy Kalya

Abstract Due to the intensive development of industry and transportation, an increasing number of harmful substances are included in the atmosphere. One of the main elements for improving the atmosphere is the greening of territories, which, along with other functions, also performs a sanitary and hygienic role. The research aims to study the phytoncide activity of woody plants used in landscaping. Phytoncide activity was determined by the chromatographic analysis method. To concentrate phytoorganic substances, an EA–I electroaspirator with a rotameter and a steel tube (10×0.2 cm) containing 0.2 g of Polysorb-I sorbent was applied. A gas-liquid chromatograph "Chrom-41" with a flame ionization detector was employed for the analysis. Instead of a column, a capillary measuring 20×0.2 cm was used. Samples were taken directly under the canopy of the studied plant. The quantity of volatile substances was calculated based on the area of the peak, determined by generally accepted methods. The analysis of the dynamics of phytoncidal activity in the studied species shows that the peak phytoncidal activity does not coincide with the flowering period in many species. All studied species possess phytoncidal activity, which varies throughout the growing season, reaching a peak from May to July. The results obtained confirm the findings of other researchers that, overall, the dynamics of phytoncidal activity are determined by the intensity of growth processes and photosynthesis.

Keywords Phytoncidal activity · Gas-liquid chromatograph · Electric aspirator · Flame ionization detector · Volatile substances · Area of the recorded peak · Vegetation period · Species specificity · Maximum phytoncidal activity · Average phytoncidal activity

T. K. Erkebaev · A. T. Tazhibaev · Z. D. Abdullaeva (✉) · Y. T. Saryeva · T. Kalya
Osh State University, Osh, Kyrgyzstan
e-mail: jypar.science@oshsu.kg

T. K. Erkebaev
e-mail: erkebaev1962@oshsu.kg

Y. T. Saryeva
e-mail: saryeva@oshsu.kg

A. B. Karbekova (ed.), *Sustainable Development of Rural Territories Based on Climate-Smart Technologies*, Climate Change Management,
https://doi.org/10.1007/978-3-032-20106-5_47

JEL Classification Q1 · Q2 · Q20

1 Introduction

According to contemporary concepts, phytoncides are substances produced by plants that possess antimicrobial properties and are one of the factors of the natural immunity of plants (Vichkanova 1981). The phytoncidal activity of various species of woody-shrub vegetation differs under the same conditions. According to the reference (Artemieva 1963), one ha of coniferous forest on the southern coast of Crimea is capable of releasing 4 kg of volatile organic substances into the air per day, while a deciduous forest releases 2 kg. Several studies have focused on the quantitative and qualitative composition of phytoncides in forest biogeocenoses (Kryuchkov 1978; Tokin 1974). The highest phytoncidal activity is found in coniferous species, warty birch, balsam poplar, Manchurian walnut, and others (Akimov 1987). Maximum phytoncidal activity is observed in July, with the minimum occurring in the fall. Phytoncidal activity of different species within the same genus can vary significantly (Kochergina and Darkovskaya 2009).

In areas of industrial pollution, the antimicrobial effect of volatile plant emissions changes (Tomchuk et al. 1976). A sharp decrease in the phytoncidal activity of tree species under conditions of increased air pollution and dustiness is associated with a weakening of the physiological state of plants due to the characteristics of the urban ecological environment (Ostrikova 1982).

Plants can exhibit phytoncidal activity regardless of their growing location, climatic, ecological, or other conditions. Simultaneously, the quantitative and qualitative composition of phytoncides may vary depending on the phase of plant vegetation and climatic-geographical conditions.

The activity of volatile substance emission by plants depends on many factors. Overall, it is determined by the intensity of growth processes and photosynthesis (Vlasyuk 1981). Volatile substances are usually most actively released during the growing season (May–August), with emissions decreasing in fall, and are almost absent during deep dormancy.

Phytoncide release intensity reaches its highest level in summer and weakens comparatively in spring and fall. In Uzbekistan, most woody-shrub species used in landscaping possess high or moderate phytoncidal activity (Berida et al. 2024). High activity is characteristic of Crimean pine, eastern arborvitae, Virginian juniper, Eldar pine, white acacia, tree of heaven, porous-branched elm, horse chestnut, bearded birch, black elderberry, small-leaved maple, walnut, and Bolle's poplar. Moderate activity is typical for species such as Canadian red oak, pedunculate oak, Japanese sophora, and Pennsylvania ash. It has been established that the phytoncidal activity of these species varies throughout the vegetation period.

The degree of phytoncide activity depends on ecological conditions. When comparing the phytoncidal properties of the same plant species from the Far North and Latvia, it was found that these properties are sometimes similar and sometimes different (Aĭzenman et al. 1973). An increase in the altitude of plant growth (up to an optimum), phytoncidal activity intensifies (Makarenko 1975).

The degree to which phytoncidal activity is detected largely depends on the methods used for its determination. The antimicrobial activity of plants observed in their natural growing conditions often does not correspond to the results of laboratory studies (Akhmedova and Annaeva 2024).

The antimicrobial activity of plants changes during the vegetation period; the pattern of these changes differs among various plants (Berida et al. 2024). Factors influencing activity include the age characteristics of the organism and its individual organs. The activity of different plant organs varies: in some plants, the roots are more active, whereas in others the leaves, needles, seeds, or other organs exhibit higher activity. Even different parts of the same organ (e.g., flowers) can differ in their activity (Pryazhnikov 1967).

Direct use of the natural phyto-organic air background created by plants is one of the simplest ways to practically apply the properties of volatile terpenoids. The creation of specialized green plantations ensures the necessary sanitizing effect and a certain impact on human body activity (Makarchuk et al. 1990). Priority is given to the use of existing plantations, especially parks. Such plantations typically have a sufficiently wide species composition. However, their layout and placement were often done without considering the possibility of utilizing volatile phytoncides. Simultaneously, the abundance of parks raises the question of the feasibility of using them to improve air quality and positively affect human health.

In recent years, studies on the hygienic importance of phytoncides and their effects on the human body and the environment have gained increasing recognition abroad (Heller 1977; Li et al. 2009).

2 Materials and Methods

The phytoncide activity was determined using the chromatographic analysis method (Akimov 1987). To concentrate phytoorganic substances, the authors applied an electro-aspirator EA-I with a rotameter and a steel tube (10 × 0.2 cm) containing 0.2 g of sorbent-Polysorb-I.

A gas-liquid chromatograph "Chrom-41" equipped with a flame-ionization detector was employed for the analysis. Instead of a traditional column, a capillary tube measuring 20 × 0.2 cm was used. Samples were taken directly under the canopy of the studied plant.

The number of volatile substances was calculated based on the area of the peak determined by standard methods. The concentration of volatile substances in the air (C) was calculated by the formula:

$$C = (1000 \times K \times S)/V, \mathrm{mg/m^3}$$

where:

S is the area of the registered peak during analysis, mm^2;

K is the coefficient relating the detector signal to the amount of substance passing through it, mg/mm^2;

V is the volume of air pumped through the tube when sampling, liters.

Measurements were performed at the Botanical Garden of the Academy of Sciences of the Republic of Uzbekistan, with triple repetition during midday hours (12:00–15:00), under calm weather conditions from May to September (12th–15th of each month). Coniferous species were also studied during winter.

Statistical processing of the data was carried out according to the reference (Lakin 1990).

3 Results and Discussion

The authors established that all studied species possess, to varying degrees, phytoncide activity, which significantly fluctuates throughout the growing season. The authors' observations confirm the findings (Pryazhnikov 1967; Starovoytova and Lakhno 1967) regarding the species-specific and seasonal variability of plants' phytoncide activity. This variability is associated with the composition of metabolites, which differ among various plant activities (Roshina and Roshina 1989) and change during the vegetative period.

The maximum levels of phytoncide activity among studied species range from 0.12 mg/m^3 in Salix alba to 0.82 mg/m^3 in *Pinus pallasiana* (Tables 1 and 2).

The coniferous species investigated demonstrate higher phytoncide activity than deciduous trees, consistent with the findings in the reference (Makarchuk et al. 1990).

As shown in Tables 1 and 2, the maximum values of phytoncide activity for most species occur in the first and second measurement years. These differences are not statistically significant ($t = 2.78$).

Based on the maximum phytoncide activity, the studied plant species can be grouped into the following three categories:

- High activity—above 0.50 mg/m^3 of phytoorganic substances: *Pinus pallasiana*, *Juniperus virginiana*, and *Biota orientalis*;
- Medium activity—from 0.25 to 0.50 mg/m^3: *Cercis canadensis*, *Quercus robur*, *Fraxinus excelsior*, *Aesculus hippocastanum*, *Robinia pseudoacacia*, *Juglans regia*, *Platanus orientalis*, *Sophora japonica*, *Albizzia julibrissin*, *Ulmus pumila*, and *Betula pendula*;
- Low activity—up to 0.25 mg/m^3: *Gleditsia triacanthos*, *Juglans regia*, *Acer campestre*, and *Catalpa speciosa.*

The results largely align with data obtained in Uzbekistan (Berida et al. 2024; Kochergina and Darkovskaya 2009). The observed differences are primarily attributed to methodological discrepancies, as it is known that the phytocind

Table 1 Phytoncide activity of studied plant species, mg/m³ first year, t = 2.78

No.	Kind	June	July	August	September	t
	1	2	3	4	5	6
1	*Pinus pallasiana*	0.78 ± 0.05	0.63 ± 0.04	0.32 ± 0.02	0.16 ± 0.01	0.80
2	*Juniperus virginiana*	0.67 ± 0.84	0.57 ± 0.04	0.35 ± 0.02	0.21 ± 0.01	0.0
3	*Biota orientalis*	0.62 ± 0.04	0.39 ± 0.02	0.28 ± 0.01	0.14 ± 0.01	0.0
4	*Fraxinus excelsior*	0.45 ± 0.03	0.25 ± 0.02	0.12 ± 0.01		0.55
5	*Quercus robur*	0.38 ± 0,02	0.31 ± 0.02	0.19 ± 0.01		0.60
6	*Robinia pseudoacacia*	0.36 ± 0.01	0.17 ± 0.01	0.10 ± 0.06		1.43
7	*Platanus orientalis*	0.21 ± 0.01	0.29 ± 0.01	0.14 ± 0.01		0.91
8	*Sophora Japonica*	0.31 ± 0.02	0.27 ± 0.02	0.11 ± 0.01		0.0
9	*Aesculus hippocastanum*	0.22 ± 0.02	0.22 ± 0.01	0.15 ± 0.01		3.57
10	*Ulmus pumila*	0.11 ± 0.01	0.26 ± 0.01	0,09 ± 0.008		2.14
11	*Gleditschia triacanthos*	0.20 ± 0.02	0.15 ± 0.01	0.15 ± 0.01		1.43
12	*Juglans regia*	0.22 ± 0.01	0.15 ± 0.01	0.14 ± 0.02		0.71
13	*Catalpa speciosa*	0.12 ± 0.01	0.13 ± 0.01	0.04 ± 0.001		2.27
14	*Acer campestre*	0.15 ± 0.01	0.12 ± 0.01	0.12 ± 0.01		0.83
15	*Populus nigra*	0.15 ± 0.01	0.13 ± 0.01	0.08 ± 0.006		0.0
16	*Salix alba*	0.12 ± 0.01	0.10 ± 0.01	0.08 ± 0.007		0.0

Note: t—significance of the differences between the maximum values of the first and second year of measurement. *Source* compiled by the authors

activity of plants under natural conditions often does not match laboratory results (Starovoytova and Lakhno 1967).

Based on the dynamics of phytoncide activity, all studied species can be divided into the following groups:

- First group—species with peak phytoncide activity in May–June: *Pinus pallasiana, Juglans regia, Robinia pseudoacacia*, and *Cercis canadensis*;
- Second group—species with maximum phytoncide activity in June: *Juniperus virginiana, Betula pendula, Gleditschia triacanthos, Sophora japonica*, and *Albizzia julibrissin*;
- Third group—species with peak phytoncide activity in July: *Quercus robur*, *Ulmus pumila*, and *Platanus orientalis*.

Table 2 Phytoncide activity of studied plant species, mg/m^3, second year

No	Kind	May	June	July	August	September	December
1	*Pinus pallasiana*	0.81 ± 0.03	0.82 ± 0.01	0.62 ± 0.02	0.27 ± 0.02	0.12 ± 0.01	0.19 ± 0.02
2	*Juniperus virginiana*	0.62 ± 0.01	0.67 ± 0.02	0.55 ± 0.01	0.32 ± 0.02	0.15 ± 0.006	0.22 ± 0.001
3	*Biota orientalis*	0.53 ± 0.02	0.62 ± 0.02	0.39 ± 0.01	0.20 ± 0.02	0.07 ± 0.01	0.19 ± 0.02
4	*Cercis canadensis*	0.50 ± 0.01	0.50 ± 0.01	0.32 ± 0.01	0.04 ± 0.003	Traces	
5	*Fraxinus excelsior*	0.35 ± 0.02	0.47 ± 0.02	0.23 ± 0.02	0.14 ± 0.01	Traces	
6	*Quercus robur*	0.26 ± 0.02	0.33 ± 0.004	0.41 ± 0.05	0.20 ± 0.01	0.03 ± 0.004	
7	*Robinia pseudoacacia*	0.36 ± 0.01	0.38 ± 0.01	0.14 ± 0.01	0.08 ± 0.007	Traces	
8	*Platanus orientalis*	0.13 ± 0.001	0.21 ± 0.01	0.31 ± 0.02	0.14 ± 0.01	0.056 ± 0.006	
9	*Sophora Japonica*	0.17 ± 0.002	0.31 ± 0.01	0.27 ± 0.01	0.11 ± 0.01	Traces	
10	*Aesculus hippocastanum*	0.33 ± 0.02	0.23 ± 0.01	0.20 ± 0.01	0.14 ± 0.006	Traces	
11	*Ulmus pumila*	0.10 ± 0.004	0.12 ± 0.018	0.29 ± 0.01	0.10 ± 0.004	Traces	
12	*Albizzia julibrissin*	0.20 ± 0.009	0.29 ± 0.015	0.25 ± 0.01	0.03 ± 0.003	Traces	
13	*Betula pendula*	0.22 ± 0.02	0.25 ± 0.01	1.10 ± 0.01	0.06 ± 0.006	Traces	
14	*Gleditschia triacanthos*	0.17 ± 0.007	0.24 ± 0.02	0.09 ± 0.007	0.02 ± 0.002	Traces	
15	*Juglans regia*	0.20 ± 0.01	0.23 ± 0.01	0.12 ± 0.005	0.02 ± 0.004	Traces	
16	*Catalpa speciosa*	0.12 ± 0.02	0.18 ± 0.01	0.10 ± 0.01	0.04 ± 0.001	Traces	
17	*Acer campestre*	0.16 ± 0.007	0.13 ± 0.005	0.11 ± 0.003	0.13 ± 0.01	Traces	
18	*Populus nigra*	0.15 ± 0.01	0.15 ± 0.01	0.11 ± 0.01	0.08 ± 0.006	Traces	
19	*Salix alba*	0.12 ± 0.01	0.12 ± 0.01	0.09 ± 0.01	0.09 ± 0.007	Traces	

Source compiled by the authors

Thus, most species studied exhibit their highest phytoncide activity between May and July (Fig. 1a). Phytoncide activity increases especially strongly during the flowering growth phase of plants and reaches its maximum in the first and second decades of June during the growth and leafing phase of shoots (Fig. 1b).

The analysis of the dynamics of phytoncide activity in the studied species shows that the peak activity does not coincide with the flowering period for most species. Only *Catalpa speciosa* and *Albizzia* exhibit maximum phytoncide activity close to their flowering phase, whereas for the other species, peak activity occurs after flowering (Fig. 1c).

4 Conclusion

Many of the species we studied show maximum phytoncide activity during active shoot growth. For example, *Juglans regia, Acer campestre, Aesculus hippocastanum, Gleditsia triacanthos,* and *Albizzia julibrissin* exhibit active growth with the highest phytoncide activity occurring in June. However, species such as *Quercus robur*, *Ulmus pumila,* and *Platanus orientalis* reach peak phytoncide activity in July, even though their active shoot growth takes place in April or June.

Therefore, all studied species have phytoncide activity that varies throughout the growing season, reaching a maximum from May to July.

Additionally, *Aesculus hippocastanum, Acer campestre, Cercis canadensis*, and *Catalpa speciosa* show peaks in photosynthesis intensity and phytoncide activity in May. For *Betula pendula, Juglans regia, Albizzia julibrissin, Juniperus virginiana, Pinus pallasiana,* and *Biota orientalis*, these peaks coincide in June. In *Platanus orientalis* and *Quercus robur*, the highest photosynthesis intensity is observed from June to July, with maximum phytoncide activity also occurring in July. In *Ulmus pumila*, both maxima occur in July. The results of this research support previously published findings that the dynamics of phytoncide activity are generally determined by the intensity of growth processes and photosynthesis.

All coniferous species exhibit high phytoncide activity, as do such deciduous species as *Aesculus hippocastanum, Cercis canadensis, Quercus robur*, *Fraxinus excelsior,* and *Platanus orientalis*. These species are highly decorative, which enhances their value for landscaping and greening projects.

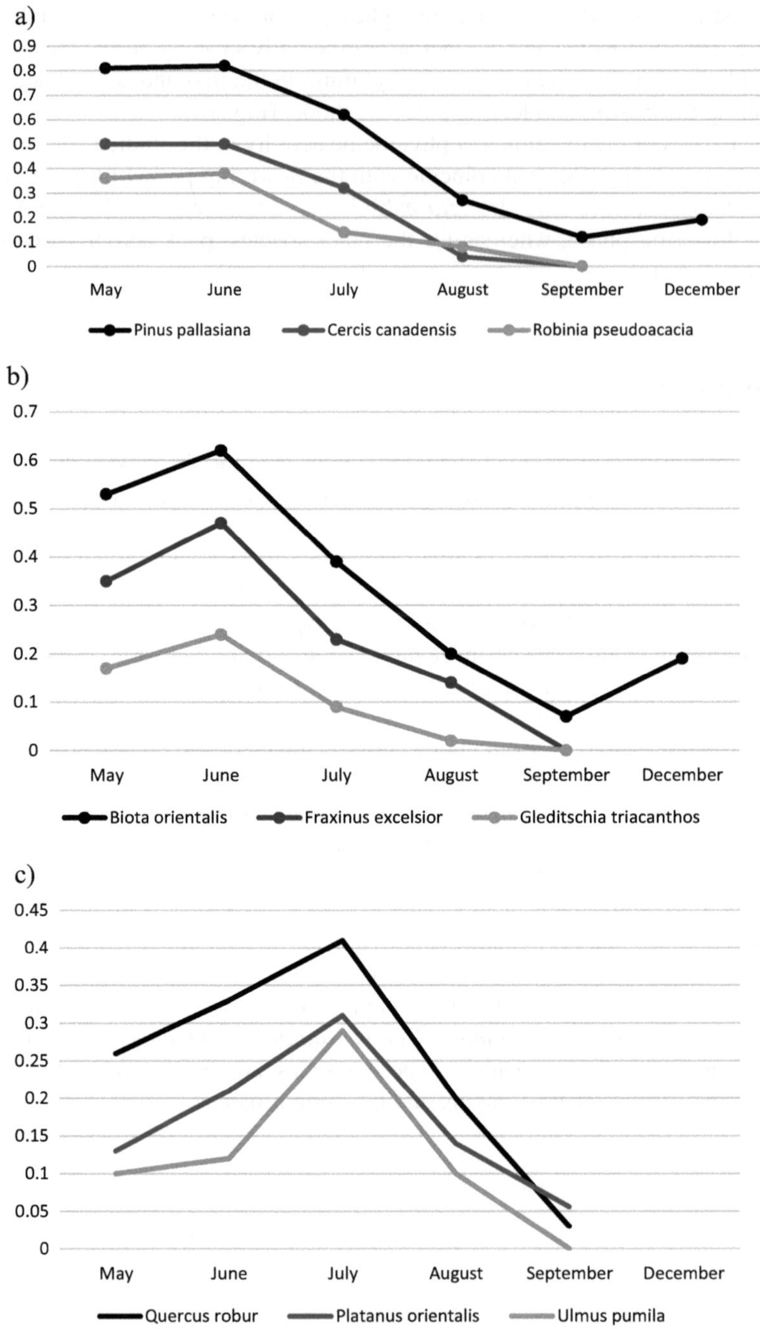

Fig. 1 **a** Maximum phytoncide activity from May to September of plants in the first group, mg/m^3; **b** maximum phytoncide activity from May to September of plants in the second group, mg/m^3; **c** maximum phytoncide activity from May to September of plants in the third group, mg/m^3. *Source* compiled by authors

References

Aĭzenman BE, Zelepukha SI, Negrash AK (1973) VII conference on the problem of phytoncides. Institute of microbiology and virology named after D.K. Zabolotny, Leningrad State University named after Andrei named after A.A. Zhdanov. Kyiv, USSR

Akhmedova G, Annaeva G (2024) Phytoncidal properties of species of the onion family (Alliaceae). Innovative Science 11-2-2:42–43

Akimov YA (1987) Ecological biochemistry of volatile terpenoids of higher plants. Nikitsky Botanical Garden, Yalta, USSR

Artemieva MN (1963) The content of phytoncides in plants of the southern coast of Crimea and their influence on oxidative processes in animals (synopsis of dissertation of the candidate of biological sciences). Taras Shevchenko University of Kyiv, Kyiv, USSR

Berida TI, Adekunle YA, Dada-Adegbola H, Kdimy A, Roy S, Sarker SD (2024) Plant antibacterials: the challenges and opportunities. Heliyon 10(10):e31145. https://doi.org/10.1016/j.heliyon.2024.e31145

Heller A (1977) Important but unknown throughout life. Phytocides—health store. Bodon Health, Langburg, Germany

Kochergina MV, Darkovskaya AS (2009). Phytoncidal properties of plantings of Petrovsky Square in Voronezh. Current Problems of the Forest Complex 23: 180–183

Kryuchkov VA (1978) Phytoncides as a factor in biosphere optimization. In: Phytoncides. Role in biogeocenoses, significance for medicine. Nauka, Leningrad, USSR, p 101

Lakin GF (1990) Biometry: textbook, 4th edn. Vysshaya shkola, Moscow, USSR

Li Q, Kobayashi M, Wakayama Y, Inagaki H, Katsumata M, Hirata Y, Miyazaki Y (2009) Effect of phytoncide from trees on human natural killer cell function. Int J Immunopathol Pharmacol 22(4):951–959. https://doi.org/10.1177/039463200902200410

Makarchuk NM, Leshinskaya YS, Akimov YA, Lebedeva AF (1990) Phytoncides and medicine. Nauka Dumka, Kyiv, USSR

Makarenko NG (1975) The effect of forest belt plants on antimicrobial activity. In: Phytocides. Kyiv, USSR, Nauka Dumka, pp 98–102

Ostrikova VM (1982) Phytocide activity of trees and shrub species used for greening Alma-Ata. In: Bespaev SB (ed) Plant protection and role of clean air in the city. Nauka, Alma-Ata, USSR, pp 77–85

Pryazhnikov AN (1967) Seasonal variation of phytocides in cedar forest areas. In: Phytocides. Kyiv, USSR, Nauka Dumka, pp 54–56

Roshina VD, Roshina VV (1989) Functional emission of higher plants. Nauka, Moscow, USSR

Starovoytova TV, Lakhno ES (1967) Antimicrobial properties of trees and shrubs used in Kyiv's greenery. In: Drobotko VG (ed) Phytocides, their biological role, and significance for medicine and the national economy. Nauka Dumka, Kyiv, USSR, pp 82–84

Tokin BP (1974) Healing plant poisons. A tale of phytoncides. University Press, Leningrad, USSR

Tomchuk RI, Spakhova AS, Konovalova VN (1976) On the influence of air pollution on the antimicrobial activity of woody plants. In: Abstracts of reports from the IV all-union conference on the physiological and biochemical foundations of plant interactions in Phytocenoses. Kyiv, USSR, Nauka Dumka, pp 128–129

Vichkanova SA (1981) BME. In: Great soviet Encyclopedia, vol 26. Soviet Encyclopedia, Moscow, USSR, pp 334–335

Vlasyuk VN (1981) The role of phytocides and forests in conservation. In: Phytocides. Kyiv, USSR, Nauka Dumka, pp 146–148

The Long-Term Outlook for Climate Change Management in Rural Areas: Strategic Sustainable Development Based on the Smart Agricultural Technologies of the Future (Conclusion)

Rural areas are particularly vulnerable to climate change, and addressing it most effectively requires reliance on climate-smart agricultural technologies—this constitutes the main authorial conclusion summarizing the results of the research presented in this book. The key specific findings can be outlined as follows:

- The most pressing challenges to the sustainable development of rural areas under climate change include declining agricultural productivity, reduced investment attractiveness, slower economic growth, and population outmigration;
- The advanced potential of climate-smart agricultural technologies includes achieving the environmental autonomy of smart farms, increasing their production capacity and productivity, and improving the quality of natural agricultural products;
- The contribution of climate-smart agricultural technologies to the sustainable development of rural areas lies in activating agricultural "growth points" within these territories. Activating such growth points requires strengthening the regulatory and infrastructural foundations of the AI-driven economy in rural areas, as well as improving their venture investment climate;
- International experience in the sustainable development of rural areas based on climate-smart agricultural technologies demonstrates the existence of different national models of climate management, including the Eurasian model, the BRICS model, and the Central Asian model;
- Promising applied solutions for the sustainable development of rural areas based on climate-smart agricultural technologies include high-productivity resource use enabled by AI-driven automation and flexible climate adaptation of natural agriculture. These solutions allow rural areas to achieve more stable economic growth amid global economic and climatic uncertainty.

Thus, this book expands scientific understanding of climate management in rural areas during the Decade of Action (up to 2030) by enhancing insight into the causal links between the technological modernization of agricultural production (associated

A. B. Karbekova (ed.), *Sustainable Development of Rural Territories Based on Climate-Smart Technologies*, Climate Change Management,
https://doi.org/10.1007/978-3-032-20106-5

with AI-driven automation), the effectiveness of climate change mitigation in rural areas, and the pace and sustainability of their socio-economic development.

Simultaneously, the dynamically changing economic and technological landscape of climate change mitigation has been identified, which leaves the long-term outlook for climate change management in rural areas uncertain. The scientific results achieved in this book have significantly enriched existing knowledge of this form of management by incorporating the author's vision of the highly efficient use of climate-smart agricultural technologies and the economic and climatic resilience of rural development during the Decade of Action, while also giving rise to new research questions related to the subsequent period.

Particularly, the materials of this book have raised the research question of what the social profile of future rural areas will look like, given that AI-driven automation leads to workforce displacement on smart farms and enables remote climate management. In this context, the sustainable development of rural areas based on climate-smart management of high-tech agricultural production in the long term implies accelerated economic growth, but does not guarantee population settlement.

Another research question emerging from this book concerns the future form of smart agricultural technologies and how to ensure their widespread dissemination and highly efficient application in rural areas. Among the research issues highlighted in this book is the need to rethink sustainable rural development from a strategic perspective after the completion of the Decade of Action, moving beyond the formulation of the Sustainable Development Goals for 2015–2030.

Future studies building on this book are recommended to focus on addressing these research questions, with particular attention to a detailed examination of the long-term outlook for climate change management in rural areas and the development of a concept for the strategic sustainable development of these territories based on the smart agricultural technologies of the future.

MIX
Papier aus verantwortungsvollen Quellen
Paper from responsible sources
FSC® C105338

If you have any concerns about our products,
you can contact us on
ProductSafety@springernature.com

In case Publisher is established outside the EU,
the EU authorized representative is:
Springer Nature Customer Service Center GmbH
Europaplatz 3, 69115 Heidelberg, Germany

Printed by Libri Plureos GmbH
in Hamburg, Germany